The Packers' Encyclopedia
A Handbook of Modern Meat Packing House Practice

by Paul I. Aldrich

with an introduction by Sam Chambers

This work contains material that was originally published in 1922.

This publication is within the Public Domain.

This edition is reprinted for educational purposes and in accordance with all applicable Federal Laws.

Introduction Copyright 2018 by Sam Chambers

Self Reliance Books

Get more historic titles on animal and stock breeding, gardening and old fashioned skills by visiting us at:

http://selfreliancebooks.blogspot.com/

Introduction

I am pleased to present another book on the curing and preservation of meat on the farm or in the kitchen.

The work is in the Public Domain and is re-printed here in accordance with Federal Laws.

As with all reprinted books of this age that are intended to perfectly reproduce the original edition, considerable pains and effort had to be undertaken to correct fading and sometimes outright damage to existing proofs of this title. At times, this task is quite monumental, requiring an almost total "rebuilding" of some pages from digital proofs of multiple copies. Despite this, imperfections still sometimes exist in the final proof and may detract from the visual appearance of the text.

I hope you enjoy reading this book as much as I enjoyed re-publishing and making it available to fanciers again.

With Regards,

Sam Chambers

PREFACE

This Packers' Encyclopedia has been compiled as the result of an insistent demand for a ready reference work on the meat packing and allied industries. It has been felt for a long time that such data on the largest industry in the United States should be brought together. It was considered appropriate that this task should fall to the official trade publication of the industry, THE NATIONAL PROVISIONER.

The work naturally divides into three parts, each of which is decribed more in detail in its own foreword.

Part I is a Hand-book of Modern Packing House Practice. The material is arranged in the simplest possible form, by classes of animals, nature of products and order of operations. An attempt has been made to approximate the latest and best American packing house practice, condensed within the space available in a single volume, and adapted to the needs of the average operator.

This portion of the work is what gave the book its name, The Packers' Encyclopedia. Its preparation would not have been possible save for the generous co-operation of the leading operating experts of the industry.

Part II is a Statistical Section which offers, chiefly in chart form, graphic comparisons covering a decade of the number and prices of meat animals and their chief products; production, exports, imports and consumption. Freight rate data and officially-adopted trade term definitions are also included for the convenience of the reader.

Part III is the first comprehensive Trade Directory ever attempted for the industry. Here is listed data of corporation information, capacity, operations, brands and trade-marks, equipment, etc., covering the meat-packing industry of the United States and Canada, together with names of packers in

PREFACE

other countries. There are also lists of wholesale meat dealers, sausage manufacturers, renderers and other allied trades. This section of the work, though more readily subject to change than the others, nevertheless will be of great practical interest and value.

The aim throughout has been to prepare a work of ready reference and strictly practical purpose, which should meet the average man's every-day need. In shaping the plan and carrying out the purpose of the work the Editor desires to acknowledge the invaluable assistance of Arnold C. Schueren. It is not out of place here to express the hope that this book is a step on the way to an adequate industry library.

TABLE OF CONTENTS

PART ONE: PACKING HOUSE PRACTICE

Chapter One: CATTLE......................................1-62
    Breeds of Cattle
    Market Classes and Grades of Cattle and Calves
    Dressing Percentages of Cattle
    Beef Slaughtering
    Beef Cooling
    Beef Grading
    Beef Loading
    Handling of Beef for Export
    Beef Cutting and Boning
    Plate Beef
    Mess Beef
    Curing Barreled Beef
    Manufacture of Dried Beef
    Handling Beef Offal
    Handling Miscellaneous Meats
    Handling and Grading Beef Casings
    Manufacture of Beef Extract
    Manufacture of Oleo Products
    Tallow
    Handling of Hides

Chapter Two: HOGS..63-119
    Breeds of Hogs
    Market Classes and Grades of Hogs
    Dressing Yields of Hogs
    Hog Killing Operations
    Hog Cooling
    Shipper Pigs
    Pork Cuts
    Curing Pork Cuts
    Smokehouse Operation
    Ham Boning and Cooking
    Lard Manufacture
    Hog Casings
    Edible Hog Offal or Miscellaneous Meats
    Preparation of Pigs' Feet

TABLE OF CONTENTS

Chapter Three: SMALL STOCK..........................120-125
    Market Classes and Grades of Sheep and Lambs
    Sheep Killing
    Calves
    Sheep Casings
    Casings from Calves and Yearlings

Chapter Four: INEDIBLE BY-PRODUCTS..............126-146
    Inedible Tank House
    Blood and Tankage Yields
    Calculating Tankage Values
    Digester Tankage
    Tallow and Grease Refining
    Manufacture of Glue
    Bones, Horns and Hoofs
    Handling Hog Hair
    Catch Basins
    Cost and Return on By-Products

Chapter Five: MISCELLANEOUS........................147-204
    Sausage Manufacture
    Meat Canning
    Animal Glands and Their Uses
    Packinghouse Chemistry
    Packinghouse Cost and Accounting Methods
    Location of Packing Plants
    Construction of Packing Plants
    Packinghouse Refrigeration

Chapter Six: VEGETABLE OILS..........................205-221
    Vegetable Oil Refining
    Compound Lard
    Winter Oil
    Hydrogenation of Oils and Fats
    Manufacture of Margarin

PART TWO: STATISTICS

United States Meat Industry Statistics.....................223-253
    Sources of U. S. Meat Supply
    Areas of U. S. Meat Consumption
    Cattle and Hog Loading Points and Slaughtering Centers
    Yearly Top Prices of Beef, Hogs and Sheep, 1910-1920
    Monthly Average and Top Prices of Native Beef Cattle, 1910-1920
    Cattle and Corn Prices Compared, 1910-1920
    Hide, Tallow and Oleo Oil Prices
    Beef Production and Consumption, 1907-1921
    Hog Population and Average Prices, 1910-1920
    Hog and Corn Prices Compared, 1910-1920

TABLE OF CONTENTS

    Pork and Lard Production and Consumption, 1907-1921
    Sheep Population and Average Prices, 1910-1920
    Mutton Production and Consumption, 1907-1921
    Veal Production and Consumption, 1907-1921
    Livestock Population in the United States, 1900-1922
    Slaughtering in the United States, 1907-1921
    Meat Packing in the United States, 1914-1919
    Exports of Meat Products, 1910-1921
    Provision Prices at Chicago, 1910-1921
Canadian Meat Industry Statistics..........................254-255
Vegetable Oil and Margarin Statistics......................253, 261
Railroad Rates on Cattle, Beef and Packing House Products...256-261
Domestic Trade Term Definitions..........................262-267
Export Trade Term Definitions............................267-272

PART THREE: TRADE DIRECTORY

Meat Packers and Slaughterers..............................274-391
    United States
    Mexico
    Cuba
    Canada
    South America
    South Africa
    Australia
    New Zealand
Wholesale Meat Dealers, Sausage Makers and Provisioners...392-416
Renderers 417-424
Refiners of Edible Oils..................................425-428
Margarin Manufacturers 428-429
Brokers in Packing House Products and Vegetable Oils.......430-439
Livestock Order Buyers...................................440-445

    Advertising Section 447-520
    Topical Index 521-525
    Index to Illustrations............................. 526
    Index to Advertisements............................527-529

FOREWORD

Part I of The Packers' Encyclopedia is a book written by practical packinghouse men, and intended for the use of practical packinghouse men. It is not theory, or a one-man book, but the result of the experiences of many.

The arrangement of material is by classes of animals, nature of products and order of operations, making it easy to follow through or to refer to any particular part. It is, therefore, in handy form for the student. In addition there is a topical index.

Methods here described represent the best American packinghouse practice, as developed in both large and medium-sized plants. Large packer practice has been used in many instances, as most experimentation and development heretofore has been in large plants. But the tendency is growing to operate packing plants in smaller units, and so-called small plant practice has been kept in line in directions and suggestions given. Where difference of opinion exists among authorities, the practice quoted is that most adaptable to the average packinghouse.

It is not expected that packinghouse operators will agree on many of the details of practice given herein. Hardly any two experts agree; each has his own methods and prefers them. The object here has been to outline the main points and emphasize the best procedure. The reader is not expected to follow blindly what he finds here, but to adapt it to his own special needs. Detailed description of all operations would have required a series of volumes instead of one. Requests for more detailed information on any subject may be submitted to the Editor, THE NATIONAL PROVISIONER, Chicago.

An effort has been made to standardize illustrations. All pictures of animals, carcasses and cuts are from official photographs of the U. S. Department of Agriculture (Bureau of Markets and Crop Estimates), published here for the first time. No attempt has been made to show machinery, as this is the province of the manufacturer's catalog. Explanatory fundamental drawings are shown instead. Construction and refrigeration details also are left to experienced packinghouse architects and engineers, who know how to lay out each particular job to meet conditions. What is written herein on construction and refrigeration will apply to everyday packinghouse practice.

PART ONE

Packing House Practice

Chapter I—CATTLE

BREEDS OF CATTLE

There are four strictly beef breeds of cattle; namely, Shorthorn, Hereford, Angus and Galloway. These breeds have been developed for the sole purpose of producing an animal which is very efficient in the production of meat. The ideals toward which the sponsors of all the beef breeds are working are a low-set animal with plenty of depth, a wide spring of rib, short neck and legs, and quarters that carry the fleshing down well. In short, a blocky, rectangular conformation that carries a maximum amount of beef.

The dressing percentage of carcass, more about which will be said later, is influenced markedly by the use of good-type, pure-bred beef sires, and it is principally from this standpoint that the packer is interested. This type of animal not only is generally a good investment for the producer, but gets high quality animals which dress out well, in both of which the packer is particularly interested.

Beef Breeds Described

The Shorthorn is roan, white, red or red and white in color. It has a quiet disposition, is adapted for farm beef making, good for grading up herds, growthy and early maturing, and dresses out well. The weight of mature bulls is from 1,800 to 2,500 pounds, and of mature cows 1,200 to 1,800 on an average.

The Hereford is red with white markings, commonly known as the "white face," is a good rustler and widely used on the ranges as well as for a farm beef animal, thrives under adverse conditions and does well in the feed lot. It matures early and fattens out well. The mature cows weigh from 1,300 to 1,700 pounds and mature bulls from 1,800 to 2,300 pounds.

The Angus is solid black in color, has soft, mellow skin and fine hair, and no horns. They fatten well on grass and respond to liberal feeding in winter. The Angus probably does best under maximum conditions, but also gives excellent returns on either the range or the general farm in any section of the country, and is increasing in popularity. The mature cows weigh from 1,200 to 1,600 pounds and the bulls from 1,600 to 2,100 pounds. These cattle are commonly known as "doddies."

The Galloway is one of the oldest breeds of cattle. The mature

Prime Killing Steer

Medium Beef Steer

Common Killing Steer

MARKET CLASSES OF CATTLE

cows weigh from 1,000 to 1,500 pounds, and the bulls from 1,400 to 1,800 pounds. They are solid black in color, have long curly hair, and are polled. They mature somewhat more slowly as compared to the other beef breeds, but they are exceptionally well adapted to climates having severe winters. They are good rustlers and winter well on roughage. They are not so well adapted to feed lot conditions as are the other three breeds. The grain of the meat is fine and high in quality.

There are some breeds of dual-purpose cattle which are designed for the production of both milk and beef. Average animals of these breeds do not reach the highest attainments, however, in the production of either. Among these are the Red Polls, Dutch Belted, Devons and some others. While these breeds do not reach the maximum of either milk or beef production, they find very wide usage on many farms where general purpose cattle are desired.

MARKET CLASSES AND GRADES OF CATTLE

Cattle and calves intended primarily for meat production fall into two general divisions; first, killing cattle and calves, which are those utilized for immediate slaughter; and second, feeder and stocker cattle and calves, which are utilized for further finish and development. The feeder animal is ready to go into the feed-lot at once, while the stock animal is thin and best adapted, economically, to be further developed on cheap feeds before being put in the feed-lot for intensive feeding designed to promote the rapid production of flesh and fat.

The market classes in which bovine animals are placed are based on sex and age. However, as the animals in any one class are not equal in quality, form and condition, they are further divided into grades to indicate their relative merit within the class. Further division of classes in the case of killing steers, killing calves and feeder steers is made according to weight, as weight is often important as a price-determining factor within these classes, but logically must be considered as a matter of selection rather than as a reliable indication of grade.

As baby beef must show some of the characteristics of veal, animals classified as baby beeves, which may be regarded as a specialty, are given maximum weight and age limits. Requirements for animals falling within this class are such, however, that by no means all of the young steers and heifers falling under the maximum weight and age limits shown in the classification properly classify as baby beeves, owing to deficiencies in quality or finish or both; and occasionally animals of somewhat greater weight and age qualify in the carcass as baby beeves.

The buyer must be a competent judge of the different grades, and be able to calculate mentally what dressing percentage may be obtained, as well as the quality of meat that will be forthcoming from the different classes and grades. Daily test sheets showing a comparison between the buyer's calculation and the actual dressing percentage and quality of the meat are valuable aids in determining whether the buyer is working along the right lines.

The following classification—in arranging which effort has been made to eliminate all class or grade nomenclature that may be considered vague

Baby Beeve

Good Fat Cow

Common Fat Cow

MARKET CLASSES OF CATTLE

or more or less misleading—has been tentatively adopted by the United States Department of Agriculture, Bureau of Markets and Crop Estimates, after extended investigation and conferences with representatives of all interests in the trade, as one suitable with perhaps minor modifications, to be used uniformly at all markets:

Killing Cattle and Calves

Class	Sub-Class	Grade	
Steers:	1. Heavy Weight (1,300 lbs. up)	Prime or No. A1 Choice or No. 1 Good or No. 2	Medium or No. 3 Common or No. 4
	2. Medium Weight (1,100 to 1,300 lbs.)	Prime or No. A1 Choice or No. 1 Good or No. 2	Medium or No. 3 Common or No. 4 Cutter or No. 5
	3. Light Weight (1,100 lbs. down)	Prime or No. A1 Choice or No. 1 Good or No. 2 Medium or No. 3	Common or No. 4 Cutter or No. 5 Canner or No. 6
Heifers		Prime or No. A1 Choice or No. 1 Good or No. 2 Medium or No. 3	Common or No. 4 Cutter or No. 5 Canner or No. 6
Baby Beeves (Steers and Heifers, 950 lbs. down, 15 mos. and under)		Prime or No. A1 Choice or No. 1	Good or No. 2
Cows		Choice or No. 1 Good or No. 2 Medium or No. 3	Common or No. 4 Cutter or No. 5 Canner or No. 6
Bulls		Choice or No. 1 (Butcher and Beef) Good or No. 2 (Butcher and Beef) Medium or No. 3 (Bologna) Common or No. 4 (Bologna) Canner or No. 6	
Stags		Choice or No. 1 Good or No. 2	Medium or No. 3 Common or No. 4
Calves:	1. Light (110 lbs. down) 2. Handy (110-190 lbs.)	Choice or No. 1 Good or No. 2 Medium or No. 3 Common or No. 4 Cull or No. 7	
	3. Medium (190-260 lbs.) 4. Heavy (260 lbs. up)	Choice or No. 1 Good or No. 2 Medium or No. 3 Common or No. 4 Canner or No. 6	

Feeder and Stocker Cattle and Calves

1. Feeders

Class	Sub-Class	Grade	
Steers:	1. Heavy Weight (1,000 lbs. up) .... 2. Light & Medium Weight (1,000 lbs. down)....	Fancy Selected or No. A1 Choice or No. 1 Good or No. 2 Medium or No. 3 Common or No. 4	
Heifers		Choice or No. 1 Good or No. 2	Medium or No. 3
Cows		Choice or No. 1 Good or No. 2	Medium or No. 3 Common or No. 4
Bulls		Choice or No. 1 Good or No. 2	Medium or No. 3
Calves		Fancy Selected or No. A1 Choice or No. 1	Good or No. 2

2. Stockers

Class	Grade	
Steers	Fancy Selected or No. A1 Choice or No. 1 Good or No. 2	Medium or No. 3 Common or No. 4
Heifers, Cows, Bulls..............	Choice or No. 1 Good or No. 2	Medium or No. 3 Common or No. 4
Calves	Choice or No. 1 Good or No. 2 Medium or No. 3	

Percentage of Classes and Grades Slaughtered

The percentage of the different classes and grades of beef animals slaughtered in the various centers varies widely, depending upon the market as well as upon climatic conditions. For instance, during droughts in the Southwest and in the Northwest many thousands of thin and immature cattle and of breeding stock, that would have been held under favorable conditions, were shipped to market because of the scarcity of feed.

The composite chart shown here illustrates very clearly the percentage of different beef animals slaughtered at various centers over a period of time.

DRESSING PERCENTAGES OF CATTLE

Depending upon the breeding, fill and condition, the dressing percentage of all cattle varies. Other conditions also affect the percentage of beef obtained from any animal; namely, the freedom from paunchiness, the type and quality. Fat steers always outdress animals of less finish. The filling of the digestive organs with feed and water is as important as the condition or degree of fatness. The broad, thick type

MARKET CLASSES AND GRADES OF CATTLE 7

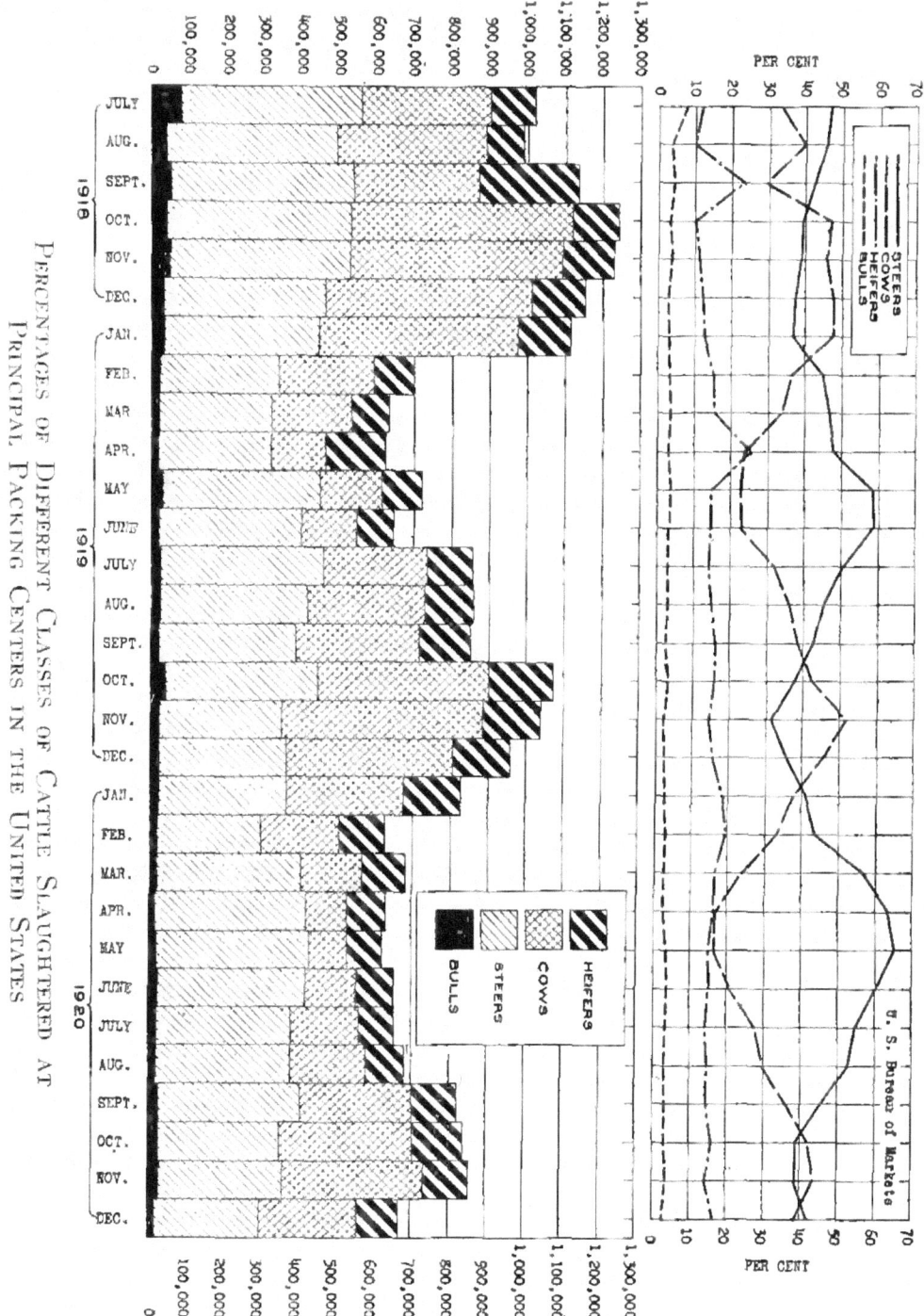

Percentages of Different Classes of Cattle Slaughtered at Principal Packing Centers in the United States

of steer will outdress the steer of dairy type, even when the condition and fill are the same, by three to five per cent.

While quality, hide, meat and bone may affect the per cent ratio by one to two per cent, the average run of steers of the type marketed today at most markets will dress out about 53 per cent. The good to choice will dress from 56 to 59, and steers of the fancy type will dress from 59 to 63 per cent.

A table showing the average dressing percentage of fair steers, baby beef, cows and canners, together with the percentage of fat and hide, is as follows:

	Avg. Live Weight Lbs.	Avg. Dressed Weight Lbs.	Yield of Beef P.C.	Fat P.C.	Hide P.C.
Fair Steers	1,050	580	.5552	4½	6½
Baby Beef	900	527	.582	5¾	6¾
Cows	1,100	572	.52	4¼	6¼
Canners	800	340	.425	1½	6¼

Aside from the fat and hide there is little, if any, difference in the yield of offal products from the various classes of cattle.

The yield of cuts from the dressed carcass from canners, No. 1 cutters and native steers is as follows, in percentage:

	Canners %	No. 1 Cutters %	Native Steer %
Loins	14.34	14.79	17.15
Ribs	10.77	9.40	9.51
Plates	13.14	16.33	12.39
Rounds	20.31	20.97	23.63
Total	58.56	61.49	62.68

From this table it will be seen that the native steer, of course, has a higher percentage of loin and round than do the poor classes of cattle, and that the percentage of plates is considerably less. As this percentage is based on the dressed carcass as 100 per cent the difference in the yield is much more pronounced when figured from the live weight basis.

By-Product Yield of a 1,000 Lb. Steer

The various by-products in pounds derived from a 1,000 lb. steer are as follows, and may be used as a general guide for the packer in estimating the yields which he should obtain:

Trimmed tongue	5.00 lbs.	Middle casing	32 feet
Cheek and head meat	5.00 lbs.	Round casing	105 feet
Brain	.90 lbs.	Weasand	1 piece
Gullet	.25 lbs.	Bladder	1 piece
Lips	1.25 lbs.	Bung	1 piece
Heart	3.50 lbs.	No. 1 oleo oil	22.00 lbs.
Liver	10.00 lbs.	No. 2 oleo oil	1.80 lbs.
Kidneys	.75 lbs.	No. 3 oleo oil	.75 lbs.
Tail	1.25 lbs.	Stearine	13.00 lbs.
Sweetbread	.30 lbs.	Prime tallow	4.10 lbs.
Suprarenal glands	.06 lbs.	No. 1 tallow	1.75 lbs.
Honeycomb tripe	1.50 lbs.	Brown grease	.16 lbs.
Plain tripe	6.50 lbs.	Hide	65.00 lbs.

Switch	1 piece	Thighs	1.45 lbs.
Sinews and pizzle	2.62 lbs.	Buttock bones	1.15 lbs.
Dewclaws	.40 lbs.	Cannon bone	1.00 lbs.
Green blood	35.00 lbs.	Neatsfoot oil	.85 lbs.
Dry blood	7.00 lbs.	Grinding bones	13.00 lbs.
Tankage	10.00 lbs.	Horns	.70 lbs.
Hoofs	1.85 lbs.	Horn piths	.90 lbs.
Shin bones	1.60 lbs.		

Offal Test on 508 Shipping Cattle

The following is an offal test on 508 shipping cattle, which averaged 1,242 lbs. live weight:

	Total	Per head
Raw fat for oleo	26,023 lbs.	51.23 lbs.
Cheek meat	2,061 lbs.	3.94 lbs.
Head meat	540 lbs.	1.06 lbs.
Ox lips	606 lbs.	1.19 lbs.
Long cut tongues	2,624 lbs.	5.16 lbs.
Brains	442 lbs.	.87 lbs.
Sweetbreads	85 lbs.	.14 lbs.
Tails	599 lbs.	1.18 lbs.
Horns	1,490 lbs.	2.93 lbs.
Hearts	2,128 lbs.	4.19 lbs.
Melts	897 lbs.	1.76 lbs.
Livers	5,422 lbs.	10.67 lbs.
Heads	7,568 lbs.	14.89 lbs.
Jaws	2,112 lbs.	4.15 lbs.
Feet	7,723 lbs.	15.20 lbs.
Sinews	1,271 lbs.	2.50 lbs.
Pizzles	274 lbs.	.55 lbs.
Tripe	7,909 lbs.	15.57 lbs.
Bladders	156 pieces	.30 lbs. per piece
Weasands	501 pieces	.98 lbs. per piece
Tankage	3,048 lbs.	6.00 lbs.
Blood	3,556 lbs.	7.00 lbs.
Neck trimmings	445 lbs.	.87 lbs.
Rendered tallow	2,602 lbs.	5.12 lbs.
Grease	170 lbs.	.33 lbs.
Export rounds	945 lbs.	.64 set
Domestic rounds	635 lbs.	.35 set
Middles	914 lbs.	.36 set
Bungs	545 lbs.	.99 piece
Switches	452 pieces	.89 piece

There was a total of 17,861 lbs. offal which went to the tanks for tallow, grease and fertilizer.

Offal Test on 499 Butcher Cattle

In comparison to the above, here is a test on 499 butcher cattle, averaging 926 lbs. live weight:

	Total	Per head
Fat for oleo	14,342 lbs.	28.74 lbs.
Tongues	1,800 lbs.	3.41 lbs.
Heads	6,447 lbs.	12.92 lbs.
Jaws	1,865 lbs.	3.73 lbs.

	Total	Per head
Cheek meat	1,810 lbs.	3.62 lbs.
Head meat	135 lbs.	.27 lbs.
Ox lips	815 lbs.	1.63 lbs.
Brains	375 lbs.	.75 lbs.
Sweetbreads	20 lbs.	.04 lbs.
Hearts	1,320 lbs.	2.64 lbs.
Melts	585 lbs.	1.17 lbs.
Sinews	955 lbs.	1.90 lbs.
Pizzles	160 lbs.	.32 lbs.
Horns	1,045 lbs.	2.09 lbs.
Tripe tanked	7,820 lbs.	15.67 lbs.
Tails	553 lbs.	1.10 lbs.
Bladders	169 pieces	.34 lbs. per piece
Weasands	449 pieces	.90 lbs. per piece
Feet	6,068 lbs.	12.16 lbs.
Livers	3,285 lbs.	6.59 lbs.
Blood	3,243 lbs.	6.50 lbs.
Tankage	3,003 lbs.	6.02 lbs.
Tallow	1,434 lbs.	2.89 lbs.
Grease	166 lbs.	.33 lbs.
Export rounds	842 lbs.	.57 set
Domestic rounds	788 lbs.	.43 set
Middles	1,025 lbs.	.37 set
Bungs	580 lbs.	1.00 piece
Switches	459 pieces	.92 piece

There was a total of 15,560 lbs. offal which went to the tanks for tallow, grease and fertilizer.

[The word "offal" formerly covered both edible and inedible products. The edible products, such as cheek meat, tails, sweetbreads, etc., are now termed by some packers "fancy meats" and by others "miscellaneous meats."]

BEEF SLAUGHTERING

Beef slaughtering as conducted in the modern packing house is a continuity of specialized butchering. As compared with the old method, where one man dressed the carcass from beginning to end, the method now employed is that of dividing the operation so that the most skillful butcher works on that part of the carcass where the most skill is required.

This is desirable not only from the standpoint of producing a well-dressed carcass at a minimum cost, but also from the standpoint of expediency. It would be almost an impossibility to procure sufficient butchers who have the ability to dress a carcass throughout, to meet the requirements of the present rate of slaughtering, not to mention the lost motion involved in attempting to do so. It is possible now to take any man and start him on one of the minor operations, such as foot skinning, and by advancement all along the line produce a man to fill any job on the floor.

A complete description of each operation on the killing floor will explain the methods used, and show how almost perfect co-ordination is obtained. Each operation is explained in the order in which it occurs.

Driving to Knocking Pen.—Care should be taken as far as possible not to unduly excite the animal, as this may subsequently show in the flushed appearance and the rupture of surface blood vessels in the carcass when it is placed in the cooler. Drivers should be equipped with a pole, having an electrical contact, to drive the cattle, rather than whips or clubs; this will prove more efficient and reduce bruises materially.

Knocking.—A four-pound sledge is sufficient, as this will produce a concussion of sufficient violence to render the animal insensible.

The entire operation to this point should be conducted with the utmost humaneness. Knocking pens should be so constructed that one side may be lifted and the bottom built on rockers, that the animal may be easily ejected. In so ejecting there should be little violence, as it

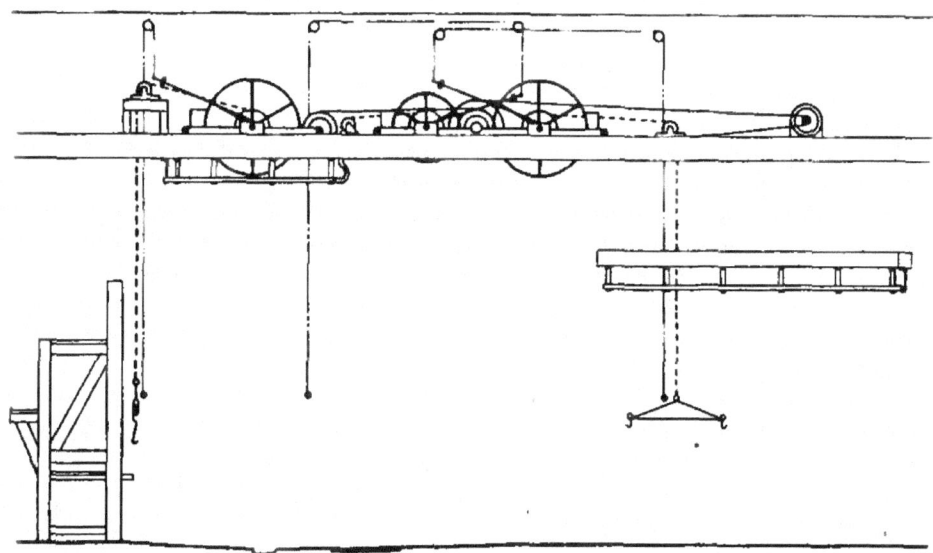

ELEVATION OF BEEF KILLING EQUIPMENT AS USED IN LARGE MODERN PLANTS

The cattle are elevated by one hoist to the bleeding rail. A dropper lowers them again to the killing floor for further dressing, and another hoist elevates them again to the dressing rail.

is not infrequent that back bones are broken or dislocated at this time by careless handling, which will make itself apparent at the time of splitting. A knocker should also be taught to gauge his blow so that when knocking thin-skulled cattle he will not cause an unsightly blood clot that will make the brain unsalable.

Shackling.—A short chain with a hook at one end and a ring at the other is used extensively. The chain is passed around both hind feet above the ankle joint and hooked tight. Particular attention should be given that both feet be shackled, for if one leg is allowed to hang loose it prevents complete draining, and in some cases by putting the entire weight on one leg it may cause a broken bone. If this style of shackle is used it will be necessary to have a man on a bridge at the rail level so that when the bullock is raised to the rail he can connect the ring of the shackle to the hook of the beef trolley and disconnect it from the hoist.

This style of shackle is being superseded in many houses by a device known as an automatic lander. The beef trolley is connected to the shackle permanently, and this device will place the trolley on the rail automatically. This method cannot be used to advantage when kosher cattle are being slaughtered, due to the weight of the trolley.

The shackling of cattle for the kosher style of slaughtering is a very difficult operation. As no knocking is permitted, it is necessary to shackle while the animal is fully alive. This can best be accomplished by slightly raising the side of the knocking pen, and with a quick throw of the chain encircle one leg. An almost instantaneous pull from the hoist is necessary or the cattle will kick the chain off. The animal is then raised so that its shoulder, neck and head touches the floor. A muzzle is placed over the nose, the neck bent sidewise to tense the muscle of the neck, and the throat cut by the Rabbi.

Kosher Killing

In certain of the large slaughtering centers, chiefly Chicago and New York, quite a number of the cattle are slaughtered according to the rights of the orthodox Jewish church, the reasons for which it is not necessary to enter into here. The chief difference in the operation, as indicated, is in the method of killing, which requires more time than the regular method, due to the fact that each animal must be shackled without being stunned and the throat cut by a representative of an authority of the church.

The Jewish law of slaughtering applies to animals and birds. This is entrusted only to people versed in the law and skilled in the work. The operation, however, cannot be done by a deaf mute, an idiot, a minor, by one who is intoxicated, nor by an old man whose hands tremble, for the reason that he may press the knife against the throat of the animal instead of moving it forward and backward, which is the prescribed method.

The length of the knife must be twice the width of the throat of the animal, the maximum being 14 finger breadths. The knife must be sharp, smooth, and without any perceptible notch. It is examined before slaughtering, first being tested on the finger and then on the edge of the finger nail, on both sides of the knife. It must also be examined immediately after slaughtering and if a notch is found afterward the animal is declared unfit for use. Before slaughtering a blessing must be pronounced to the Lord, as commanded. When many animals are slaughtered at the same time one blessing is sufficient for the whole lot.

The act of slaughtering consists of cutting through the windpipe and the gullet. The principles which must be observed are as follows:

1. There should be no delay by interruption while the slaughtering is being performed.

2. The knife should be kept in continuous motion forward and backward until the organs are cut through. A delay of a moment makes the animals unfit.

3. The knife must be drawn gently across the throat without any undue exertion on the part of the killer.

4. The killer has no right to lay a finger on the blade while killing as the slightest pressure renders the animal unfit.

5. The knife must be drawn over the neck of the animal; if it is placed between the windpipe and the gullet or under the skin so that any part of the knife is not visible while the act is being performed, then the animal is unfit for food, even though the other actions may have been correctly executed.

6. The limits within which the knife may be inserted are from the large ring of the windpipe to the top of the upper lobe of the lungs when inflated. Slaughtering by the insertion of the knife in any part above or below these limits is called "slipping," and renders the animal unfit for food.

If either the windpipe or the gullet is torn out or removed from its regular position during slaughter, the animal is branded unfit for food. This is called "tearing." Soon after the slaughtering the killer must examine the throat of the animal and ascertain whether the windpipe and gullet are cut through according to the requirements of the Jewish law.

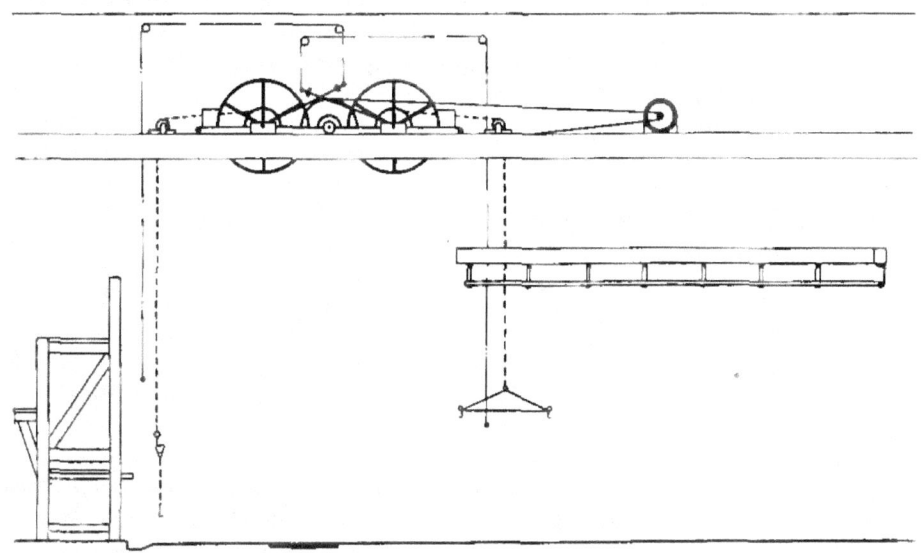

ELEVATION OF BEEF KILLING EQUIPMENT AS USED IN SMALL AND MEDIUM-SIZED PLANTS

It consists primarily of a double hoist, on which the cattle are raised and bled. The other side of the hoist is used for elevating the cattle to the dressing rail.

In New York 90 per cent of the cattle and calves are koshered and nearly 50 per cent of the New York markets carry Kosher beef only.

It is important to have the killing house arranged so that the animal may have plenty of time on a bleeding rail, as it is a prime requisite that the animal be well bled.

Regular Sticking

Under the regular slaughtering method the knife should be inserted in the dewlap and rib upward to the confluence of the jugular vein. In opening the jugular vein the knife point should touch; then move the knife in such a manner that the heel will make the actual incision. It is to be desired that the vein be opened on one side only, as in the event of cutting through the animal will, when bleeding, cause a certain amount of blood to regurgitate into the lung cavity, which is very undesirable for the following reasons: Loss of blood, discoloration of the pleura, and to a certain extent a flushed condition of the entire carcass, due to faulty bleeding. Cattle should be bled for at least six minutes before any further operation is attempted, and longer if possible, up to twenty or twenty-five minutes. This will allow a thorough bleeding and yield a whiter carcass.

Heading.—Insert knife back of horn and draw over to the left side. Cut on a straight line from the left horn down alongside of the left eye to the snout. Remove the skin from the face. Continue around right jaw to the center of neck. On a line with the incision made by the sticker the hide is then opened to the lip. Next remove skin from left jaw.

In this operation, as well as in all subsequent skinning operations, the butcher must be impressed with the danger of damage by cutting the hide. While a head cut in front of the ears is not as serious damage to the hide as it would be behind, nevertheless it is anything but desirable.

The head is now removed by cutting through the button or atlo-occipetal joint. If long-cut tongues are to be made the trachea, frequently called gullet, should be cut four rings behind the tongue; if short cut, one ring behind. When skinning along the front of the neck, a header may injure the tongue, unless great care is used that the ball of the tongue is not removed with the hide, thereby exposing the lean tissue of the tongue and a consequent loss in yield. Further precaution should be taken that the greater portion of the lip be left on the hide.

At this point the ear hair is usually removed. This hair has a very fine texture and is used in making high-grade brushes. When the head is removed, some means must be devised to make the identification of this head possible until the viscera of the carcass has been inspected. This precaution must be taken so that in the event of total condemnation the correct head may be located. Numbered racks or head chains usually are employed.

All of the sticking and heading work is done as the carcass moves along the rail, usually propelled by an endless chain in the larger packing plants. When the carcass reaches its designated bed it is lowered from the rail and "pritched up" on its back.

Foot Skinning.—Front foot.—First cut around the hoof so that the hide will present a straight edge. The dewclaws are then cut off and a straight cut is made on the inside from the hoof to the knee joint. The foot is then skinned on either side and the knee joint disconnected. Taking the shin bone in the left hand, the hide on the front of the foot is removed by one cut from the knee to the hoof.

Leg Breaking.—Hind foot.—This operation is practically identical with the front foot operation, but precaution must be observed that the fell of the web is not cut or broken.

Ripping Open.—A straight incision is then made from the original incision made by the sticker to the pizzle butt. While it is necessary to open up the carcass into the abdominal cavity, at the same time the paunch must not be molested.

Raising Gullet.—Cut down alongside the gullet, using extreme care that the sweetbread thymus gland is not injured. This must be accomplished in such a manner that the sweetbread is left entirely on the left side of the neck. Next the weasand or oesophagus must be separated from the pluck and gullet; this is accomplished to meet U. S. meat inspection requirements from a sanitary standpoint, as follows:

The weasand is separated from the gullet with a knife midway between the pluck and the neck end to the extent of about three inches. A rod with a worm, similar to a corkscrew, is next used. This is inserted into the incision between the weasand and gullet and screwed around the weasand. By pushing forward, the weasand may then be separated from the gullet and pluck to the mouth of the paunch. Next, the rod is drawn forward to within two rings of the end of the gullet. These two rings are then cut from the gullet, left attached to the weasand, and a knot is tied to prevent any paunch manure from being expelled. By simply reaching up through the abdominal cavity the weasand may be drawn through and pulled back into the abdominal cavity.

Floorsman.—The brisket on the high side—that is, the side opposite to that on which the bullock is "pritched up"—is first skinned, then the belly is skinned on the same side to the cod fat. This operation is called "rim-over." The cod is then cleared and the rim-over is carried forward on the pritch to the pizzle. The brisket on the pritch side is next skinned and the rim-over is carried on down to the pizzle. The rim-over then is completed to the ribs on both sides of the carcass.

A straight cut is now made on a line with the incision made by the leg breaker, to meet the open-up incision about four inches behind the cod. Both hind legs are then skinned on the inside. The front shank is opened up on a line with that made by the foot skinner to the center of the shoulder and then on a line to a point about two inches in front of the beginning of the brisket bone, where the original opening incision is met. The high side of the carcass is skinned over the ribs until the flank and nose are entirely cleaned. The pritch stick is changed to the high side and the pritch side is cleaned in like manner.

The floorsman has now completed his task, and this is held to be the most difficult and skillful of the entire slaughtering operation. Essentially his primary consideration was to save the fell from mutilation, but at the same time equal attention was given that the hide be neither scored nor cut.

Breast Bone Sawer.—This operation should be carried on from the neck end backward, as by so doing there is less likelihood of cutting the diaphragm.

Aitch Bone Opening.—Heavily-lactating cow udders are removed, while those on yearling heifers are cut in two at the center, or the pizzle is removed to the pizzle butt and the scrotum cut off, as the case may be. The cartilage of the aitch bone is then cut with a knife at the center line. An incision is made in the gam chord and a spreading device on a hoist is connected. The carcass is pulled forward now to the dressing bed rails, and is elevated to such an extent that the rump is about four feet from the floor.

Fell Cutting.—This operation lies between that of the leg breaker and floorsman. The most simple explanation is that it consists in skin-

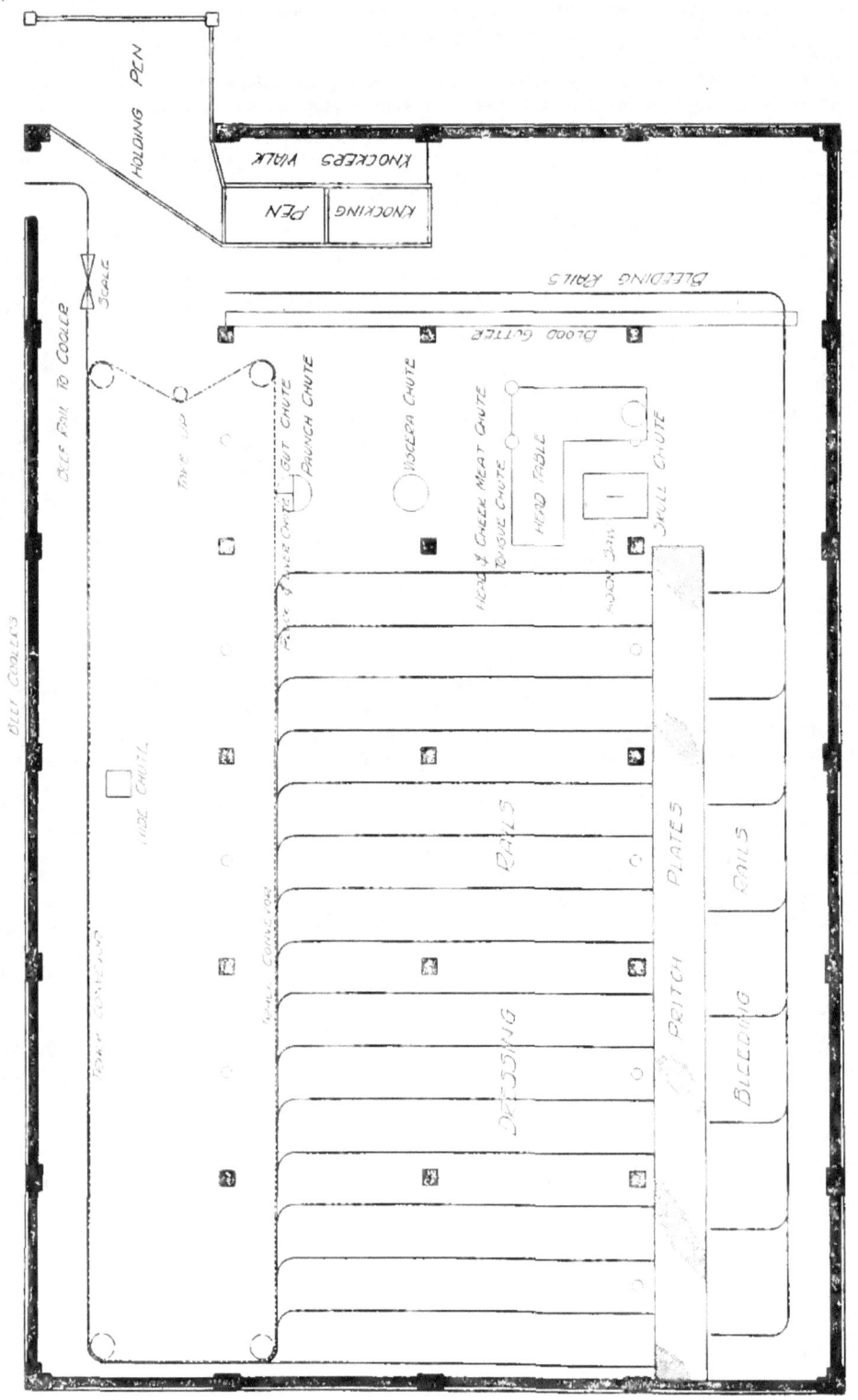

Lay-Out of Modern Beef Killing Floor With Eight Dressing Beds

ning the outside of the hind leg or round. To start this operation, cut upward from the point where the floorsman left off, whereas on the left leg the start is made downward from the point where the leg-breaker discontinued. The most important feature of this operation is the preservation of the fell covering of the round.

Rumping.—Continuing immediately after the fell cutter, the rumper begins at the tail butt and clears out around it. From there the left or low leg is skinned with a downward cut to the fell covering of the hip bone. The right or high leg is worked in the same manner, but the butcher must use his left hand. Because of its subsequent bearing on operations which are to follow, the most important feature of this operation is the absolute necessity of leaving the hip fell covering intact.

Bung Dropping and Tail Ripping.—These two operations are usually assigned to one man. Carefully cut around the bung in such a manner that very little crotch fat is left attached to the bung, at the same time using extreme care that the neck of the bladder is not cut. The tail is now ripped from the tip to the bung opening. By holding the tail securely the skin can be readily pulled off.

Aitch Bone Sawing.—This bone must be sawed upward, and it should be remembered that the bladder is immediately beneath and very liable to be cut. The carcass is now raised entirely from the floor and hung off on the trolleys, which will carry it through the remaining operations and into the cooler.

Fell Pulling and Beating.—The hide is pulled down half way from the hip fell by collaboration of the pulley and beater. It is then removed from the remainder of the hips. If the rumper has done his work well this usually can be accomplished without breaking the hip fell. If, on the contrary, the hip fell has been torn away, it detracts greatly from the appearance of the carcass. Special attention should be given yearling cattle, as in this case it is exceptionally easy to mutilate this fell.

Backing.—This is a continuation from the point where the floorsman, fell puller and rumper stopped. By downward cutting, the hide is removed from the back between the hips and shoulders. As in all foregoing skinning operations, the fell must be left intact. This is doubly true at this point, as any imperfection here is extremely noticeable; the back is the most conspicuous part of the cattle when hanging in the sales cooler.

Clearing Out.—Cut with a downward motion around neck and shoulders from the point where the floorsman discontinued. Extreme care must be used that the hide is not cut, as experience has shown this to be the point where hides are frequently scored.

Hide Dropping.—This operation consists in removing the hide from the back of the neck after the clear-out has been completed, and the hide is entirely removed. The cord in the back of the neck is split on the center line. A skillful hide dropper is able to rectify many of the miscuts of the header and produce a well-evened neck. The hide is now entirely removed. In large establishments it is usually considered best to put the carcass on the moving chain rail at this point, completing all remaining operations while moving toward the cooler

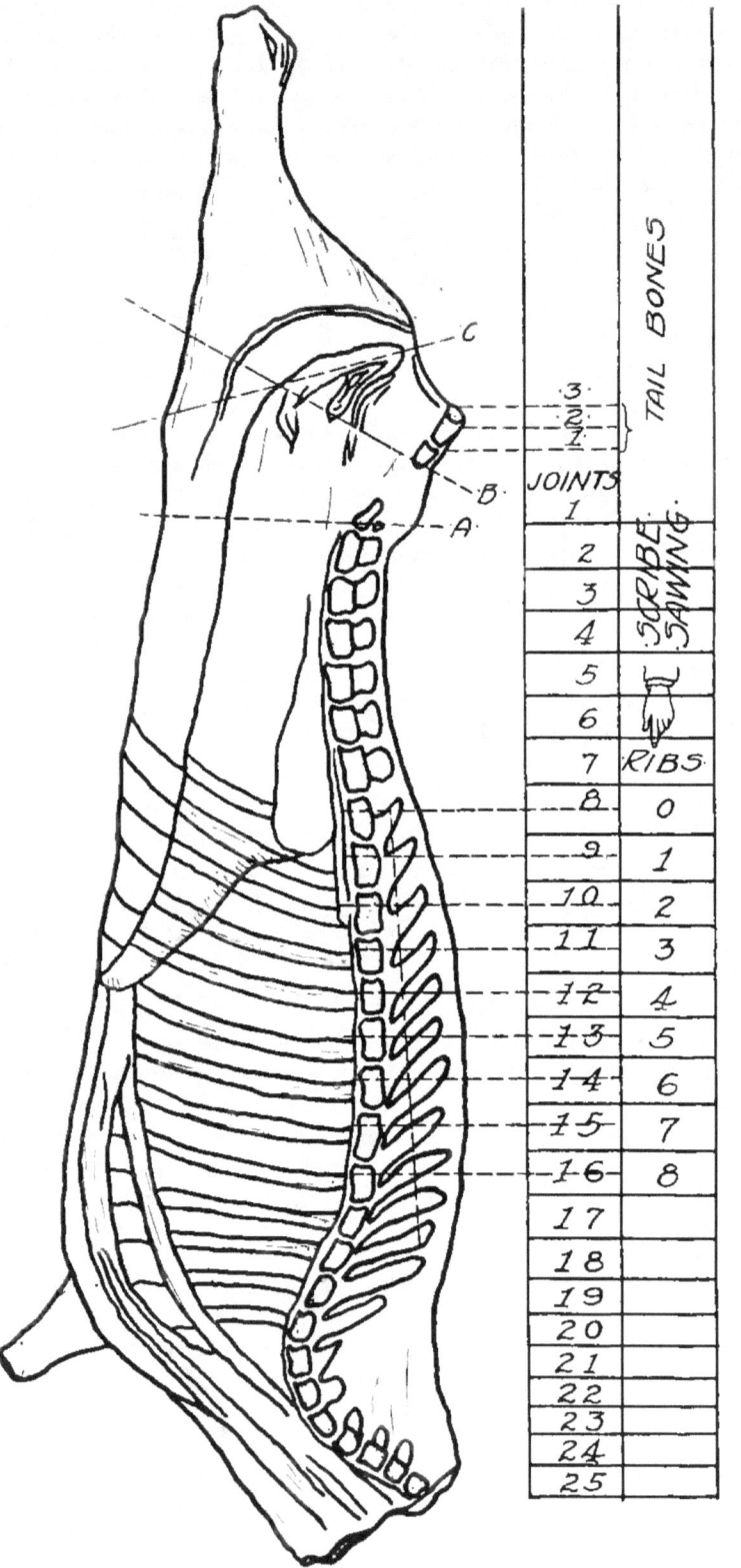

STANDARD METHOD OF DRESSING BEEF CARCASS

Showing location of joints, ribs and place for scribe sawing. C, short round; B, short round and rump; A, full round. A proper round has three tail bones—1, 2 and 3.

Gutting.—In the event that a cow is being handled, the uterus is first cut off, the ovaries recovered and the remainder immediately removed to the inedible section of the house. The peritoneum is cut on both sides, thereby disengaging the bung gut and bladder. The guts, along with the paunch, will drop down of their own weight. The liver is separated from the jejunum or round gut, by cutting through the pancreatic gland, and then is lifted out after disengaging it from the vertebrae. The diaphragm is now cut and the heart, lungs and trachea (pluck) removed.

The yield on the dressed carcass will greatly depend on the ability of the gutter; first, from the standpoint of not detaching the pelvic fats, kidney fats, and diaphragm from the carcass; secondly, that he does no damage to the viscera.

Tail Sawing.—The tail is now disjointed at the third joint. The saw is placed in the center of the tail butt and an equal division of the tail bone is made to the first lumbar vertebra.

Splitting.—Inasmuch as the sale of beef is made chiefly on the appearance of the carcass, the splitting operation holds a very prominent place in the dressing. Inattention to splitting may render all previous efforts along this line futile. A poorly split backbone, very conspicuous on the carcass, will probably do more injury to the appearance of the carcass than any other imperfection.

The splitter should hold the cleaver close to the blade, as he will have better control by so doing and thereby reduce the possibilities of shattering the bone. In splitting the skin the blade should come into contact with the backbone at right angles; that is to say, horizontally. After completing the loin, the blade should be so directed that the contact with the bone will be at an angle of forty-five degrees, with the handle down and the blade up. This method makes it possible to split the fine bones to better advantage. The splitter should continue until he reaches the cervical vertebrae.

Neck Splitting.—Continuation of the splitting from the beginning of the cervical vertebrae through the neck, thereby separating the carcass into halves. This completes the actual slaughtering operation. The carcass is now trimmed and washed as follows:

Scribe Sawing.—With a scribing saw the "fin" bones are sawed almost through and then bent outward to an angle of about forty-five degrees. This gives a broadening effect to the back, and adds considerably to the appearance. All outside bruises are trimmed off. The pizzle butt is trimmed, flushed with the aitch bone, and the cow bags evened to present a uniform appearance. In the case of cuts or scores on the fell, the surrounding fat may be pulled over to cover and set with skewers. After cooling, the skewers must be removed, and the fat will lay as placed.

The tail is then taken off and sent to the chill room. The spinal cord is pulled from the spinal column and the diaphragm is trimmed. The aorta and jugular are removed and heart fat trimmed to present a good appearance. The sweetbread is cut off, trimmed and sent to the cooler.

Good results are obtained, and it is very desirable to pump the shoulder by moving it up and down vigorously, as this will induce any blood remaining in the carcass to drain out.

A very great saving in edible fats may be effected if the trimming is closely supervised. All trimmings should be immediately collected and guarded against contamination.

The entire carcass is now thoroughly washed in water of about 120° Fahrenheit. It is then scraped and dried with warm cloths throughout. This drying prevents carrying moisture to the coolers. Towels or napkins are placed under the kidneys and at the neck to absorb any dropping blood from the large arteries and veins. The U. S. Government inspection legend is branded on the primal cuts, and the carcass is sent to the cooler.

Beef Killing Costs

It is very difficult to set a standard as to the cost of slaughtering per head of cattle, for it is rare, if ever, that the same conditions are found in two plants. Much depends upon the general layout, the equipment used, and the speed of the operators.

For illustration, however, there is given here a schedule of the labor used in a cattle-dressing gang in a large packing plant which is considered a very efficient organization:

Labor in Beef Washing Gang.

Position	50	70	90	100	120	130	150
Foreman	1	1	1	1	1	1	1
Push back beef	½	½	1	1	0	0	0
Switchman	½	½	½	2	2	1½	2
Cut off cords	½	½	½	0	1	1	1
Trim skirts	½	½	½	1	1	2	2
Trim bruises	1	1	1	1	2	2	2
Spraying kidneys	½	½	½	1	1	1	1
Trim. on ladders	1	1	1½	2	2	2	2
Scrub hind shanks and rounds	1	1	1	1½	3	3	4
Pump and wash kidneys	½	½	½	1	1	1	1
Pumping shoulders	½	½	1	1	1	1	1
Scrubbing outsides	1	2	2	1½	2	3	3
Scrubbing shanks and shoulders	0	0	0	1	1	1	1
Scraping outsides	1	1	1	1	1	2	2
Pumping skirt veins	0	0	½	½	1	1	1
Wiping outsides	1	1	1	1	2	2	2
Put in neck and kidney rags	½	½	1	½	2	2	2
Scrub necks and chines	1	1½	1½	2	2	3	3
Sewing necks and breast veins	½	½	1	1	1	1	1
Wiping hind shanks and rounds	½	1	1	1	2	2	2
Put on Govt. stamps	½	½	1	1	1	1	2
Scribe sawing	½	1	1	1	1	½	2
Wiping ribs	0	0	0	0	1	1	1
	14	16½	20	24	32	35	39

Labor in Cattle Dressing Gang

Number of Men for Different Gangs, from 50 to 150 Cattle per Hour.

Position	50	70	90	100	120	130	150
Foreman	1	1	1	1	1	1	1
Clerk	0	0	0	1	1	1	1
Knocker	1	1	1	1	1½	2	2
Shacklers	1	1	1	1	1½	2	2
Hoisters	1	1	1	1	1½	2	2
Stickers	1	1	1	1	1	1	1
Headers	1	1	1	2	2½	3	3
Squeegee	0	0	0	0	1	1	1
Cutting off heads and putting in racks	0	0	0	0	1	1	1
Dropping cattle	½	½	½	½	½	1	1
Pritching up	½	½	½	½	½	1	1
Carrying over shackle	0	0	0	0	½	1	1
Skinning feet	1	2	2	2	2½	3	3½
Raising gullets and cutting out sweetbreads	½	0	0	½	1	1	1
Ripping open	0	0	½	½	1	1	1
Breaking hind legs	2	2	2½	3	3½	4½	5
Floorsmen	3	4	5	6	7	8	9
Caul pullers	1	1	1	1	2	2½	3
Crotch washer	0	0	½	½	1	1	1
Opening notches	½	½	½	1	1	1	1
Sawing breasts	½	1	1	1½	2	2	2
Put in spreaders and hoisting	1	1	1½	2	2	2½	3
Pulling hides on shank hind	½	½	½	1	1	2	2
Break legs on hoist	½	½	½	1	2	2	2
Ripping tails	½	½	½	½	½	1	1
Cut out bladders	½	½	½	½	½	½	1
Fell cutting	2	2½	3	3½	4	5	5½
Rumping and cutting out bungs	1	1½	1½	2	2½	2½	3
Scrub hind shanks	½	1	1	1	2	2	2
Hold out hide for fell beater	½	1	1	1	1½	2	2
Fell beater	½	½	1	1	1½	2	2
Backing	1	1½	1½	2	2½	2½	3
Cutters	1	1½	1½	2	2	2½	3
Sawing tails	1	2	2½	3	4	4½	5
Pulling tails	1	1	1	1	1	1	1
Splitting	2	2½	3	3½	4	5	6
Hoisting and hanging off	1	1	2	2½	3	3½	4
Pulling back cattle	0	0	0	0	1	1	1
Squilgee	1	1	1	1	1	1	1
Put up travelers	1	1	1	1½	2	2	2
Clearing out	1	1	1½	2	2½	3	3½
Dropping hides	1	1½	2	2	2	3	3½
Splitting necks	1½	1	1½	1½	2	2	3
	35½	42	49½	60½	79	93½	103

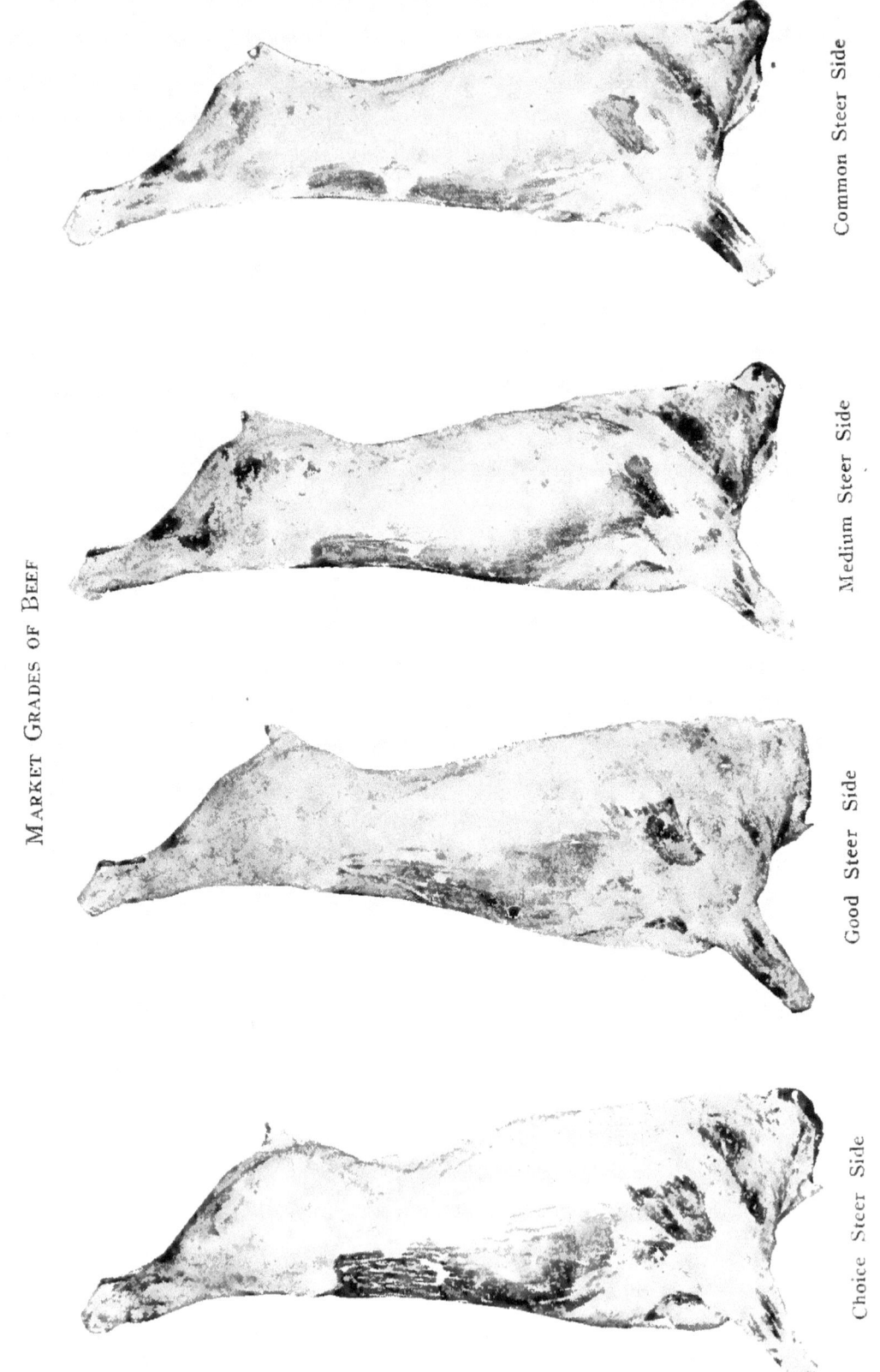

Market Grades of Beef

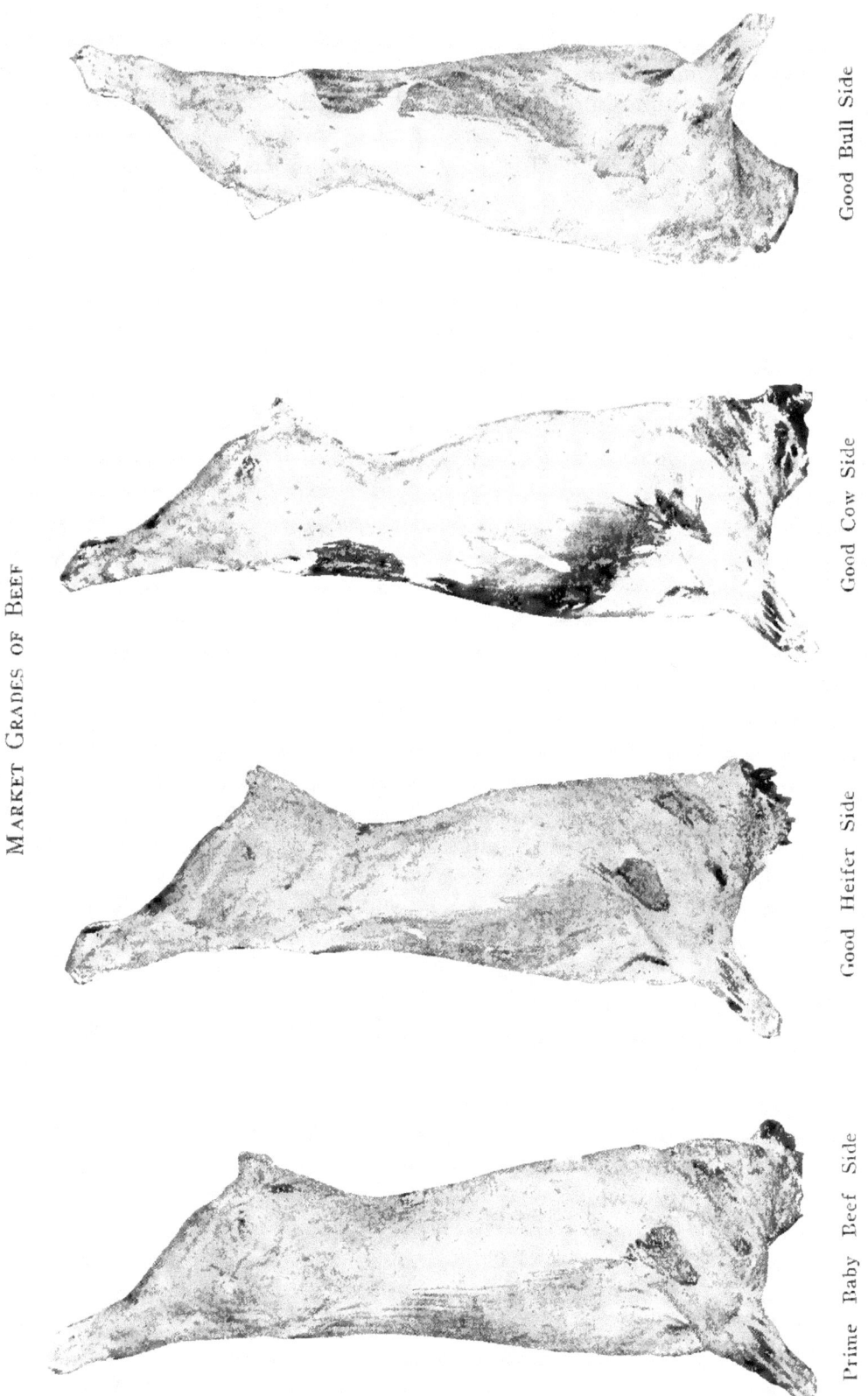

Market Grades of Beef: Good Bull Side; Good Cow Side; Good Heifer Side; Prime Baby Beef Side

BEEF COOLING

There are many different kinds of refrigeration for cattle coolers. Originally the medium of refrigeration was the old-fashioned ice box, in which the ice would be filled in the ice loft by conveyors. It was only a step from this to the ammonia expansion or brine pipes. About twenty years ago the use of brine sprays, or the curtain system, was introduced. In later years the nozzle brine spray system seems to have become the popular method, and is no doubt the most economical and the most efficient method. (See chapter on "Refrigeration" for illustrations.)

In filling the beef cooler with cattle it is always advisable to have the cattle spaced about six inches apart. In other words, do not let the sides come in contact one with another, as this prevents proper circulation and is frequently the cause of sour spots.

The following is a schedule of the standard temperatures used in beef coolers in some of the largest packing plants:

Temperature of cooler at time of filling	30-34° Fahr.
High point during filling, not over	44° Fahr.
Ten hours after filling	38° Fahr.
Twenty hours after filling	36° Fahr.
Thirty-six hours after filling	35° Fahr.
Forty-eight hours after filling	34° Fahr.

Heavy carcasses should remain in this temperature at least 48 hours, and it is advisable to leave them in 72 hours. The temperature of the meat in the hip socket and in the shoulder on heavy carcasses, twenty-four hours after the carcass is in, subject to the above temperature, will show from 51° to 54°. Forty-eight hours after being placed in the cooler the meat temperatures will show around 42°. In seventy-two hours the temperature of the meat will be down around 38°, which is the ideal temperature and indicates clearly that all the animal heat has been removed.

Light cattle and thin cows can be shipped out in twenty-four hours, but even on this grade it is hardly satisfactory, because in ribbing the cattle on the loading dock, unless the beef is thoroughly chilled and set, the meat will slip from the rib bones and have an unsightly appearance.

It is very essential in the beef cooler to put down fresh sawdust at frequent intervals and keep the cooler painted up, so that it will always have a clean, fresh smell.

The average shrinkage of beef in the cooler forty-eight hours, under proper conditions, is about 1½%, but sometimes runs as high as 1¾%.

BEEF GRADING

The grading of dressed beef is practiced in every packing plant, and if the buyers are skillful the grades will correspond closely to the quality of the purchase.

There is one chief consideration in beef grading: the quality of the meat. Carcasses weighing 1,050 lbs. and up are considered "extra heavy"; those from 900 to 1,050 lbs. are "heavies." The other gradings according to weight range as follows: 800 to 900 lbs.; 700 to 800 lbs.; 600 to 700 lbs.; 500 to 600 lbs.; under 500 lbs. The designation of each grade varies according to the packing plant, either letters or numbers being used.

The grading as to quality in most plants is as follows: Prime, choice, good, medium, fair and poor. In addition to this there are old cow carcasses, cows which have had but one calf, heifers, bulls and stags. In the case of cows, bulls and stags, the terms "common," "cutter" and "canner" are frequently used. Some packers go into much more detail than others in grading.

The U. S. Bureau of Markets and Crop Estimates has prepared a good deal of detailed material on this subject, published in bulletin form, which may be secured if a complete description of each particular grade according to Bureau standards is desired.

"Spotters" is a term applied to carcasses having small dark blood spots, apparently caused by the rupture of minute blood vessels and the coagulation of small quantities of blood which did not escape at the time of bleeding. They are more frequent in well-finished carcasses. These spots, which often develop in the muscular tissue, vary in size from small specks to areas one-half inch or more in length and one-fourth of an inch or more in diameter. The exact cause for this is not fully known.

BEEF LOADING

That beef may reach its destination in the same prime condition as when hanging in the packing house cooler, refrigerator cars, equipped with brine tanks and beef rails, have been built. While these make the transportation of highly-perishable meats possible, this can only be successfully accomplished by the utmost care in loading.

Beef will never reach the firm condition necessary for proper loading in less than thirty-six hours from the time of slaughter, and then only in the case of light cattle. Every effort should be made to see that the carcass is allowed to remain at least forty-eight hours in the cooler at 35° F. before it is loaded.

The cattle that have been selected and ordered for a given car are assembled on the rail. For the purpose of identification they are numbered; tickets bearing the carcass number are placed on each of the two forequarters and on each of the two hindquarters.

Ribbing.—The forequarter and hindquarter are usually separated between the twelfth and thirteenth rib; that is to say, one rib is left on the hindquarter. While this is the usual practice, there are many exceptions, where two, three and even four ribs are left on the hindquarter. This, of course, is governed entirely by the demand.

The actual operation requires careful and skillful butchering, that the exposed lean tissue will present a well-evened appearance. The knife is inserted eleven inches from the backbone, and with a single cut the carcass is opened from that point toward the belly side. About three inches of the flank is left intact to support the forequarter when the backbone shall have been cut. The knife is then reversed and again with a single stroke toward the backbone the remainder of the lean tissue is separated. The backbone is sawed and the forequarter left suspended by the flank until loaded.

The knife blade should be directed as closely as possible to the rib on the forequarter, thereby allowing the lean tissue between the twelfth and thirteenth rib to remain on the hindquarter. Care must be taken also that the kidney fat is not mutilated.

Refrigerator Cars.—The car in which the beef is to be loaded must be washed thoroughly with hot water and soap throughout, after which the doors should be allowed to remain open. Fresh air and sunshine are the best agents to sweeten and dry the car. It is imperative that the car be pre-chilled to a temperature below 40° F. before loading is started.

To accomplish this, after washing and ventilating the car is closed and well iced with salt and crushed ice. Standard specifications are 12 per cent of salt in summer and 8 per cent in winter. In some cases cars are equipped with brine sprays, and may be connected with the house brine refrigerating system, thereby hastening the pre-chilling operation.

A new type of refrigerator car is equipped with a patented system of refrigeration by means of brine circulation, which is accomplished by connecting the brine tanks in each end of the car with brine pipes, three on a side. Each brine tank is divided by a partition in which is located check valves which operate in opposite directions in each end of the car. The drain pipe is located high enough in the brine tank so that there is always some brine left in the tank, and when the ice and salt is dumped in the brine rises to the level of the pipes, and the swaying motion of the car causes it to circulate from one end of the car to the other. The claims made for this system are that less ice is required, the refrigeration is more complete in all parts of the car, and there is less wastage in meats at destination.

It is unfortunate that all refrigerator cars are not equipped with brine tanks, but it frequently becomes necessary to load beef in those equipped with large wooden tanks. These cars are iced by filling the tank to one-third capacity with lump ice, then about one-third of the salt to be used is distributed over the top of the ice. The remainder of the tank is filled with a mixture of lump ice and salt. Crushed ice in this style of tank would form into a large cake, giving poor circulation and refrigeration, and frequently causing serious damage by the sliding in the bunkers due to the train movement.

Loading Platform.—When any chilled object is brought into contact with the warm air, the moisture of the air will be condensed on it, causing what is known as "sweating." This is true of chilled beef, and any carcass that has been allowed to sweat will subsequently become slimy and moldy in transit. To guard against this, loading platforms should be closed in, all doors leading to the cars vestibuled, and the temperature carried as closely as possible the same as that of the cooler.

Loading the Car

The carcass having been ribbed in the cooler, is now brought forward on the rail to the car door on the loading platform.

The hinds and fores are usually loaded into opposite ends of the car. As only one man is used to assist the luggers in hanging off the beef, it is considered good practice to load a given number of fores and then the corresponding hinds.

In hanging in the car rights and lefts must alternate, so that the quarters will hang bone to bone and flesh to flesh in even rows. By so hanging, mutilations of the fell covering that would be caused by the bone tearing into the flesh when swaying in the car, due to train movement, are avoided. To prevent accidents in lugging beef, the lugger should carry the right quarter on the left shoulder and the left quarter on the right shoulder, thus turning the bone away from himself, which on account of its sharp, ragged edges has caused many serious cuts.

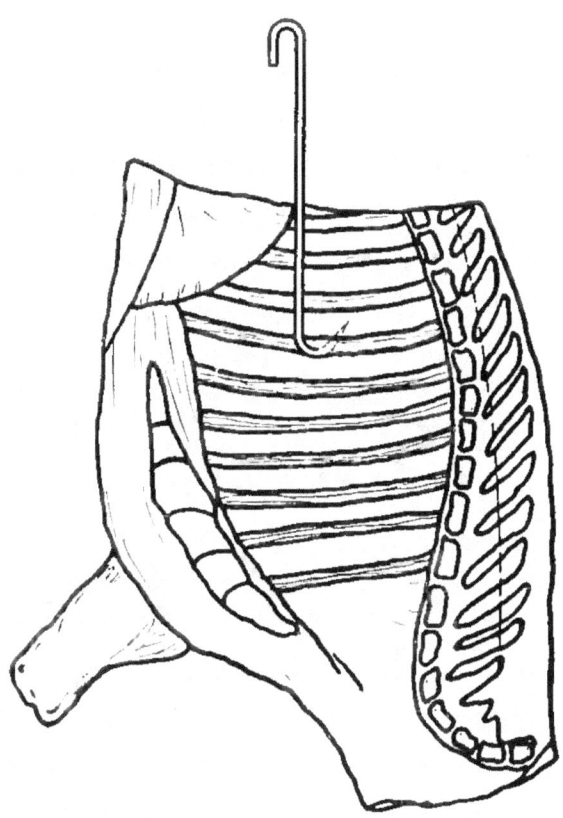

Proper Method of Hanging Beef Forequarters in Refrigerator Car

In hanging the hindquarter the car men should be closely supervised to see that the car hook be placed through the gam chord, in the same hole where the beef trolley hook was originally inserted, thereby eliminating a second hole and preserving the web of the gam. The forequarter should be hung between the ninth and tenth rib, that is to say, three ribs forward from the plate rather than through the more expensive meaty portion of the ribs.

The tallow covering of beef is very susceptible to dirt, rust, and grease marks. To avoid this rails should be well painted and free of rust; the beef trolley should be wiped free from oil after lubricating, and all engaged in the lugging and handling of the beef should be well supplied with clean frocks.

With careful regard for loading along the lines as herein set forth, and re-icing every twenty-four hours while in transit, the beef may be transported to its destination in the clean, sweet, and wholesome condition essential to food products.

HANDLING BEEF FOR EXPORT

Export beef is selected from well-finished native cattle, usually steers, although fat cows are sometimes utilized, but in very few cases is bull beef shipped.

In dressing, extra and particular care is used. In the first place, export cattle must be stuck as soon as possible after knocking. In ripping open the cattle, care should be taken not to reach the clod and crotch, only rip through the hide. Have the caul puller open the clod and crotch and properly wash with a cloth, using hot water. The hind shanks should be scrubbed on the first hoist with a brush, using hot water; also in this case the hind shanks and rounds should be wiped with a hot cloth. Two loose joints should be left on the tail. After the backer has finished his work, the backs of the cattle should be wiped with a hot cloth which has been dipped into a pail of hot water and thoroughly rinsed, allowing very little water to remain in the cloth. The flank should be dropped slightly at a high point. Do not use scribe saw in dressing; simply beat back the loin.

Care should be taken in backing the cattle, so as not to peel any loins, as this is quite objectionable. The breast fat should be trimmed out as soon as hide is dropped and the neck should be wiped immediately on the outside with a hot cloth.

No water should be used to cleanse the carcasses that are not bruised; but instead, dry, hot cloths should be used to cleanse thoroughly. Should there be bruises, trim them well, then scrub with a fountain brush and as little water as possible. They should then be scraped and wiped thoroughly with dry, hot cloths. The inside and outside of the neck is an exception to this rule, and may be washed with a cloth and hot water; also hot water may be used on the ribs and possibly on the outside of the brisket. All these parts should be wiped thoroughly after washing, with dry, hot cloths.

The shoulder should be pumped in order to drain the blood from the neck and skirt veins. Then the skirt veins should be tied and shoulder veins skewered before washing. Neck cloths should be left in for five hours.

In the cooler, the handling is much the same as with custom cattle, excepting that the carcasses are given a thorough wiping under the skirts and inside generally. The kidneys are handled in the same way as custom cattle. The crotch fat should not be destroyed.

Export cattle should enter cooler at a temperature of 45° Fahr.; a lower temperature destroys the bloom. Allow two feet of space between each carcass. The following morning the temperature of the cooler should be 36°, and within 48 hours it should be 31°. If at all possible, the beef should be in the cooler 72 hours before shipping; but if it should be necessary to ship within 48 hours after killing, arrangements should be made to spread the cattle as well as possible in the cooler, so as to improve their condition. The best way to stop complaints on blood-stained backs in transit or at destination is to get the cattle firm and hard before shipping.

In loading, care should be taken to see that the luggers do not handle the sack of beef with dirty gloves or greasy frocks. In hanging in the cooler, it is best always to load on the fourth rib.

The car in which the beef is to be shipped should be iced a day before shipment and the temperatures preparatory to loading must not exceed 34°, keeping doors of the cars closed as much as possible, so that it will not exceed a temperature above 44° when loaded. The car should be iced again just before shipment, and, of course, kept iced in transit to seaboard. In initial icing 15 per cent salt should be used; in re-icing, 12 per cent salt may be used; but in winter this may be reduced to approximately 6 per cent.

It is best to rack both sides and ends of all cars used for export beef. Refrigerated steamers equipped for handling chilled beef should be able to hold a 29° summer temperature and a 30° to 31° winter temperature. This may vary according to the condition of the beef upon arrival.

For direct shipment to foreign houses a heavy muslin or beef cloth is used, the amount required being four yards per quarter. This is sewed all around the beef with four-ply cotton twine. Should it be necessary to ship to domestic seaboard freezers for storage before shipment, or if shipping frozen to distant ports, the beef first should be covered with the best grade of cheesecloth, which again is enclosed with burlap as described above. The weights should be distinctly marked on the inside of each quarter. Shipments consist principally of steers, weighing from 600 to 1,000 lbs. If a part of the carcasses are cows, they are marked as such.

BEEF CUTTING AND BONING

The boning of beef is done in cases where the carcasses are not suitable for regular wholesale cuts, or where only a portion of them are eligible for this purpose. Boneless beef is sold very largely to the trade, and at certain times there is a heavy demand for this class of product for canning purposes and the manufacture of sausage. Following is a description of the various cuts generally made by boning out cattle:

Beef Roll.—Beef roll "old style" is cut from the back of a beef rib and is entirely boneless. It is used for both roasts and steaks and has quite a wide range in weight, the average being about 5 to 6 lbs.

Spencer Roll.—This is a cut with the back covering of the rib left on the roll for about three-quarters of the length, making it more desirable for roasts.

Rib Ends.—These are pieces usually 2 to 4 inches long, cut from the plate end of a regular beef rib. Rib ends are used for roasting.

Beef Flank.—A cut from the hindquarter just below the loin and in front of the round. When trimmed it has the flank steak and the tallow removed; untrimmed, it is known as a rough flank with the flank steak and tallow left in.

Beef Fore Shank.—The lower part of the fore leg cut with a piece of the clod on or off.

Beef Skirt.—The fleshy part of the diaphragm.

Beef Sirloin Strip, "Boneless."—This is cut from the back of a short loin and used for sirloin steaks, the average weight being about 5 to 6 pounds.

Beef Strip Loin, "Bone In."—Practically the same cut as a sirloin, except with the bone in, and weighs about 30 per cent more than a boneless sirloin strip.

Beef Tenderloin.—The beef tenderloin is cut from just under the backbone of a full loin of beef. It is used principally for steaks, sometimes for roasts. The average weight is about 3 to 5 pounds, and anything under 3 pounds is called a "T" strip. The "T" strip is practically the same cut as the beef tenderloin, except that it is of lighter weight.

Carcass Cuts, Chicago Style

The following cuts are usually designated as "Chicago style" cuts. Methods used in other centers are described elsewhere.

Beef Rattle.—The forequarter with the seven-rib roast removed.

Beef Chuck, Square Cut.—The front end of the fore quarter, including the first five ribs, the seventh rib roast, brisket, plate, neck and shank removed. Manhattan style or cross-cut chuck is the front end of the fore quarter and includes the brisket, neck and shank.

Beef Round.—That part of the hind quarter that remains after removing the loin and flank, or the entire hind leg after removing the rump and shank. When the shank is removed and the cod fat left on it is commonly known as a buttock.

Beef Loin.—This is the front part of the hind quarter with the flank removed. It has one rib and ranges in weight from 25 pounds for the lightest up to 100 pounds for the heaviest. No. 1 native steer loin has a correspondingly wide range in price.

Beef Loin End.—This is the butt of a full loin which is cut off in making a short loin for high-class hotel trade.

Beef Rib Roast.—That part of the fore-quarter back of the chuck, and has seven ribs. The weight ranges from 16 pounds for the lightest to as high as 55 pounds for the heaviest native steers, with a wide range of price.

Beef Short Loin.—This is made from the regular loin, principally for high-class hotel trade, and is the small or rib end of the loin cut in different lengths, from a pin-bone cut to a flat-bone cut. A pin-bone short-cut loin is the shortest and most expensive, and is about 50 per cent of the full loin.

Beef Sirloin Butt, Boneless.—The large end of a full loin "loin end" bones out. It is used for both roasts and steaks, and has a large range in weights, the average being from 6 to 8 pounds.

Beef Rump Butt.—This is cut from the hip or top end of a round, and practically boneless, only an insignificant piece of the tail bone being left in.

Beef Clod.—Beef clod is frequently called shoulder clod and is cut from the shoulder part of the chuck or the upper part of the foreleg, entirely boneless. The average weight is about 8 to 10 pounds.

Beef Hams.—The entire meaty part of beef round split, from which the bone has been removed. The set consists of knuckle, outside and inside, all boneless with the exception of a very small piece of bone, hardly enough to consider, left in the knuckle. This can be sold in either regular sets or each piece separately, and is used principally for dried beef, but is very desirable for pot roasts or steaks.

Beef Plate.—This is cut from the forequarter, the bottom part below the chuck and the rib. The full plate contains navel end and brisket.

Beef Brisket.—The front part of a full plate.

Beef Navel End.—The rear part of the full beef plate.

Beef Neck.—The rough part of the neck removed from the chuck.

Hanging Beef Tenderloin, "Boneless."—This grows directly under the middle of the backbone, is about one foot long and weighs about 2 pounds.

Beef Hind Shank.—Lower part of the hind leg or round.

Yields in Boning Out a Carcass

Following is a test which indicates the percentage of meat, bones and by-products obtained when "boning out" six carcasses:

Meat

Beef hams	356 lbs.	
Canning meat	580 lbs.	
K. or black meat	196 lbs.	
Flank steaks	12 lbs.	
T. strips	15 lbs.	
Shoulder clods	108 lbs.	
Boneless briskets	44 lbs.	
Boneless chucks	300 lbs.	
Shank meat	124 lbs.	
Weight	1,735 lbs.	71.84%

Bones

Plate and rib bones	151 lbs.	
Forequarter shank bones	79 lbs.	
Hindquarter bones	99 lbs.	
Chuck bone	72 lbs.	
Rump and sirloin bone	123 lbs.	
Shoulder blade bone	19 lbs.	
Weight	543 lbs.	22.48%

Oleo Tallow

Beef tallow for oleo .... 43 lbs.   1.78%

Tankage

Scrap tallow .... 39 lbs.
Scrap tankage .... 42 lbs.
                                                        ─────
Weight .... 81 lbs.   3.36%

Kidneys

Kidneys .... 13 lbs.   0.54%
Total weight .... 2,415 lbs.
Percentage 100%

Boneless Cuts Compared to Carcass Cuts

The following test shows the per cent of boneless cuts in comparison to regular carcass cuts.

Loins 17%
- Tenderloins 12%
- Butts 24%
- Strips 20%
- Bones 21%
- Trimmings 13%
- Tallow 10%
                                                        ─────
                                                         100%

3" Pin Bone..... { 55% loin end / 45% short loin } 100%

Flat bone loin.. { 36% loin end / 64% short loin } 100%   { 88% boneless butts / 12% bones } 100%

Ribs 6%
- Rolls 32%   41%
- Bones 30%   30%
- Trimmings 38%   29%
                                                        ───── ─────
                                                         100% 100%

1½" pin bone.... { 50% loin end / 50% short loin } 100%

Strip loin........ { 76% boneless strip / 20% bones / 4% trimmings } 100%

S.C. chucks..26%
- Boneless chucks 62%
- Clod 16%
- Bones 22%
                                                        ─────
                                                         100%

Round ...23% { Rump butts ... 14%
Canning meat ... 3%
Shank meat ... 6%
Tallow ... 3%
Bones ... 19%
Insides ... 26%
Outsides ... 16%
Knuckles ... 13%

100%

Full plate ... 13% { Navel end ... 60%
Brisket ... 40% { Boneless brisket ... 67%
Bones and trim'gs ... 33%

100%　　　　　　　　100%

Shanks ... 4% { Shank meat ... 50%
Bones ... 50%

100%

Flank ... 4%
Suet ... 4%

100%

Yields in Cutting "Canner" Cattle

Here is a sample test made in canning cattle which indicates the yield of the various cuts:

Sirloin butts	3.903%
Strips	4.204%
Tenderloins	2.552%
Boneless chuck	13.813%
Rolls	2.552%
Plates	12.162%
Insides	7.957%
Outsides	5.555%
Knuckles	5.555%
Clods	5.105%
Rump butts	2.402%
Flank steak	.600%
Tenderloin	.450%
Front shanks	7.207%
Hind shanks	4.650%
Soft bones	6.906%
Trimmings	8.108%
Tallow	1.200%
Kidneys	.600%
Tankage and marrow	4.519%
	100.00 %

Wholesale Beef Cuts and Their Uses

The wholesale or primal cuts into which a beef carcass is divided, and the percentage of each, are as follows:

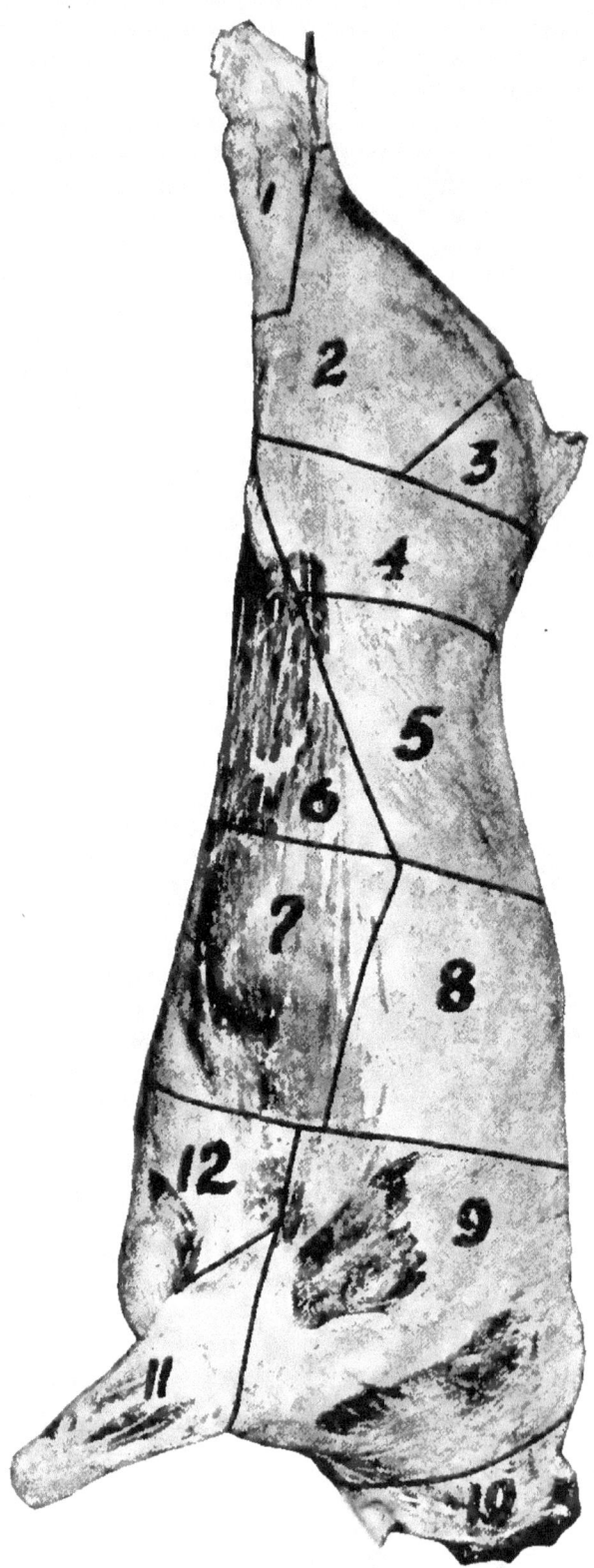

Standard Wholesale Beef Cuts (Chicago Style)

1. Hind shank
2. **Round**
3. Rump
4. Loin end
5. Short loin
6. Flank
7. Plate
8. Rib
9. Chuck
10. Neck
11. Foreshank
12. Brisket

Round with rump and shank	23%
Full loin	17%
Flank	4%
Full plate	13%
Ribs	9%
Chuck	26%
Foreshank	4%
Suet	4%
	100%

Round.—The above round, which makes up 23% of the carcass, is removed at the point of the ball joint, expert cutters being able to strike this joint so that where it is sawed the surface of the bone will be about the size of a dollar. The rump and shank can now be removed from the round, the former being used for steaks and pot roasts and the latter for soup meat.

Loin.—The full loin, which makes up about 17% of the weight of the carcass, contains the best beef. It may be divided into a short loin and a loin end, this being done by cutting at the joint of the pin bone. These two cuts will each weigh approximately 50% of the entire loin. The short loin, or smaller end of the loin, is considered to be the better, and is worth more money than the loin end. The first cut on the short loin, or the pin bone cut, is the choicest of all.

Flank.—While the entire flank of the carcass comprises four per cent of its weight, the flank steak, which is left after the full flank has been trimmed, weighs less than one-fifth of the entire flank. The balance of the flank is used for sausage meat usually, for the flank steak itself is considered by many as a choice piece of meat.

Plate.—A full plate contains twelve ribs, and consists of the lower side of the carcass from a point just back of the brisket, extending to the flank. The full plate is generally divided between the fifth and sixth ribs, the front end being called "brisket" and the back the "navel end." Short ribs may be cut from the upper side of the plate by cutting off a strip about four inches wide, upper or rib side, the same being used for roasting.

Ribs.—Beef ribs consists generally of seven ribs cut just back of the chuck and ahead of the loin. The quality of this beef is very high, being used for roasts. The percentage of ribs run about 9% of weight of the carcass.

Chuck.—The chuck or forequarter contains five ribs and figures 26% of the carcass. In fancy cattle, cuts off the chuck may be used for steak, but in lower-grade cattle the chuck is used for stewing beef.

The clod or the immediate portion that lies directly over the shoulder blade is sometimes removed separately for stewing. The foreshank is used for soup meat chiefly.

While beef may be cut in many various ways, this is a brief description of the usual wholesale cuts.

Variations in Beef Cutting

From the standard method of cutting carcasses into the wholesale cuts given here, a variation is resorted to in certain sections, especially in the East, such as Philadelphia, Boston and New York. These cuts are made for the purpose of including a larger percentage of either the round or chuck in with the loin and rib cuts, so that more steaks will result therefrom. Top and bottom rounds, for instance, are well known in the Boston and New York districts.

As an example of the yields which may be obtained from this practice, here are three tests showing the percentage of each cut obtained, first by the New York and second the Philadelphia method, compared to the Chicago method:

New York Cuts

Ribs	9.55 per cent
Loins	15.74 per cent
Flanks	5.55 per cent
Suet	3.62 per cent
Navel	8.61 per cent
Rounds	23.27 per cent
New York chucks	33.66 per cent
Total	100.00 per cent

Philadelphia Cuts

Rumps and rounds	34.00 per cent
Ribs and loins	22.00 per cent
Rattler, chuck, plate, brisket, and flank	44.00 per cent
Total	100.00 per cent

Chicago Cuts

Chucks	28.00 per cent
Rounds	23.00 per cent
Navels	8.00 per cent
Flanks	2.00 per cent
Flank steaks	.50 per cent
Kidneys	.25 per cent
Ribs	10.00 per cent
Loins	15.00 per cent
Suet, No. 1	3.00 per cent
Suet, No. 2	.50 per cent
Shanks	4.00 per cent
Brisket	5.00 per cent
Necks	.75 per cent
Total	100.00 per cent

PLATE BEEF

The beef plate portion of the carcass is one which is frequently used for curing, so that it may be stored and sold at a future time, or shipped a great distance for consumption.

As has been described heretofore, the plate usually contains twelve ribs and is divided for curing between the fifth and sixth ribs, the brisket end and the navel end.

Various grades of plate beef are produced by the different packers, selections being made on the basis of the quality of the plates; that is, whether they come from first, second or third-grade cattle. This is usually gauged by the weights. No. 1 plates weigh in the neighborhood of 55 to 60 lbs. with the brisket off; navel ends cut into three pieces, and rib strips into two equal parts. In packing this grade usually ten ribs are put in each barrel and the barrel completed with navel ends. This style of packing is used principally for Great Britain.

The second grade of plate beef is made from plates weighing from 40 to 50 lbs. The assortment in packing is the same as on No. 1 plates. In this case, about twenty to twenty-five pieces per barrel are required.

A third grade is made from plates weighing from thirty-five to forty-five pounds average.

Still another grade is made from plates which average less than forty pounds, but which do not include the brisket.

MESS BEEF

There is another classification of beef cuts used for barreling and curing for export. This is known as Mess Beef, and generally consists of the following assortments:

One plate with brisket off, cut in to four pieces.
One rump butt.
One chuck, cut into four pieces of about 10 lbs. each.

In addition to the above, these other cuts may be used the same way: Flanks, briskets and necks. In cases of this kind the chucks are cut into smaller pieces, usually six or seven.

CURING OF BEEF

The curing of barrelled beef is simple. The beef is packed in tierces, pickled with 100 degree plain brine, and at the time of packing two ounces of nitrate of soda per 100 lbs. of meat are added. The tierces should be rolled on the 5th, 10th and 20th days. The shipping age is 25 days. Shipments to the United Kingdom are usually packed 202 lbs. to the barrel. Sixty pounds of capping salt should be added per barrel when repacking for shipment.

MANUFACTURE OF DRIED BEEF

The curing of dried beef is somewhat more difficult than the curing of most pork products. This is due to the fact that there is a greater variation in the quality of the meat itself, and the length of time of curing and other factors have to be varied accordingly.

An average pickle for curing beef hams may be made up as follows: To 100 gallons of water add 245 lbs. of salt, 20 lbs. of brown sugar and 5 lbs. of saltpeter or its equivalent of nitrate of soda. The length of time to cure depends on the factors mentioned above, usually from 6 to 8 days to a pound is sufficient, depending on the temperature. Many packers use a temperature of 36° Fahr. and find it satisfactory. Others cure at a temperature of 28° to 30°, retarding the cure but allowing the curing time to extend 25 per cent longer.

After curing, the beef hams should be soaked from 20 to 24 hours in fresh 60° F. water, and then washed in warm water at a temperature of 120° to 130° before being hung in the smokehouse. The time required for smoking is again dependent upon the weight of the hams. A good rule to follow is to smoke until the hams are dry. Sometimes four days may be sufficient, again it may require seven or eight days. Ninety-six hours at a temperature of 135° is usually sufficient time to smoke.

BEEF OFFAL

The packing industry has gained an enviable reputation by utilizing by-products. It is in the offal and casing departments where, in a large measure, this reputation was acquired. The edible division of this department receives the head and viscera from the killing floor, and these constitute the raw product. In the following pages will be described how these are subdivided and processed. The inedible division receives the front and hind feet, condemned viscera, cow udders and uterus.

The Head.—After a thorough inspection of the glands by the government inspectors, the tongue bones are cut and the sides disengaged from the jaw. The pharynx is opened and the tonsils removed and sent to the inedible tanks.

The tongue is then thoroughly washed and hung in the cooler. In removing the tongue care must be taken that no cuts are made, as this will depreciate its value. When a hair pocket is present, it must be cut out. After the tongue has chilled it is trimmed into long cut, short cut and canning tongues, as desired.

The trimmings are retrimmed; the lean tissue is used in the sausage department and the fats sent to the edible tallow tanks.

The horns are now removed from the head with the circular saw and delivered to the bone house for processing.

The cheek meat is loosened from the lower jaw, the joint chiseled and the lower jaw separated from the upper. Cheek meat, lips and head meat are entirely removed from the skull. Considerable lean tissue may be saved by giving close attention that all has been removed from the head.

The skull is now split, the brain covering cut, the cords disconnected and the brain lifted out. This product should be placed on pans so constructed as to allow complete drainage, and placed in the cooler as soon thereafter as possible. If the pituitary and pineal glands are to be saved for pharmaceutical purposes they are recovered at this time, as they lie at the base of the brain. (See Chapter V for description of animal glands.) The skull and the jaw bones are sent to the glue house for processing.

The Viscera.—After the government inspectors have completed the inspection, the intestinal organs are separated from the remainder of the viscera at the end of the rennet, and sent to the gut table.

The Bladder.—This is removed and all fats detached. Any ammonia smell may be avoided by promptly draining the urine, very thorough washing and quick handling into ice water.

When the animal heat has entirely disappeared, the bladders are inflated with air and allowed to dry. It is highly important that all operations on the bladder be accomplished in such a manner as to eliminate criticism. The government inspectors will reject for sausage stuffing any bladder defective in the slightest manner.

The Bung Gut, Middle Gut and Round Gut.—The rounds, middles and bungs are now separated, fatted, turned and slimed, and are processed as described in the section following, on "Beef Casings."

The casing operation is now complete. The remainder of the viscera is handled next; that is, the four stomachs of the ruminant, along with the liver, heart and lungs. This operation is carried on simultaneously with the casing operation.

The Liver.—The operator removes the gall bladder from the side of the liver. The bladder is cut open and the bile drained through a screen to recover gall stones, if any are present. The drained bladder is sent to the inedible tank, the portal vein or hepatic glands and any portion of the pancreatic glands which may have been left attached are trimmed off and tanked. An incision is made in the liver at the distal end; that is, the thin end, to allow for drainage, and it is hung in the cooler.

The Pluck.—The pluck (heart, lungs and trachea) is hung by the throat end of the trachea and the lobes of the lungs are removed. The heart sack (pericardium) is slit, the pulmonary veins and the common aorta are cut, and the heart removed. All adhering fats are then stripped from the trachea, washed and sent to the oleo department. The giblet meat, which is a small portion of the hanging tender that has been left attached to the gullet when the viscera was removed, is now trimmed off and is an edible product. The lower part of the trachea is cut from the remainder of the trachea and rendered into prime tallow. The upper part of the trachea is sent to the glue house for processing.

If the lungs are to be used for edible purposes, all bronchial tubes must be laid open to expose any fæcal matter that may be present, in which event they are rejected. If not required for edible purposes they are rendered in the brown grease tank. The heart ears (auricles) are separated from the ventricle, and the fat and arteries removed from the lean tissue. The remainder of the heart, having been previously opened by the federal inspectors, is now washed and sent to the cooler.

All fats that have been removed from the pluck are put through a hasher into a tank of cold water. The specific gravity of fats, being less than that of water, they will float and are recovered by skimming, while the lean tissue will precipitate and is subsequently recovered and rendered.

The Paunch.—If the floor arrangement makes it possible all fats, including the caul, should be removed immediately after the paunch is taken from the abdominal cavity. If this is not possible, the operation should be accomplished before any incisions are made. If this policy is not adhered to, these fats are liable to contamination from paunch content, which would cause their rejection as an edible product.

The paunch of the ruminant is divided into four distinct stomachs, viz: First paunch (rumen); Honey comb paunch (reticulum); Peck (omasum); and Rennet (abomasum).

After all the fats have been removed the melt (spleen) is cut from the side of the paunch. While the melt is an edible product, the demand is very limited. As an edible product it is washed thoroughly after removal and hung in the cooler. If it is not desired as edible, it is rendered in the brown grease tank.

The peck, with the rennet attached, is then cut from the honey-comb paunch, and a division is at once made thereafter, separating the peck and rennet. The rennet may be regarded as an edible product, in which case it is flushed, then laid open from end to end, to permit a thorough washing. The food value of a rennet is hardly sufficient to warrant its use in the sausage room, hence it is almost universally diverted to the tallow tanks after flushing.

The peck may be hashed and the manure washed out; in which case it may also be rendered for tallow. The yield, however, in tallow does not favor this operation, and more frequently it is merely slashed crosswise and rendered in the brown grease tank.

The operation on the remainder of the paunch is carried on simultaneously as follows: A blunt hook is inserted in the opening of the honey-comb paunch, where the peck has been removed. The hook is attached to a trolley by means of which the paunch is conveyed over a receptacle to receive the paunch content. The paunch is laid open from the point where the hook has been inserted to a point diametrically opposite; opening in a direction so as to avoid cutting through the honey-comb portion. The paunch content having been expelled, the tripe is placed on a conical table, usually called an umbrella washer, where by spraying and scrubbing with coarse brushes all adhering paunch content is removed. Further washing and processing is carried on in the tripe department.

Inedible Offal Department

Condemned Viscera.—Fæcal matter is removed from the guts and paunch, to make the attached fats available for tallow. This may be accomplished by hashing and washing all intestinal organs or by merely laying them open with the aid of a knife and then washing. The liver and pluck are rendered in the brown grease tank.

Condemned Carcasses.—Kidney knobs, pelvic and heart fats are removed and rendered as tallow; the remainder is diverted to the brown grease tank.

Uteri.—The uterus is opened and the slunk (fetus) removed. If the fetus is sufficiently matured so as to make it desirable to save the skin, this operation then follows: The feet, head and shoulders are skinned with the aid of a knife; the remainder of the skin can then be removed by pulling. The skin is sent to the hide cellar for curing, and the remainder to the rendering tank.

Feet.—All sinews are removed and put into glue stock. The hoof can readily be removed after steaming for a short time. The shin bone is then sawed out, the nerve holes indicating the point where they should be sawed. Hoofs and shin bones are further processed in the bone house. The foot bones and that portion of the knee joint which has been sawed from the skin are sent to the glue house.

The offal and casing operation is now complete. As has been indicated the successful operation of this department depends on constant vigilance so that no product be lost or reduced in value by careless handling. Astonishing results and yields may be attained by watchful and intelligent supervision.

Handling Edible Offal or Miscellaneous Meats

Hearts.—On the killing floor all hearts are cut in half. This is done in order to make possible the examining of the same for measles, and also to let out the blood. The fat and heart ears are trimmed off, and the hearts may be used for fresh products at the market, or in sausage, for extract, canning, mince meat, or may be pickled.

Livers.—The gall bag is first removed from the livers. They are then washed and may be sold fresh, or frozen, especially if there is an oversupply. If necessary to tank livers, the fat should be trimmed off and rendered separately.

Kidneys.—Kidneys are used for extract; some are sold fresh; others may be trimmed and frozen; those not salable are tanked.

Brains.—Brains are removed carefully immediately after the skull is split, are placed in shallow pans and put in the chill room. Those not offered for immediate sale may be frozen in a sharp freezer directly after they have been packed in tin pails ready for marketing.

Tongues.—Tongues are first washed and then chilled. In the chill room they are hung up by the gullet end, because if hung up by the narrow end the tongue will stretch, which makes it look very thin and unattractive. Tongues are cut in two ways—a long-cut tongue and a short-cut tongue. In the former case the gullet is left on. Short cut tongues are generally sold fresh.

Ox Tails.—A small per cent of ox tails are sold fresh. The surplus is used in soups, or is sometimes cured in salt and pickled.

Tripe.—The preparation of beef tripe is as follows: The paunch, after being ridden of its contents, is washed on an umbrella spray, being scrubbed with brushes. The dirty pieces are then trimmed off, as well as any fat which may be used for oleo. The fresh tripe may then immediately be put into a rotary tripe washer and thoroughly agitated in hot water at a temperature of not over 140° Fahr. A small amount of soda in the water is necessary to remove the scurf and whiten the tripe. After removing from the washer, any excess scurf or mucus remaining is removed by scrapers. The tripe may then be cooked in a sheet steel or wooden vat cooker at a scalding temperature for about three hours, care being taken that the steam does not come in direct contact with the tripe. To test when it is done, place the finger through the heavy seams of the tripe, and if it is soft the tripe is done. Cold water is then turned on, and when the tripe is chilled it is ready for the finishers.

The finishers should take care not to throw good pieces of tripe away with the skin. Also, the tripe should not be trimmed too wide and wasted. One-fourth inch trim is sufficient. After finishing, the tripe is inspected for quality and cleanliness, then it is put into ice water and chilled thoroughly before pickling. From the grading vats, if not sold or used fresh, it is placed in vinegar pickle of 55 grain strength and

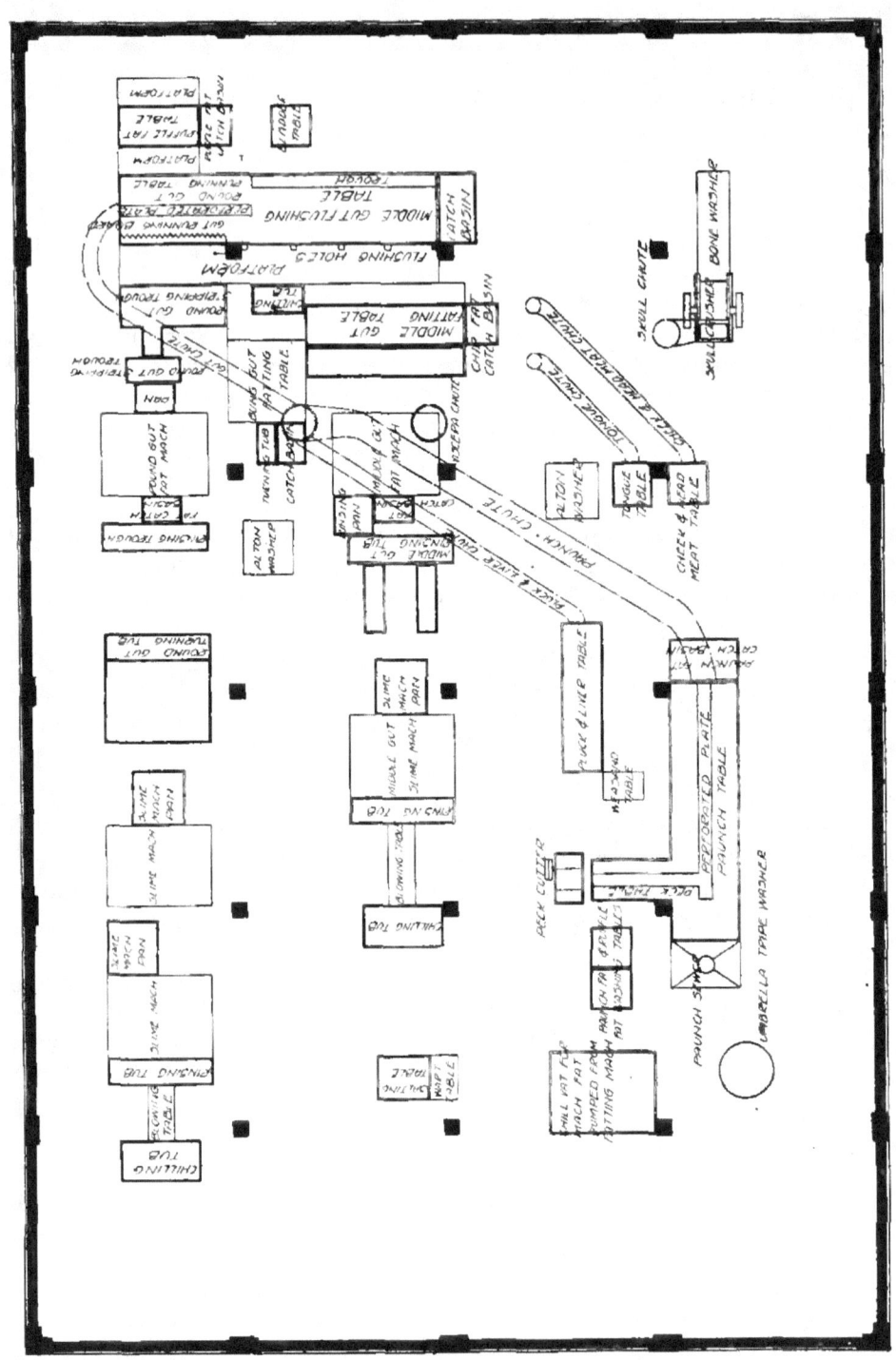

Lay-Out of a Well-Arranged Offal Floor

left for at least 3 weeks, during which time it will take on 20 per cent in weight. The temperature of the curing room should not exceed 40° F. When packed fresh in small packages for shipment 45 grain strength vinegar pickle is used for packing.

In order to give an idea of the yield and weight of tripe during its various stages of processing, the following test is given:

100 pieces of unclean tripe, 1,429 lbs.
100 pieces of scraped weight, 1,270 lbs.
100 pieces of cooked weight, 768 lbs.
100 pieces of finished weight, 567 lbs.

BEEF CASINGS

Packers in years gone by saved beef casings primarily for use in their own sausage kitchens, and as they were utilized almost as fast as produced, very little attention was given to the method in which they were put up. But since a considerable portion of the casing output of a modern packinghouse is now packed for the trade, certain rules, measurements, etc., must be observed in order to standardize this merchandise and improve its keeping qualities.

Beef Rounds, Middles and Bungs, after being cleaned and graded, but before salting, should be placed in a tub of clean cold water, never exceeding 50° F., so as to thoroughly chill them and remove all animal heat. In winter, when temperature of water is sufficiently cold, no ice need be added; but in summer, when water is of higher temperature than 50°, keep tub well iced. Allow casings to remain in this chilling tub from 30 minutes to one hour, stir them around occasionally so that all of them receive proper chilling.

After thoroughly chilling, the casings are salted with bulk provision salt and allowed to lie in bins for 24 to 36 hours in order to drain, when they will be ready to pack in tierces for storage. In order to get the required quantity in each tierce, it will be necessary to press the excess moisture out of the casings, which is done by boring holes about 1 in. in diameter in a glucose tierce, packing the casings in this tierce and subjecting them to pressure for several hours, when they can be packed into a regular glucose or casing tierce.

After packing, fill the tierce with 100 degree strength pickle, and head up water tight, storing in cellar or cooler of 45° F. temperature.

Beef Rounds

After separating the paunch from the viscera the round casing, or small intestine, should be run off with the aid of a sharp knife, beginning at small or stomach end. By allowing the edge of viscera holding the round casing to hang over the end of the table, this operation is easily accomplished.

Modern packinghouses, however, are installing sanitary steel tables which are equipped with fluted iron arms, especially constructed for the holding and handling of beef casings, which, from a sanitary standpoint, the ease with which casings are handled and the minimum number of men required, will more than save their cost in a few months' operation.

After the full set of round casings is run off, strip, wash and put them through a fatting machine to remove small particles of fat; then put them in tub of warm water of about 80° F. and wash off any loose fat.

Then turn the casing—this can be done easily by means of a funnel; put one end through the funnel, stripping the end back over the spout, put the rest of the set in the hopper of funnel with a dipper full of water, and it will run through very quickly without making a hole in the casings. Another method is to cut the casing crosswise about the middle of the gut and run both pieces at the same time.

After casing is turned, tie five or six sets in loops in center and put casings through the sliming machine. These sets are held by a looped string fastened in the middle, and casings are passed through machine about three or four times, so as to thoroughly remove all slime. After sliming, put in tub of cool water until ready for inspection.

The inspection is made to see that the casings are free of tallow and slime, to determine the size and quality, and to separate Export Rounds from Domestic Rounds. The casings can be inspected either by flooding with water or blowing with air. After inspecting and separating the two grades, place the rounds in ice water for 30 minutes or more to remove all trace of animal heat.

Export Rounds.—These must be well fatted and refatted (and hand-inspected after machine fatting), so that they are free of tallow, well slimed, free of stains, prime quality, good color and smell, absolutely free of all warts, pimples or knots and reasonably free of holes. Not more than five holes (turning holes included) are permissible in a set. Export rounds must measure 106 to 108 ft. green, and must not contain to exceed five pieces to the set. Both ends must be cut off square. If export rounds are not graded for size they can be put up 180 to 200 sets to the tierce, and should be branded "House Run Export Rounds." If sufficient cattle are killed to warrant grading for size, then they are to be packed as follows:

Narrow Export Rounds, to measure 1¼ in. and under, and packed in glucose tierces 225 or 250 sets per tierce and branded "Narrow Export Rounds."

Regular Export Rounds, to measure 1¼ in. to 1½ in. and packed in glucose tierces 200 to 225 sets per tierce, and branded "Regular Export Rounds."

Wide Export Rounds, to measure 1½ in. and over and packed in glucose tierces, 160 to 180 sets per tierce, and branded "Wide Export Rounds."

In order to get the required number of sets in a tierce it will be necessary to put the rounds in a press to squeeze out moisture, as already described.

Domestic Rounds.—Same general rules govern as apply to the export grade, with this exception: The domestic grade of rounds permit of a few small knots or warts, but judgment must be exercised in saving these rounds, so that the extremely heavily-infested sets, or those containing large green warts, or those profusely covered with black pimples, are not saved. These should be tanked.

Domestic rounds are not permitted to be used in U. S. inspected plants, and on this account are kept separate from the export grade. They can, however, be used in non-inspected places. Domestic rounds are to measure

106 to 108 ft. green, five pieces to the set, although six pieces are permissible occasionally, with not more than five holes (turning holes included) and packed as follows:

> Regular Domestic Rounds, packed 125 to 140 or 160 sets to the tierce, are graded 1¼ in. to 1½ in. in diameter and packed in glucose tierces, branded "Regular Domestic Rounds."
>
> Medium Domestic Rounds, graded 1¼ in. and under, and packed 180 sets to the tierce, and branded "Medium Domestic Rounds."

Stump Rounds.—Stump rounds or short pieces are not to be included in either the export or domestic grades, but are to be packed separately for sausage room use, or put into sets of 106 ft. green, and packed 180 sets to the tierce, and branded "Stump Rounds."

Calf Rounds.—Calf rounds are to be cleaned in the same manner as beef rounds. Calf rounds are to be practically free of holes, carefully fatted and slimed, to measure green 84 to 86 ft., and cured about 80 ft. No specified number of pieces to the set, but this to be held down to lowest minimum possible. No piece less than 3 ft. 4 in. to be included. Grading to be as follows:

> Wide Calf Rounds to measure 1⅛ to 1¼ in. in diameter.
> Medium Calf Rounds to measure 1 to 1⅛ in. in diameter.
> Narrow Calf Rounds to measure 13/16 to 1 in. in diameter.

Each grade to be packed in separate packages and packed as many as possible in a tierce, branding "Choice Select Calf Rounds."

Beef Middles

After separating the middle gut from the bung it should be flushed with warm water (not hot), and care must be taken in flushing so as not to fill too full of water, as this has a tendency to burst the casing.

Modern packinghouses are equipping offal floors with the latest improved sanitary steel tables, especially designed for running casings, and properly flushing them. These tables are so arranged that one end of the middle casings is slipped over a water pipe, the other end directed into a manure chute and the contents of the casing flooded into sewer, avoiding all danger of stains and contaminating the valuable fat.

After flushing, the middles are passed to the fatting bench, where the fat is removed by trimming with a sharp knife. The latest improved fatting bench is a marvel of efficiency, also of steel, fitted with perforated pipes which play a constant stream of water upon the back or "catch board," and in this manner, wash, chill and conduct the fat into the drain gutter, thence into the chill tub, without handling and without any danger of contamination. The thick end or "fat end" is to be cut off about 9 to 12 in., then the middle is put through the fatting machine to remove all particles of fat, then put into a tub of water about 85° F. to wash off loose fat, then inspected for manure stains and spots. If any are found, remove them, as they are very objectionable.

Then turn the middle and send to the sliming machine, putting them through in bunches of five or six sets held by looped string fastened in center, and pass through machine two or three times to remove all traces of slime.

After thoroughly sliming, put middles on grading table and inspect by either flooding with water or blowing with air to determine size, holes, etc. After selecting, put casings in ice water to chill and remove all traces of animal heat, allowing to stay in chilling tub 30 minutes to one hour.

The middles are then measured into sets of 61 to 63 ft. green, or 57 ft. when cured, not more than five pieces to the set, and no piece less than 3 ft. 4 in. in length is permissible. Middles must be practically free of holes, although a small hole in a long piece is permitted. In making quality tests, inspectors will pass ten small holes or five large holes in a total of ten full sets of middles. All ends should be cut off square; must be prime quality, good color and smell.

After casings are measured, pack in bulk salt and put on drain bins for 24 hours, then shake out and pack in tierces, filling tierce with 100 degree pickle and place in cold storage.

Middles are graded for size and packed as follows:

> Narrow Middles, to be graded only in localities where there is a demand for this grade. Size 1⅜ in. and under, and packed 130 sets per tierce, and branded "Select Narrow Beef Middles."
>
> Regular Middles, to contain all middles under 2 in. in diameter, packed 110 sets to the tierce, and branded "Regular Beef Middles."
>
> Wide Middles, to be graded 2 in. to 2½ in. in diameter, packed 90 sets to the tierce, and branded "Wide Beef Middles."
>
> Extra Wide Middles, to contain all middles measuring 2½ in. in diameter and over, packed 80 sets to the tierce, and branded "Extra Wide Beef Middles."
>
> Kosher Middles, to be put up from kosher-killed cattle, same specifications as "regular" middles. Tallow to be left on these. Also have kosher mark or tag placed on middles.

Beef Bungs

Beef bungs average 3 to 4 ft. in length, and are removed after the round guts have been run from the viscera. The open end is then placed over water pipe, filled with water and stripped of all contents; then place bung in cold water for 20 or 30 minutes to harden the tallow and fat, which is trimmed off with a sharp knife much easier when properly hardened.

After fatting, wash with a brush to remove stains or loose fat, then turn the bung and remove the slime by means of a wooden scraper, or put in washing machine, which is an excellent slimer. Bungs are then put in tub of cold water and all warts or knots are trimmed off with a regular surgeon's shears, care being taken not to trim too closely so as to score the bung, which weakens them and causes breakage in stuffing.

Bungs are then inspected with water or air to discover scores, holes and for grading, which is done as follows:

> No. 1 Beef Bungs, to be full length; that is, 3½ ft. or over, and the nipple or round hole must be in center; or in other words, the open end must be as long as the cap end. Must be 3 in. and over in width, prime quality, good color and smell, free of stains and tallow, well slimed, reasonably free of scores and no holes, although a small hole within 2 in. of nipple hole is allowed. Absolutely free of all warts and pimples. All skins left in bung. Tie in bundles of five bungs and pack 80 bundles or 400 pieces to the tierce, branding "No. 1 Beef Bungs."

No. 2 Beef Bungs are bungs measuring 3 in. and under in diameter, prime quality, good color and smell, reasonably free of scores and manure stains, absolutely free of knots or pimples, either naturally or brought to this condition by trimming. Tie in bundles of five bungs and pack 100 bundles or 500 pieces to the tierce, branding "No. 2 Beef Bungs."

No. 3 Beef Bungs consist of bungs too heavily infested with warts or pimples. These cannot be used in U. S. inspected houses, and should be branded "Domestic Bungs."

Cap Ends.—In the event the open end of the bung is badly scored or cut, or should the bung be cut off too short, which would prohibit it being placed in the No. 2 grade, then the cap end only should be saved, provided it is not damaged and is free of warts or pimples. They are to be packed 750 to the tierce, and marked "Cap Ends."

Beef Bung Skins.—Beef bung skins should be left in the bung, except when market conditions warrant their being removed. However, the removal of the inner skin weakens the bung about one-third. When these bung skins, or gold-beater skins, as they are called, are to be removed, they are selected as follows:

No. 1 Bung Skins must be taken off with great care, pulled to the seams on both sides, 30 in. and over in length, free from holes ¼ in. from edge. All fat or tallow trimmed off the edges and holes in edges to be trimmed out. Any skins with rust, manure stains or fatty adhesions are to go in the No. 2 grade. Pack No. 1 skins 25 pieces to the bundle, and pack 3,500 pieces to the tierce.

No. 2 Bung Skins are all skins measuring 18 in. to 30 in., and those having too much tallow, knots, rust and manure stains are to be included in this grade. Pack 25 pieces to the bundle, 4,500 pieces to the tierce.

No. 3 Bung Skins are to include all skins 12 in. to 18 in. in length, packing 25 pieces to the bundle, 5,000 pieces to the tierce.

All tierces containing bung skins should be thoroughly washed, paraffined and lined with muslin and parchment paper.

NOTE—Wherever possible, beef casing packing should be done in a cooler or cold cellar, and casings should be started in process of packing not later than three days after production, and all packing, heading of tierces, etc., should be completed within a week. Always store casings on bilge in cold storage, and see that tierces are well coopered and water tight.

Beef Weasands

The weasand is cut from the paunch and immediately flushed. The two rings of the gullet that have been left attached are cut off and any fat that may be adhering is removed. A thorough washing is now given and the outside covering of lean tissue is stripped off with a sharp knife. Care must be taken that the weasand is not cut in this operation. This meat so recovered is edible and used for sausage purposes. Any worms found attached to the weasand under the meat must not be removed, as this impairs its strength, causing it to burst. If weasands show more than four worms, send to the rendering tank.

The weasand is now turned inside out and one end tied. Inflate with air, tie up the other end and allow to air dry. All weasands for sausage stuffing must be prime quality, properly dried, thoroughly cleaned, absolutely free from bruises and worms.

Weasands are graded as follows:

> No. 1 Weasands are to consist of those which are 24 in. and over in length and 2¾ in. and over in width, free of grubs, with a few blood stains permissible. Pack 25 pieces to the bundle, tieing ends and center, packing 100 bundles in box or sugar barrel. Mark package "No. 1 Weasands."
>
> No. 2 Weasands are to be 24 in. and over in length, any width, may contain blood stains, grubs and weasands under 2¾ in. wide which would otherwise be graded as No. 1. No. 2 weasands to be packed 50 pieces to the bundle, 60 bundles to box or barrel. Mark package "No. 2 Weasands."
>
> No. 3 Weasands are to be from 18 in. to 24 in. in length, may contain grubs, blood stains and be of any width. Packed 50 pieces to the bundle, 70 bundles to box or barrel. Mark package "No. 3 Weasands."

Beef Bladders

Beef bladders are to be well trimmed, necks free of fat, and bladder free of manure or blood stains. To be trimmed, blown and dried daily as produced. After drying and before grading, soften them by holding in steam for a few moments, or hang in moist cooler for a short time, which will prevent cracking when folding and measuring. Measurement of bladders to be taken across the widest part, and not the length. Grading to be as follows:

> Large Bladders, neck on, to be 12 in. and over in width, 24 pieces to the bundle, tied both ends, and packed 25 bundles or 50 dozen per barrel. Mark "Large Beef Bladders."
>
> Medium Bladders, neck on, to measure from 10 in. to 12 in. in width, 24 pieces to the bundle, tied both ends, and pack 40 bundles or 80 dozen per barrel. Mark "Medium Beef Bladders."
>
> Small Bladders, neck on, to measure from 7 in. to 10 in. in width, 24 pieces to the bundle, tied both ends, and packed 50 bundles or 100 dozen per barrel. Mark "Small Beef Bladders."

Only bladders with neck on are used for sausage purposes. Those with the neck off are not to be included. The ones with neck off are to be graded same as above, packed separately and designated as "No Neck Bladders." These are used principally by putty manufacturers.

Calf Bladders

The method of handling calf bladders—that is, trimming, washing, blowing, etc.—is exactly the same as outlined for beef bladders; grading, however, to be as follows:

> Small Calf Bladders, to measure 4 in. to 6 in. across widest part, 24 pieces to the bundle, tied on both ends, packed 125 bundles or 3,000 pieces to the barrel, and marked "Small Calf Bladders."
>
> Large Calf Bladders, to measure 6 in. and over across widest part, 24 pieces to the bundle, tied both ends, packed 100 bundles or 2,400 pieces to the barrel, and marked "Large Calf Bladders."

In packing dried casings, such as bladders and weasands, be particular to use only first-class sugar barrels which are sound in every way, in order to keep out rats and flies. It is well to line packages with tar paper, or newspaper, as printer's ink helps to repel bugs. Always store dry casings in a dry, cool storage.

MANUFACTURE OF BEEF EXTRACT

Beef extract is made in most of the large plants from the following waters:

Waters from the cooking of beef for canning purposes.
Waters from the cooking of beef hearts.
Waters from the cooking of beef cutting bones.
Waters from the cooking of corned beef.
Waters from the cooking of beef livers.

The general practice is to make a separate grade of extract from the separate raw materials, but many plants mix the above ingredients in certain proportions to make a standard formula. Following is a list of certain formulas:

No. 1 Extract.—75% roast beef water; 25% heart cooking water; 3 pounds of sugar to 100 pounds of finished extract.

No. 2 Extract.—50% water from cooking beef livers; 50% water from cooking beef hearts; 3 pounds of sugar to 100 pounds of finished product.

When making fluid extract, add 40% water to 100 pounds of finished extract.

Process and Equipment.—The ordinary process is to cook the meat in regular cooking vats from four to six hours at a temperature of 210°, although some packers make extract from the soaking waters in thawing out frozen beef. The list of machinery required is as follows: 1st, soaking vats; 2nd, cooking vats; 3d, evaporators; 4th, concentrating tanks; 5th, finishing kettles; 6th, mixers.

In general practice 1 pound of finished extract per 100 pounds of raw material, such as beef hearts, livers, etc., is obtained. A general description of the manufacture of beef extract follows:

Regular Beef Extract.—Formula: 30% weasand meat; 70% beef hearts.

Use fresh weasand meat and hearts; grind through an Enterprise hasher, using ¼ inch plate on hearts and ⅜ inch plate on weasand meat. Soak in water at temperature not less than 40° Fahr. in the following manner:

Fill 800 pounds of meat in vat and cover same with about 200 gallons of water. Allow to soak three hours. Fill 800 pounds of beef in an adjoining vat and pump water from the first vat into the second vat. Allow to soak three hours. Fill 800 pounds of meat in the third vat and pump water from the second vat over meat in the third vat. Allow to soak three hours. This water is now strong enough to go to the receiving storage.

Cover meat in the first vat with another 200 gallons of water; allow to soak three hours and pump into second vat; allow to soak three hours and pump into third vat; allow to soak three hours and pump into the fourth vat. From here pump to the receiving tank.

This process is continued until each vat of meat has been covered by at least four changes of water, and each change of water has been used on at least one vat of fresh meat. Then the soaked meat is delivered to the extract room to be cooked for extract, water being delivered to the storage tank preparatory to heating.

Pump water from the receiving vat into heating vat. Heat up to temperature of 210° Fahr. Hold at this temperature long enough to solidify the albuminous matter that will rise to the surface of the water during the process of cooking. This usually takes about ten or fifteen minutes. Agitate again in order to break the hardened albumen on the surface of the water, and mix it throughout the vat of water, so as to make a filtering pad during process of filtering.

This water is now pumped through the filter and all foreign material caught in the filter cloth. From the filter press, the water goes into a double effect vacuum pan. Water should be at a temperature of about 200° when going into the pans, and should be evaporated in the pans to 15° or 18° Beaumé, or reduced to about 40% solids.

When operating the vacuum pans, carry a 15-inch vacuum on the first effect and a 27-inch vacuum on the second effect. From the vacuum pans, the extract now goes to the finishing kettles. Here it is boiled down nearly to a solid, and afterwards dipped out and put into receptacles for storage. Carry a vacuum of 27 inches on these finishing kettles.

Hold extract thirty days in order to give it a chance to work. This is *crude extract*. Then add 6 pounds sugar, 9 pounds salt, 1 pound caramel and 55 pounds of water to every 100 pounds of crude extract, boiling in an open steam kettle, until the above ingredients are thoroughly mixed. Put into settling jars. Allow to stand over night. Syphon off clear liquor, filtering settlings so as to remove all albuminous matter. Boil together in the finishing kettle to about 82% solids; pass through an agitator in order to give it a good color and make it smooth and glossy. Fill in packages for trade.

Extract (Second Grade).—Take soaked hearts, after same are used for regular extract. Put in open vat. Cover with cold water. Cook one hour at temperature of 180°; agitate constantly during the process of cooking. Pump to storage cooking vat. Boil 12 hours with closed coils. Pass through filter press. Handle in vacuum pans the same as regular extract; put through finishing pans. Boil to about 82% solids and fill in cans as crude extract. This is now ready to mix with other extracts.

Extracts (Third Grade).—Made from beef ham pickle. Pickle is pumped into heating vat; cooked at temperature of 210° for one hour, in order to allow the albumen to rise to the surface. Skim off all the albuminous matter and send to tank. Allow the water to remain in the vat over night. Syphon off clear water and pass through vacuum pans the same as the regular extract. Send settlings, which are usually about 10%, to the tank. Draw the extract from the vacuum pans, when it is reduced from 36 to 38 Beaumés; from vacuum pans pump into settling

BEEF EXTRACT

pans and allow to remain about five days. Syphon off the clear liquor from the top when using, putting the settlings into the next batch of extract that is cooked.

Yields on Beef Extract

Every packer must judge for himself if it will be profitable for him to go into the manufacturing of beef extract. The following yields and costs may be of interest:

Extract made from corned beef cooking water: 14,228 gallons of cooking water produces 1,205 pounds of beef extract.

Extract made from roast beef cooking water: 1,783 gallons of cooking water produces 113 pounds of beef extract.

Extract made from ground beef hearts: 3,110 pounds of ground beef hearts produces 122 pounds of beef extract.

It is usually safe to figure that for every 100 pounds of good beef, one pound of finished extract can be produced.

Methods of figuring beef extract packed in jars are also given here. These prices, of course, will have to be changed to correspond with current market prices, and are simply shown as an example of how to figure the cost:

(1) Test on Beef Extract

Debits:

3,110 lbs. beef hearts, $0.02½ per lb.	$ 77.75
Curing material	.60
Labor	8.45
Administrative, 122 lbs., .2012 per lb.	24.55
	$111.35

Credits:

122 lbs. beef extract, $.7724 per lb.	$106.43
35 lbs. tallow, $0.04 per lb.	1.40
1,408 lbs. heart pulp, $0.00¼ per lb.	3.52
	$111.35

Cost per lb. including administrative, 0.8724 cents.
Cost per lb. not including administrative, 0.6712 cents.

(2) Test on Beef Fluid

Debits:

100 lbs. beef extract, $0.6712 per lb.	$67.12
40 lbs. water	0.00
Curing material	.29
Labor	.20
Administrative, 140 lbs., $0.1437 per lb.	20.12
	$87.73

Credits:

140 lbs. beef fluid, $0.6267 per lb.	$87.73

Cost per lb., including administrative, .06267 cents.
Cost per lb., not including administrative, .4829 cents

Statement Showing Cost Per 100 Lbs.

(3) Packed in Different Style Packages

(2-oz. jars—12 2-oz. boxes)

100 lbs. extract	$0.8724	per lb.	$ 87.24
800 2 oz. jars	.14	7/12 doz.	9.72
800 corks	.75	per gr.	4.16
800 cork covers	.16	per M	.13
800 caps	3.72	per M	2.98
800 labels	1.64	per M	1.31
800 wrappers	1.18	per M	.95
800 circulars	1.40	per M	1.12
3 oz. sealing wax	.30	per lb.	.06
66⅔ sets packing	10.00	per M	.67
66 boxes	5.00	per hund.	3.34
Labor			1.86

Cost per 100 lbs. $111.54

(4) Packed in Different Style Packages

(4-oz. jars—12 4-oz. boxes)

100 lbs. extract	$0.8724	per lb.	$ 87.24
400 4 oz. jars	.225	per doz.	7.50
400 corks	.69	per gr.	1.92
400 cork covers	.35	per M	1.40
400 caps	6.10	per M	2.44
400 labels	2.67	per M	1.07
400 wrappers	1.36	per M	.55
400 circulars	1.40	per M	.56
2 oz. sealing wax	.30	per lb.	.04
33⅓ sets packing	14.00	per M	.47
33 boxes	7.75	per hund.	2.58
Labor			1.48

Cost per 100 lbs. $107.25

MANUFACTURE OF OLEO PRODUCTS

Oleo products as produced in the modern packing house are the second most valuable by-product produced—the hide, of course, being the most important.

The fats which go to make up oleo products are as follows: caul, ruffle, intestinal, machine fat, heart fat, paunch, peck, rennet, tripe trimmings, beef cut trimmings, kidney fat and suet.

The beef cut trimmings and machine fat are graded into a No. 2 product, whereas the other fats may be used for manufacturing a No. 1 product.

Yield of Fat.—The amount of fat coming from the carcass depends upon the degree of finish that the animal has. Real good steers will produce an average of 55 lbs. of fat per head; canners produce 20 to

25 lbs. per head. On the average, all classes of cattle will produce between 40 and 45 lbs. of raw fat per head.

Handling of the Fats

Washing and Chilling.—These fats, which are produced on the killing and offal floors by removing same from the carcass and viscera, are immediately washed and weighed and conveyed to a vat of cold water, passing first, however, through a fat cutting machine. This water, which is kept at a temperature of around 38° to 40° F., either by ice or artificial chilling, gradually chills the fat, removing the animal heat, which puts it in good condition for further processing. The chilling requires usually not less than twelve hours, but in some instances fats are processed further after being chilled only six or eight hours, although this is not advisable.

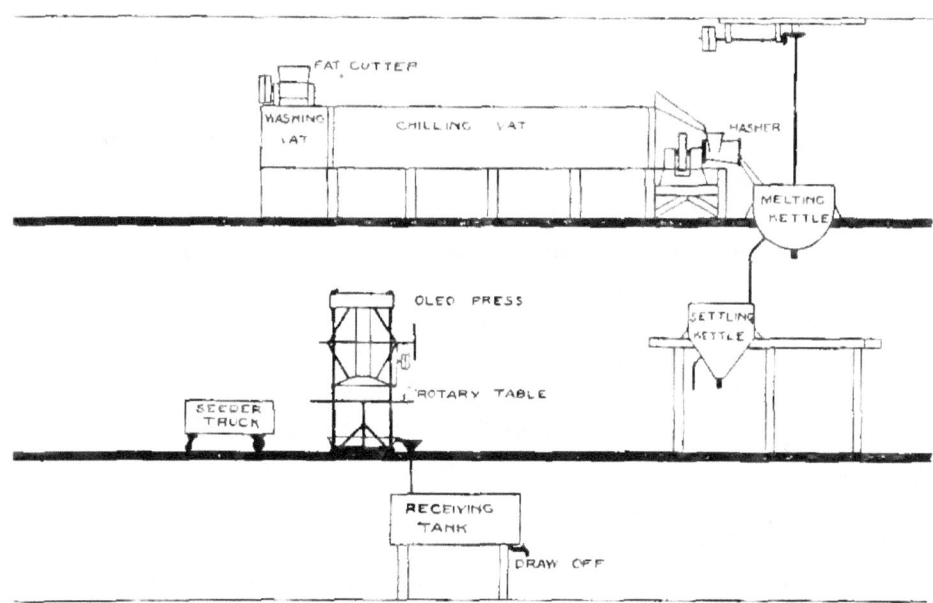

ELEVATION OF A TYPICAL OLEO PLANT

The oleo equipment shown here consists of a fat-cutter, a chilling vat, hasher, melting and settling kettle or clarifier, oleo press, seeder truck or seeding machine, and the receiving tank for the oleo oil.

These fats should not be held over 48 hours. The water should be changed twice a week, or as often as necessary to keep them sweet, and all sinkers removed, washing the vats thoroughly after each draining.

Hashing, Melting and Settling.—After the fats are well chilled, the next step is the hashing, which is accomplished by means of an ordinary power hasher of the Enterprise type, a ⅜" plate being used. If the fat is well chilled, this hasher will thoroughly disintegrate the fibre and make the separation of the oil easy.

The hashed fat is conveyed by gravity or by means of a spiral conveyor to cast-iron melting kettles, which range in capacity from 2,000 to 3,000 lbs. These kettles, which are equipped with a water jacket and siphon pipe, have slowly revolving paddles. In this receptacle the melting takes place, about two hours' time being required. The temperature to which the fat is heated is 155° for No. 1 and 160° for No. 2.

After the melting is completed—that is, a separation of liquid fat and the fibre has taken place thoroughly—the agitators are stopped and the whole is allowed to settle for about twenty minutes, coarse salt being used (10 lbs. to 20 lbs. per kettle) to help the settling operation, making a brine of the water which is in the kettles and at the same time carrying down the fibre. Each kettle is skimmed and the skimmings reworked into the next kettle.

After a thorough settling the liquid fat, which is now known as oleo stock, is siphoned to a clarifying or settling tank on the floor below, great care being taken to prevent any of the bottoms getting in with the oleo stock. After the oil is removed from the melting kettle, the bottom is then dropped through another opening to tanks below, where it is reheated and more oil recovered for the No. 3 grade, the remaining fibre being used for either tankage or cracklings.

The oleo stock which is now in the clarifying tanks is allowed to stand from four to five hours in order to further rid it of fine meat fibre and moisture. Ground rock salt is again used to aid the process by sifting it over the surface of the stock.

Old and New Seeding Processes.—After clarifying, the usual process has been to draw the oleo stock off at a temperature not below 140° into the seeding trucks, which are usually hardwood trucks holding from 800 to 1,000 lbs. each. In these trucks the oleo is wheeled into a seeding room, which is kept at a temperature of 90°. In the seeding room the stock must be held for from 72 to 96 hours, during which time the stearine or oil of a high melting point becomes crystal in character, settling to the bottom, allowing the oleo oil to remain in a liquid form.

When this crystallizing process is complete, the stock is broken or mixed by means of a mechanical stock breaker, and is then ready to press.

A more modern method of seeding oleo stock which is just beginning to be adopted is the use of a seeding machine, which consists of a jacketed tank of proper shape, equipped with cooling coils, stock breaker and discharging equipment. The oleo stock from the clarifiers (or settling kettles) is transferred directly into the machine, and by the circulation of water of the right temperature the seeding process is accomplished in one-quarter to one-third the time required by seeding trucks. A direct-connected motor furnishes the power which mixes the stock, and it emerges ready for the press. As this process saves time, labor, steam and product, it will no doubt be very widely used in the future by packers manufacturing oleo oil.

Pressing Oleo Stock.—Pressing is the next process. The stock should be 90° to 94° in temperature. From six to eight pounds of mixed stock is put in each press cloth by means of a filling table. No. 10 duck; 26" wide, is the best press cloth. Eight cloths per plate is the usual number, and at this rate each press will hold approximately 2,000 lbs. of stock.

Slow, steady pressure, peculiar to the knuckle joint type of press, is then applied. The draining of the oil or separation of the liquid oil from the solid oil or stearine requires about one hour and three-quarters. At this rate, figuring the time for loading and unloading, it takes from two and one-quarters to two and one-half hours for a press to repeat.

Handling Oleo Stearine.—After the pressing, the stearine cakes are shaken out of the cloths into a bin or directly into a melting tank. The eventual use of almost all stearine is for making compound lard. If the stearine has to be stored, it is packed by means of a rotary stearine packer in 500-pound slack tierces, and kept in a cool, dry place at an even temperature, 45° to 55° being suitable for the purpose. It is very important that the stearine storage be dry.

Handling Oleo Oil.—The oleo oil which comes from the presses is gravitated into large holding tanks, the larger the better, so that a uniform oil will result. This oil is heated very carefully to a temperature of 110°, and is then drawn into tierces and immediately put into storage. It may be carried in storage at 58° indefinitely, although it is much better to sell the product when it is fairly fresh. Oil six months old and over is, in general, more undesirable than the fresher product. After being put into storage it should not be removed for a day or two, for by moving it may break the crystal form which oleomargarine manufacturers like to see.

There seems to be a general belief among oleomargarine manufacturers that a well-crystallized oil denotes quality, but it has been well proven that this is a matter of temperature, and has little relation to the original quality of the raw fat.

Oleo oil ready for the market should have a clean, bland, nutty flavor, with no suggestion of greasiness or tallowness. The melting point of oleo oil should be below that of body temperature, a melting point of not above 80° to 85° being desirable. The average oleo stearine should be clean in odor and should have a melting point of approximately 128° F. and titre of 50°.

Grades of Oleo Oil.—Most manufacturers of oleo oil make two grades or more, the usual grades of the larger makers being as follows:

>No. 1 oleo oil, made from the highest-grade fats, as described heretofore.
>
>No. 2 oleo oil, made from cutting, machine fat and poor killing fats.
>
>No. 3 oleo oil, made from the further cooking of bottoms from melting kettles and poor cutting fats.

Some manufacturers produce a special yellow oil which is made by selecting the yellow fats in the raw state and processing separately. The yellow oil frequently brings a special price on the market. The color of such oils may best be judged by physical examination when they are cold.

Mutton Fats.—Mutton fats should be handled separately. The operation is the same as when processing beef fats, except that the melting is done at a temperature of 165° F. Mutton stock may be reworked into regular oleo stock by adding 5% of mutton stock to the regular oleo stock. This is best done by adding about 75 lbs. of mutton stock to the beef fat in the melting kettle. If the mutton stock has a very pronounced odor and flavor it is best to use it in No. 2 oleo only.

Yields from Oleo Fats.—One hundred pounds of oleo fats will produce from 65 to 70 lbs. of oleo stock, more or less, depending upon the quality of the fat. Fats from canners will produce much less, while fats from shipping cattle will yield up to 80 per cent. The yield of oleo oil from the oleo stock will be about 60 per cent, and the stearine will run 40 per cent, approximately.

Whether or not it will pay to make oleo stock or oil instead of edible tallow depends both upon the relative market prices and the amount of oleo fats which are available. At present market prices (1921) there is a difference of 2¼c per pound between oleo stock and edible tallow, and considering the cost to manufacture oleo oil and stearine complete, which depending upon the plant runs from 1½ to 2¼c per pound, and the cost to manufacture edible tallow, which roughly is one-half of one cent per pound, it may be calculated whether or not it would be profitable to adopt the oleo process. There are a good many packers who would do well to make oleo stock, even though they do not press it. This, however, is something which must be determined locally, depending upon the conditions present.

The oleo process should be handled with the utmost cleanliness throughout, in order to produce a high quality product. The delicate oils absorb flavors readily, especially when warm, and when in combination with the meat fibre decomposition takes place very quickly unless the utmost care and cleanliness is exercised throughout.

Tests.—Those packers who kill cattle, sheep and calves will be interested in the following test which gives an idea of the amount of oleo stock and inedible tallows which may be derived from beef, sheep and calf fat. This test was made on 2,013 shipper cattle, 2,405 sheep and 1,616 calves. The raw fat obtained from the above animals was as follows:

Beef caul and ruffle fat	38,682 lbs.
Beef gut fat	54,840 lbs.
Sheep caul fat	2,614 lbs.
Sheep ruffle fat	2,128 lbs.
Calf fat	287 lbs.
Total	98,551 lbs.

The yields obtained from this fat were as follows:

No. 1 oleo stock	61,145 lbs., or 62.04%
No. 2 oleo stock	4,416 lbs., or 4.48%
No. 3 oleo stock	1,886 lbs., or 1.92%
Total	67,447 lbs., or 68.44%

The prime tallow yield was as follows:
3,439 lbs., or 3.49%

Pressed tankage yield:
135 lbs., or .14%

MANUFACTURE OF TALLOW

All fats coming from the beef department which are not utilized for oleo are made into tallow. There is, of course, a small production of brown grease from this source. Theoretically, if all the fats and offal could be sliced and thoroughly cleaned, the production of the highest grade tallow would result. That, however, has been found to be impractical, because of the amount of labor and expense necessary. Accordingly the tallows are graded, depending largely upon the condition of the raw product which enters the tank, and secondly, upon the government regulations in regard to certain products which cannot be used for edible tallow.

In a large number of packing plants there is considerable loss, however, due to the fact that they do not utilize and care for many products which may be used in higher grades of tallow economically. Herewith is a list of the various products which go to make up the different grades of tallow. This is compiled from experience in larger packing houses, and is meant for general guidance only. Smaller packing houses may not find it practical to carry out the process to as fine a degree as the larger packers do.

Raw Products for Edible Tallow.—Cheek trimmings, beef rennets, head trimming fats, skirt trimmings, gall bag trimmings, beef trimmings from sausage, beef gullet trimmings, tongue trimmings, bones from passed for sterilization carcasses, beef aortas, scrap tallow from beef cutting, tail joints, sheep gullet trimmings, trimmings from calves and sheep, pancreas glands.

Products for Prime Tallow.—Broken clean guts, fat end of guts, pancreas glands, rough paunch trimmings, beef windpipes, washed skulls, jaws and skins from killing, all clean catch-basin skimmings, sheep tripe, cheek and tongue trimmings, trimmings from sheep ruffle, sheep windpipes, stripped sheep guts, beef ear drums, dirty tongue trimmings, beef rennet trimmings, condemned beef and sheep carcasses, skinned slunks, tonsils, shipper cow bags, bruised pieces from beef killing, oleo floor scrapings, beef cutting floor pickings, bladder trimmings, sinkers from tallow chill vats, round stumps of large guts, liver sweetbreads, scraps from oleo kettles, condemned viscera, cooked and raw tripe trimmings, sheep pelt trimmings, scraps from market, canner cow udders; sheep pecks, tripe and rennets, hashed and washed; nodulated sheep guts, hashed and washed; sheep lungs and pancreas glands, hide cellar trimmings.

All raw products entering the tanks which are not clean of course are rendered into a No. 2 tallow or brown grease, more about which will be mentioned later.

After rendering, the tallow is drawn off into vats, the usual custom being to run it through a small separator or settling device attached to the vat, so that all moisture and other fibre material can be removed. This is important in order to keep the tallow in good condition.

The yield of tallow per head is approximately as follows in well-regulated houses:

Edible	.30 lb.
Prime tallow	3.50 lbs.
No. 2	1.00 lb.
	4.80 lbs.

Tallow Grading Standards.—The standardization of tallows has been given considerable thought, and generally speaking the standard used by most packers for grading tallows is as follows:

Grade and Color	Titre	Free Fatty Acid	Moisture and Impurities
Edible tallow, pure white	42½ to 43½	½ to 1%	not over ½ of 1%
Prime tallow, almost white, slightly gray	42½ to 43½	½ to 1%	not over ½ of 1%
No. 2 tallow, darker than prime	42 to 43	5 to 8%	.6 to .75%
Brown grease, brown	no standard	not over 25%	not over 1%

The rendering of tallows is fully described later on, under the by-product and tankhouse sections.

HANDLING OF HIDES

In the beef slaughtering operation it was shown that next to the importance of leaving the fell covering intact, the elimination of cuts and scores on the hide was the highest consideration. The very first requisite in the tannery for a number one hide is the square feet of surface free from cuts, scores or blemishes. After the hide is dropped on the killing floor a close inspection is made for cuts and scores, and the responsibility therefor determined. The hide is then sent to the curing cellar.

Classes.—The first operation is reinspection, grading and weighing. The classification is as follows:

Spready native steers, not less than 6 ft. 6 in. in width.
Heavy native steers, 60 lbs. and up.
Light native steers, 50 lbs. to 60 lbs.
Extra light native steers, 25 lbs. to 50 lbs.
Kosher native steers, same as above.
Heavy Texas steers, 60 lbs. and up.
Light steers, 50 lbs. to 60 lbs.
Extra light steers, 25 lbs. to 50 lbs.
Butt branded steers, 60 lbs. and up.
Butt branded steers, 50 lbs. to 60 lbs.
Butt branded steers, 25 lbs. to 50 lbs.
Kosher butt branded steers, 60 lbs. and up.
Kosher butt branded steers, 50 lbs. to 60 lbs.
Kosher butt branded steers, 25 lbs. to 50 lbs.
Colorado steers, 60 lbs. and up.
Colorado steers, 50 lbs. to 60 lbs.
Colorado steers, 25 lbs. to 50 lbs.
Kosher Colorado steers, 60 lbs. and up.
Kosher Colorado steers, 50 lbs. to 60 lbs.
Kosher Colorado steers, 25 lbs. to 50 lbs.
Heavy native cows, 55 lbs. and up.
Light native cows, 25 lbs. to 55 lbs.
Kosher heavy native cows, 55 lbs. and up.
Kosher light native cows, 25 lbs. to 55 lbs.
Branded cows, all weights over 25 lbs.
Block cows, usually 55 lbs. and over.
Native bulls, all over 25 lbs.
Branded bulls, all over 25 lbs.

The above are cured weights. Hence in the green weight sufficient allowance must be made to cover shrinkage and attached dirt.

Trimming.—The ears are split to allow them to flatten out and thereby prevent faulty curing in the pack. The switch is removed at a point fourteen to sixteen inches from the tail butt.

From the inspection floor, the hide is taken to the pack of that quality that has been determined by the grading. Any hide showing a cut, score, sores, mutilation or excessive grubs must be considered a number two hide. It is considered good practice to carry each pack to accommodate the kill by months, as far as possible.

Salting the Hide.—The greatest possibility of damage to the hide is in this operation. A hair-slipped hide is the inevitable result of careless salting. The usual method is as follows: A sprinkling of good coarse rock salt is thrown on the floor and the corner hide laid down with the flesh up. From 40 to 50 pounds of salt is used on each hide. To make a straight edge on each side of the pack the belly is folded over sufficiently to make a straight edge and the butt folded over to make the corner rectangular. Considerable precaution must be observed that the salt is well distributed throughout the fold. The hides are then laid so that they will overlap, as in shingling. An absolutely flat surface must be maintained on each layer to prevent the drainage of the pickle. All water leaks or steam condensation must be guarded against.

A steer hide is thicker than a cow's hide and is wider than a bull's in the hindquarter. A bull's hide is wider than a steer's in the neck and shoulders and very thick in the butt and plate. Hides are thicker in the winter than in the summer and the hair is longer. Winter hides are packed close, having only about one foot lap. Short-haired hides, or summer hides, are generally packed wider apart; namely, about three feet. Packed in this way they preserve the moisture better.

Bull hides are given special attention in salting on account of their greater thickness. Hides carrying excessive manure must also be given extra salting to prevent decomposition. Long-hair hides are more readily cured, due to the greater capillary attraction. It is obvious, therefore, that short-hair hides should have closer attention in salting.

Temperature Requirements.—The temperature should not be lower than forty degrees, as the curing will be retarded if allowed to go below this point. Higher temperatures are perfectly safe up to 75° Fahr., but increase the shrink materially. The average of most hide cellars is 55° to 60° F.

Assuming that the salt has been thoroughly and uniformly distributed throughout the pack, no further overhauling or resalting is required. Overhauling adds nothing to the cure and should always be avoided, as any additional handling is detrimental to the hide. It is, however, frequently resorted to, to overcome excessive shrinkages or because of excessive air circulation that has dried out the pack. It would be far better to attempt to overcome the physical condition that has caused the excessive shrink rather than overhaul.

Building the Hide Pack.—The size of the pack, as to area, is usually determined by existing conditions, bearing in mind that the larger pack has less exposed edge, in comparison to the number of hides in the pack, than the smaller pack. If, however, it is built too large, the working conditions will become more difficult, salting not infrequently slighted and the level of the pack may suffer. The height of the pack should not exceed three and one-half feet, although the packs are sometimes built to 4½′, but anything above 3½′ causes shrinkage. The ears, pates, legs and tails should be well salted to prevent hair slips and deterioration. Build the pack so there is a gradual slope to the center, which keeps the pickle from draining away.

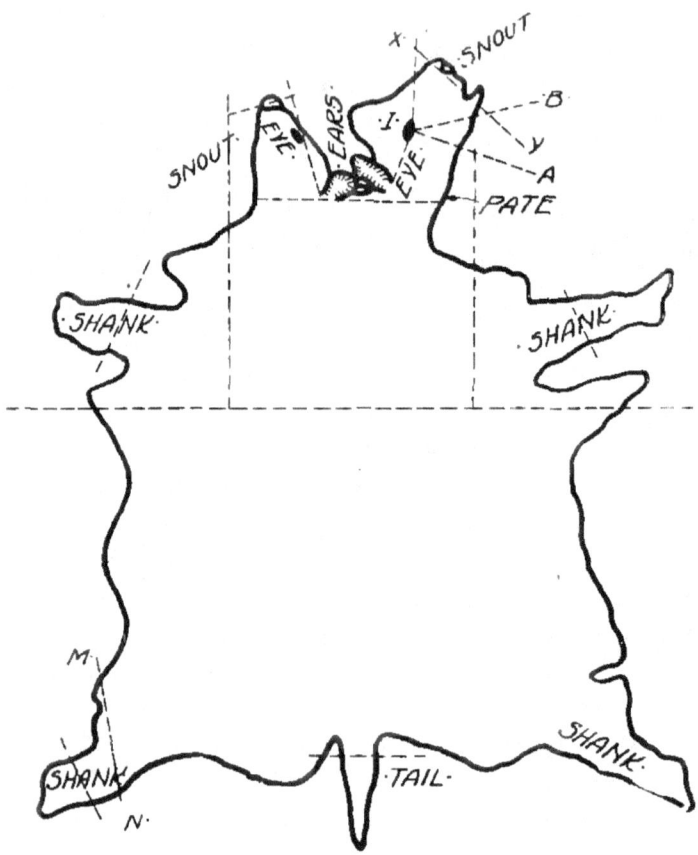

Proper Layout of a Packer Hide

Hides are thoroughly cured after thirty or forty days, but frequently they are left in cure for a much longer period. It has been demonstrated that no damage is done up to two years and possibly longer, with the exception that after the first year discoloration or salt stain begins to be apparent. Hides cure in less time in summer because of the higher temperature, and summer hides are more porous and quicker to absorb pickle. The shrinkage also is greater in summer than in winter, the summer averaging around 18 per cent and the winter shrinkage from 12 to 15 per cent. Hides pickled 24 hours will be cured enough for shipment if tanned immediately. Slipped hair on hides is caused by insufficient pickle or undersalting.

Many discolorations of hides are due to impurities in the salt. The salts of iron or copper are the worst offenders. In the event of extreme

HIDES

discoloration, samples of the salt should be sent to the chemical laboratory to determine the quantitative and qualitative analysis of such impurities.

Taking Hides Out of Pack.—The hides are taken up as required, put over a tosser to remove adhering salt and wrapped into bundles. Usually this bundling is done in such a manner that the skin will be exposed rather than the hair. This is the better method, as the hair carries sufficient moisture to keep the hide damp and in better condition. Some tanneries object to this method. Hides are usually taken up and bundled the day before weighing and shipping. It is with this in view that causes the tanneries to order the hides bundled with hair out, as it is expected that the hair will have dried out, showing a corresponding loss in weight. Tare allowance is regulated by taking ten hides from the lot, and by sweeping to determine the percentage of allowance.

Hides with more than five grub holes or with slipped hair are number two hides, and sell for less per pound. It is not uncommon at certain seasons of the year to receive a large percentage of grubby hides. From 25 to 55 per cent of grubby hides are found at certain seasons. Most of these are received in the months of January, February, March, April and May.

The allowance made for grubby hides is determined in much the same manner as tare allowance is regulated, ten hides each being selected by the buyer and seller from a pack, the average of the twenty being used as a compromise. In case the buyer is still dissatisfied, he can demand the selection of a second 20.

Hides from condemned cattle are sterilized by soaking in a 20 per cent salt solution, which contains one one-thousandth part of mercuric chloride.

Hide Left on Feet, Head, Etc.—It is very important in removing the hide for the skinner to be sure to mark the hoofs, heads, etc., so that no more hide will be left on these parts than is absolutely necessary, for the hide is worth much more as such than for fertilizer. The following is a test on the amount of hide left on, which is considered to be a good standard to approach. Should a greater weight of hide be left on, it is an indication that the skinners are not doing the most efficient work.

```
100 feet ............................ 2 lbs. hide
100 heads ........................... 4 lbs. hide
100 pairs horns ..................... 5 lbs. hide
```

Curing Switches.—The switches which are removed from the hides are washed in cold water and are put into saturated salt pickle over night. They are then dried and salted down in fine salt and left for about a month, or until such time as the accumulation is large enough for selling purposes.

Slunks.—Hairless slunks are put into light salt water over night and then taken out and hung to dry for a half day. They are then put down in fine salt. It should be the endeavor to dispose of these within a month after the curing process is started.

Calf Skins.—Calf skins may be handled the same as cattle hides, but it is best to use new salt. On regular cattle hides 50 per cent of second-hand salt may be used after it is washed.

General Information

Packer Hides:

Average Weight, Green	60	lbs.
Average Weight, Cured	51	lbs.
Average Length	7	ft.
Average Width	6	ft.
Space Required in Pack	1	cu. ft.
Space Required on Floor	4	sq. ft.
Average Height of Pack	3½	ft.
Salt Used in Curing per Hide	40	lbs.
Time Required to Cure	30	days
Average Shrink	15%	
Temperature of Cellar	55	Degrees Fahr.

Grubbing Season: Branded Cows, Nov. 1 to June 1; Texas, Nov. 1 to June 1; Colorados, Dec. 1 to June 1; Bulls, not grubbed; balance, Jan. 1 to June 1.

Spread is measured just behind the front legs; it refers to heart girth.

Chapter II—HOGS

BREEDS OF HOGS

Uniform, smooth meaty hogs of proper type are of paramount importance to the American meat packer. The breeders of purebred hogs have developed these characteristics to a marked extent. During the last 50 years, especially, great progress has been made in improving hog types. With the continued purchase of breeding stock in different sections of the country, numerous types or strains with distinguishing characteristics have been developed, and packers should be familiar with these.

In general there are two main types of hogs—the bacon hog and the fat or lard hog. The bacon hog is essentially a European product. It is usually characterized as being narrow of body and long, with deep sides, smooth but comparatively light hams and shoulders, and with a characteristic "leanness" throughout, with marked refinement or quality.

The fat or lard hog is almost exclusively a product of the United States. In the past it has been almost the opposite of the bacon type. It was thick of body throughout, with heavy hams, shoulders, and jowl; of short, squatty conformation, carrying an excessive amount of fat and giving the appearance of being rather "roly-poly." Today, however, the breeders of the lard type are selecting towards a hog which more nearly approaches the bacon type than that which they were producing a number of years ago, in order that the product may be a hog which gains weight rapidly and one which will be prolific.

The lard hog of today has great size, deep body, strong back, good feet and legs, smoothness-throughout, and as much refinement or quality as is consistent with vigor or ruggedness.

The important breeds of hogs in the United States at the present time are: Berkshire, Chester White, Duroc Jersey, Poland China, Hampshire, Spotted Poland, Large Yorkshire, and Tamworth. There are also a number of other breeds which are found in smaller numbers in various sections of the country, such as Kentucky Red Berkshires, Cheshire, Essex, Large Black, and Mulefoot.

Trade Requirements in Hogs

While the packer is interested in the breeds of hogs, he is not so much interested in what one particular breed can do, as in the market type which hog breeders are working toward to produce a hog which has quality, proper weight and which will finish in good condition, producing a desirable product.

In considering quality, hogs should not be overfinished. An extremely lardy meat is too fat for the best trade. At the same time, thin unfinished pork lacks the substance and maturity which is preferred. Moderate to

Typical Bacon Hog

Typical Lard Hog

good finish is most desirable, and as a general thing will bring the best prices.

Killing hogs should show smoothness, uniformity and fullness throughout. Heavy shoulders and shallow, narrow hams yield too great a proportion of bone and shank to suit the consumer. Low, narrow backs cut out a pinched loin, which produces a small amount of pork chops. Soft, wrinkled and seedy bellies never yield the trim, firm attractive strips of bacon.

In considering weight, the packer should remember that the best trade desires pork from good hogs which have been finished at a weight of under 250 lbs. A 16-lb. loin cuts chops weighing around one-half pound apiece; an 8 to 9 lb. loin will yield four chops to the pound. This naturally puts a premium on the lighter loins. The 6 to 10-lb. bacon strips carry a more popular portion of fat and lean than do the 14 to 20-lb. bellies which come from 300 to 400-lb. hogs. The 10 to 14-lb. ham is more nearly family size than is the 20-lb. average.

MARKET CLASSES AND GRADES OF HOGS

The following classification has been tentatively adopted by the United States Bureau of Markets and Crop Estimates for use in describing and quoting upon the various killing classes, sub-classes and grades of hogs:

Class	Sub-Class	Grade
Butcher and Bacon Hogs:	1. Heavy Weight (250 lbs. up)	Choice Good Medium
	2. Medium Weight (200-250 lbs.)	Choice Good Medium Common
	3. Light Weight (150-200 lbs.)	Choice Good Medium Common
	4. Light lights (130-150 lbs.)	Choice Good Medium Common
Pigs (130 lbs. down)		Choice Good Medium Common
Packing Sows		Smooth Rough
Stags		
Boars		

Butcher and Bacon Hogs.—This general class embraces a very large percentage of the hogs marketed, and includes all hogs the fresh and cured products of which are most sought after by the consuming public, which dictates in large measure the kind of hogs preferred by the packer, and which therefore command the best prices on the market. The

Choice Hog

Good Fat Hog

Medium Butcher Hog

Common Hog

Skip or Rough Hog

Rough Packing Sow

higher grades cannot be materially deficient in condition, form and quality. There are four-sub-classes, divided as to weight, as in deciding upon relative market values of this general class weight is usually the most important factor, since the weight of the live animal determines the approximate weight of the various dressed cuts.

The heavy weight sub-class embraces hogs weighing from 250 pounds up, and grading from medium to choice, that are usually referred to by the trade as "heavy butchers," and frequently, when weighing over 300 pounds, and of excellent quality, condition, and conformation, as "prime heavies." The number of prime heavies found in our markets today is limited.

The light and medium weight sub-classes embrace hogs weighing from 150 to 250 pounds, that are commonly known in the trade as "light" and "medium weight butchers." These are usually the most popular weight hogs, since they furnish chops, hams and bacon (the most valuable cuts of the hog) of the weights most desired by consumers. The percentage of low-priced cuts from these hogs is small. These sub-classes also furnish the popular export cuts known as "Wiltshires" and "Cumberlands."

The fourth sub-class, "light lights," weighing from 130 to 150 pounds, consist chiefly of immature hogs that are utilized almost exclusively for the fresh meat outlet. Excepting at periods of partial or total crop failures, the percentage of "light light" hogs marketed is small, and because of the uncertain supply demand for them is relatively less stable than for hogs of other weights. They are frequently used by killers as a substitute for strong weight pigs, or for the lighter hogs of the light weight sub-class, when supplies of such are inadequate.

Pigs.—Killing pigs include all swine used for slaughter weighing under 130 pounds, excepting "roasters." As demand for such small cuts as pigs supply is limited, pigs are usually dressed "shipper style." (See special section on "Shipper Pigs.")

Packing Sows.—In the classification given here sows that show evidence of having had pigs are classed as packing sows, in contrast to gilts and barren sows, which fall into the butcher and bacon class. Rough and plain barrows, usually of excessive weight, are generally bought and classed with packing sows . Packing sows are used chiefly for packing purposes, meaning that the products have to be processed before being offered for sale. They are graded as "smooth" and "rough."

Stags.—Stags are boars which have been castrated. They are bought with dockage of 70 pounds, because they are very wasty in dressing.

Boars.—This class is seldom found in the stock yards, as they are practically worthless for food on account of the odor of the meat. Boars are usually condemned by the government inspectors and are used for fertilizer and grease only.

Miscellaneous.—"Skips" are very inferior throw-out hogs, usually of light weight, and can be bought for a lower market price than rough packing sows. "Roasting pigs" usually weigh from fifteen to thirty-five pounds. They are in demand chiefly during the holidays and on special occasions, and as a class therefore are not important to the packer. Any grade of hog that is very lame, or for any reason is unable to walk, is called a "cripple." "Dead hogs" are those which arrive dead in the car, and are valuable only to those interested in the manufacture of grease and fertilizer.

Advantages of Different Grades to the Packer

The kind of hog for the packer to buy depends upon his market outlet and what his customers want. With lard high in price, the heavier, extremely fat lard type of hog, of course, can be used to best advantage. On the other hand, should a packer have considerable export business, he can use to best advantage lighter hogs of bacon type, but with good conformation. In general hogs of the butcher class are most in demand.

Buyers look for hogs which have smooth backs and sides, free from wrinkles, and for hogs that carry down well. Roughness over the shoulders and long necks are avoided. The demand of the general public has been for the lighter cuts, especially of ham and bacon. The heavier cuts sell at a discount, due to the fact that the buyers have to put more money in the heavier cuts, and they are usually coarse in fibre and do not have the quality which the lighter hogs carry. Of course, the packer must consider his market and act accordingly.

DRESSING YIELDS OF HOGS

The yield of hogs depends considerably upon the quality, condition, age, weight, fill and breeding. Barrows of the same weight of sows will dress out a higher percentage of carcass. Prime hogs will dress about 78 to 80%, and lighter hogs not well finished around 65%, head off and leaf out. In the percentages given here the hair, blood and viscera are not included.

The average dressing percentage of good, fair and common hogs, showing the yield, and percentages of the various cuts, is shown in the following tests:

Cut	Good %	Fair %	Common %
Hams	12.50	13.00	13.25
Shoulders	10.50	11.00	11.00
Sides: Bacon Belly	11.50	10.00	10.00
Fat Backs	9.95	6.25	4.25
Loins	9.75	10.00	10.00
Rendered lard, from cutting	8.00	7.75	7.50
Cutting trimmings	6.50	6.00	5.50
Total dressed carcass	68.70	64.00	61.50
Rendered lard, from killing	4.00	3.50	3.25
Leaf fat	3.00	2.60	2.35
Killing products: Livers, hearts, greases, tankage, etc.	12.00	13.75	13.75
Total yield	87.70	83.85	80.85
Loss in moisture, etc.	12.30	16.15	19.15
	100.00	100.00	100.00
Percentage of barrows	70.00	55.00	40.00

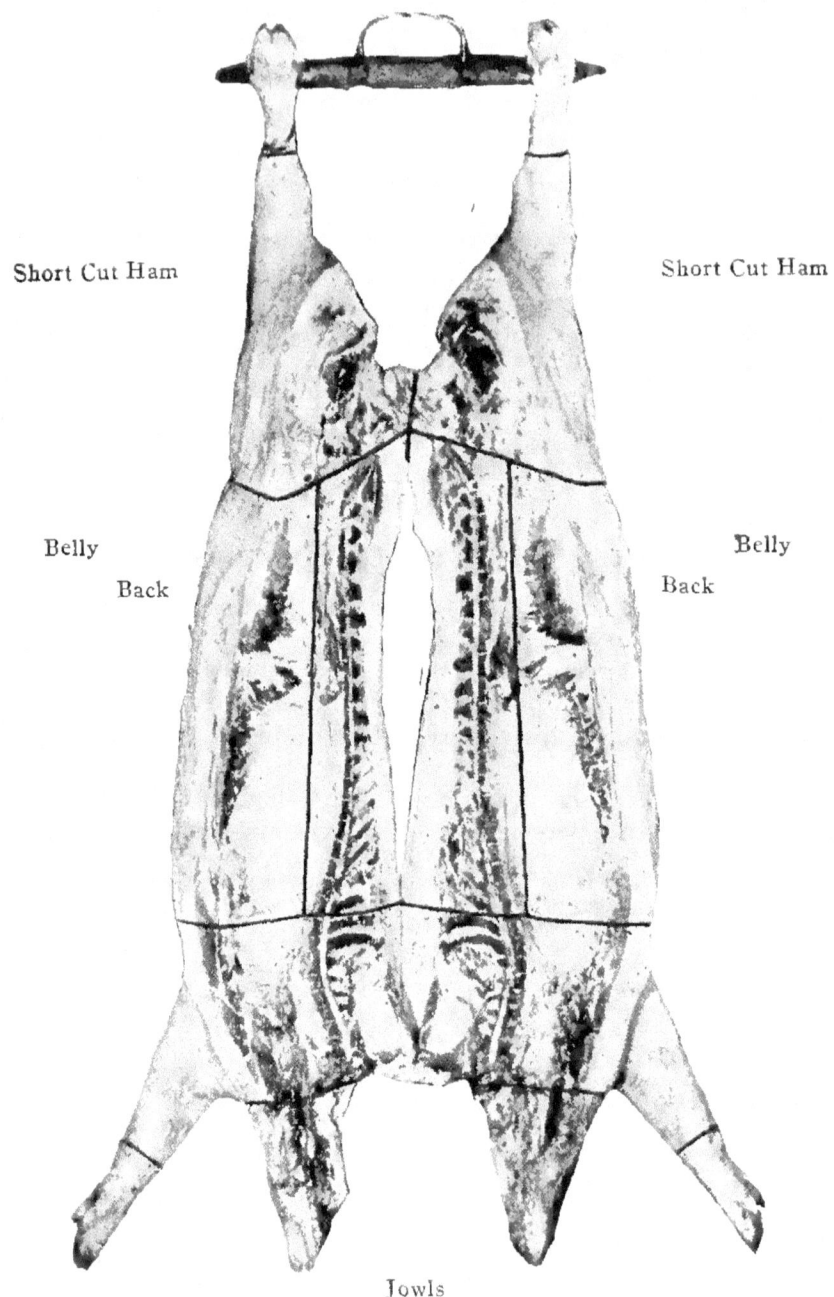

Packing Hog Carcass, Showing Domestic Cuts

Special cuts for export or other purposes do not change the figures on total percent, but affect only the percentage of each cut, the amount of trimmings, etc., based on the figures given.

As a general guide each Superintendent will, of course, have his own tests, from which he can figure the yield of the carcass on various cuts which he usually makes.

Here is a second test made on two heavy, rough hogs, which will serve as an example of the various cuts which may be expected from a carcass of this type.

Cut	Weight	Per Cent
Skinned hams	51 lbs.	16.72
Ham skins	1½ lbs.	.49
Boston butts	23 lbs.	7.54
Jumbo loins (tenderloin out)	47 lbs.	15.41
Tenderloins	2 lbs.	.65
Rib bellies (30-40 lbs. avg.)	65½ lbs.	21.48
Trimmed neck bones	3½ lbs.	1.15
Shoulder bones	3 lbs.	.98
Front feet	3 lbs.	.98
Hind feet	3 lbs.	.98
Tails	½ lb.	.16
Trimmings	22½ lbs.	7.41
Neck bone trimmings	1 lb.	.33
Fat backs (18-20 avg.)	38 lbs.	12.46
Fat backs skinned	9 lbs.	2.95
Shoulder fat and skin (65% P. S. L.)	15½ lbs.	5.08
Ham fat (70% P. S. L.)	14 lbs.	4.57
Fat from tenderloins (30% P. S. L.)	1 lb.	.33
Loss in cutting	1 lb.	.33
Total	305 lbs.	100.00

Here is another representative test on hogs weighing 209 lbs. average, as a guide for what may be expected from the ordinary runs coming on the market today (1922).

Cut	Per Cent
Hams	13.48
Shoulders	7.60
Sides	19.07
S. P. Bellies	6.43
Loins	6.76
Leaf Lard	2.82
P. S. Lard	11.37
Total cuts	67.53
Cutting offal	6.47
Grand total	74.00

The percentage yield of cuts in this case is calculated from the live weight of the hog as 100%.

HOG KILLING

In driving hogs to the killing floors they should never be hurried, nor should the drivers be allowed to use anything as a persuader in the form of a stick or club, because when a hog is struck, either on the shoulder or flank, there is left from the force of the blow a bruised spot, rendering that part of the meat unfit for No. 1 quality. The driving is best done by making some sort of a noise or by the use of a flat canvas strap.

Another serious result from pushing the hogs too rapidly is that they become overheated, and when killed in this condition the meat hanging in the cooler does not chill properly, and may result in sour meats. Should they become overheated they should be allowed to stand for some time before killing, but if this is not possible a water sprinkler may be used, but with great precaution, as death may result from the effects of the cold water on the overheated hogs. Modern practice turns the hose under the hog, never over his back.

Hogs that are kept in the pens should not be allowed to become too hungry, as they will get very nervous. On the other hand, they should not be given too much food, nor should they be overfed just previous to their being killed, as that renders the lining of the stomach of little value, especially if it is to be utilized for the manufacture of pepsin.

Shackling.—The shackler should be given specific instructions never to strike the hog with the shackling chain, to avoid bruises. The hogs should be driven close to the shackling hoist, and the chain put around the hind leg nearest the hoist. In this way there is very little chance of a joint being thrown out. If the latter occurs, the ham is damaged.

Sticking.—When the hogs slide down the rail to the "sticker" they should be stopped squarely in front of him, so that when making the cut the knife will go straight and make a clean cut about 3½ inches long, so the blood will flow freely and rapidly. As in the case of the steer, the jugular vein is severed. Should the sticker not run the knife straight in, the result will be a "stuck" shoulder, which will fill with blood and leave an ugly clot, which must be trimmed out later when making California or picnic hams, in order to make them salable.

Bleeding.—The regulations of the U. S. Bureau of Animal Industry specify that a hog shall hang on the bleeding rail for at least 6 minutes before being dropped into the scalding tank. This will give ample time for the blood to drain thoroughly.

Scalding and Dehairing.—A little soap or weak alkali is generally added to the water in the scalding vat, in order to loosen the scurf and cleanse the carcass thoroughly. The temperature of the scalding vat water in the larger plants—and, in fact, it is advisable for all plants—is best regulated by a thermostat, which allows just enough steam to enter to keep the water at a temperature of 146° to 147° F. The temperature should not go above the 150° mark, as that will result in cooked carcasses.

From three and a half to four minutes is usually long enough to scald a hog, depending upon the season of the year, the hardness of the hair and the individuality of the hog. The man looking after this work by means of hooks should test each hog, and as soon as the scalding is complete allow it to pass on through the hog-dehairing machine. Should a carcass not be thoroughly scalded the hair and bristles will break off at the roots, and the result will be a dark-appearing carcass, which will make only second grade stock.

From the hog-dehairing machine, the carcasses pass on to a scraping table, or in some plants on to a chain where any hairs that may remain can be shaved off. Modern dehairing machines are rapidly approaching a high state of efficiency, however. Before passing on to the chain, a slit is made exposing the gam cords, so that the gambrel stick may be inserted. Most packers find it advantageous, also, to pull some of the toes on the moving table, immediately after the carcass is ejected from the dehairing machine, as this may be done very easily with hand hooks at that time.

Heading.—After shaving, and burning the hair from the ears and snout by means of gas torches, the slaughtering operation commences. Heading is the first, care being taken to make the cut very close to the head bone, leaving no more meat attached to the head than possible. The head is left attached to the carcass by means of a small strip until after the Government veterinarian has inspected the neck glands.

Gutting.—The opening of the carcass the entire length and exposing the viscera is the next step. This must be done very skilfully by men of experience, in order to make a straight cut and avoid cut casings. Irregular cut bellies have to be trimmed later to a uniform shape, which means a loss in first-class meats, as the trimmings only bring sausage prices. Should the casings or gall sack be punctured the contents flow over the meat, discoloring it and often causing it to be rejected by the Government inspectors.

Just before the viscera is completely removed from the carcass the bung is dropped, care being taken when passing the knife around it not to include too much fat, which if left on the bung is apt to get dirty, and will have to go to the grease tank instead of being utilized for lard. After the bung is dropped and tied, the entire viscera is removed for inspection by the Government veterinarians. Modern viscera inspection tables, either stationary or moving on an endless belt, and divided into compartments for the different viscera, are now installed in the most up-to-date plants.

The hams are then faced, care being taken not to cut too near the lean meat.

Splitting.—The splitting of the carcass requires great care, in order that the cut may be made directly in the center, allowing an equal amount of bone to appear on each side. Before the splitting is complete side hooks are used to spread the carcass, and the leaf lard is pulled out. The puller should remove every particle possible, as this fat can be used for a higher grade product than prime steam. After the leaf is removed the remaining particles of fat are scraped out, and are caught in a clean receptacle and saved for use in prime steam lard. Any bruises or ulcers which are cut from the carcass should be dropped into a separate inedible pan, and utilized for white grease purposes.

Before the carcass enters the cooler it is washed thoroughly, inside and out, with a spray of luke-warm water; then the carcasses are scaled or weighed before they pass on to the cooler. Many houses weigh hogs as

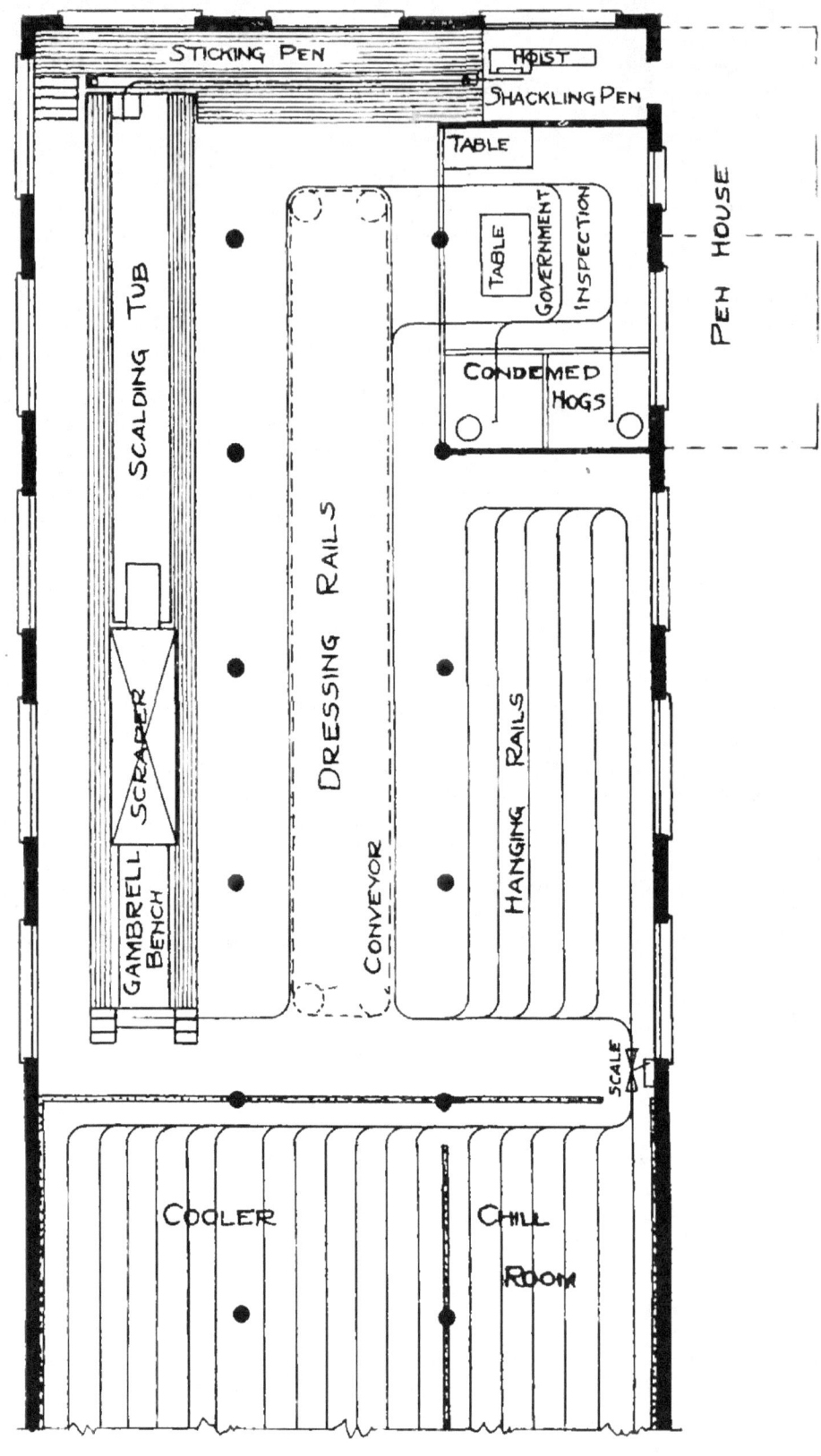

Layout of a Typical Hog-Killing Floor

soon as viscera are removed; weighing with head on, leaf in and ham facings on.

The cost of killing hogs varies according to the size of the plant, the labor cost, and the degree to which the by-products are utilized for manufacture. The cost for labor alone will vary from 10c to 20c per hog.

Typical Hog Killing Floor

The arrangement shown in the illustration is a typical layout of a hog killing floor, capacity from 100 to 400 hogs per hour. The size of the machines, scalding tubs and rails will vary according to the capacity.

Hogs are driven from the penhouse into the shackling pen, which should not be too wide. If this pen is made too wide, considerable delay will be experienced in shackling hogs, as one man will have to keep constantly after the hogs, throwing shackles around the hind legs.

Rails.—The hoists deliver the animal on to the sticking rail, where the hog slides towards the dropper, located at the end of the sticking room. The sticking rails are usually 7 feet to 9 feet above the floor, depending entirely upon the kind of hoist used. It should be borne in mind that the height of the scalding tub, sticking rail and hoist is governed entirely by the height of the rail above the gambrelling bench. Upon this dimension depends the height of the equipment coming before it.

The U. S. inspection authority prefers hog-killing floor rails not to be lower than 9 feet, but in existing buildings as low as 8 feet has been allowed.

Scalding Tub.—The scalding tub should be of sufficient length to allow for proper scalding of the animal, and should never be less than 14 feet long, which is practically the minimum capacity to take care of any kind of hog scraper. Experts consider it is necessary to have scalding tubs of the following lengths for the following capacities: 100 hogs, 16 feet to 20 feet; 200 hogs, 20 feet to 30 feet; 300 hogs, 30 feet to 40 feet, and 400 hogs, 40 feet, to 50 feet.

Dressing.—After thorough scalding, the hogs are delivered into the dehairing machine, from which they are discharged, either on a stationary or moving gambrelling bench. Gambrel storage is provided overhead for the convenience of the gambrelling men, who deliver the hogs on to the rail, from where they go to the dressing rails, which are usually provided with a track conveyor to allow for steady movement of the hog through the various operations.

Two separate enclosures, either with a low concrete wall, or fenced in with wire, are provided for the Government inspector, with two extra compartments for condemned animals.

Hanging.—As soon as the hogs leave the conveyor they may be put into the cooler direct. In cool weather some packers find it advisable to leave them on the hanging rails on the killing floor for some time, although good practice recommends that the hog be put in the chill room immediately after being cleaned.

The height of the hog-killing floor is governed primarily by the type of hoist used. A wheel hoist requires much greater heights than a vertical or triangle type hoist. This is due to the fact that the shackle hook which engages the bleeding rail may be delivered right from the top of the vertical hoist, and from the highest point, on to the bleeding rail. The wheel hoist, however, requires additional height for the shackle itself, and for the chain which is attached to the wheel. For this reason, from two to four more feet of height are required on a wheel hoist than on the triangle type of hoist.

HOG COOLING

At one time it was thought that a hog should hang for two or three hours on a hanging floor at ordinary temperature before entering the cooler, but this practice has now been abandoned in most plants. If the hogs are fairly dry before entering the cooler, they may be run in at once. Care should be taken to see that there is ample space between each carcass, i. e., they should not touch. The temperature of the cooler when starting to fill should be 30° to 32°, and when filled high point should not exceed 44°. Ten hours later it should be down to 36°; the following morning the temperature should be not over 32° and the second morning 30° to 32°.

The brine spray system, as well as the direct expansion system, are both used for chilling hog coolers. The temperature, as given above, will in 40 to 48 hours thoroughly prepare the carcass for cutting. Only in very exceptional cases should a carcass be cut under the 40-hour period. If good results are to be obtained in the following process of curing, etc., it is very important that these cooler temperatures be watched very closely.

Hog carcasses will shrink an average of 2.6 per cent after hanging in the cooler for 48 hours, and about 2.7 per cent if left in the cooler for 72 hours. After this period, the shrink will be practically nil. In a series of tests made on 60,000 hogs over a period of a month, the various lots killed each day showed the following shrinkage in a cooler chilled by the direct expansion method:

48 hours .... 2.55 per cent	48 hours .... 2.61 per cent
48 hours .... 2.60 per cent	48 hours .... 2.73 per cent
48 hours .... 2.44 per cent	48 hours .... 2.68 per cent
48 hours .... 2.62 per cent	72 hours .... 2.40 per cent
48 hours .... 2.35 per cent	72 hours .... 2.54 per cent
48 hours .... 2.77 per cent	72 hours .... 2.83 per cent
48 hours .... 2.40 per cent	72 hours .... 2.92 per cent
48 hours .... 2.56 per cent	

If green hog products are to be stored for a period of 2 to 3 days, they will carry well at a temperature of 20° to 25°, but if necessary to store for a longer period they should be sharp frozen at a temperature of zero or lower, and then transferred to and held at a temperature of 12° to 15° above zero Fahr. (Also see Section on "Packinghouse Refrigeration.")

SHIPPER PIGS

Shipping hogs are dressed with head on and leaf lard in.

When possible these should be bought separate from the regular drove, as this gives a clear record as to live weights and cost, and enables the packer to figure a regular test to ascertain the dressed cost.

When "shippers" are selected out of the regular packer drove and the live weight is not known, use the following formula: Shrink the hot weight of the hogs selected 2½ per cent and figure an arbitrary yield of

78 per cent, thereby obtaining an approximate live weight for the shipper pigs, using as live cost the average live cost of the drove for that day's kill.

The hot dressed weight of the pigs should be shrunk 2½ per cent to bring to chilled basis.

Offal Test on Shipper Pigs

The offal credit on "shippers" is as follows: Pigs under 90 lbs. alive, 25% per head; pigs over 90 lbs. alive, 40% per head. These are based on the following tests:

	Heavy Pigs Per Hd.	Light Pigs Per Hd.
Hearts	.61	.35
Giblets	.37	.31
Stomachs	....	....
Gullets	1.33	.50
Lungs	1.90	.92
Livers	4.44	1.86
Gut fat, 55%	2.67	....
Caul fat, 55%	1.55	3.00
Bung gut, 55%	.56	....
Tankage	.03	.01
Dry blood	.02	.01
Hair and bristles	.02	.01
Casings	.03	.02

Test on 16 Shipper Pigs Cut Up

For the packer or butcher who does not slaughter, but who buys dressed hogs, the following cutting test on shipper pigs with the head on will be of interest:

Avg.			
6 lbs.	188 lbs.	Skinned shoulders	15.27%
6 lbs.	189½ lbs.	Rib bellies	15.39
5½ lbs.	179½ lbs.	Loins	14.57
8 lbs.	246½ lbs.	Hams	20.02
	72½ lbs.	Back fat	5.89
	22½ lbs.	Neck bones	1.87
	35½ lbs.	Lean trimmings	2.88
	120 lbs.	Fat trimmings (65%)	9.75
	2½ lbs.	Tails	0.20
	14½ lbs.	Front feet	1.18
	18 lbs.	Hind feet	1.46
	5 lbs.	Kidneys	0.41
	35 lbs.	Leaf lard	2.84
	17 lbs.	Cheek meat	1.38 ⎫
	3 lbs.	Head meat	0.24 ⎭
	5½ lbs.	Snouts	0.45
	4½ lbs.	Ears	0.36
	28 lbs.	Head fat (30%)	2.27
	9 lbs.	Jaw bones	0.73
	31 lbs.	Skulls	2.52
	4 lbs.	Ear drums	0.32
	1,231 lbs.		100.00%

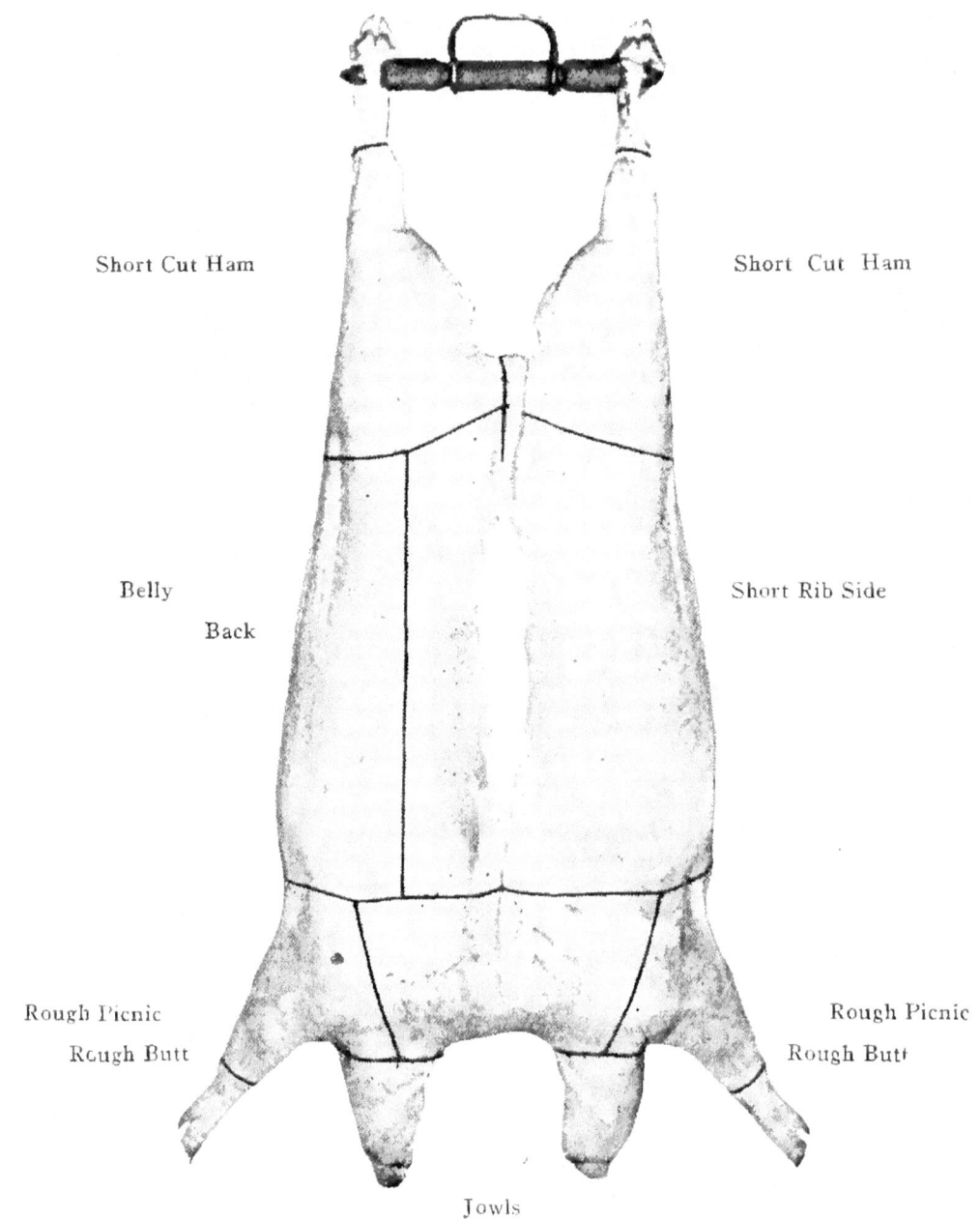

PACKING HOG CARCASS
Lined to show typical domestic cuts.

PORK CUTS

Following are the revised regulations (Sept. 15, 1921), adopted by the Board of Directors of the Chicago Board of Trade, setting forth the requirements in the cutting and packing of hog products. These regulations are accepted in both domestic and export trade as official.

Barreled Pork

Mess Pork.—Standard mess pork should be made from sides of well-fatted hogs, split through the center, cut into strips of reasonably uniform width, properly flanked and not backstrapped.

Between October 1st and the last day of February, inclusive, one hundred and ninety (190) pounds, and between March 1st and September 30th, inclusive, one hundred and ninety-three (193) pounds of green meat, numbering not over sixteen (16) pieces, including the regular proportion of flank and shoulder cuts, placed four layers on edge without excessive crowding or bruising, shall be packed in each barrel, with not less than forty (40) pounds of coarse salt, and barrel filled with brine of full strength; or forty (40) pounds of coarse salt, and, in addition thereto fifteen (15) pounds of salt, and barrel filled with cold water.

Back Pork.—Back pork should be made from the backs of well-fatted hogs, after bellies have been taken off, cut into pieces of about six (6) inches in width, and in all other respects to be cut, selected and packed in the same manner as mess pork.

Extra Clear Pork.—Extra clear pork should be made from the sides of extra heavy well-fatted hogs, the backbone and ribs to be taken out, and in all other respects to be cut, selected and packed in the same manner as mess pork.

Clear Pork.—Clear pork should be made from the sides of extra heavy, well-fatted hogs, the backbone and half the rib next the backbone to be taken out, and in all other respects to be cut, selected and packed in the same manner as mess pork.

Clear Back Pork.—Clear back pork should be made from the backs of heavy well-fatted hogs, after bellies have been taken off, and backbone and ribs taken out, cut into pieces of about six (6) inches in width, and in all other respects to be packed in the same manner as mess pork.

Fat Back Pork.—Shall be made from well-fatted hogs, after the loin, blade bone and belly have been removed, cut into about six (6) inch pieces of uniform thickness, packed on edge, and placed in four (4) layers. The barrel to be filled with full strength pickle, and at least thirty (30) pounds of coarse salt, and shall weigh at time of shipment, two hundred and four (204) pounds net, or two hundred and six (206) pounds out of pickle.

Ham Butt Pork.—Shall be made of pieces cut from the rump end of the side in squaring, after a short ham has been removed; somewhat of a triangular piece in shape. The barrel to be filled with full strength pickle, and at least thirty (30) pounds of coarse salt, and shall weigh at time of shipment two hundred and four (204) pounds net, or two hundred and six (206) pounds out of pickle.

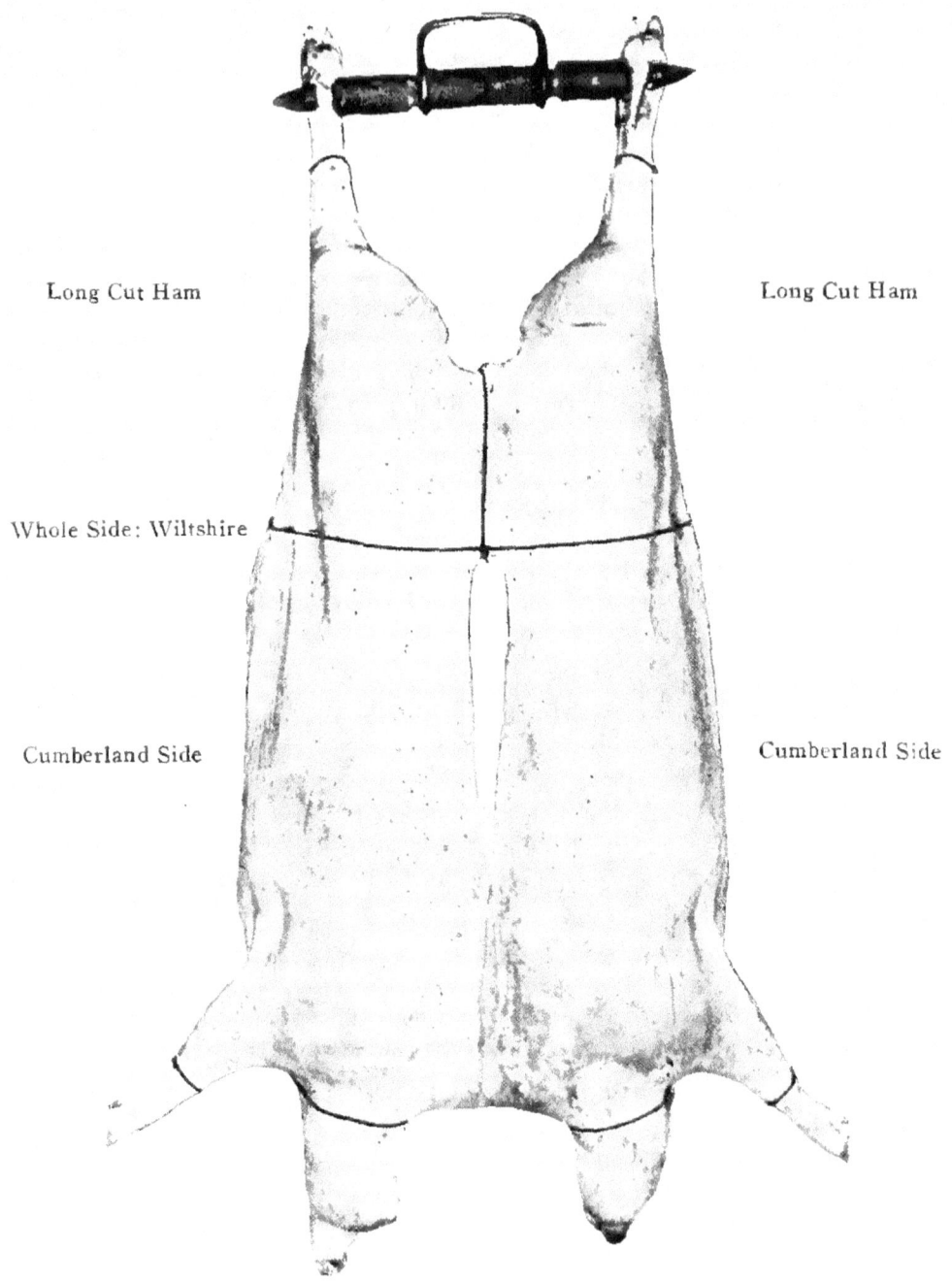

BACON HOG CARCASS
Lined to show export cuts.

Bean Pork.—Shall be made from the jowl, neatly trimmed on the face, and cut into square pieces. The barrel to be filled with full strength pickle, and at least thirty (30) pounds of coarse salt, and shall weigh at time of shipment two hundred and four (204) pounds net, or two hundred and six (206) pounds out of pickle.

Jowl Pork.—Shall be made from the jowl, trimmed on the face and edges, blood clots and loose pieces removed, and the side next to the shoulder squared. The barrel to be filled with full strength pickle, and at least thirty (30) pounds of coarse salt, and shall weigh at time of shipment two hundred and four (204) pounds net, or two hundred and six (206) pounds out of pickle.

Clear Plate Pork.—Shall be made from the fat end of the shoulder, free of bone, reasonably free of lean, and squared on the neck side. The barrel to be filled with full strength pickle, and at least thirty (30) pounds of coarse salt, and shall weigh at time of shipment two hundred and four (204) pounds net, or two hundred and six (206) pounds out of pickle.

Plate Pork.—Shall be made from the fat end of the shoulder, part of the blade bone left on, reasonably free of lean, and squared on the neck side. The barrel to be filled with full strength pickle, and at least thirty (30) pounds of coarse salt, and shall weigh at time of shipment two hundred and four (204) pounds net, or two hundred and six (206) pounds out of pickle.

Shoulder Butt Pork.—Shall be made from the butt end of the shoulder, after the picnic is cut off, the neck bone and part of the blade left in, neck squared, and the flaring fat at the butt end squared off. The barrel to be filled with full strength pickle, and at least thirty (30) pounds of coarse salt, and shall weigh at time of shipment two hundred and four (204) pounds net, or two hundred and six (206) pounds out of pickle.

Clear Shoulder Butt Pork.—Shall be made the same as shoulder butt pork, except that the neck bone shall be removed. The barrel to be filled with full strength pickle, and at least thirty (30) pounds of coarse salt, and shall weigh at time of shipment two hundred and four (204) pounds net, or two hundred and six (206) pounds out of pickle.

Loin Pork.—Shall be made from loins, tenderloin out, cut into two three or four pieces, and packed on edge without excessive crowding. The barrel to be filled with full strength pickle, and at least thirty (30) pounds of coarse salt, and shall weigh at time of shipment two hundred and four (204) pounds net, or two hundred and six (206) pounds out of pickle.

Green or Sweet Pickled Meats

Standard Hams.—Shall be cut off about two and one-half (2½) inches from the exposed end of the aitch bone, properly faced, shank cut off in or above the hock joint, loose and gut fat removed from the face, and the ham well rounded.

Skinned Hams.—Shall be cut, in all respects, the same as standard hams, except that the skin must be removed down to within, at most, four (4) inches from the shank, the fat to be leveled back at least three (3)

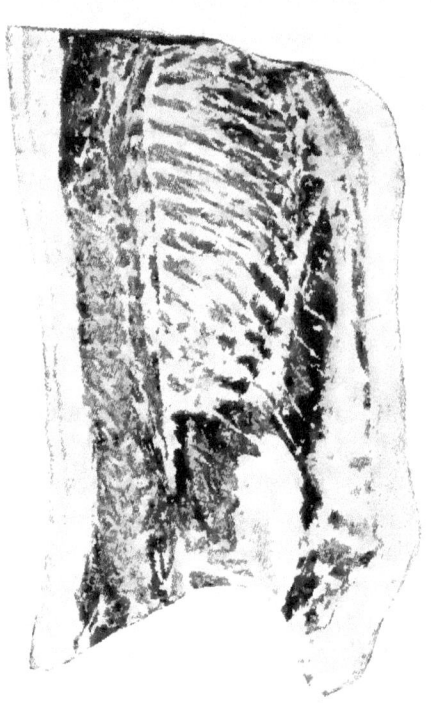
D. S. Short Clear

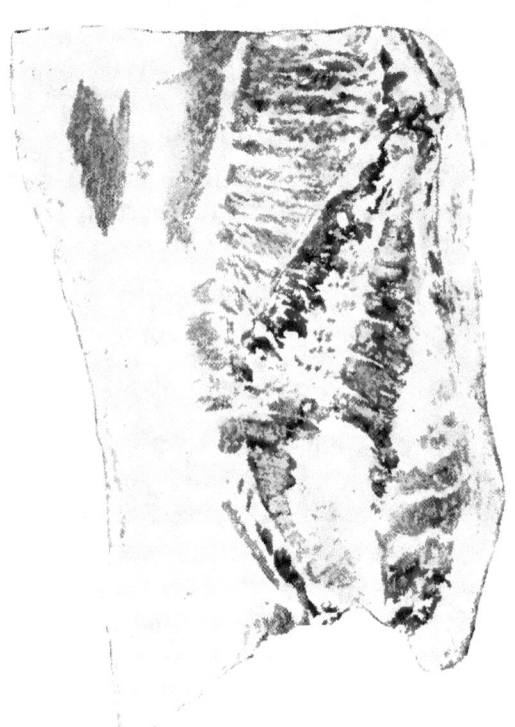
D. S. Extra Short Clear

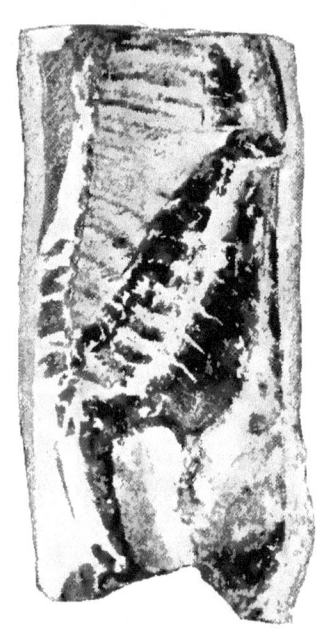
D. S. Rib Belly

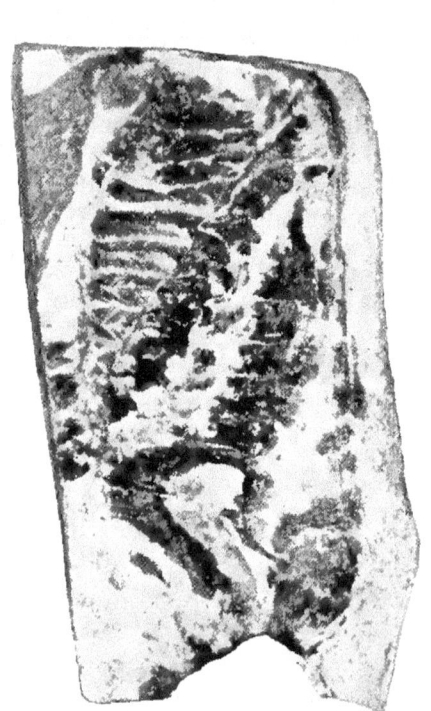
D. S. American Clear Belly

inches from the lean at the butt, and to be neatly rounded and beveled on flank and cushion, not over one and one-quarter (1¼) inches of fat to be left on any portion of the ham from which the skin has been removed.

Standard Picnics.—Shall be made from the shoulder, after the butt has been cut off, to leave not less than one (1) inch, nor more than two and one-half (2½) inches of blade bone in the picnic. The shank cut off in or above the knee joint, breast flap, loose fat or lean removed, trimmed full and the fat beveled on the butt end, which must be well rounded.

New York Shoulders.—Shall be made from smooth hogs, shank cut off one (1) inch above the knee joint, butted about one (1) inch from the blade bone, neck bone and breast flap taken off, neck removed and trimmed close.

Rib Bellies.—Shall be made from nice, smooth sides after the back has been removed, reasonably square cut and trim, and the breast bone removed. No scribed cut bellies shall be classed as standard.

Clear Bellies.—Shall be made from nice, smooth sides, after the back has been removed, reasonably square cut and trim, and free of bone. No scribed cut bellies shall be classed as standard.

Clear Bellies, Square Cut and Seedless.—Shall be made from nice smooth sides of barrow hogs, after the back has been removed. Sows will be acceptable, provided that they are cut down, until the seed, if any, is removed. To be free of bone and trimmed square on all edges. No scribed cut, extremely long and narrow, or wide and short belly will be classed as standard.

Branding.—The packer's name, location, number of pieces and date of packing shall be branded on the head of each package of pickled meats, at time of packing. Also, on each package of lard shall be branded the month and year of packing such lard.

Uniformity of Pickled Meats.—All pickled meats shall be sized when packed, the light, medium and heavy separately, as nearly as practicable.

Dry Salt Meats

Short Rib Sides.—Shall be made by splitting the hog through the backbone. The ham, shoulder and loose lard or fat to be taken off. The feather of the blade bone should not be removed, and no incision (pocket) shall be made in the side.

Extra Short Clear Sides.—Shall be cut reasonably square at each end, the loin and spareribs to be removed, the breastbone cut out or sawed down level with the face of the side. The blade bone may be left in or removed.

Extra Short Rib Sides.—Shall be made the same as extra short clear sides, except that the spareribs and breastbone shall be left in.

Short Clear Sides.—Shall be reasonably square at each end, the backbone and spareribs to be taken out, henchbone and breastbone sawed down smooth and even with the face of the side.

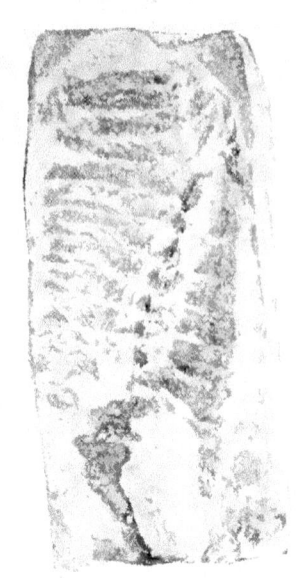 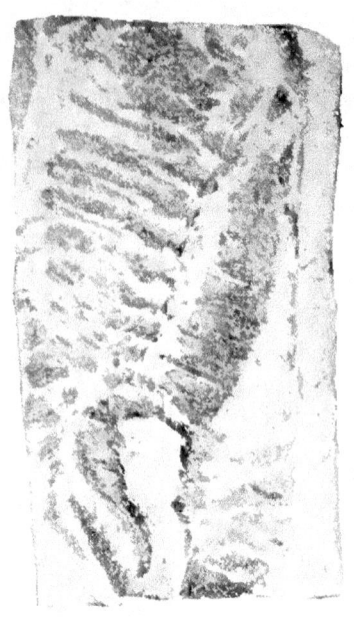

LIGHT ENGLISH BELLY   HEAVY ENGLISH BELLY

D. S. RIB BACK   D. S. FAT BACK

Long Clear Sides.—Shall be cut reasonably square at the tail and shoulder ends, the neck taken off and smoothly trimmed, the backbone, ribs, bladebone, shoulder bones and leg bone removed, hench bone and breast bone sawed off or cut down level with the face of the side.

Short Clear Backs.—Shall be made from the sides of smooth hogs, from which the bellies have been cut, backbone and ribs taken out, and the lean left on, tail bone sawed off even with the face of the meat, trimmed smooth, and reasonably squared on all edges.

Rib Backs.—Shall be made the same as short clear backs, except that the spareribs shall be left in.

Rough Backs.—Shall be made from short rib sides, from which the bellies have been removed; all bones left in.

Rib Bellies.—Shall be made from the side, after the back has been removed, reasonably square cut, and trimmed on all sides. Ribs and breastbone left in, and free of loose fat. No scribed cut bellies shall be classed as standard.

Clear Bellies.—Shall be made from the side, after the back has been removed, reasonably square cut, and trimmed on all sides, and to be free of bone and loose fat. No scribed cut bellies shall be classed as standard.

English Bellies.—Shall be made from nice smooth sides of barrow hogs, after the back has been removed. Sows may be used, however, provided the seed is cut out, and the width of the belly is in proportion to its length. All edges shall be trimmed square and all bones removed. No scribed cut bellies shall be classed as standard. Note.—Barrow bellies are preferable.

Short Fat Backs.—Shall be made from the sides of well-fatted hogs, from which the bellies and loins have been removed. Bladebone cut off and practically free of lean. All edges to be reasonably squared, with the exception of the tail end, which shall be squared sufficiently to leave not more than a two (2) inch bevel on the corner. The width and thickness of the back shall be reasonably uniform its entire length.

Regular Shoulders.—Shall be cut fairly close to the back part of the forearm joint, fat end butted, neckbone and ribs taken out, neck squared, breast flap trimmed off and foot to be cut off in or above the knee joint.

New Orleans Shoulders.—Shall be made the same as D. S. regular shoulders, except that they must be cut from one (1) inch to two (2) inches narrower, part of the neck left on, and the leg cut off below the knee joint.

Regular Plates.—Shall be made from the fat end of the shoulder, with part of the bladebone left in, and the neck side squared.

Clear Plates.—Shall be made from the fat end of the shoulder, with the bladebone removed and the neck side squared.

Jowl Butts.—Shall be made from the jowl, slightly faced, and the loose pieces cut off.

Wiltshire Sides.—Shall be made from nice smooth selected hogs. The shoulder, side and ham left together in one piece. The foreleg to be cut off at or above the knee joint, and the hind leg at or above the hock joint. The shoulder ribs, neckbone, backbone, aitchbone, skirt and

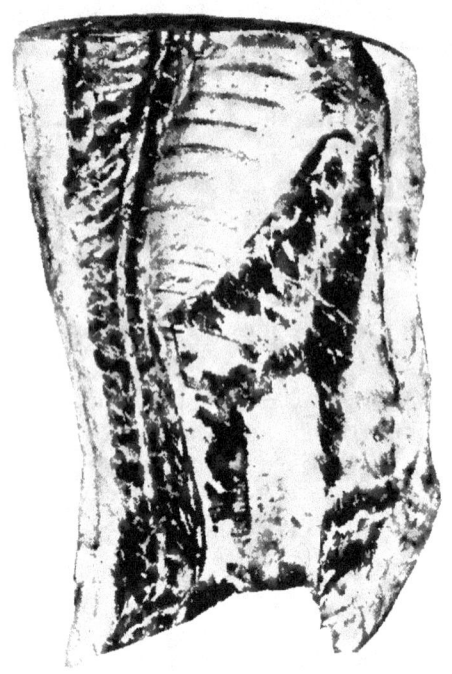
D. S. Short Rib

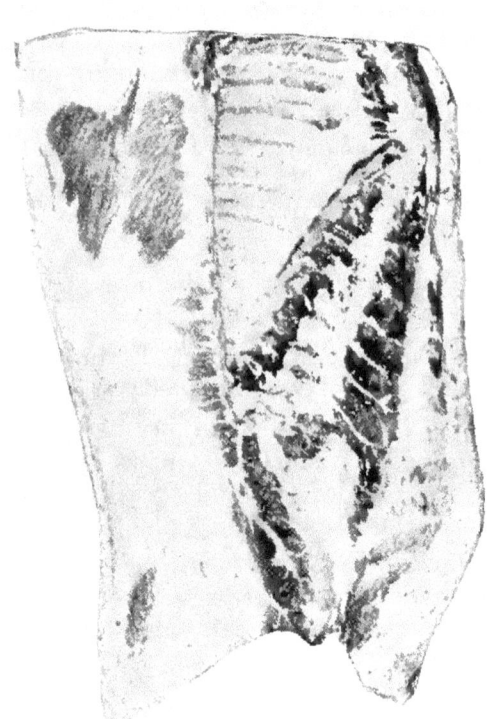
D. S. Extra Short Rib

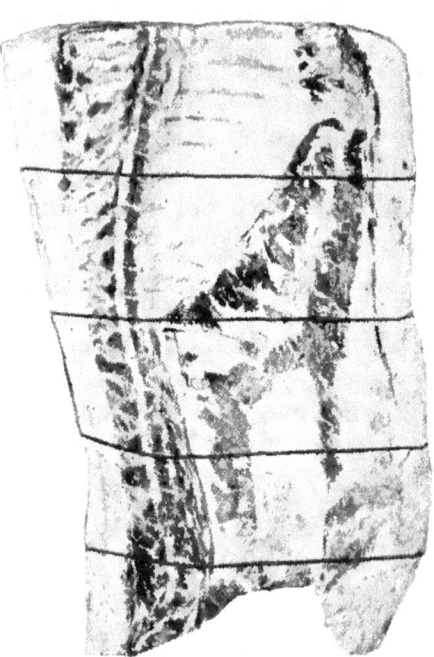
D. S. Short Rib Lined for Mess Pork

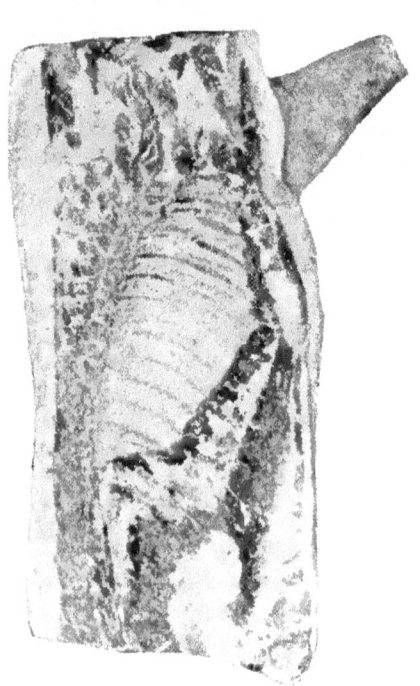
D. S. Cumberland Side

loose fat to be removed. The breastbone to be sawed, or cut down, even with face of side. Neatly trimmed on belly and squared on the neck.

Cumberland Sides.—Shall be made from nice smooth selected hogs, after the ham has been taken off. The leg cut off in or above the knee joint. The shoulder ribs, neckbone, backbone, skirt and loose fat removed. The breastbone sawed, or cut down, smooth and even with the face of the side. Neatly trimmed on the belly, and squared on ham and neck ends.

Dublin Middles.—Shall be made from smooth light hogs. The side must be thin; made the same as Cumberlands, except that the leg shall be cut off close to the breast.

Long Rib Sides.—Shall be taken from smooth light hogs; made the same as Cumberlands, except that the bladebone must be removed, and the leg cut off close to the breast.

Birmingham Sides.—Shall have the backbone, ribs and bladebone removed, pocket piece cut out and pocket nicely rounded, knucklebone left in, and leg cut off close to the breast.

South Staffordshire Sides.—Shall be made the same as Birmingham, except loin taken out full to top of shoulder blade, leaving only a thin strip of lean along the back, knuckle left in, and leg cut off close to the breast.

Yorkshire Sides.—Shall be made the same as Cumberlands, with ribs out.

Irish Cut Sides.—Shall be made the same as long clears, with the knucklebone left in.

English Short Clear Backs.—Shall be made from the sides of smooth hogs, from which the bellies have been removed, squared on all edges, lean left on and free from bone.

English Rib Backs.—Shall be made the same as English short clear backs, except that the spareribs shall be left in.

Long Cut Hams.—Shall be cut from the side by separating with a knife, the hip bone from the rump. Properly rounded out and foot taken off at, or above, the hock joint. The ham shall not be faced.

Square Shoulders.—Shall be made from nice smooth hogs, and cut fully three (3) ribs wide through the gristle of the bladebone. The neckbone, breast flap and ribs shall be removed, foot cut off at or above the knee joint, and squared on the neck and butt.

South Staffordshire Hams.—Shall be cut short, hipbone taken out at socket joint, back unjointed at first joint below the hock joint.

Manchester Hams.—Shall be made in all respects like the South Staffordshire hams, except that the hipbone must be left in.

Backstrapping.—Backstrapping of side meats is not permissible.

Uniformity of Boxed Meats.—In packing meats in boxes, the pieces shall be uniform in average, as nearly as practicable, in packages made to suit the different sizes.

Green Meats

In the sale of green meats, it is understood that "chilled" weights shall be delivered.

Skinned New York Shoulder Skinned Regular Shoulder New York Shoulder

Picnic New York Shoulder Rough Shoulder

D. S. Regular Plate D. S. Clear Plate—Two Averages

On sales of green hams, it is understood that hams from which the lean ones have been taken will not be considered a fair tender.

Range of Green, Sweet Pickled and Dry Salt Meats

No piece lighter than	Average	No piece heavier than	No piece lighter than	Average	No piece heavier than
3	4	5	18	23	28
4	5	6	19	24	29
5	6	7	20	25	30
6	7	8	21	26	31
6	8	10	22	27	32
7	9	11	23	28	33
8	10	12	24	29	34
9	11	13	25	30	35
9	12	15	25	31	37
10	13	16	26	32	38
11	14	17	27	33	39
11	15	19	27	34	41
12	16	20	28	35	42
13	17	21	29	36	43
14	18	22	29	37	45
15	19	23	30	38	46
16	20	24	31	39	47
16	21	26	32	40	48
17	22	27	33	41	49

It is understood that the above range is the limit from the actual average of the lot delivered.

Packages

Barrels.—Pork barrels shall be made of well-seasoned white, red or burr oak, white ash or birch, and may be coated on the inside with refined paraffine wax, or other substance acceptable to the United States Department of Agriculture, a sufficient quantity only to be used to fill up the pores of the wood for the purpose of preventing soakage and leakage.

Staves should be about 5/8 inch thick, 29 or 30 inches long; head about 18 inches, and the head about 5/8 inches thick in center. Six galvanized iron hoops to be used, three on each end as follows: Head hoops, 1½ in. x 19 gauge; quarter hoops, 1¼ in. x 19 gauge; bilge hoops, 1½ inch x 19 gauge.

Wood-bound barrels shall be hooped not less than eleven-sixteenths (11/16) of the surface, with hickory or white oak.

Tierces for Pickled Meats.—Tierces for pickled meats shall be made of well-seasoned white, red or burr oak, white ash or birch, and may be coated on the inside with refined paraffine wax, or other substance acceptable to the United States Department of Agriculture and suitable for use in the packing of pickled meats. In other respects the same specifications as for lard tierces.

Boxes.—Shall be made of good, sound lumber, dressed on both sides to a thickness of not less than 13/16 of an inch. To have good, strong hardwood, whitewood, or pine cleats outside, nailed and clinched to the top, bottom and sides of each end. Straps to be of iron or steel, about

⅝ of an inch in width, to overlap about 6 inches on the side of the box, and nailed with ten-penny nails.

Grading of Pork Meats

Pork meats are graded usually according to weight. For instance, No. 1 hams are graded into 8, 10, 12, 14, 16-pound averages and sometimes into 16, 18 and 20's. Skinned hams are graded from the following averages: 12/14, 14/16, 16/18, 18/20, 20/22, 22/24 pounds. Calas are graded in the following averages: 4/6, 6/8, 8/10, 10/12 pounds. No. 1 bellies for box cure are graded from 4/6, 6/8, 8/10, 10/12. Regular bellies are graded as follows: 6/8, 8/10, 10/12, 12/14, 14/16.

Grading depends on sales policy and demand, and varies according to the locality and outlet for the product.

In shipping sweet pickle and dry salt meats, the cars should always be bottom racked, and in warm weather side racked as well. The meat should be laid skin down, except for the top layer, which is laid skin up.

CURING PORK CUTS

The curing of pork cuts, though it has been practiced for a good many years, is still to a considerable extent an unknown science. This topic alone is big enough to occupy the subject matter of an entire book. In this treatise, however, brief mention will be made of the essential facts in the following order:

First—Function of the curing ingredients.
Second—Ordinary formulas for sweet pickle cure.
Third—Formulas for dry curing.
Fourth—Dry salt methods.
Fifth—Formulas for export cures.

In following the directions given it must be borne in mind that the formulas quoted may not meet with the approval of every curer; in fact, many different curing formulas are in use. Those given here are standard with some of the best packers, and are reliable as to percentages, etc.

Function of the Curing Ingredients.—Salt or sodium chloride is the oldest and most commonly used mineral salt to preserve meats. It is readily dissolved in water and penetrates muscular and fat tissues quickly. Besides having valuable keeping qualities, it adds a flavor to the product and supplies a mineral salt which the human body craves. The using of ground rock salt for making sweet pickle is the predominant practice among American meat packers. The question of the kind of salt to use hinges solely on the cost and the purity of the article.

Sugar is a second curing ingredient, which is used generally on the higher quality cuts. Sugar is used principally to temper the saltiness of the meat and to add to the flavor of the product. Sugar provides the medium for certain chemical processes—the starting of fermentations and the production of acids and other by-products which produce the desirable flavor.

If the price is right, refiners' sirup may be used to advantage to replace sugar. This sirup usually runs 56% or under in sugar content, and the amount used should depend upon this. Care should be taken to see that it does not run too high in ash content, as over six per cent will generally produce a bitter flavor.

Either saltpeter or nitrate of soda, closely-related mineral salts, are used to perform the following functions: First—Preserve the red of the meat. Second—There is a current opinion that they act to preserve proper condition of the meat. The most important function of either of these salts, however, is to preserve the red color of the meat.

Nitrate of soda is 16% stronger than saltpeter; i. e., 84 lbs. of nitrate will do as much curing as 100 lbs. of saltpeter. Nitrate of soda gives satisfactory results if care is used as to proportions. Saltpeter, however, is still being used by many packers as a curing agent.

Borax is a meat preservative which is used on export meats only, a Government regulation forbidding its use on meats to be consumed in the United States. Details of the use of borax are given elsewhere in this chapter.

Formulas for Sweet Pickle Cure

This cure is used on all fancy cuts of ham and bacon and also shoulders. The method of making sweet pickle solution is as follows:

Pass pure water through a salt leaching bed, best located under the dock adjacent to the salt bins, the usual size of which is 5x5x4 ft. From here it goes to a settling vat of 2,000 to 5,000 gallons capacity, where it should be settled over night. From the settling vat it is pumped or gravitated to the formulating vats above the curing cellar level. For the amount of salt required to make given strengths of pickle, refer to salinometer table later on in this chapter.

The solution thus made will give approximately 100 degree pickle, which is reduced to 70 degrees as tested by the salinometer. To this solution, for the fancy sugar-cured meats, may be added 27 lbs. of brown sugar, although for the lower grade meats approximately one-half this amount may be used, or 14 lbs. Four pounds of nitrate of soda or its equivalent in saltpeter should also be added per 100 gallons, care being taken, if nitrate of soda is used, to see that it has dissolved thoroughly, as it does not dissolve as quickly as saltpeter. The nitrate and sugar are usually dissolved together in brine. Finished pickle should show 70 to 75 degrees on the salinometer. Some packers use 80 degree pickle.

As a safeguard for preserving meats, and in order to hasten curing, shoulders, hams, Calas, etc., should be pumped with 85 degree pickle containing 25 lbs. to 35 lbs. of nitrate per 100 gallons, or its equivalent in saltpeter, as follows:

```
Hams ....................12-14 lbs. averaging 2 strokes
Hams ....................14-16 lbs. averaging 3 strokes
Hams ....................16 lbs. up averaging 4 strokes
Shoulders ...............10-12 lbs. averaging 3 strokes
Shoulders ...............12-14 lbs. averaging 4 strokes
Shoulders ...............16-22 lbs. averaging 6 strokes
Shoulders ...............22-37 lbs. averaging 7 strokes
Calas ...................  5-6 lbs. averaging 2 strokes
Calas ...................  6-8 lbs. averaging 3 strokes
Calas ................... 8-12 lbs. averaging 4 strokes
```

The needle should be inserted so as to deliver the pickle in the region of the joint and down the shank. Three ounces of pickle should be discharged at each stroke.

Overhauling.—A standard wooden curing vat will hold approximately 1,450 lbs. of meat, which requires about 75 gallons of pickle to cure, except for such cuts as are cured in 30 days or less, in which case the overhauling should take place on the third, tenth and twentieth days. Overhauling usually is done on the fifth, fifteenth and thirtieth days. Overhauling makes thorough pickling certain and increases the weight.

The curing, of course, is done at a temperature of 36° to 38° F. Hams will gain from five to six per cent in curing, and bellies as much as ten per cent in pickle. It usually requires from 55 to 100 days to thoroughly cure hams, depending on the weight, from 35 to 45 to cure Calas, and approximately 20 to 25 days for bellies.

Dry Curing and Dry Salt Methods

Dry curing is the using of mineral salts direct upon the meat, the cure being effected by absorption of these mineral salts by the native moisture in the meat. This plan is usually carried out when curing fancy bellies, brisket pieces and jowl butts in boxes, and is known as the box cure.

A satisfactory formula for high grade sugar cured meats (box cure) is as follows: Salt, 70%; granulated sugar, 25%; saltpeter, 5%. Cover the pork with this mixture, using 5%, or 5 lbs. per 100 lbs. of meat, and pack tightly in a galvanized iron lined box.

The cure on high-grade bellies will be effected in twenty-five days. The dry cure is not ordinarily used on hams and shoulders.

Dry Salt Methods.—Long-cut hams for export, ribs, sides and fat backs are cured in dry salt. This is done merely by rubbing salt well into the cuts and packing them in the curing cellar, well surrounded by this ingredient. Bellies will cure in this way in from 18 to 20 days. A gain of 2% may be made by pickling the cuts ten to fifteen days before dry salting.

Overhauling.—The bulk should be broken and the cuts re-rubbed in salt and repiled at about twenty days, with the possible exception of fat backs, which may be allowed to go unturned for a somewhat longer time. It is important during the period of storage, which, in the case of D. S. meats, may extend to six months or longer, that frequent inspection be made, and where the meats are bare of salt another overhauling given, or where there are any signs of sliming or purging at the ends, a thorough rubbing of salt be applied.

Dry salt meats, if smoked, are soaked from one to six hours, depending on the age, before going to the smoke house.

Formulas for Export Cures.—On sweet pickle cuts which are to be exported, the very best sugar cure is generally used. The dry salt cure is used on many export cuts, particularly side meats.

In curing for export, figure a day to a pound on dry salt meats, except on Wiltshires, which run one-half day to the pound, and for sweet pickle meats figure two days to a pound.

Curing Wiltshires.—On Wiltshire sides special care must be exercised. The following is a good guide to use in curing this particular export cut:

Pump with the same pumping pickle used for domestic cuts, using at each stroke two ounces of pickle as follows: Shoulders, 3 strokes; loin, 3 strokes; ham, 4 strokes. Curing formula: Refined salt, 95%; saltpeter or nitrate of soda, 5%.

Before applying the curing formula pass the product through a 100° brine. Apply 6% of the curing mixture, rubbing well over face and side. Pile the sides twelve to fourteen pieces high. The temperature of the curing room should be 36° to 40°. Overhaul in ten to fifteen days. If the meat is not to be shipped as soon as cured, postpone overhauling to fifteen days. Apply straight salt at the time of overhauling. Use the curing ages given above.

Packing English Meats.—Borax is used in the packing of English meats. The usual plan is to rub five to six pounds of borax on the meat which will be put into one box, which usually runs from five to six hundred pounds of meat. Before shipping cuts of this kind, they are generally allowed to age a little longer than domestic meats. It is the aim of the British Government not to allow an excess of $\frac{1}{3}$ of 1 per cent of borax in the meat; so if the meats are aged a little longer there will be no danger of too much absorption of the borax in transit, as the meat will have already absorbed the salt used in the curing.

General Curing Information

Generally speaking, four days per pound of ham is a rough guide to use in figuring minimum curing ages and dates for smoking. For instance, a 15-lb. ham will cure in sixty days in sweet pickle cure. This may vary in different sections of the country, depending on local conditions and methods of handling, which must be determined by the superintendents.

In dry salt curing, two days per pound of meat is a good guide to follow on domestic meats.

Holding Green Meats.—Should the time of marketing curing meats be delayed, the various cuts may be held in cure by back-packing. For instance, in the case of hams, they may be pulled from the curing vats five days under cure, put into tierces, covered with 40° to 50° pickle, and then put into freezer temperatures of 5° above zero or lower.

Back-packing, however, is a custom which has died down somewhat, the modern method being to freeze surplus green hams 72 hours at a temperature of five to twelve degrees below zero; at the end of that time transfer to a freezer at 12° above zero. When the trade demands make it necessary to take them out of the freezer, put them into cure. Freezer hams will cure in 30% less time than hams from the block. This means that a 14 to 16-lb. averaging ham put into cure requires 65 to 70 days, while the freezer ham of the same weight will cure in 40 days.

If a packing plant has not sufficient freezer space, but considerable space around 12° to 15° above, the practical thing to do is to back-pack these hams, cover them with 40° pickle and put them into this space. It must be remembered, however, that meats will continue to cure at 15 deg. Fahr.

Curing Periods.—A useful table showing the time required for curing different cuts is as follows:

Sweet Pickle Meats:

Extra Light Hams, 10 lbs. and down	50 days
Extra Light Hams, 10 to 14 lbs.	55 days
Extra Medium Hams, 15 to 18 lbs.	65 days
Extra Heavy Hams, 18 to 23 lbs.	75 days
Fancy Skinned Hams	65 days
Heavier Skinned Hams	65 to 70 days
Light Shoulders	35 days
Cala Hams, 9 to 11 lbs.	40 days
Cala Hams, 11 lbs. and up	55 days
New York Shoulders, 10 to 12 lbs.	45 days
Sweet Pickle Ribs	25 days
Extra Light Bellies, 10 lbs. and down	20 days
Light Bellies	25 days
Heavy Bellies	30 days

Dry Salt Meats:

Fancy Bellies	20 days
10 to 12 lbs. average Bellies	28 days
14 to 16 lbs. average Bellies	28 days
16 to 18 lbs. average Bellies	32 days
18 to 20 lbs. average Bellies	32 days
20 to 25 lbs. average Bellies	35 days
25 to 30 lbs. average Bellies	40 days
English Bellies	25 days
Boneless Backs	30 days
English Shoulders	50 days
New Orleans Shoulders	40 days

Calculating Weights of Meats.—In calculating the weights of sweet pickle meats to fill an order for a definite amount of pounds, one should allow 5 per cent increase as shown by the following example:

Considering 20,000 lbs. as the known net weight, to find the gross weight, multiply by 1.05, equals 21,000 lbs., or gross weight.

Considering 21,000 lbs. as the gross, to find the net divide by 105, equals 20,000 lbs., or the net weight.

Drainage Allowances.—Following are the usual drainage allowances on meats coming directly out of pickle: Hams, 4 per cent; calas, 5 per cent; S. P. Bellies, 5 per cent. If drained for twenty-four hours or longer, an allowance of 2 per cent only is the usual practice.

Recovery of Pickle.—Sweet pickle remaining in the vats after the curing process may be recovered and used again if proper care is exercised. Whether or not this will pay depends upon the amount of pickle available; in other words, the size of the plant, as it would not pay the very small packer to put in the equipment to handle it unless the amount to be cared for was of sufficient quantity.

Just how much can be saved by this process can best be determined by each curer in his own peculiar situation, as it will vary according to the value of the ingredients used. Briefly, the process is as follows:

The pickle is pumped from the curing vats by means of an electrically-driven piston pump connected to a header to which is attached armored or wire-wound hose with a 2-inch brace, goose neck and strainer attached. The pickle is concentrated at some central point in the cellar of the building, and then pumped to a point higher than the curing cellars, where it should be pasteurized at a temperature of 190° to 200° F. for an hour, in order to kill off the wild yeasts or other organisms that may have developed, and which may later cause sour pickle.

The operator will find that used or second pickle will not be as clear as first pickle unless the trouble is taken to pump it through a filter press, which is not absolutely necessary.

The used pickle will vary in density. It may have a salinometer reading of 50° to 70° or possibly more. Any coagulated albumen which may arise to the surface after pasteurization should be skimmed off. It should then be filtered, this latter operation being necessary in order to remove the coagulum.

A chemical analysis should be made of this used pickle on every occasion after the recovery process has been started, in order that experience may be gained as to just how much salt, sugar and saltpeter or nitrate of soda must be added in order to bring it up to normal.

As a rule, used pickle is not utilized for curing the best grade of meats, but it is very good for curing all other grades. By calculating the amount of salt, sugar and saltpeter in the used pickle and finding its value at current market prices, and then deducting the interest and depreciation on the equipment necessary, the saving for each particular plant can be calculated very closely. Some of the larger packers also find that a saving can be made by having pans underneath the point where the various cuts are pumped, so that the overflow also may be recovered.

Brine Calculating Table

For the convenience of curers the following table is presented, showing the specific gravity, per cent of salt, weight of a gallon of brine in pounds, and pounds of salt in a gallon of brine, compared to salimeter degrees from 40 to 100 inclusive:

Salimeter degrees	Baume degrees	Specific gravity	Per cent of salt	Wt. of a gal. of this brine in lbs. of 7,000 grains each	Lbs. of salt in gal. of brine of 231 cubic inches
40	10.40	1.073	10.600	8.939	.947
41	10.66	1.075	10.865	8.955	.973
42	10.92	1.077	11.130	8.972	.998
43	11.18	1.079	11.395	8.989	1.024
44	11.44	1.081	11.660	9.005	1.050
45	11.70	1.083	11.925	1.072	1.075
46	11.96	1.085	12.190	9.039	1.101
47	12.22	1.087	12.455	9.055	1.127
48	12.48	1.089	12.720	9.072	1.154
49	12.74	1.091	12.985	9.089	1.180
50	13.00	1.093	13.250	9.105	1.206
51	13.26	1.095	13.515	9.122	1.232
52	13.52	1.097	13.780	9.139	1.259

Salimeter degrees	Baume degrees	Specific gravity	Per cent of salt	Wt. of a gal. of this brine in lbs. of 7,000 grains each	Lbs. of salt in gal. of brine of 231 cubic inches
53	13.78	1.100	14.045	9.164	1.287
54	14.04	1.102	14.310	9.180	1.313
55	14.30	1.104	14.575	9.197	1.340
56	14.56	1.106	14.840	9.214	1.367
57	14.82	1.108	15.105	9.230	1.394
58	15.08	1.110	15.370	9.247	1.421
59	15.34	1.112	15.635	9.264	1.448
60	15.60	1.114	15.900	9.280	1.475
61	15.86	1.116	16.165	9.297	1.502
62	16.12	1.118	16.430	9.314	1.530
63	16.38	1.121	16.695	9.339	1.559
64	16.64	1.123	16.960	9.355	1.586
65	16.90	1.125	17.225	9.372	1.614
66	17.16	1.127	17.490	9.389	1.642
67	17.42	1.129	17.755	9.405	1.670
68	17.68	1.131	18.020	9.422	1.697
69	17.94	1.133	18.285	9.439	1.725
70	18.20	1.136	18.550	9.464	1.755
71	18.46	1.138	18.815	9.480	1.783
72	18.72	1.140	19.080	9.497	1.812
73	18.98	1.142	19.345	9.514	1.840
74	19.24	1.144	19.610	9.530	1.868
75	19.50	1.147	19.875	9.555	1.899
76	19.76	1.149	20.140	9.572	1.927
77	20.02	1.151	20.405	9.580	1.956
78	20.28	1.154	20.670	9.614	1.987
79	20.54	1.156	20.935	9.630	2.016
80	20.80	1.158	21.200	9.647	2.045
81	21.06	1.160	21.465	9.664	2.074
82	21.32	1.163	21.730	9.689	2.105
83	21.58	1.165	21.995	9.705	2.134
84	21.84	1.167	22.260	9.722	2.164
85	22.10	1.170	22.525	9.747	2.195
86	22.36	1.172	22.790	9.764	2.225
87	22.62	1.175	23.055	9.780	2.256
88	22.88	1.177	23.320	9.805	2.286
89	23.14	1.179	23.585	9.822	2.316
90	23.40	1.182	23.850	9.847	2.348
91	23.66	1.184	23.115	9.864	2.378
92	23.92	1.186	24.380	9.880	2.408
93	24.18	1.189	24.645	9.905	2.441
94	24.44	1.191	24.910	9.922	2.471
95	24.70	1.194	25.175	9.947	2.504
96	24.96	1.196	25.440	9.964	2.534
97	25.22	1.198	25.705	9.980	2.565
98	25.48	1.201	25.970	10.005	2.598
99	25.74	1.203	26.235	10.022	2.629
100	26.00	1.205	26.500	10.039	2.660

SMOKEHOUSE OPERATION

The smoking of sweet pickle and dry salt meats has long been a practice in order to preserve the meat and add a desirable flavor to it. The result is accomplished in two ways; first by reducing the moisture content; second, the meat absorbs some of the smoke fumes, which act as a preservative.

Careless handling and smoking cause excessive shrinkages, while too hasty smoking may cause too much moisture to be retained in the meat. It is therefore difficult to set a definite standard as to the percentage meats should shrink in smokehouses. Many packers prefer to have a

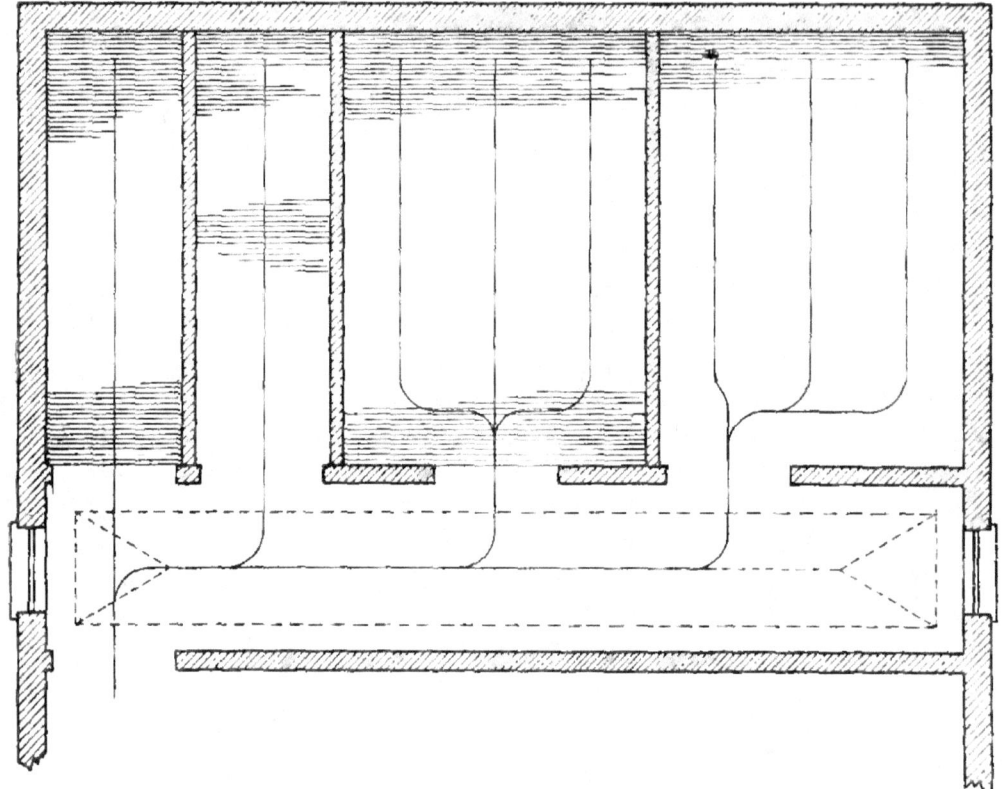

"A"—Layout of a Standard Packinghouse Smoke House

firm piece of meat, and therefore do not watch the shrinkage too closely; while other packers have a trade which will be satisfied only with meats with a lot of moisture content. Therefore, a reduction in shrinkage on smoked products is not always desirable. It depends entirely upon the trade, and the final destination of the product.

The sweet pickle meats just coming from the curing vats are soaked in 60 degree temperature water three minutes for each day they have been in pickle. For instance, hams that have been in cure from 60 to 80 days should be soaked from three to four hours. Hams that are 100 days old or older, and that have been held in the original pickle, should receive a change of water, two hours in the first and three hours in the second. After soaking, the meats are washed in lukewarm water of 90 to 100 degrees temperature, and are scraped so as to remove all traces of slime. They are then strung, branded and conveyed to the smokehouse to dry, overhead carriers of suitable design being used to avoid unnecessary handling.

Depending upon the destination of shipment and the kind of product, various lengths of time are used for smoking. Bacon is ordinarily smoked from 18 to 24 hours, hams from 24 to 30 hours, and sausage for a shorter

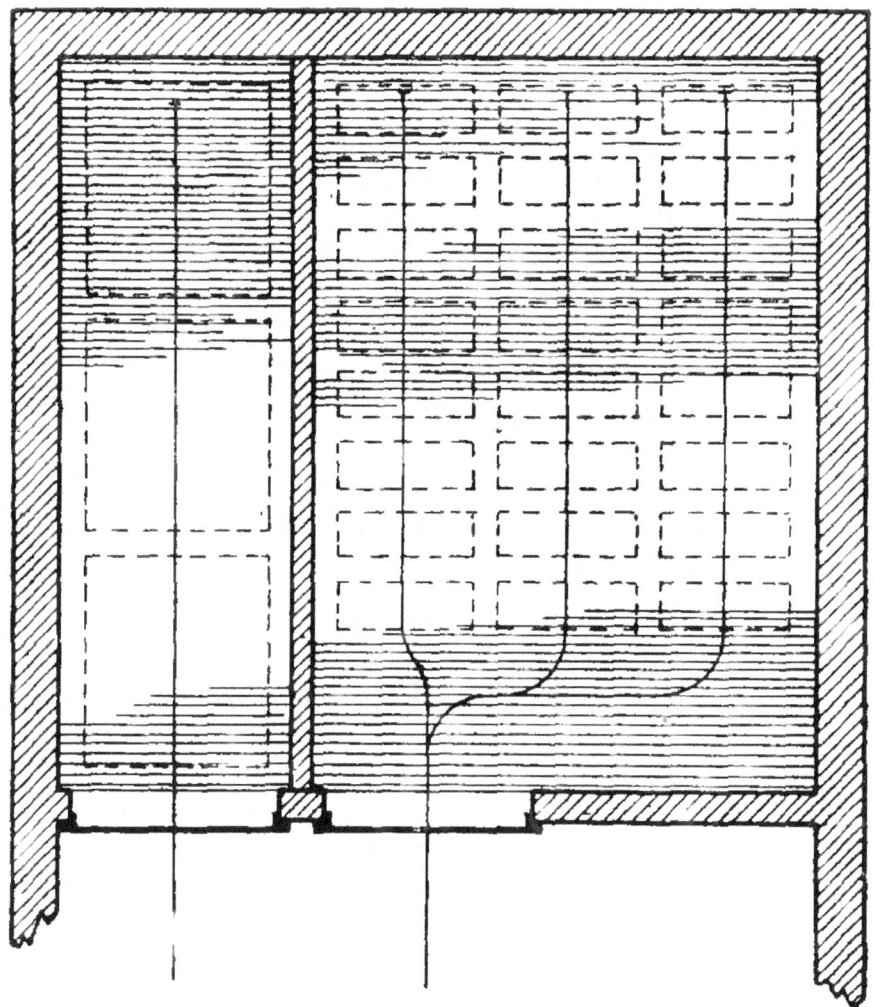

"B"—Another Smoke House Layout, Showing Bacon and Ham Carriers in Place

time, generally 10 hours, while dried beef is smoked at a higher temperature, namely, about 135 degrees Fahrenheit for 85 to 90 hours.

After smoking the meats should be handled as little as possible, as they will become greasy-looking and will lose the smoke color. The longer the meats are smoked, the greater will be the shrinkage; therefore, the time of smoking and temperature should be watched very closely.

Should the meats be desired for immediate use, or if they are to be sent to a comparatively cool climate, a lighter smoke may be given. On the other hand, if they are to be subjected to more severe conditions, a longer smoking will be required. The shrinkage on smoked meats will vary from 6 to 15 per cent, depending a great deal on the gain in the curing process.

If smoked meats upon being removed from the smokehouse are allowed to hang in a room of steady temperature, they will frequently gain slightly in weight, but if left any length of time, say over 10 days, they will be apt to become moldy. As a rule, smoked meats are wrapped, packed and shipped immediately, according to the particular desires of the individual packer.

Smokehouses

The size and height of the smokehouse depends upon the capacity of the plant, and it should therefore be designed to meet the requirements of the individual plant, and the design and construction should be under the supervision of competent packing house engineers or architects, as it is very important that certain principles be followed in the construction.

Walls of smokehouses should be at least 13 inches thick, so as to retain the heat as long as possible. Doors and openings should be of

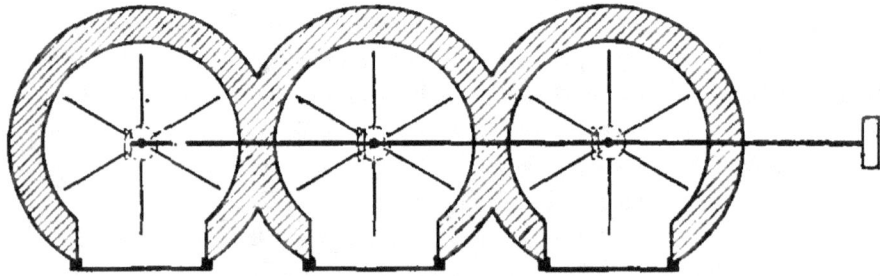

"C"—Circular Type of Smoke House

sufficient width to allow the widest smokehouse carrier to pass easily through them. Double roof or a false ceiling is also required in climates where snow may rest on the smokehouse roof, as condensation may collect on the roof of the smokehouse. This can be prevented by a false ceiling.

Wooden timbers should be avoided as supports for track hangers; "I" or channel beams should be provided for this purpose. The tracking in smokehouses also should be arranged for the convenient handling of carriers. This is illustrated in Drawing "A."

The narrow smokehouse is equipped with one track and dotted lines show the outlines of sausage carriers. Carriers hold from 600 to 1,200 pounds of meats and since the time of smoking is known, the number and capacities of smokehouses required may be easily figured out.

Another smokehouse, shown in Drawing "B," is filled with bacon or ham carriers, from which capacity can be quickly computed.

Types of Smokehouses.—The first illustration shows what may be termed standard packing-house smokehouses. They are divided from the main building by a corridor, which assists greatly in confining the smoke escaping from doors, and prevents it from entering the manufacturing rooms. When the smokehouses are only one story high, a large skylight in the roof and windows on each end of the corridor remove the smoke quickly. The drawing also shows the different track arrangement which may be used.

Fire pits are usually from four to six feet below the first floor grate, and suitable draft openings should be provided in the fire pit doors to allow for proper regulation of draft.

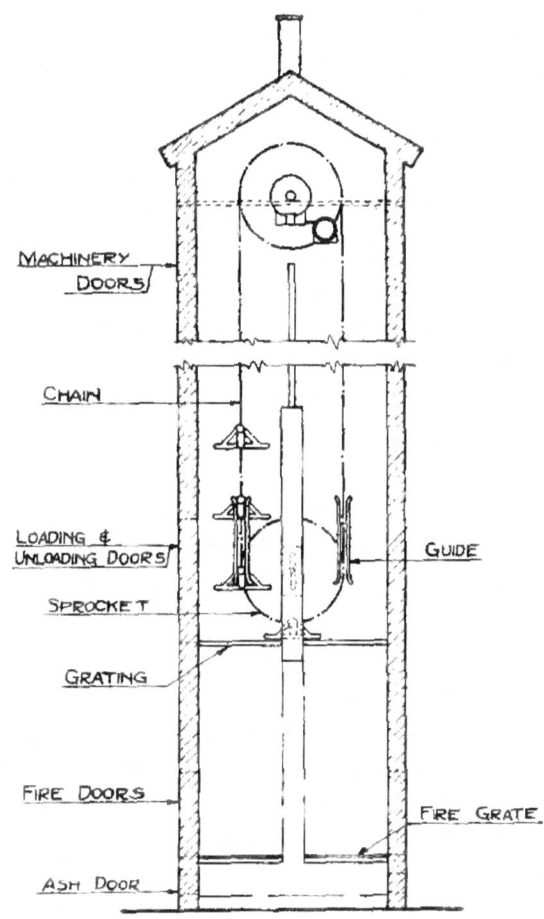

"D"—Continuous Operation Type of Smoke House

In various parts of the country one finds different types of smokehouses. One type is illustrated in Drawing "C." These smokehouses are of a circular type, with a stationary carrier on the inside, which is constantly revolving during the process of smoking.

Another type of smokehouse is shown in Drawing "D." This type, it is claimed, is a great saver of floor space. It may be operated continuously, as it is loaded on one side and the finished product taken out on the other side. This patented type of smokehouse may be built several stories high, depending upon the capacity of the plant, and one house twenty carriers high will have a capacity of approximately 1,000 pounds of sausage per hour.

Smoking Material.—Packers once considered hickory sawdust the best for smoking purposes, but for many years all kinds of sawdust have been used successfully, such as white wood, maple, birch, bass wood, cedar, cypress, etc. Wood that contains pitch or rosin, such as Southern pine, is unfit for smoking meats.

Gas has become a great factor in meat smoking. The easy control of temperature, often done automatically, the elimination of labor, and the saving of wood storage space, makes gas an ideal smoking element. In fact, gas-fired apparatus is coming into general use in many departments of the packing house.

The great majority of smokehouses are also equipped with steam coils to assist in cold weather in maintaining an even heat of 120 degrees Fahrenheit.

Smoking by electricity is also being developed, but at this time (1922) the process is not far enough advanced to have been put into practice to any extent.

HAM BONING AND COOKING

Ham cooking is an important part of the operation of nearly every packing plant, and it is usually necessary for every packer to bone and cook a certain number of hams, according to the demands of his trade. Success in producing a first-class cooked ham depends upon three functions: Curing, boning and cooking.

Curing.—Only newly-cured hams should be used for cooking, and many packers use a special mild cure for cooking hams. It is also good practice to soak the hams in water not over 80° F. temperature for 3 to 5 hours.

Boning.—The next operation is the boning, which is very important, as the retail trade desires a ham which will give even and uniform slices on the slicing machines.

In making the insertion on the face of the ham, to break the joint below the aitch bone, care should be used by the boner in not making the insertion too large, and in not making any bad cuts in the meat around the sucele bone or joint. Most ham boners think it is necessary to sever the lean meat just below the aitch bone in order to remove the small bone after the joint has been broken. If a thin boning knife is used, this lean meat can be raised from the bone and the bone pushed out where the insertion is made on the face of the ham, instead of being pulled out at the butt of the ham. The same care should be exercised in breaking the top joint and removing the shank bone.

Many ham boners will cut the heavy toe sinew running along the shank bone with the chisel, then follow the shank bone toward the center joint. If the boner, in using the chisel, will not cut this sinew, but instead follow the shank bone closely down to the joint, he will find this heavy sinew grows out from the joint in each shank and is tender at the base, and can be split very easily without lacerating the good lean meat in the shank.

Grading or Sorting.—After boning the next important step is the sorting of the hams, as the weight of the ham determines the time of cooking. It also saves time in handling, so that all hams of uniform weight may be cooked together.

Cooking.—The cooking of the hams is done either in cooking vats equipped with live steam lines, or in cooking boxes with steam or vapor. It was once customary to cook hams which weighed when boned, fatted and wrapped over 12 pounds, for 30 minutes per pound at 165° F., and allowing 25 minutes for hams weighing less. The temperature of the water at the start would be 175° to 180° F., and was brought down to 165° F.

It has been proven, however, that fast cooking has a tendency to increase the shrinkage. Tests under this method of cooking show the following results:

6 hams, skin on, weight 57 lbs. 6 hams, skin on, weight 79 lbs.
Cooked, weight .... 45 lbs.   Cooked, weight .... 56 lbs.
Shrinkage, 21 per cent.   Shrinkage, 20 per cent.

By using lower temperature and lengthening the time of cooking shrinkage can be reduced to 10 to 15 per cent. It is claimed by manufacturers of vapor or steam cooking boxes that this method will save considerable shrinkage.

It must also not be overlooked that in case skins are rendered into prime steam lard, proper credit should be given to the hams. It is also well to bear in mind, wherever temperatures of cooking are given, that they are based upon sea level, and in higher altitudes they should be changed accordingly.

Pressing.—In wrapping hams for pressing and cooking, the ham should be placed lengthwise and the cloth pulled tightly together, so that the ends overlap and prevent the ham lengthening when under pressure. The pressing is done under either hand, air, steam or hydraulic pressure.

Cooked hams are put in either round, square or oval containers, depending upon the demand of the trade. These containers are manufactured in a variety of styles, and their use, together with that of steam or vapor cooking appliances, has worked a revolution in ham-cooking practice.

LARD MANUFACTURE

The three principal kinds of lard manufactured in American packing plants are: first, Prime Steam Lard; second, Kettle Rendered Lard; third, Neutral Lard.

The first requisite in making any kind of lard is absolute cleanliness in the handling of the product. This fact seems to be underestimated by a great many packers, but experience has proven that the very best product is turned out where it is handled in the most sanitary places, and by the most sanitary methods.

Prime Steam Lard

Ingredients.—Prime Steam Lard is made up of all edible killing fats not used for any other purpose. Some sweet pickle and dry salt fat scraps are also used in small quantities, at times.

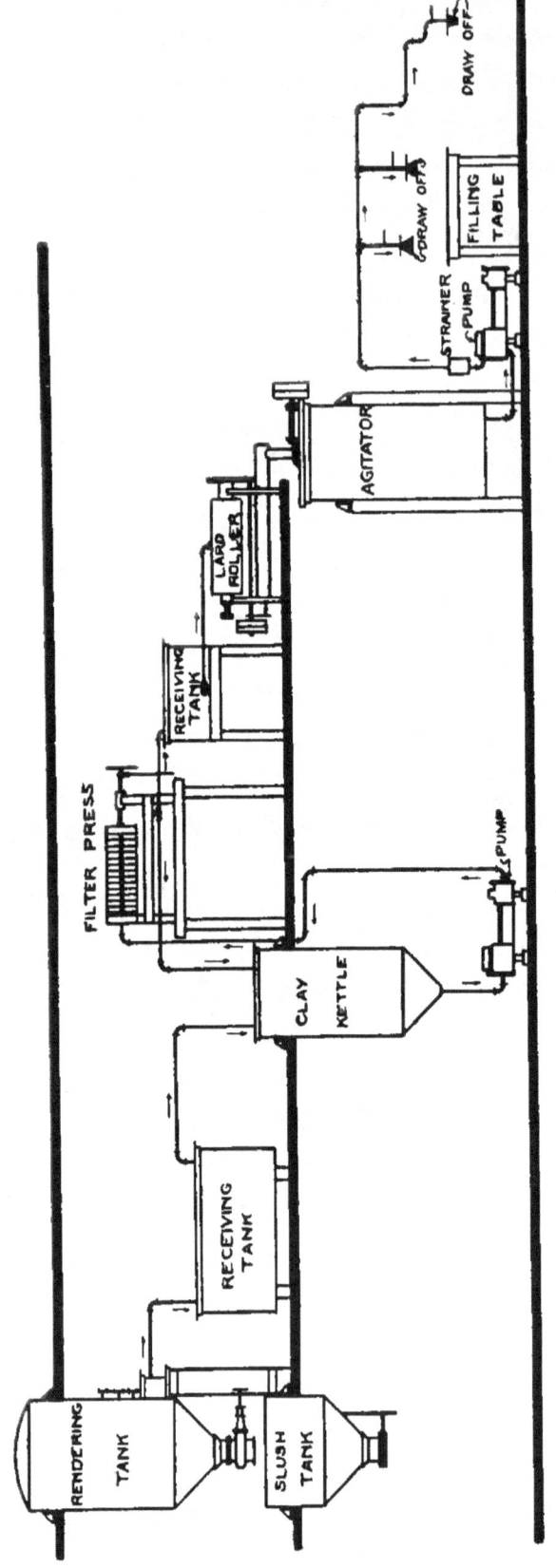

ELEVATION OF A TYPICAL LARD REFINERY

This shows the equipment required in a standard lard refinery, consisting of rendering and slush tanks; receiving tank to receive the rendered lard; clay or refining kettle for bleaching; filter press for removing the fullers' earth; receiving tank to hold the lard from the filter press, whence it rolls by gravity over a lard roller, and from thence into an agitator, or vice versa.

It is well to bear in mind, however, that whenever sweet pickle or dry salt fats are used they should be washed thoroughly in cold water before rendering, so as to take out as much of the salt as possible. Too much salty material will discolor the lard, as during the rendering process the salt will precipitate the fibre to the bottom of the tank and the steam will burn it, which naturally leaves a dark-colored effect on the product.

The principal products which enter into prime steam lard are given here, and the list contains items which are classified according to large packinghouse practice. The small packer should be reminded that in his plant several of these items might be included in one, as the separating of fats in the small packing plant is, of course, not carried on to the same extent that it is in the large plant, and therefore the list is for general guidance only. It is as follows:

1. Bellies, snouts, lips and ears for sterilization.
2. Bones for sterilization.
3. Trimmings for sterilization.
4. Inside hog scrapings.
5. Head bones and jaw bones.
6. Sweet pickle ham trimmings.
7. Outside trimmings.
8. Hog stomach trimmings.
9. Caul fat.
10. Ruffle fat.
11. Head fat.
12. Fat from black guts.
13. Fats from sterilization heads.
14. Ham boiling vat skimmings.
15. Pigs' feet cooking skimmings.
16. Skimmings from lard storage vats.
17. Ham trimmings from skinning hams.
18. Tongue trimmings.
19. Ham facings.
20. Kidney trimmings from canning department.
21. Bung gut fat.
22. Hashed gullets.
23. Ears.
24. Liver trimmings.
25. Melt trimmings.
26. Pancreas gland.
27. Fresh pigs' feet bones from canning and sausage.
28. Neutral lard scrap.
29. Fat from fleshing machine.
30. Fresh and pickled ham bones.
31. Pickled fat and bones from canning.
32. Heart trimmings.
33. Back fat when tanked.
34. Leaf lard when tanked.
35. Dry salt trimmings.
36. Neck bones when tanked.
37. Back bones when tanked.
38. Trimmings from hams, shoulders, sides, jaws and heads.
39. Cracklings from refinery when tanked.
40. Skimmings from back fat skin washer.

Rendering and Cooking.—The size of the steam rendering tanks used depends entirely upon the size and capacity of the plant. Standard sized tanks are from four to six feet in diameter, and from eight to sixteen feet high on the straight side. These tanks are equipped with cone bottoms and a ten or twelve-inch gate valve.

All material which now enters the rendering tank should be thoroughly washed. Unless the material for rendering has a lot of natural water in it, it is very advisable to admit clean water through the bottom water connection so that the cone is about half filled. It is standard practice in American packing plants to fill the bottom, or the cone of the tank, with green bones. This will prevent any fats from packing in the bottom of the tank and interfering with the free flow of steam.

While the tanks are being loaded, and especially if the tanks are held open for any length of time, as is customary in small plants, the steam should be turned on just enough to keep the water hot.

The tank should never be filled more than about four-fifths full of raw material, so as to allow plenty of room for "rolling" during the cooking process.

After the tank has been charged, the man-hole must be secured tightly, and the tank is now ready for cooking. Steam should be turned on gradually until the proper pressure is on the tank.

A reducing valve is very advisable on all rendering tanks, and it is common practice to cook under 40 pounds steam pressure. This seems to be the standard in the majority of houses, but some prefer 35 pounds, while some cook under 45 pounds pressure.

Although there is not a great variation on the pressure carried on rendering tanks for cooking, there is, however, a great difference of opinion among packing-house superintendents as to the time required for cooking lard. Many packers will cook a tank in three to four hours, while others recommend seven to nine hours for the same material. Actual tests, however, prove that the best results are obtained by cooking a longer time and at lower pressure—about 35 to 40 pounds.

The valve on the exhaust pipe in the top of the tank is kept closed until the gauge shows the proper pressure on the tank. Then it is opened, slowly, for a few seconds, to allow a free exhaust of steam. Then it is closed down so as to allow just enough steam to exhaust into the air to carry away the gases freely.

It is very important that a constant pressure be maintained during the cooking process; and after the lard has cooked the required time the steam is shut off and the tank allowed to stand for a few minutes, and then the exhaust steam pipe should be slowly opened, which is in the manhead, being very careful that no lard passes off with the steam. After all the pressure has been released, the manhead can be taken off.

Now the lard is allowed to stand in the tank for at least three hours. Fine salt is sprinkled over it while it is standing, to further settle and clarify it, and the lard is allowed to settle for about thirty minutes after the salt has been added.

Drawing Off.—In drawing off the lard, the operator will try the bottom outlet valve on the side of the rendering tank. If clear lard flows from the valve, he goes ahead and runs off the lard through this valve into a small separator, from where it flows into the lard receiver.

Some packers allow the lard to flow from the separator to a storage tank immediately, while others let it run into a large receiving tank equipped with steam coils to drive off any additional moisture which may be contained in the product. The latter method is recommended as being the most practical.

Recent experiments and tests have proved that considerable lard can be wasted during this draw-off process. A lot of raw material, such as hog stomachs, neck pieces, etc., are usually found floating on the surface. These, of course, cannot pass through the small draw-off valve, with the result that they are usually dropped into the slush vat, where operators will try to recover additional lard from the dropped tankage. This, however, is very difficult—especially as some operators raise their tanks with cold water.

This has a tendency to congeal the fats, which makes it practically impossible to press them out. A patented apparatus, which is called a separator, has recently been placed on the market for the purpose of overcoming this obstacle. It has the advantage of preventing the lard from getting into the slush vats, thus decreasing the lard contents in the tankage and giving a larger yield all the way round.

The lard in the receiving tank is kept there for several hours for further settling. These tanks are equipped with two outlets, near the bottom. The bottom outlet is used to draw off the bottom, or settlings, of the lard, which are kept in a cold place and added to the next tank of fat to be cooked.

Prime Steam Lard has a very distinct cooked flavor, which is, of course, difficult to describe, but which is very familiar to all lard men.

The quality of any finished product depends, to a large degree, on the care taken in handling the raw fat, and also in drawing off the finished product so that no tank water is mixed with the lard, the careful cooking of the rendering tank, and the elimination of all possible gases during the process of cooking.

Refining.—The finished product is now drawn by gravity, or pumped, into the bleaching, or refining tank. The bleaching operation is accomplished by the use of fuller's earth, which is a peculiar form of clay which has the properties necessary to absorb coloring matters in fats, providing such fats are in prime condition so that they will give up this coloring matter.

This method of bleaching for edible purposes has supplanted practically all chemical bleaching, and is used exclusively in the United States. English fuller's earth is used more than domestic fuller's earth, but recently large domestic deposits of fuller's earth have been developed in the United States, and this earth is becoming an active competitor of the English product.

After fuller's earth has been used once it is of no further value in a small plant, as its absorbing qualities are gone, and in small plants the earth is usually thrown away. Large refining plants, however, have found it profitable, if the price of lard and oil will justify it, to extract fuller's earth by means of a solvent process. The advisability of doing this can only be determined when the quantities of earth to be extracted are known.

Refining kettles, or clay kettles, as they are commonly called, are heated by means of a double bottom and steam coils, to a temperature of 130 to 140 degrees, and the lard is violently agitated by compressed air. The pipe which supplies the air should run down to the bottom of the tank within a few inches of the lowest part.

The quantity of fuller's earth to be used for refining depends upon the quality of the raw product. One-half to one per cent of fuller's earth is considered a good average for lard.

After the fuller's earth has been added, the lard is thoroughly agitated for ten to fifteen minutes. This is accomplished by air or mechanical agitator; some operators pump the lard out of the bottom of the clay kettle and circulate it constantly.

After thorough agitation, the line which flows from the pump to the filter press is opened and some of the product run through the filter press. Tests, which are usually taken in a glass bottle, will show the operator if the lard is now thoroughly refined. It is customary, however, to allow the first lard which comes from the press to run back into the refining tank, until such time as the lard has come absolutely clear, and of the proper color.

When the color is about suitable, further valves are opened and the refined product can be drawn either to the receiver, storage tank, or agitator.

Filtering.—The filter press, in this instance, is used for no other purpose than to take out the fuller's earth from the lard. It is usually equipped with double press cloths, sewed together at center holes.

In putting up the filter press for refining, the operator should watch carefully to see that the press cloths are all smooth where the plates come in contact with each other, as otherwise there may be a leak, no matter how tight the press may be adjusted. It is also well to bear in mind that filter press cloths should always be dry, clean, and that they should be free from holes.

After all lard has been pumped through the press, the valves that control the air pressure are opened slowly, to blow out as much of the refined lard as possible. This lard can be put with the finished product.

Now the air is shut off and steam turned into the press, but the lard which now blows out should be re-cooked, and not be turned into the finished product, as it may be scorched by the steam.

The press is now allowed to stand and cool. Then the screws are loosened, the plates separated, so that air can circulate freely and the apparatus will dry. When the press is dry, the earth can be shaken out of the cloths very easily.

Cooling Lard.—One of the most important points in the manufacture of lard is the proper handling of it after the raw product has been turned into lard; that is, the drawing off of the product from either the storage tanks or the receiver. There are two standard methods used.

One method is to run the lard over steel cylinders, with the lard leaving the storage or receiving tank at a temperature of from 100 to 130 degress Fahr. It may then be drawn from the picker box of the lard-cooling cylinder directly into the package.

Some operators, however, prefer to draw off the lard from the cooling cylinders into the agitator, so that the man who draws off has a constant supply of lard to draw from, for unless there is a ready supply of lard which keeps the picker box constantly full of lard, there is a possibility that the operator may draw off lard too fast from this picker box, with the result that he may cause a lot of air to be mixed with the finished product in the packages, which is, of course, detrimental.

There are various kinds of lard-cooling cylinders used today. In a few plants the lard is first cooled by running it over a roll cooled by well or lake water. Then it is run over a second roll, which is a brine roll. This method, however, is practiced only by those having very large capacities.

The majority of rolls is use today are of the single or double cylinder type, and are usually equipped for brine circulation. Another type of roll is equipped with direct expansion coils, cooling the brine which is stationary in the cylinder; and still another type which has recently been placed on the market, is the direct expansion lard roll, in which the ammonia expands directly into the cooling cylinder. Advantages are claimed by the manufacturer of each type, which the purchaser may investigate for himself.

Large refiners usually have methods of their own in handling lard, and it is rather difficult to describe a standard method for producing a product, because it must be turned out to suit the demand of the trade in a certain locality. Lard as it is demanded in certain states, must be of an entirely different color than what may required in other states.

The lard roller is used primarily to give firmness and grain to the product. It is also commonly known that agitation has a great bearing upon the grain, but primarily upon the color of the product, and it is for this reason that various operators have different speeds on their picker box shafts. Wherever a fluffy or light lard is required, more rapid agitation is employed to accomplish this.

The prime object in giving refined lard a sudden chill on the lard roll is to chill the lard before the stearine in the product has a chance to crystallize. The stearine, however, does slightly crystallize on the roll. But these crystals are so small that the grain is much finer than it would be if the lard had been allowed to cool slowly; and if the refined lard wasn't given a sudden chill, it would be as coarse in grain as an unrefined product.

Lard which eventually will be pressed should never be run over a lard roll, as the lard roll is intended to prevent the separation of the oil and stearine which form the component parts of lard.

Brine rolls usually circulate from zero to ten above zero Fahr. and the lard, if run over from 110 to 130 degrees Fahr., will fall into the picker boxes at 40 to 50 degrees Fahr. Higher temperatures are employed; as, for instance, if lard drops off of the rolls at 70 degrees, it is not only customery, but it is advisable, to use an agitator.

The lard is now ready to be drawn off into the various packages, which range from one pound cartons to tierces.

Stiffening Lard.—In some sections of the country considerable trouble is encountered in making lard of sufficient hardness to stand up under the climatic conditions. It is not permissible, under the regulations of the United States Bureau of Animal Industry, to add anything to lard except lard stearine, which may be used up to five per cent. This, of course, refers to inspected houses only. In certain sections of the country uninspected houses are now adding as high as fifty per cent of tallow and beef fats to lard, which they either market under a trade name, or sell as a compound. However, these smaller packers have difficulty in disposing of their beef fats in any other way.

Yields from Various Fats.—Lard yields from various averages as follows, depending upon the quality of the fat:

Products	Per Cent
Fresh ham fat	80
Fresh ham bones	20
Outside trimmings	70
Giblet fat	12
Mixed fat from shipping pigs	60
Pork loin bones	20
Scrap leaf lard	90
Sweet pickle ham fat	75
Sweet pickle belly trimmings from curing cellar	45
Fresh pork tongue trimmings	70
Dry salt fat backs	75
Dry salt pork trimmings	55
Whole heads	30
Back bones	15
Blade bones	11
Knuckles	13
Pork cheek trimmings	39
Heart trimmings	15
Pig tails	18
Feet, hind and front	12
Leaf lard	94
Ham facings	75
Brisket bones	32
Frozen belly trimmings	60
Pork lips	10
Shoulder bones	20
Bung gut fat	35
Trimmed neck bones	15
Skins	10
Pork loin fat	90
Head bones	9
Head fat	35
Snouts	18
Sweet pickle outside fat	45
Lard skimmings	60
Jowl trimmings	63
Sweetbreads	55
Snout trimmings	18
Sterilized pork fat	70

Kettle Rendered Lard

Ingredients.—The products entering into kettle rendered lard are pure leaf, back fat, pork trimmings, caul and ruffle fat. The first two are used much more frequently than the others.

Chilling.—The leaf or back fat may be chilled before rendering; in fact, hashing and rendering are facilitated considerably if chilling is practiced. The hasher generally will disintegrate the fiber more readily, and rendering will take place more quickly and a higher yield result, this being due, of course, to the fact that the fat shrinks upon chilling.

Hashing.—A power hasher of large type is used for hashing. This hasher is generally equipped with a series of knives, so that disintegration will be thorough, and the fat is forced by means of the hasher through a half-inch plate, although a five-eighths inch plate is sometimes used.

Rendering.—The hashed fat is then gravitated directly to the rendering kettle, equipped with agitators which revolve about 16 to 20 times per minute. The rendering kettle may be of any size, the usual range in this respect being from three to six thousand pounds capacity. The kettle has a steam jacket which is capable of being subjected to 100 pounds steam pressure. The steam is turned on and the rendering takes place in two to three hours. Sometimes a longer time is required, depending upon the nature of the material.

A good method of operation is this: Turn on steam to jacketed kettle with reducing valve set for 45 lbs. steam and start the center shaft revolving. Begin to hash the fat into the kettle, using any suitable hasher. The hashed fat should be about the size of hickory nuts. The more uniform the pieces to render the better will be the color of the lard, because the cracklings will cook uniformly and will brown together.

When the kettle has cooked about two hours there will be a rise in temperature to about 230 degrees F. Increase the temperature of the steam by changing the reducing valve to 60 lbs. steam pressure and cook right up until the temperature of the lard reaches 255 degrees F.

Settling and Filling.—After rendering a method generally used is this: The steam is turned off, the agitators are stopped, and the contents of kettle are allowed to settle for an hour or so, until the cracklings are well bedded in the bottom of the kettle.

An objection to this method is that the lard loses its color and flavor because of remaining in the high heat of the rendering kettle. Another method is to drop the lard from the rendering kettle as soon as the cracklings turn yellow (usually at a temperature of 255 degrees to 260 degrees Fahr.) into a shallow receiver, so that the high heat of rendering may be dissipated as soon as possible. In 20 to 30 minutes after running into this receiver, lard can be drained off clear and white, with good flavor.

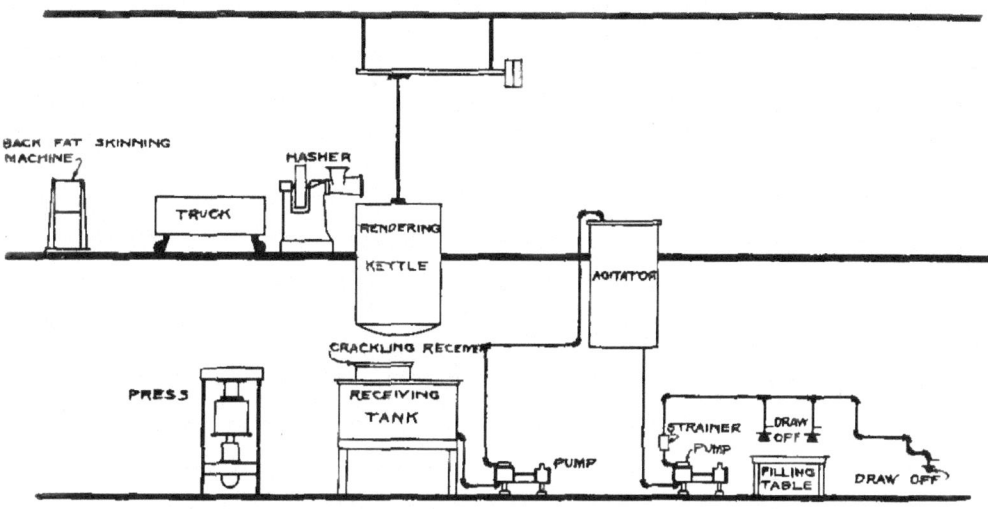

ELEVATION OF A TYPICAL KETTLE-RENDERED LARD PLANT

The lard is then carefuly siphoned off through two strainers—one ⅛ inch and one of cheese cloth—into containers, and is filled directly into the final package without any further processing. The filling in the final packages is generally done in a refrigerated room, and the lard is chilled as quickly as possible.

Yields.—The yield of kettle-rendered lard depends, of course, upon the raw material used. If made from pure leaf lard, a total yield of about 92% may be expected. If back fat and pork trimmings are used, the yield will vary from 80 to 85 per cent. Kettle-rendered lard has a flavor distinct and separate from prime steam lard.

Lard Cracklings.—The remaining cracklings contain a large amount of lard which packers find it advantageous to recover.

The majority of packing plants press their cracklings by either screw pressure or hydraulic pressure in curb presses. This method, as usually practiced, leaves anywhere from 15% to 35% of lard in the cracklings.

Another method used chiefly by large operators has been developed whereby the lard contents of the cracklings are reduced to a small percentage. In this process cracklings are spread on an iron pan and allowed to drain and cool. This pan has a small steam coil under it for convenience in keeping the cracklings warm while draining. The cooling is continued until the cracklings lose their gummy, plastic condition. Usually it is convenient to leave them over night.

Cracklings are then slightly warmed and run through an Anderson expeller. The cracklings, under enormous pressure, emerge from the press in a continuous plastic sheet quite warm and pliable, which become brittle immediately upon cooling. In this condition the crackling is ground to any condition desired, either coarse for chicken feed, or it can be ground still finer and made into flour. Cracklings thus handled contain about 6% to 7% of lard.

Lard crackling flour can be used as an edible food product, such as potted meats and in sausage, or it can be made into a palatable bread with ordinary flour, by making a mixture of 1 pound of lard crackling flour with 6 pounds of wheat flour. Crackling flour has 84% to 86% protein; wheat flour contains about 11% protein.

Butchers' Lard

Butcher's lard is made from all edible lard scraps which are produced in meat shops. It is made generally by the butcher himself, by rendering in an open kettle. Such lard, of course, varies in quality and hardness, depending up the kind of raw material used, and the care which is taken in preparing. The method of operation differs, depending upon the individual.

Neutral Lard

Neutral lard is made for use in oleomargarin. Two raw products, namely leaf and back fat, are used. Leaf lard will make what is known to the trade as a No. 1 Neutral, and back fat what is known as No. 2 Neutral. In either case the raw fat is chilled over night by hanging it over sticks at a temperature of about 34 degrees before hashing.

Hashing and Rendering.—The hashing is accomplished in a way similar to that used in hashing lard for kettle rendered purposes, although packers generally like to have the fiber disintegrated more thoroughly in the case of neutral lard, because the rendering is done at a much lower temperature.

Neutral lard kettles are either tin, galvanized iron, or tank steel, and vary in size. The paddles have about 20 revolutions per minute, and the kettles are often equipped with bumpers on the side so that the larger pieces of fat may be broken up and melted easily.

The only difference between the operations in making neutral and kettle lard is the fact that neutral lard is melted in a water-jacket kettle at low temperature. In kettle-rendered lard steam is turned on directly into the jacket and rendered at a very much higher temperature. Steam of about 5 pounds pressure, usually from an exhaust line, can be utilized to heat the water in the neutral lard kettle.

In making No. 1 Neutral a temperature of 120 degrees F. is used. If the previous steps are carried out correctly and with care, the lard will be fairly free from the fiber in about an hour and a half's time. When the rendering is finished, the scrap will come to the top in a foamy mass, and after settling for 20 to 25 minutes, the neutral lard can be drained off from beneath, and after straining through cheese cloth may be filled into tierce and put into storage at 40 degrees Fahr. until ready for use by the oleomargarin manufacturer.

The making of No. 2 Neutral, using back fat as a raw product, is accomplished in much the same manner, running the rendering temperature, however, up to 125 to 130 degrees.

To Get the Neutral Quality.—Neutral lard is clear in color and has no lard taste, but should be distinctly sweet and contain no foreign flavors or odors. It is especially desirable that no porky odor be present, and to accomplish this many packers have practiced the removing of the bungs during the first night in storage, allowing the tierces to lie on their sides over night, thus subjecting the greatest amount of lard surface possible to the air.

Some makers of neutral lard render at a higher temperature, and salt the scrap to the bottom, drawing the lard off at the top, but this does not produce the same high quality neutral as the other method described.

A very good grain can be obtained by rolling tierces of neutral into a temperature of 70 degrees, leaving them there twenty-four hours and then into a cooler at temperature of 41 to 42 degrees. When handled in this way the neutral lard can be shipped in seven days. Never roll a tierce of neutral lard after standing three days, as it ruins the grain. It can be moved by a truck. Good neutral lard should be grainy and taste like hickory nuts.

Yields.—Leaf lard made into neutral by the first method described will yield from 80 to 82% of neutral, and 10 to 12% of prime steam, the latter originating, of course, from the remaining scrap which has not been thoroughly rendered. No. 2 Neutral made from back fat varies greatly in yield, depending largely upon the quality of the fat, 70 to 75% being an average yield.

Wet Neutral.—Some oleomargarin manufacturers make what is known as Wet Neutral. While this is not generally practiced by packing plants, should a packer have a call for wet neutral, the method of manufacture is as follows:

The chilled leaf lard is hashed and rendered in the usual way, but at a temperature of about 135 degrees, and the neutral is run from the settling kettle into a vat of ice-cold water. It is then broken with a paddle until thoroughly chilled. When this is accomplished it is forked on to a drain table so that the excess water may be thrown off. It may then be packed in tierces; or better, it should be used immediately for incorporating in oleomargarin.

Lard Oil and Stearine.—There is very little difference in the pressing of lard in comparison with the other products mentioned here. The lard is usually run through at a temperature of 40 to 45 degrees, and pressed in cloths. The lard stearine thus recovered is used frequently for stiffening of lard of low titre.

There are usually two qualities of lard oil made, No. 1 and No. 2. No. 1 is made from the better grade of fats, while the latter is usually made from condemned hogs; the only difference is that No. 1 is edible and No. 2 is inedible. (See also Chapter IV, "Tallow and Grease Refining.")

General Lard Information

Operating Precautions.—Lard, like all other fats which in the processing are heated, is susceptible to rapid deterioration unless everything is kept absolutely clean. It is also important that no moisture or fiber be left in the lard, as it will sour very quickly. Thorough settling at all stages of the process where settling occurs is important, and care should be taken in siphoning off the lard so that no part of a kettle bottom will be mixed in with the general run. Choice of materials, cleanliness, and proper temperature are the chief things to remember in connection with lard manufacture. In storing lard it may be held very nicely at a temperature of 35 to 40 degrees.

Definition of Prime Steam Lard.—The Chicago Board of Trade (official) requirement is as follows:

"Standard prime steam lard" shall be solely the product of the trimmings and other fat parts of hogs, rendered in tanks by the direct application of steam, and without subsequent change in grain or character by the use of agitators or other machinery, except as such change may unavoidably come from transportation. It must have proper color, flavor and soundness for keeping, and no material which has been salted must be included. The name and location of the renderer, the month and year of filling, and the grade of the lard shall be plainly branded on each package at time of filling. Each tierce shall be properly filled.

Prime Steam Lard of superior quality as to color, flavor and body may be inspected as "prime steam lard, choice quality" and shall also be deliverable on contracts for "prime steam lard."

Board of Trade Regulation for Lard Tierces.—Lard tierces shall be made of well-seasoned white, red or burr oak or white ash, and may be coated on the inside with silicate of soda, or other substance acceptable to the United States Department of Agriculture, a sufficient quantity only to be used to fill up the pores of the wood for the purpose of preventing soakage and leakage. Staves should be about 34 inches long with a head about 20½ inches, or about 33 inches long with a head about 21 inches. The tierces should contain not less than 360 pounds net, nor more than 392 pounds net. Staves to be chamferred at the head and about ¾ inch thick, head about ¾ inch thick in center and ¾ inch at bevel. At least 6 iron hoops to be used, three on each end 1¾, 1½ and 1¾ for head, quarter and bilge. Head hoops, 1¾ in. x 18 gauge; quarter hoops, 1½ in. x 19 gauge; bilge hoops, 1¾ in. x 19 gauge.

Wood-bound tierces shall be hooped not less than eleven-sixteenths (11/16) of the surface, with hickory or white oak.

HOG CASINGS

The hog offal department is a very important point in any packing plant, and it is here that large savings can be made by the proper handling of the product.

Handling the Viscera.—The complete viscera which comes to the puller's table is treated as follows: First, the bung gut is pulled out, care being exercised when pulling not to break it. The paunch is then cut off, also the pluck. The pullers then strip off the long guts carefully so as not to break them. The ruffle fat is separated from the gut by breaking it with the left hand. Otherwise the fat will roll down on the casing, causing it to break. The ruffle fat and middle gut, which is used for chitterlings to some extent, are then separated, the ruffle being sent to the prime steam lard tank, and the middle guts are washed and fatted by means of a fatting machine, if they are to be used for chitterlings. There is 8 feet or so which may be used for sausage casings if handled correctly.

Stripping and Soaking.—When the pullers are about half way through a set, the gut is thrown over a rod to mark the middle of the set, then nine or ten sets are caught up about the middle of each, tied in a knot, stripped, put on a stick, and 10 bundles or 100 sets are put into a tierce or barrel.

They are now placed in the cleaning room and allowed to soak in water at a temperature of about 75° in the winter; in the summer some ice is used. They are allowed to remain in the water for 24 hours, but should the water be too warm the casings will spoil, and will not have strength enough to withstand the pressure of the sliming machines. On the other hand, if the casings are too cold they will be brittle, and will be apt to break also.

After the soaking process they are stripped again and allowed to remain in water at a temperature of 100 to 110° for three or four hours. A third stripping is then given in water at the same temperature, allowing them to stand for an hour and a half, and then the process is repeated just before they are put into the sliming machine.

Sliming.—When feeding the machine each set should be untied, and the portion of the gut which is somewhere near the center should be fed through the machine first, so that the helper can put the sets over a peg and still retain the center of the bunch. Feeding is an important work, and the spreader should watch his knives to see that they are not set too close to the drum, which may cut the casings. Also while the casings are passing through, the operator should keep stripping them so as to remove as much of the manure as possible, which prevents the weight in the gut causing it to break.

Overcleaning.—The 100 sets are now caught on the other side of the machine and tied into one bundle, and put into cold water until such time as the overcleaners can get to them. The overcleaners take two or three bundles of 100 sets each, put them into a tub of water at a temperature of 95 to 100° F., and then the cleaning is done with the back of a knife. This is necessary in order to remove all of the dirt left on the inside of the casing by the machine, also to cut out all holes larger than the size of a lead pencil, and to cut the ends square.

Grading.—The next step is to put the casings in a grading tub at a temperature of 90°. Here they are graded into three classes: No. Ones, which are all lengths over 15 feet; No. Twos, which vary from 6 to 15 feet in length; No. Threes, which run from 2½ to 6 feet in length. The extra wides are sorted out also, and after being graded they are sent to the bleaching tubs.

Bleaching and Salting.—Clear water is used, the best results being obtained when the water is around 55° F. Here they are left for 24 hours, each grade being tied in a separate bundle. After bleaching is complete they are sent to the salters, where the casings are measured in a small receptacle 3½ inches deep and 7 inches in diameter. With this pan full of casings the bunch should weigh, including water, close to 2½ lbs.

The casings are then thoroughly salted in rough salt and packed in pans for from 6 to 10 days. If left longer they may become too dry and the salt will fall off, or if taken out too soon they may not be cured properly. When the curing is complete they are taken out, the rough salt is shaken off, the knots are taken out and they are re-salted in fine 3X salt. At the time this is done the casings should have no odor, but should smell fresh and clean. If there is a disagreeable odor it is caused by either improper cleaning, insufficient curing, or from the fact that they were not bleached long enough.

When re-packing for export shipment, it is well to pour a small quantity of water over the barrel to be repacked, so as to dampen the bunches, in order that they may hold the salt. This helps to keep the casings moist in transit.

Specifications for Grading and Packing

Export Hog Bungs should be 1 13/16" and over in diameter when inflated and measured by gauge approximately 18 to 20 inches from the crown. The general appearance of the bung, or size from that point to the crown, is a guide for the inspectors as to what kind of bung it should make. The bungs should be perfect as regards crown scores, reasonably perfect as regards tail scores, and should be free from holes 32 inches from the crown. In case there is a hole 32 inches from the crown and the tail of the bung is still on, it is satisfactory to put same into exports, but no bung is to be put into this selection which is less than 36 inches in length, and it is understood that bungs be full length whenever possible.

Prime Hog Bungs.—The bungs should be 1 10/16" to 1 13/16" in diameter and 18 to 22 inches from the crown. The same rules as to selection, quality, etc., apply as for export bungs.

Small Prime Hog Bungs.—These bungs should measure 1 7/16 to 1⅝ inches in diameter and be 18 to 20 inches from the crown. The same rules govern as in exports.

Skips.—These bungs to be 1 4/16 to 1 7/16 inches in diameter and 18 to 20 inches from the crown. Same rules apply as in case of other bungs.

Broken Export Hog Bungs.—These bungs consist of all the export bungs which have been broken in handling, and should be at least 14 inches and up to 35 inches in length. They also include all the exports that have been cut near the crown so badly that they cannot be put into the regular export selection. The greatest care must be exercised on the killing floor to keep the percentage of cut exports down to a minimum. Every cut export hog bung means a loss of about 7 to 8 cents per piece.

Broken Prime Hog Bungs.—These bungs consist of all the prime hog bungs which have been broken in handling. They must be at least 14 inches, and up to 35 inches in length They also include all primes that are cut near the crown so badly that they cannot be put in the regular prime selection. The percentage of cuts should be kept down to a minimum.

Broken Small Prime Hog Bungs.—These bungs consist of all the small primes which measure from 14 to 35 inches in length, and all cut small primes. The percentage of these, too, should be kept at a minimum.

Packing.—Exports, primes, small primes and skips to be packed in second-hand ham or lard tierces, which are full eight-hoop, iron-bound. Broken hog bungs may be packed in glucose tierces. All tierces must be thoroughly cleaned, overhauled and put in first-class condition before being used. Second-hand vinegar, whiskey or wine barrels must never be used as containers for hog bungs.

The goods are to be put up as follows:

```
Exports ........................400 pieces to the tierce
Primes .........................500 pieces to the tierce
Small primes ...................600 pieces to the tierce
Skips ..........................700 pieces to the tierce
Broken exports .................550 pieces to the tierce
Broken primes ..................650 pieces to the tierce
```

All bungs should be so handled and packed that the tierce when opened will present a good appearance. This cannot be done unless, in pulling, the full length bung is saved and the bundles are well salted and thoroughly pickled when packed.

Hog Bladders

Hog bladders should have necks cut as long as possible, blow and trim all fat off as close as possible. Dry in a temperature about 110° to 115°. It is well to soften bladders before letting out air. Then grade as follows, taking measurement across widest part and not lengthwise:

Small Hog Bladders measure 5 in. to 7 in., are tied 25 pieces to the bundle and packed 200 bundles or 5,000 pieces to the barrel. Mark "Small Hog Bladders."

Medium Hog Bladders measure 7 in. to 9 in., are tied 25 pieces to the bundle, and packed 100 bundles or 2,500 pieces to the barrel. Mark "Medium Hog Bladders."

Large Hog Bladders measure 9 in. and over, are tied 25 pieces to the bundle and packed 60 bundles or 1,500 pieces to the barrel. Mark "Large Hog Bladders."

For export trade hog bladders are packed only in boxes; the small and medium grades are packed 5,000 pieces to the box, and the large size 3,000 pieces to the box.

Rules governing manufacturing and packing of hog casings. It is of the utmost importance that there be sufficient pullers to pull the gut carefully and properly, so as to save as many casings as possible and avoid breaking the gut. This often makes a difference of one or two hogs to the pound free-of-salt in the yield, so the great importance of having the work done efficiently will be appreciated.

Next in importance is the cleaning of the casings. This must be done carefully, so that there are no dirty strings in any of the finished product.

The heavy or stomach end (about a foot long) should not be left on, as buyers seriously object to it.

The regular production of hog casings put up for sales stock must be in as long lengths as possible; no piece should be less than 6 feet in length, and there should not be more than two 6-foot lengths in any one bundle.

Hog casing stumps. For sausage room use from 2 to 6-foot lengths are put up and packed with 40 to 45% salt. It is very desirable that the percentage of stumps be kept down to a minimum. Stumps put up for selling stock should be repacked with 60 to 65% salt.

Packages.—All hog casing packages must be strictly cleaned and thoroughly paraffined before being used, so as to avoid the casings being stained black by the natural acid contained in the sap of the lumber in the packages.

Specifications for putting up hog middle guts. These should be put up at least $7\frac{1}{2}$ feet to a set, each set to contain not more than 2 pieces. The narrow end measures $1\frac{3}{4}$ inches wide, and the wide end not over 4 inches The gut must be closely fatted and free from holes. It is absolutely necessary that all guts be thoroughly cleaned, as even the smallest particle of dirt will render them unfit for use in sausage rooms. After the dirt has been thoroughly removed they are to be chilled in ice water and packed in tierces of 150 to 200 sets each. Each tierce must be stenciled "Hog Middle Guts," and the contents shown on the other end of the tierce. After the hog middle guts are put up in tierces, as mentioned, they should be placed in cold storage so as to keep them in good condition.

EDIBLE HOG OFFAL OR MISCELLANEOUS MEATS

Following are specifications on edible hog-killing products coming from the offal floor:

Trimmed pork cheek meat is the trimmed pork cheek meat with 35 per cent of the fat trimmed off and 5 per cent of the No. 2 pork cheek meat trimmed off.

No. 2 pork cheek meat is the gland taken from the untrimmed pork cheek meat and should be free of fat.

No. 1 jawbone trimmings consist of small lean pieces of meat taken from the lower jawbones.

No. 2 jawbone meat is a small piece of meat taken from the lower jawbone and containing about 35% of fat.

Hog bung trimmings are the untrimmed bung gut trimmings.

Untrimmed pig snouts are snouts containing the lean meat.

Trimmed pig snouts are the snouts with the lean meat removed.

Snout meat, which is self-explanatory, should contain about 20% of fat.

Hog lips are lips taken from the lower jaws and must be free from hair.

Pig ears should be free from hair and must not contain any part of the eardrum.

Trimmed pork giblet meat is the trimmed pork giblet with 40% of tankage product trimmed off.

Trimmed hog gullet trimmings is the pork gullet with 75% of tankage product removed.

Jowl meat is a fat piece taken from the jowls, but which has streaks of lean running through it. These lean streaks are called "seam meat."

Pig hearts are untrimmed and slit and with 11% of the blood and tankage product removed.

Trimmed hog weasand meat is from weasands which have been split, thoroughly cleaned and with about 35% of the tankage product removed.

Hog tripe scalded is the hog stomach split, turned, with lining trimmed off. This is put through a washer and then through a second washer with brushes. The fat is trimmed off and the stomach is then scalded for 15 minutes at boiling temperature. It should then be put in cold water and all the slime scraped off before using.

PREPARATION OF PIGS' FEET

Both hind and front feet sometimes come from the killing floor with hair and toes on. To remove the toes, put them in water and bring to a temperature of 130° to 140°, agitate during heating, then take out a small quantity at a time and the toes easily can be pulled off. The feet should be shaved in the shaving machines, then singed, which removes the fine hair missed by the shaver; then the feet are put in ice water. After thorough chilling the glands between the toes must be removed, as per U. S. Government requirements. Now they are ready for pickling in a salt pickle cure of 80°—in which there is 4½ oz. of saltpeter to each 100 lbs. of feet—for 20 days, at 40° F. The feet are then ready to be cooked, in a vat or a steam cooking box, the usual method being to turn on the steam slowly for about 3 hours, then chill them off by turning on cold water. They are then split in a splitting machine, the bone dust is washed off and they are packed in tierces in a 45-grain vinegar pickle. After 20 days they are ready to repack for shipment, fresh vinegar being added, also a few coriander seeds and bay leaves being used in the top of the tub or other package.

The following test on a lot of front pigs' feet shows the yield which may be obtained in preparing them for vinegar pickle. The clean weight of the feet in the salt pickle was 3,635 lbs.

	Lbs.	Per Cent
Split feet	3,162	86.99
Bone dust	50	1.38
Prime steam lard	110	3.03
Shrink in cooking	313	8.60
Total	3,635	100.00

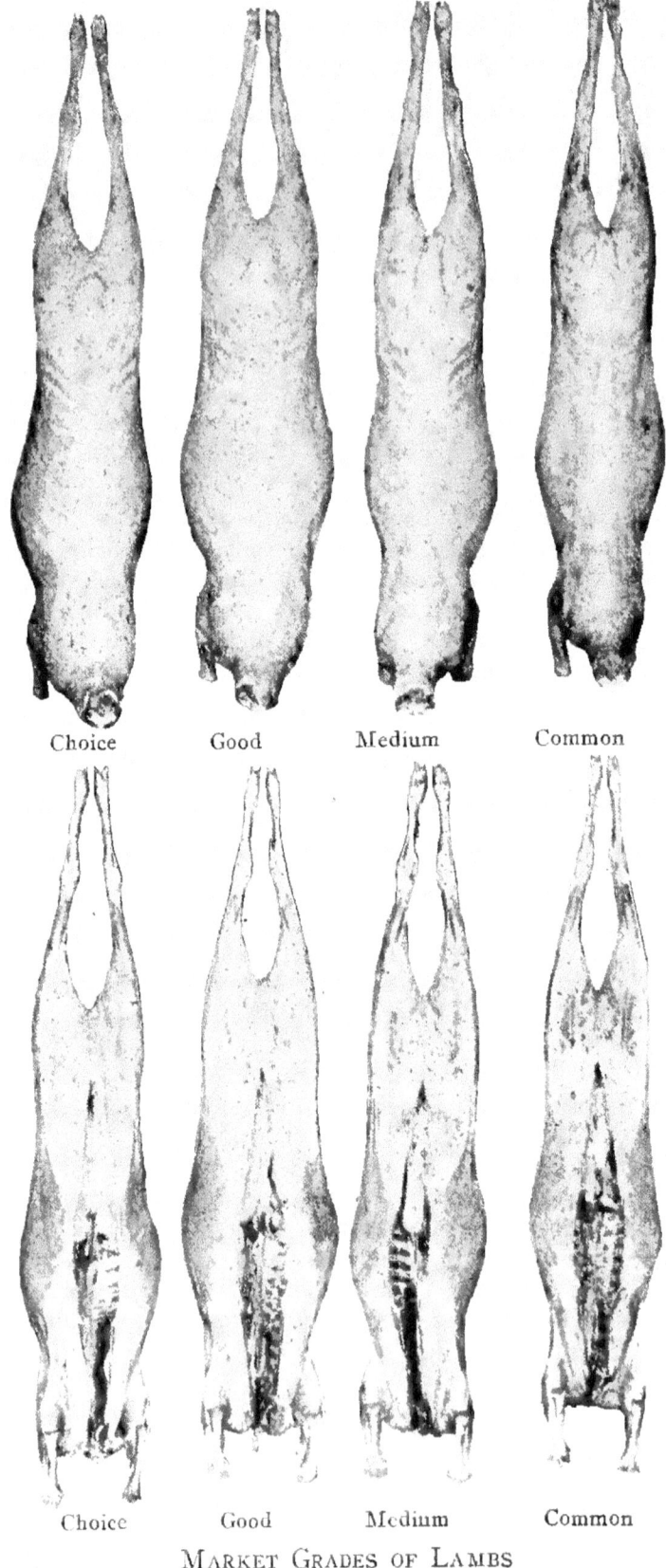

Market Grades of Lambs

Chapter III—SMALL STOCK

MARKET CLASSES OF SHEEP AND LAMBS

Sheep and lambs are classed according to their age and sex, and graded upon their condition, quality and conformation. Weights and fleece values are also very important factors in determining relative market values of animals falling within any one class. Other conditions—such as pelt values, grade and weight—being equal, the mutton breeds, adhering more closely to the approved mutton type and being of higher average dressing ability, usually outsell for slaughter the types produced primarily for wool. But at times, especially when wool is high, the condition of the wool market merits buyers' careful consideration of the relative pelt values of the two distinct types. Price differentials between shorn and wooled stock of the same class and grade are figured on a basis of pelt values.

The classification for killing sheep and lambs tentatively adopted by the United States Bureau of Markets and Crop Estimates provides the following classes: Lambs, spring lambs, yearling wethers, wethers, ewes, and rams. As an aid in properly showing market values of various grades, the lamb and yearling wether classes are further divided as to weight. The terms used to define the relative merit of the different grades within the several classes are prime, choice, good, medium, common, cull, canner. On and after June 15, all offerings born in the spring of the previous year are classified as yearlings, and those in the current year as lambs.

SHEEP KILLING

While many small packers are not engaged in sheep killing, an outline is here given of the principal operations necessary for slaughtering and handling this class of livestock.

Sheep are generally put over a hog hoist in pairs. The jugular vein is stuck, inserting the knife on the right side of the neck. After thorough bleeding, the face is skinned from forehead to nose, then the right leg is skinned outside and the foot removed. The tendons of the front legs are cut and the joints in the front leg are broken just above the pastern. The animal is then hung by the left leg, the right being released, which gives opportunity for the right to be skinned outside and the foot removed. A little more of the skin is then removed from the inside of the hind legs, and after this operation both hind and front legs are spread on the chain. A partial skinning of the neck comes next, after which the throat and neck are opened and the esophagus loosened, cut and tied. The front feet are then severed and the spreader removed.

"Ripping down facing" is the next operation, which is the splitting of the pelt the full length of the belly. "Rumping" follows, which is the

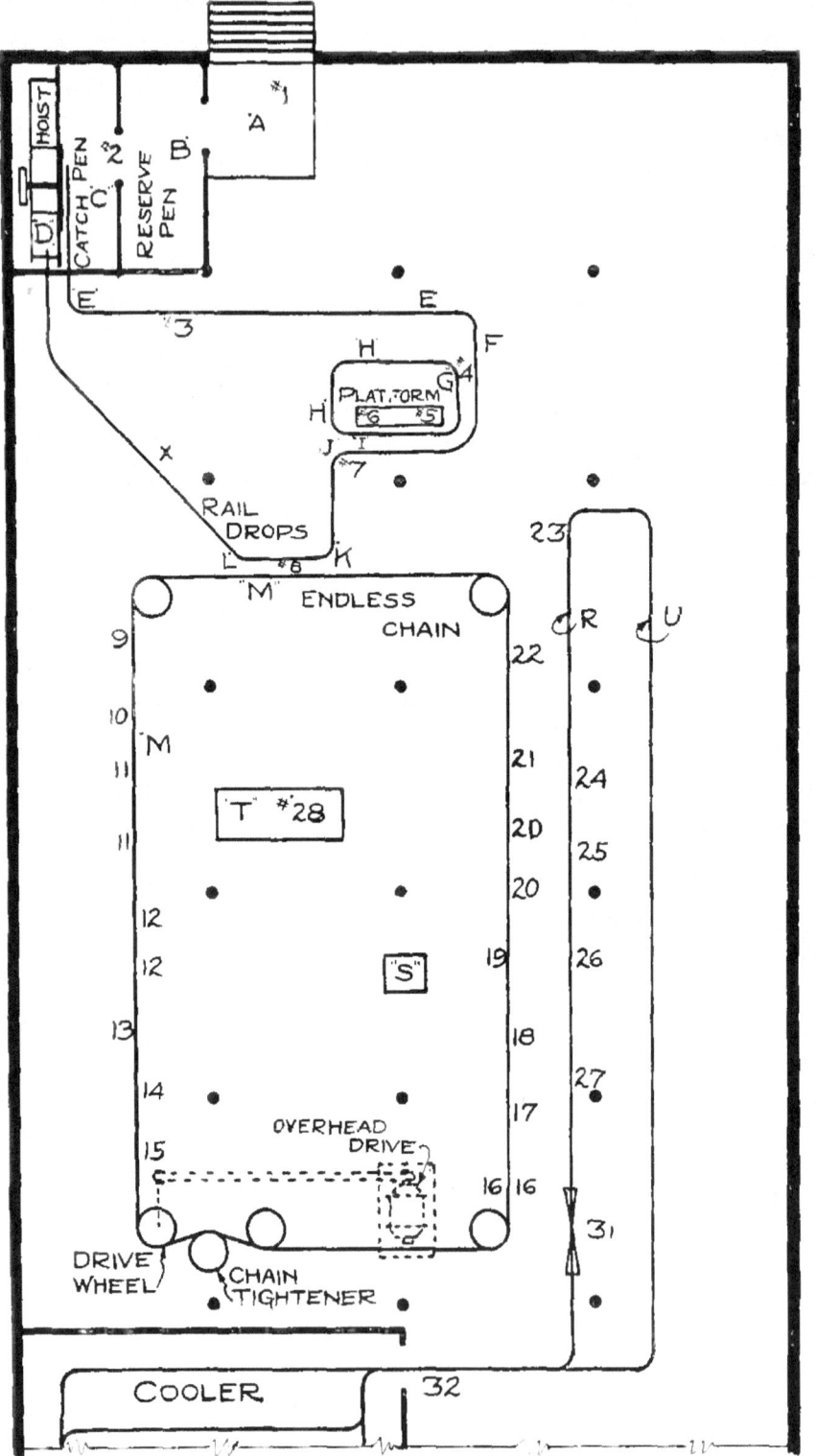

Layout of Modern Sheep Killing Floor (Using the "Sheep Ring").

Labor in modern sheep-killing ring: 1, driver-up; 2, shackler; 3, sticker; 4, hooker-up; 5, forequarter legger; 6, breast puller; 7, first hind-legger; 8, swinger-off; 9, second hind-legger; 10, hanger-up; 11, two pellers; 12, rumper and back-pullers; 13, breast splitter; 14, skinner; 15, pelt remover; 16, washers and wipers; 17, caul puller; 18, gutter; 19, header; *20, back sets; *21, dressers; 22, legger; 23, legger; 24, rack man; 25, neck trimmer; 26, skewer boy; 27, scrubber; 28, tagger; 29, trimmer; 30, floor cleaner. *If caul is dressed.

This method of operating a sheep-killing gang is used only in large plants, where labor-saving is a big item.

removing of the pelt from the hind legs, after which the pelt is loosened a little more from the belly. The tail is then skinned and the anus loosened and dropped. "Backing" follows, the pelt being pulled from the back and neck by hand. Finally the pelt is dropped completely by cutting from the head.

Butchering.—The carcass is then washed and wiped thoroughly before the butchering proper commences. The breast bone is split with a heavy knife and wooden mallet, the abdominal cavity opened, and pizzles or mammary glands are removed, the caul pulled from the viscera and put into a separate box on the chain. The intestines are then pulled down, the bung gut fat pulled off and the casings removed. These casings are generally placed on a clean, water-washed table or a moving conveyor. The pluck and gullet are then cut out, after which the uterus and bladder are removed. A string about five inches long is inserted in the left hind leg; this is called "beating block."

The carcasses are then tagged and the legs hung together and tied, and the head is severed. If the carcass is to be caul-dressed this is the point at which it occurs. The sweetbreads and thymus glands are then pulled out and the neck trimmed off. A galvanized iron rod, 12 inches by $\frac{1}{4}$ inch, is used for pinning down necks. The forequarters are then washed and wiped and the carcass stamped. Next the front legs are skewered up, all skewers are then clipped and the carcasses weighed and sent immediately to the coolers. After the carcasses are thoroughly chilled the neck pins are removed.

The cooler temperatures are the same as those used for chilling beef.

Sheep Dressing

Dressing.—There are two principal methods of sheep dressing: One is round dressing, in which the pluck is generally out, unless ordered otherwise, and no caul fat or back-sets or spreaders are used. A round-dressed sheep or lamb must be in very good condition. The second style, caul dressing, is used to a considerable extent, special styles of caul-dressing being practiced, depending upon the market to which the carcass is to be shipped.

During the world war it became a general practice to dress everything round, in order that the caul fat might be saved for edible purposes, instead of finding its way to the butcher's scrap box. However, it is possible that caul dressing may become popular again, and for this reason a listing is here made of the various methods employed.

Chicago style caul-dressing means that the carcass has the pluck in and two wooden back-sets or spreaders used. The Philadelphia style is the same as Chicago style, except the shanks are pinned close to the shoulders. The New York style is the same as Chicago, except the caul is put on in three pieces. The heavy end of the caul is put around the hind leg. In dressing Baltimore style, only one back-set is used and the pluck is left in. The Boston style consists of placing one back-set and cutting the ribs. Springfield, Mass., style is a combination of a round-dressed carcass with caul on. Light sheep are sometimes caul-dressed and the joints chopped off so that they will appear like lambs. These are called "choppers."

Offal.—The offal from the sheep killing, after having the head meat, hearts and livers separated (these products generally being used for sausage or sold fresh, or sometimes being frozen for future use), is graded as follows before being sent to the tankhouse: For prime tallow, paunches, peck and rennets are hashed and cleaned; cleaned sinkers, condemned sheep carcasses, condemned livers and plucks are also put into prime tallow. The pancreas glands, gullets, lights and trimmings, clean white guts, tongue trimmings and bung gut ends generally fall into a second grade of tallow. Lamb bags, bladders, floor pickings, condemned guts and catch-basin skimmings from the sheep killing go to make up brown grease. The feet are put directly into a tank, which is to be cooked for tankage only and no grease removed.

Tests on Sheep Yields

Sheep carcasses will shrink about the same in the cooler as beef carcasses. From 2 to 2½ per cent shrinkage with a 48-hour chill is not uncommon. Here is a test on 16,216 sheep, which will give the reader a very good idea of the offal yield which may be expected:

	Total Lbs.	Per Head Lbs.
Sheep heads	34,104 pounds	2.103 pounds
Sheep livers	11,266 pounds	.682 pounds
Tongues for sweet pickle	4,324 pounds	.265 pounds
Hearts	4,497 pounds	.277 pounds
Cheek meat	5,040 pounds	.311 pounds
No. 1 tallow for oleo	29,225 pounds	1.802 pounds
No. 2 tallow for oleo	17,293 pounds	1.066 pounds
Casings	Approximately 90% saved	
Brains	2,299 pounds	.142 pounds
Plucks	939 pieces	.058 pieces
Lamb fries	81 pieces	.005 pieces
Tripe	71 pieces	.004 pieces
Sheep pelts	9,969 pieces	
Lamb pelts	4,940 pieces	
Broken pelts	1 piece	
Fall clipped pelts	1,306 pieces	
No. 2 rendered tallow	11,380 pounds	.702 pounds
Brown grease	12,875 pounds	.794 pounds
Dry tankage	28,291 pounds	1.745 pounds
Dry blood	6,130 pounds	.378 pounds
Thyroid glands	32 pounds	.002 pounds

This test does not satisfactorily record the yield of inedible greases which should be obtained, as the greater percentage should go to prime tallow.

Following is a test on one lamb, which weighed 76 lbs. before killing:

	Lbs.	Per Cent
Dressed carcass, edible	40	52.63
Head, edible	2½	3.29
Ruffle fat, edible	¾	.99
Pluck, edible	3½	4.60
Casings, edible	2	2.63
Skin, inedible	11½	15.13
Toes, inedible	1	1.31
Stomach, inedible	1½	1.98
Raw blood, inedible	4½	5.60

Scrap, inedible	1	1.31
Loss in dressing	8	10.53
Total	76¼	100.00
Total edible		64.14%
Total inedible		25.33%
Loss in dressing		10.53%
Total		100.00%

As with cattle and hogs, the cost to kill varies, depending upon the capacity of the plant, labor conditions, etc. The labor cost at present (1921) in most sheep killing houses will run from 10 to 20c per head.

CALVES

The only difference in the dressing of veal calves from that of regular beef is the fact that the hide is not taken off, the carcass is not split and care is taken to see that the neck is neatly trimmed. In order to make the veal carcass appear to good advantage, it is a good plan to give the carcass a thorough bath in warm water, using scrub brushes on the hide until it is thoroughly clean; then drying well before putting in the cooler.

The offal from calves is handled in a manner similar to that of beef, which has been described in the first chapter of this section.

Blood and Tankage Yields.—A standard for the blood and tankage yields from sheep and calves is as follows:

Sheep, average live weight, 75 lbs.: Blood, .65 lb.; tankage, .55 lb.; concentrated tankage, .75 lb.

Calves, average live weight, 150 lbs.: Blood, 1.20 lb.; tankage, .50 lb.; concentrated tankage, .75 lb.

SHEEP CASINGS

Sheep casings are the most important casings, because of the demand for them for musical strings, surgical ligatures, etc., causing them to be much higher in price than beef and hog casings. Their manufacture is similar to the process described under beef casings, except that sheep casings are much more susceptible to breakage, and considerable care must be used at every step.

Following are specifications for manufacture and packing: Sheep casings should be pulled as carefully as possible and all gut under 6 feet in length should be tanked. They should be put up in bundles as follows:

Wides, 15-16 inches and over wide, 90 yards long.

Mediums, 12-16 to 15-16 inches wide, 120 yards long.

Narrows, under 12-16 inches to be put up for gut strings.

Wide and medium sheep casings should be put up in bundles, as yields can be more easily compared.

CASINGS FROM CALVES AND YEARLINGS

If casings from calves and yearlings are saved, they must be put up separately and sold for what they are. The trade does not relish receiving calf rounds in the regular run of beef round casings. This manufacture is similar to that of beef casings.

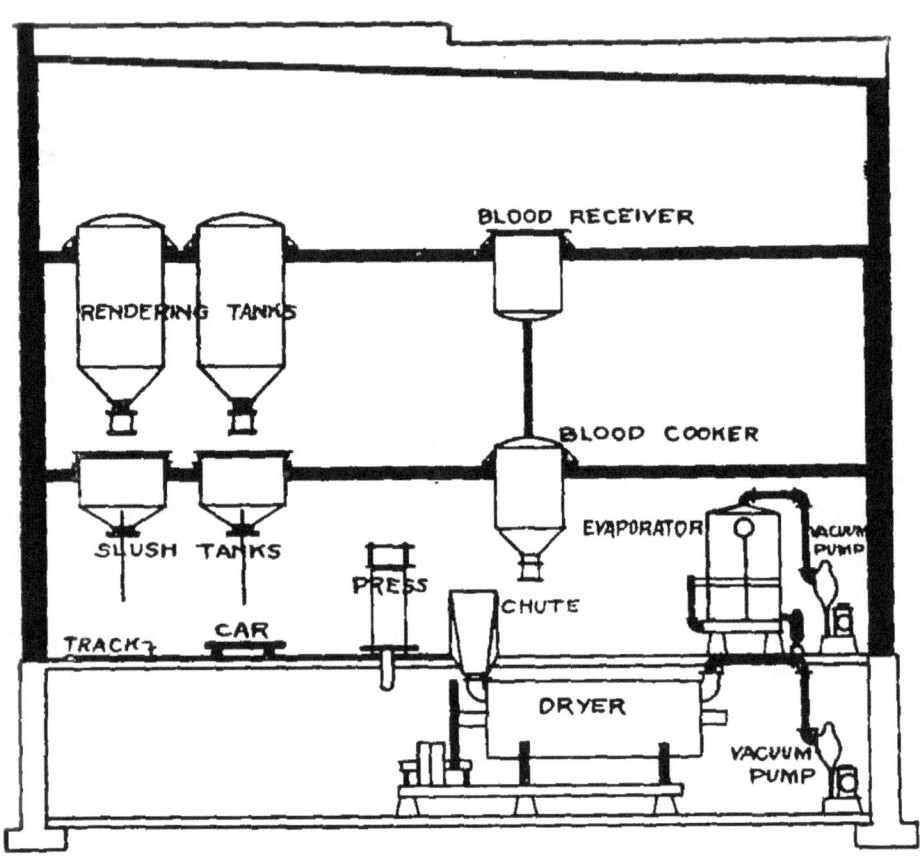

Elevation of a Typical Tank House

Chapter IV—INEDIBLE BY-PRODUCTS

INEDIBLE TANK HOUSE

The inedible tank house is not only one of the most interesting and vital parts of a packing plant, but it is one department which is frequently neglected.

Equipment.—The tanks are made in various sizes, ranging from 4 to 6 feet in diameter, depending upon the size of the packing plant. Tanks with cone-shaped bottom, equipped with 8 or 10 inch gate valves, is the standard construction. The direct steam inlet should be just above the gate valve in the bottom of the tank, some packers preferring to have three or four small inlets instead of one large one. At the top of the tank the piping is arranged so that steam may be constantly ejected into the exhaust line through a check valve which is weighted; also it is equipped so that the steam may be discharged direct when "blowing off" the tank.

The other equipment consists of a slush box, or receptacle beneath the tank, to receive the remaining tankage and tank water after the tallow or grease has been drawn off; tank water storage vats, where the tank water may be cooked and grease recovered; and tallow or grease vats, to receive those products; and hydraulic presses, to press the tankage so that it may be dried in a steam or direct-heat dryer.

In all except very small packing plants the evaporation of the tank water has been found to be a profitable source of income, provided exhaust steam and water are plentiful. Therefore many tank houses are equipped with evaporators.

Removing Tank House Odors.—If the packing plant is located in or near a city, condensers have been found necessary in order to eliminate tank house odors. Some packing plants have even found it necessary to go to the extent of washing the air from the tank house by drawing it out with fans and forcing it through a scrubber or spray.

As mentioned later on in this book, the location of a packing plant has considerable bearing upon the construction of the tankhouse so as to eliminate the objectionable odors which otherwise would arise from this building. Numerous methods have been devised, such as flues in the walls for removing obnoxious odors, special condensers and burners for the non-condensible gases, so as to make the plant as sanitary as possible.

Recently a new process, which the inventor claims is very effective, has been developed. The principle is based on the introduction of a sterilizing agent into the tankage before and after cooking. The agent used is sulphur-dioxide gas, produced by burning yellow sulphur in a rotary burner. By means of a specially-designed mixing device the gas is mixed with live steam, and this mixture is passed directly into the tankage. The exhaust from the tank is thus made odorless, it being claimed that all the sulphur gas has been absorbed in passing through the tankage. It is claimed that when the hot sulphur gases mix with live steam it gives a superheating effect, so that the entire mass of tankage is heated up to a cooking temperature in a short time.

The elimination of odors by the gas treatment is due to the fact that the majority of the putrefying bacteria in the tankage are destroyed by the germicidal sulphur gas. The mixture disseminates readily throughout the tankage, and the entire mass is well impregnated and sterilized in from 15 to 20 minutes. As stated, the tank is "gassed" before and after cooking, so as to insure a permanent elimination of odors before the tankage is dropped.

Blow System of Handling Tankage.—Several packing plants have found it advantageous to use what is called the blow system of handling tankage. Long sweep bend 8-inch pipe is connected to the tank at the bottom of the cone just below a quick-opening gate valve. This pipe may lead to a department a few hundred feet away from the rendering plant, this department usually containing slush and tank-water receiving receptacles, tankage presses and tank-water holding vats.

After the lard, tallow, or grease has been removed from the rendering tanks, the heads are placed in the tanks, and either steam or air pressure is then turned into the tank at the top. When the pressure reaches from 25 to 40 lbs., depending upon the distance which the tankage and remaining tank water is to be blown, the gate valve is opened quickly and the entire contents are discharged a few hundred feet away within a few seconds time.

The advantages of this system are: First, the rendering tanks may be placed at convenient levels with the offal floors, making charging easy; second, when placed at these levels it does not crowd the head room where the presses, etc., are located, which is a fault in very many packing plants; third, the cooking of the tank water and the pressing of the blood and tankage is removed some distance from the packing plant, and obnoxious odors in the vicinity of the food departments are thereby avoided; fourth, the method of transfer is rapid and economical.

The whole tank house should be arranged, as far as possible, to accommodate the gravity system, the charging floor generally being on a level with or beneath the offal floors. The elimination of conveyors and other such equipment is desirable.

Grading.—The grading of the products going into the tank is very important, in order that the largest amount of high-grade tallows and greases may be recovered. Careless grading may spoil a large quantity of high-grade grease or tallow, thereby decreasing the sale price three or four cents per pound.

The grading of inedible beef products has been described in the Cattle Chapter under Tallows. The grading of inedible hog products here outlined is that which is followed in a general way by the larger packers, and which is, of course, for general guidance only.

For white grease the following products may be used: Condemned fats; tonsils; bung guts; skips or thin small casings; eye lashes; bad kidneys; trimmings from ham facings; pizzles; scrap from splitters; vaginas; bladder fat; gut and paunch fat skimmings; skimmings from sterilizing machines; ground bones from head, ear trimmings; skull bone trimmings; skull bones and lard heads from retaining room; ear tubes; broken gall bags; floor scrap and condemned pieces from cutting floor; scrap from curing cellars; condemned hogs; dead hogs; condemned head bones; pork tongue trimmings; condemned tongues; rejected sausage casings; stomach trimmings; clean floor pickings; and fresh and clean catch-basin skimmings.

All products not fresh or clean enough for white grease must go into a lower grade; sometimes the lower grade is good enough for yellow, and other times it will classify only a brown grease.

The usual chemical standards for the greases are as follows, the standards for tallow having been given in the Cattle Chapter:

A White Grease.—Titre 39.5; free acid not over 3 per cent.
B White Grease.—Titre 39.5; free acid not over 5 to 7 per cent.
Yellow Grease.—Free acid, 10 to 15 per cent.
Brown Grease.—No standard on titre; free acid 25 per cent or under.

Rendering

The various products after being graded as suggested are put into the rendering tank, the head is closed and the steam turned on. The pressure should be maintained at 35 to 40 lb. for from 7 to 9 hours for most products, depending upon the material which is being cooked. When cooking livers alone, or something of that kind in which there is no grease, only three-fourths of an hour under pressure is necessary.

After complete rendering, let the tank settle well. It is sometimes necessary to blow off the tank—that is, open the pet cock in the head—when the pressure gets very low, should one be in a hurry to empty the tank and make it available for more raw material.

The tallow or grease, as the case may be, is either raised or lowered to the draw-off valve by the addition to or removal of tank water from the bottom of the tank. It has been found that a 4-inch or 6-inch floater draw-off valve is a valuable addition to tank house equipment, as it gives the operator a chance to keep separate any floaters on the top of the grease which may not be thoroughly rendered so that they can be reworked in the next tank. Tallow or grease which is drawn off is first passed through a water-settling device before entering the storage tanks.

The tankage and tank water is dropped to slush boxes where it is re-cooked and is skimmed before it is pressed, so as to recover the maximum amount of grease before pressing. Just previous to pressing, any remaining water in the slush box is drained to the tank water storage vats. There should be sufficient tank water storage to allow cooking to go on for at least 24 hours, so that the maximum amount of grease may be recovered. It is lost entirely, of course, for grease purposes once it passes into the evaporators.

Pressing.—The pressing of tankage is an important part of the process. The tankage should be kept very hot throughout. The cheeses should be made thin, so that there will be plenty of draining surface. A 2-inch thickness of cheese when the press is filled will press down to a half inch in thickness.

After the hot tankage is put into the hydraulic press the pressure is turned on slowly, and while the ram is traveling up or down a small stream of very hot tank water, which has been thoroughly cooked and skimmed, should be played on the edges of the cheeses, in order to wash off any grease or tallow that may have congealed, thereby keeping the drainage surface open and giving in all the best pressing results. Complete pressure should be applied finally before the press is emptied.

The pressed tankage should run 45 per cent in moisture and 5 to 7 per cent of fat on a 10 per cent dry basis. The raw material which is put into the tank determines the quality of the tankage. Ordinary beef tankage containing little bone will run from 9 to 10 per cent of ammonia, while hog tankage, bone excluded, will run 7 to 8 per cent ammonia. Should bones be in the tankage there will be a considerable percentage of bone phosphate of lime, although this runs rather low in ordinary tankage.

The drainage from the presses runs very high in grease, and this material should not be allowed to escape into the sewer, but should be saved and recooked in the rendering tank in order to recover the grease. Or better, if first run through a catch basin the grease may be collected and put into the tank, and the fine solids will be trapped so they can be pressed with the coarser tankage, and will not keep up continuous circulation because of the method used.

If the tank water is evaporated in the same room, or in the basement of the tank house, it is important that there be no sewer outlet whatever. No fresh water should be used in a tank house, as all the floors can be washed down and thoroughly cleaned with cooked tank water. If there is a sewer outlet, some of this valuable material is apt to escape, either by accident or otherwise.

Inedible Tankage Yields

The following tests on the yields of various products which may be obtained by rendering in the inedible tanks will be of interest to all superintendents and tankhouse men.

No. 1 White Grease Test.—Test on 5,600 hogs to find the yield of No. 1 white grease. In addition to the regular inedible offal, the following pieces were rendered: 1 dead hog, 28 condemned hogs, 30 condemned heads. The white grease recovered in this test amounted to 15,300 lbs., or 2.73 lb. per hog.

Yellow Grease Test.—The following is a test on 3,432 hogs to find the yield of yellow grease:

Products to Tank	Total Lb.	Lb. per Hog
Pig bags	6,243	1.819
Lungs	5,141	1.498
Catch basin skimmings	4,257	1.240
Floor scrap	662	.193
Total green product	16,303	4.750
Yield of yellow grease	1,320	.39

Tallow Yields.—The following test is one which indicates in a general way the yield of tallows which are obtained from cattle:

Edible tallow	1.13 lb. per head
Prime tallow	4.41 lb. per head
No. 2 tallow	.95 lb. per head
Brown grease	1.23 lb. per head
Total	7.72 lb. per head

The tallow from calves is indicated by the following test:

No. 1 tallow	.45 lb. per head
Brown grease	.65 lb. per head

The tallow yield from sheep is indicated by the following test:

No. 1 tallow	.19 lb. per head
No. 2 tallow	.28 lb. per head
Brown grease	.19 lb. per head

Tests on Condemned Offal.—The following yield of condemned viscera has an important bearing, as it is necessary to know this in order to calculate the losses on condemned cattle:

Green product tanked:

14 sets viscera,
8 paunches,
16 sets guts,
68 pieces liver.
Total weight 9,563 lb.
Yield No. 2 tallow 701 lb. or 7.33%.
Pressed tankage (10% moisture basis) 769 lb., or 8.04%.

In addition, of course, there was a yield of tank water, which, depending on the individual case, should be analyzed and its value calculated in determining the value of the offal.

Following is a test on condemned hog viscera:

192 sets.
Weight 4,430 lb.
Yield No. 2 white grease 340 lb. or 7.67%.
Pressed tankage (10% moisture basis) 204 lb. or 4.60%.

The value of the tank water in this case should be determined and enter into the calculation also.

Tank Water Evaporation

Following is a table prepared for superintendents and tankhouse men showing the Beaumé, the per cent of solids, the per cent of water, the weight per cubic foot of water, and the weight per gallon of tank water at a temperature of 150° Fahr.

Table for Tank Water at 150° Fahr.

Beaumé	% Solids	% Water	Wt. Cubic Foot H_2O	Wt. Gal
1	1.9	98.1	62.14	8.3
2	3.83	96.17	62.46	8.34
3	5.77	94.23	62.79	8.39
4	7.77	92.23	63.11	8.43
5	9.85	90.15	63.43	8.47
6	11.93	88.07	63.91	8.54
7	14.04	85.96	64.40	8.60
8	16.19	83.81	64.89	8.67
9	18.38	81.62	65.38	8.73
10	20.6	79.4	65.87	8.8
11	22.71	77.29	66.39	8.87
12	24.84	75.16	66.92	8.94
13	26.97	73.03	67.44	9.01
14	29.11	70.89	67.97	9.08
15	31.27	68.73	68.49	9.15
16	33.58	66.45	69.05	9.23
17	35.88	64.12	69.61	9.3
18	38.22	61.78	70.17	9.37
19	40.59	59.41	70.72	9.45
20	42.98	57.02	71.28	9.52
21	45.11	54.89	71.89	9.6
22	47.24	52.76	72.79	9.68
23	49.37	50.63	73.09	9.77
24	51.5	48.5	73.7	9.85
25	53.63	46.37	74.3	9.93
26	56.31	43.69	74.95	10.01
27	59.04	40.96	75.6	10.1
28	61.8	38.2	76.25	10.19
29	64.61	35.39	76.9	10.27
30	67.54	32.46	77.55	10.36
31	70.34	29.66	78.35	10.46
32	73.27	26.73	78.94	10.55
33	76.24	23.76	79.64	10.64
34	79.25	20.75	80.33	10.73
35	82.31	17.69	81.00	10.83
Water	1.00	100.00	62.425	8.345

It rarely pays to evaporate any cooking water which runs under 1° Beaumé. This is something which must be determined upon each class of cooking or blood water which may be available, and depends not only upon the concentration of the liquor, but upon the nature of the solids in the liquor; namely, whether or not those solids run high in ammonia. A liquor containing mineral salts or other solid matter in solution may run high in Beaumé, but unless the solid matter consists of nitrogenous material it will probably be a waste of money to endeavor to handle such liquor. The cost of coal, oil or whatever fuel is used in each particular locality also figures very largely in making this calculation.

Tank water before evaporating, of course, should be thoroughly cooked, and all of the grease which rises to the surface skimmed off and put back into the rendering tank. The water should also be well settled, so that the fine solids will not be drawn into the evaporator. Drawing these solids into the evaporator causes the evaporators to cake, and work with low efficiency, and it also causes a large percentage of fat to appear in the liquid "stick," or concentrated tankage.

In order to avoid this, it is a good idea to have the intake to the evaporator in a corner of the tank-water vat, shielded from the rest of the vat by two perforated plates, one with about two one-half inch openings per square inch, and a second with about three one-quarter inch openings per square inch. The intake should also be on a swinging joint, so that it will not drop clear to the bottom of the tank; in other words, so that the operator may level it to a point just above where the solid material accumulates.

No attempt will be made here to enter into the technical details of operating an evaporator, as these are supplied by the manufacturers, the directions varying with the type of machine used.

Tankwater Value Table

		% Solids	% Ammonia in Solids
"A" Tallow	Tankwater	7.49	17.86
"A" Tallow	Presswater	5.13	15.00
"A" Tallow	Parboil-water	1.87	10.58
Prime Tallow	Tankwater	5.05	17.18
Prime Tallow	Presswater	4.97	15.98
Prime Tallow	Parboil-water	2.57	12.31
Beef Offal	Tankwater	7.79	15.45
Beef Offal	Presswater	6.65	15.68
Oleo Scrap	Tankwater	6.11	9.71
Oleo Scrap	Presswater	5.52	10.64
Oleo Scrap	Drainwater	7.00	7.66
Bones	Housewater	3.71	18.84
Tripe	Cooking-water	1.50	13.62
Blood	Presswater	1.57	6.93
Grease	Tankwater	1.53	10.73
Grease	Presswater	1.87	10.32
Hog Killing	Tankwater	5.60	17.67
Hog Killing	Presswater	6.68	16.85
Hog Killing	Parboil-water	1.22	12.32
Hog Cutting	Tankwater	3.80	18.02
Hog Cutting	Presswater	3.70	17.54
Hog Cutting	Parboil-water	0.85	11.03
Hog Offal	Tankwater	3.45	10.25
Hog Offal	Presswater	5.14	13.04
Condemned Hogs	Tankwater	5.65	16.45
Condemned Hogs	Presswater	5.68	15.16
Hog Scald Tub Water		0.42	8.25
Average		4.00	13.50

Drying Concentrated Tankage.—Liquid "stick" under chemical test should not show more than 2 to 3 per cent of fat on a 10 per cent moisture basis if properly handled. The liquid stick should be evaporated to from 25 to 30° Beaumé, depending upon the amount of coarser tankage which is available to absorb the liquid stick when drying.

A method which was in common use until very recently has been the drying of the liquid stick on rolls. Ten per cent of copperas (iron sulphate) was added to 24° stick to preserve it and add to the drying qualities before running it over pressure steam-heated rolls. Evaporation took place almost instantaneously, the result being a dry powder running about 4 to 5 per cent moisture. This is a rather expensive method of drying concentrated tankage.

The most economical way is to partially dry the regular tankage in a steam-heated dryer, then add about 50 per cent of liquid stick by weight, running it in slowly while the dryer is moving, the more concentrated the stick the better. After drying for an hour or two more stick may be added, and if care is taken and the dryer is of correct type, the mass will not ball up, but will be fairly uniform and all parts will be thoroughly dry. In this way the tankage is made to absorb the liquid stick, and the cost of drying is thereby greatly reduced.

When this method is used and the bones are kept separate, the result is a high-testing tankage which may be used for feeding tankage when ground and prepared. Feeding tankage is usually dried in a steam-jacketed dryer.

Some packing houses are finding it advantageous to press and dry paunch manure, using this as a base for absorbing the liquid stick. Paunch manure contains a small amount of ammonia and this is, of course, saved and increases the yield to that extent.

Tankage Method for Small Plants

The method of parboiling is an important preliminary step to the rendering process. Especially is it valuable in small packing plants where sufficient machinery has not been installed to thoroughly disintegrate and wash the raw products before putting into the tanks. The purpose of parboiling is to wash away the manure and coloring matter, thereby reducing the acid and producing a product that bleaches better.

Parboiling is generally used on the following material:

Beef fat ends
Beef pancreas glands
Rough paunch trimmings
Beef gullets
Sheep pecks
Trimmings from hogs and sheep ruffle
Sheep rectums
Hog gullets and stomachs
Hog melt trimmings
Sheep sweetbread trimmings
Dirty tongue trimmings
Bruised beef pieces
Beef gut scraps

Round stumps of large guts
Condemned paunches
Sheep tripe
Rennet
Middle guts
Sheep cheeks and tongues
Calf gullets
Beef rennets not clear enough for edible
Cow bags
Beef pecks which have been sliced and the manure shaken from them
Floor pickings.

Method of Procedure.—Sheep heads or any bones of similar shape are washed and dumped into the bottom of an empty tank. These act as a strainer over the drain valve. Before putting in the heads 5 feet of warm water is run into the tank. Then the tank is charged, care being taken to keep the material covered with water. Steam is kept on slightly while loading, and this keeps the temperature around 175° Fahr.

When the tank is loaded the steam is shut off and the tank is allowed to stand for 20 or 30 minutes; then the water is drained off into the sewer. When the tank is empty the hot water hose is used to spray the material in the tank until the water which comes out at the bottom is fairly clear. The tank may then be headed and cooked in the usual way.

BLOOD AND TANKAGE YIELDS

The careful packinghouse superintendent will watch the yields of blood, tankage and concentrated tankage very carefully, for the efficiency of packinghouse operation depends very largely upon the total recovery of these by-products. The following figures show results which have been obtained over long periods of time on blood, tankage and concentrated tankage calculated on a basis of 10 per cent moisture:

Blood

Shipping cattle	7½ lb. per head
Canning cattle	6 lb. per head
Calves	9/10 lb. per head
Sheep	½ lb. per head
Packing hogs	1¼ lb. per head
Shipping pigs	¾ lb. per head

Concentrated Tankage

Cattle	5.80 lb. per head
Hogs	2.35 lb. per head
Sheep	.60 lb. per head

Regular Tankage

Cattle	9.25 lb. per head
Hogs	3.60 lb. per head
Sheep	.90 lb. per head

Cooking Blood.—In preparing blood—after collecting all of it in a blood tank located below the killing floors, and seeing that all of the blood is thoroughly flushed and drained from the floor down the drain pipe—the blood is pumped to the cooking tank. This tank is generally located in the by-products department, possibly in the tank house. Or if there is volume enough the blood is pumped to a separate point.

The blood is cooked in an open tank by introducing direct steam, care being taken that the tank is not more than half full, as the blood will expand and run over the sides very quickly under heat. Should there be a limited amount of blood only, a deep narrow tank is advisable in order that the steam may thoroughly agitate it while cooking. If there is a larger volume it is well to employ mechanical agitators, so that the whole will be thoroughly cooked and the albumen coagulated.

After cooking for 30 minutes coagulation should be thorough, and the tank is then allowed to settle and the lower or liquid portion drained away. Some plants make a practice of evaporating this blood water, but as it runs very low in ammonia if cooking is properly carried out, it is generally not advisable to do so, on account of the low solid content and therefore the high cost of the evaporation. Blood water is also somewhat hard on the evaporators, it having a tendency to eat the tubes more than other cooking waters. This, however, is a problem which should be determined at each local point.

Pressing.—After draining, the coagulated blood may be gravitated directly to the press and placed in the cheeses as hot as possible. The pressing operation is similar to the pressing of tankage, except since there is no grease present it is not necessary to wash down the sides with hot tank water. After pressing the blood should run 45 per cent in moisture, and it should be dried and kept separate, as it runs very high in ammonia, 17 units being a good average.

Dried blood may be used in the preparation of special feeds, or for bringing up feeding tankage to the proper protein content.

Calculating Weight of Blood per Head.—The following is a test on various classes of livestock showing the liquid weight of blood per head, its percentage of moisture and solid material, the dry weight per head, and the weight calculated on a moisture basis of 8.74%. The test also includes the factors for calculating the weight of blood per head, using the bullock basis as a standard.

	Liquid wt. per head	% of Moisture	Absolute dry basis % of Solids	Absolute dry wt. per head	On Moisture basis of 8.74	Equals
Cattle	57.83	87.3	12.7	7.34	8.04	1
Sheep	2.46	82.2	17.8	.4373	.48	16.75
Calves	4.14	82.2	17.8	.7369	.81	9.93
Shipper hogs	4.75	82.2	17.8	.8455	.93	8.65
Packing hogs	9.16	88.06	11.94	1.0884	1.19	6.76

Specifications for Animal Feeds.—The following specifications for the manufacture of feeds from animal ammoniates are set forth, with the usual methods employed:

Sixty per cent feeding tankage is generally beef tankage free from bone and containing enough liquid stick to bring up the test to standard. It is used for hog and poultry feeding.

Poultry bone consists of granulated raw bone in three grades—coarse, medium and fine.

Blood meal is pure blood, dried and put through a 40-mesh screen. It is used for cattle, horses, and hogs.

Poultry blood consists of coarse dried blood.

Tankage from condemned carcasses or tankage containing considerable amounts of bone, also steam bone tankage, are used for fertilizer purposes generally.

Steamed Bones.—This product is derived generally in the larger plants from bones which have been previously cooked for glue manufacture, which will be described hereafter. In the small plants they consist of edible and inedible bones which have been cooked separate from the tankage. Here are a few tests made on steamed bones of various kinds:

Test on 1,165 lbs. shank bones:
- Result: 710 lbs. dry bone stock or............60.94%
- 120 lbs. of No. 1 tallow...............10.30%

Total: 830 lbs. 71.24%

Test on 10,106 lbs. steamed plate bones:
- Result: 2,510 lbs. dry bone stock or............24.83%
- 1,214 lbs. No. 1 tallow.................12.01

Total: 3,724 lbs. 36.84%

Test on 10,102 lbs. steamed rib bones:
- Result: 2,800 lbs. dry bone stock..............27.71%
- 1,194 lbs. No. 1 tallow.................11.82

Total: 3,994 lbs. 39.53%

Test on 7,900 lbs. steamed cattle skulls:
- Result: 2,100 lbs. dry bone stock..............26.58%
- 974 lbs. No. 1 tallow.................11.98
- 161 lbs. brown grease................. 2.04

Total: 3,235 lbs. 40.60%

Test on 9,500 lbs. steamed sheep heads:
- Result: 3,100 lbs. dry bone stock..............32.63%
- 726 lbs. No. 1 tallow................. 7.63

Total: 3,826 lbs. 40.26%

Test on 23,000 lbs. mixed steamed bones:
- Result: 7,400 lbs. dry bone stock..............32.17%
- 2,204 lbs. No. 1 tallow................. 9.58
- 831 lbs. brown grease................. 3.61

Total: 10,435 lbs. 45.36%

Chemical Analysis.—The following table gives briefly the chemical constituents of these various animal products, at the same time showing the availability of the chemical elements when used for fertilizer purposes:

	Ammonia	% Total Phosphoric Acid	Amount Available
Raw bone	4½	22	⅓
Steamed bone	2	29	⅓
Hog house tankage, 10% moisture	7	20	½
Beef house tankage, 10% moisture	8½	10	½
Concentrated tankage, 4% moisture and under 2% fat	14		
Low grade tankage	3 to 7	20	½
Dry blood	18		
Calcined hoofs	17		

Screening.—In preparing these various feeds and fertilizers for the market, after grinding in one of the standard grinders, they should be screened through the following size mesh screens before being bagged for shipment:

Commercial fertilizers, batching .... 6 mesh
Commercial fertilizers, shipping................................ 7

Tankage 7
Commercial blood 7
Bone tankage 5
Raw bone 7
60% feeding tankage.. 10
Blood meal 40
Poultry blood 7

Calculating Analytical Results.—The following table is given for calculating analytical results:

Per Cent Given of	Multiply by	Per Cent Obtained of
Phosphoric acid	2.183	Bone phosphate
Bone phosphate	.458	Phosphoric acid
Nitrogen	1.215	Ammonia
Ammonia	.823	Nitrogen
Nitrogen	6.250	Protein
Ammonia	5.14	Protein

Casing Slime.—In passing, it should be mentioned that slime from the cleaning of casings is generally saved and cooked with the blood or added to the pressed tankage in well-regulated plants. Slime runs 12 to 13 per cent in ammonia on a 10 per cent moisture basis and it is therefore worth while to save this product in this manner.

CALCULATING TANKAGE VALUES

The cooking of the offal and the rendering process has been described under "Inedible Tankhouse." Tankage is used primarily for two purposes:

First. As an ingredient for fertilizer.
Second. For feeding purposes.

The value, whether used for feeding purposes or for fertilizer, depends upon the proportion of nitrogen which it contains. Commercially, and when used for fertilizer purposes, the percentage of nitrogen in fertilizer is expressed as ammonia, while if it is used for feeding purposes it is expressed as protein.

It has been commercial practice to base the calculation of ammonia contained in tankage on a basis of units of ammonia per ton. For example: One unit of ammonia means 20 lbs. per ton; and tankage which contains, for instance, ten units of ammonia contains 200 lbs. of ammonia per ton. While the calculation is made by the chemist in terms of ammonia, the direct determination is made on the amount of nitrogen present, and the calculation is made in terms of ammonia merely because the trade has used that term to designate the quality of the particular kind of ammoniates marketed for feeding and fertilizer purposes.

As ammonia is the chief ingredient, it is sold on the basis of so much per unit. For instance, ammonia in ground tankage may be quoted at $3.00 per unit, and if the tankage analyzes 8 per cent ammonia, the tankage would sell at 8x3, or $24.00 per ton, plus an additional small amount for the percentage of bone phosphate.

DIGESTER TANKAGE

What is known as Digester Tankage, or Hog Feed Tankage, is a product which has been in great demand for the past few years, and it is estimated that 60 to 70 per cent of the tankage produced in the large packing plants is used for feeding purposes. Inasmuch as this is sold on a protein basis, it is well to give here the method by which it is calculated.

When nitrogen is expressed as ammonia, the nitrogen is multiplied by 1.215. In order to express nitrogen as protein, the percentage of nitrogen is multiplied by 6.25. For example, 10 per cent of nitrogen is equivalent to 12.15 per cent ammonia, or 62.5 per cent of protein.

It is, of course, the desire of every packer in producing tankage for feeding purposes to get the protein content as high as possible. Since the ordinary packing house tankage will only analyze on an average between $33\frac{1}{2}$ to $41\frac{1}{2}$ per cent of protein, some products must be added to this low grade tankage in order to bring it up to the desired percentage of protein. In selecting the ingredients for feeding tankage, however, care should be used not to put in hair or other ingredients which, though they are high in ammonia content, have a low feeding value.

The methods and products used by packers differ, according to the manner in which they handle their other by-products. Blood which will analyze 15 or 16 per cent ammonia will help a great deal to bring up the protein content. So will concentrated tank water, or dry stick, as it is commonly called, help greatly in bringing up the protein percentage. Other packers buy cracklings which run very high in protein and add them to the tankage.

In order to get the desired percentage of protein it is also customary in packing plants which produce tankage for feeding purposes to eliminate all the paunch contents, and other very low grade material which may have a tendency to lower the protein contents.

Whenever the concentrated tankage is used, it is well to remember not to treat it with copperas, as it would be injurious to the animal.

It is also very important, in order to turn out a product of high quality, that the product be handled very fresh. Feeding tankage and cooking material must be drained, pressed and dried with the greatest possible speed. Rendering, of course, sterilizes the product, and destroys all causes of decomposition, but it is also a fact that tankage will immediately become contaminated as soon as it is brought into contact with floors, presses, etc.

A good many states have laws governing the percentage of protein in feeding tankage. Practically all of the packers turn out a digester tankage showing 60 per cent protein and 6 to 8 per cent of fat. Why this 6 to 8 per cent of fat is retained in the tankage has been a puzzle to many in the trade, as its value as grease on a normal market is greater than its feeding value in the tankage. Whether it should be left in or taken out, however, depends on the cost and method of extraction.

TALLOW AND GREASE REFINING

Tallow and grease refining is practiced in only a few of the larger plants. It is therefore relatively unimportant in this discussion. The process is very similar to the refining of lard, with the exception that more fuller's earth is used to remove the excess color which is present. In order to market these products the tallows, but more especially the greases, are pressed in order to produce marketable oils. The tallow and grease stearine recovered in this case is generally used for soap manufacture.

The graining and preparation for the pressing may be done either in trucks kept in a cold room at 50° or 60° Fahr., the length of time required varying, of course, with the temperature employed. The grained stock is put into press cloths and pressed in large screw presses, which are best located also in a chilled room. By pressing tallows so that the stearine will have a titre of 47½, good results in getting the right consistency of tallow oils is possible. Grease stearine will run about 43½ titre. Sometimes very brown or dark-colored grease is treated with an 18° Beaumé caustic soda solution, using one per cent of it for every per cent of acid in the oil. This is done to reduce the acidity, so that the finished oil when sold will meet the requirements of the buyer.

Lard Oil and Greases

Many grades of lard oil are put on the market, depending entirely on the grade of stock used for pressing.

Special Lard Oil.—This is made from pigs' foot stock or from a good grade of white grease chilled to 40° F. and then pressed in a temperature of 38° to 40°. This will yield about 60 per cent oil and the oil will show a cloud test at 30° F. Contains less than 2 per cent free fatty acid.

Extra Winter Strained Oil.—This is made from a good grade of white grease pressed at a temperature of 44° F., and will usually show a cloud test at 32° F. Contains 2 to 4 per cent free fatty acid.

Extra Lard Oil.—This is made from a choice fair grade of grease, pressed around 46° F., and will usually show a cloud test of 35° and a yield of 50 per cent. Contains 5 per cent free fatty acid.

Extra No. 1 Lard Oil.—This is usually made from the poorer greases, and is pressed at a temperature of 46°, and will usually show a cloud test at around 40° F. Contains 7 to 10 per cent free fatty acid.

No. 1 Lard Oil.—This is made from yellow grease, pressed at 52° F., and usually shows a cloud test of 48° to 50°; 15 to 20 per cent free fatty acid.

No. 2 Lard Oil.—This is made from brown grease, pressed at 52° F., and will show a cloud test at around 50° F.; 20 to 25 per cent free fatty acid.

Lard Stearine.—The stearine from the above pressings is usually used in the manufacture of soap.

Neatsfoot Oil

Neatsfoot oil is probably of more interest to the small packer than the other classes of inedible oils. Neatsfoot oil is made by cooking cattle feet first at a temperature of 185° for 8 hours. Other cookings follow, but little tallow is recovered on the second cooking, although the glue liquor gained therefrom is valuable.

The resulting oil is known as neatsfoot stock, and before neatsfoot oil can be produced it must be grained and pressed the same as tallow, except that the graining is done at a lower temperature, depending upon the class of neatsfoot oil which is to be manufactured.

The usual calls are for 20° and 30° neatsfoot, the requirements of which are self-explanatory. In other words, 30° neatsfoot oil is neatsfoot oil which should flow readily at a temperature of 30°. In order to accomplish this the graining should take place at a temperature ranging between 25°

and 30° Fahr., and the pressing should occur in a chill room, in order to prevent the stearine from melting and raising the melting point of the neatsfoot oil. This is something which can be determined very readily by each packer. It is not necessary to bleach the neatsfoot product.

Twenty-Degree Pure Neatsfoot Oil.—This is made from neatsfoot stock, and is pressed at a temperature of 36° F. It will not solidify or even show a cloud test. It has a temperature higher than 18° F. after pressing. This is made from pure neatsfoot stock, just as it comes from the cooking of shin bones, as described elsewhere in this book.

Thirty-Degree Pure Neatsfoot Oil.—This is pressed from the stearine produced in making twenty-degree neatsfoot oil, and is usually pressed at a temperature of 42°. It shows a cloud test at around 27°. The pressed stearine from this is afterward used in tallow oils.

Tallow Oil.—This is usually pressed from a good grade of tallow at 76° F., showing a yield of 65 per cent and a cloud test of 74°. In case it is desired to manufacture acidless tallow oil, containing less than .2 per cent free acids, the oil produced above is treated with caustic soda to have the fatty acids removed.

MANUFACTURE OF GLUE

The manufacture of glue is rather a new development in many packing plants. It is only recently that a few of the small packers have found ways and means of manufacturing this product in a practical way, and it is yet capable of great development.

There are two principal methods of manufacturing glue, one the bleaching of the raw product before cooking and evaporating the liquor; and second, the bleaching of the liquor after the cooking and before evaporation. The first method is that which is used in the larger glue manufactories extensively. The second is the method which can be adopted by the small packer with the addition of considerably less equipment than is required by the first method. For instance, glue liquor vats placed in a rendering plant, together with separate evaporators and the setting aside of a chill room to chill the jelly, and the providing of a forced warm air drying room, is all that is required.

The following raw products are generally used for glue manufacture. After listing them this discussion will follow through the manufacture of glue as it may be found practical in a small or medium-sized packing establishment.

Raw Product.—Knuckles, extract bones, hog snouts, pigs' feet, ear drums, cattle snouts, cattle feet, cattle tails, heads with snouts on, upper part of beef gullet; calf heads, feet and tails; sheep heads and feet; dewclaws, green and cured hide trimmings, horn piths, hog back bones, and neck bones when tanked.

The hog and beef products should be kept separate, because of U. S. Government regulations, of course, and because of the fact that there is a difference between greases and tallows.

Preparation.—The first step, and one which is very important, is the thorough washing in cold water of the raw product in order to remove all blood, dirt, etc. The crushing of the bones, if a crusher is available, is also advisable for two reasons; first, they will cook more readily, and second, less tank room is required. Whether this is done or not, most of the raw products are washed before entering the tank. However, if this is not possible cold water may be added to the tank and a pressure air line placed to the bottom, thereby providing means of agitation. After thorough washing, the water may be drained from the tank, and fresh hot or cold water added. The cooking then follows, with the head of the tank out, using a temperature of about 190°.

Cooking.—Should it be desired to make two grades of glue, a lower cooking temperature may be first employed, the liquor resulting therefrom being kept separate, and a higher temperature used afterward. After the first cooking, of course, the major portion of the grease or tallow is removed, as well as the glue liquor. Then follow four other cookings, in order to recover the sum total of the glue material which is present.

The glue liquor, after being received in the holding vats, is well skimmed to remove any excess oils that may have escaped, and the liquor is then bleached. This is done by forcing sulphur smoke from a small oven under air pressure through perforated coils in the bottom of the glue vats. After 2 to 5 hours, depending upon the nature of the glue liquor, which is governed by the raw product used, a white flaky coagulum will be seen in the glue liquor. It is important to watch this point very closely and stop the blowing at that time, for if the process advances too far the desired result will not be obtained. The glue liquor is then allowed to remain quiescent, so that the coagulum will settle to the bottom. After this is accomplished the liquor may be drawn immediately into the evaporators, care being used to avoid the pulling in also of part of the coagulum.

Evaporating.—The evaporators which are used for liquid "stick" will not do for glue liquor, as it is almost impossible to clean them thoroughly enough for glue purposes. Therefore it is desirable to have a separate set of evaporators.

The density to which the glue liquor is evaporated again depends upon the quality of glue it is desired to make. The recommendation here is that the liquor be evaporated to about 20° Beaumé, and from the last effect of the evaporator it may be drawn into small rectangular pans about 8 inches deep, 6 inches wide and 12 inches long. These pans are immediately placed in a chill room at a temperature of about 35° Fahr., whereupon the substance becomes glue jelly after a thorough chilling for 15 hours or more.

When a sufficient quantity of the glue jelly has accumulated, it may be taken out and loosened from the tins by dipping in warm water and turning them over. Then the glue is sliced thinly with a knife, placed on chicken-wire screen racks, and put into an alleyway through which is forced a current of warm air by means of a large rotary fan working in connection with a bank of steam-coils. Thirteen to 20 hours are required to dry the glue properly. The temperature of the air should be about 90° or 95° Fahr. Overheated air or air very high in humidity will cause the glue to melt and run down through the screens.

After the glue is properly dried it may be either sold as it is or ground and packed in barrels ready for shipment.

The glue should be amber in color and fairly transparent. A glue which is opaque or very cloudy indicates that it has not been properly bleached and that it is therefore not of the same quality as the transparent glue.

Yields.—Whether or not it will pay a packer to manufacture glue depends very largely upon his local conditions and his market outlet. Yields of one-half to three-fourths of a pound per head, bullock basis, are not infrequent. Figuring this against the kill, and the difference in the value of glue and concentrated tankage, taking into consideration the increased cost due to equipment and labor required, are the determining factors.

No mention is made here of processes employed in the larger glue establishments, except to say that some of the manufacturers use pressure tanks in cooking, some steam and some hydraulic pressure, in order to withdraw the maximum amount of glue from the raw product. As these heavy cookings, however, extract also a considerable amount of the mineral element which is undesirable in the glue, our advice is for the small manufacturer to avoid this practice.

BONES, HORNS AND HOOFS

Bones.—The bone room in the packing plant should be light, airy and dry. Most bone rooms are equipped with a set of steam coils covered with a perforated plate or screen, on which may be placed horns, hoofs and other bones which may require artificial heat. Shin bones for manufacturing purposes are cooked in an open vat for 5½ hours at a temperature of about 185° Fahr., and then allowed to cool for about 4 hours in water before being removed. The tallow recovered is a No. 2 neatsfoot stock, or it may be added to prime tallow.

Thigh bones are also handled in the same manner, and are often saved for manufacturing purposes. These bones, after two or three cookings, during which time the glue liquor and grease are thoroughly removed, are placed on racks and air dried at ordinary temperatures, and are then graded into No. 1 and No. 2 grades.

The manufacturing bones are graded usually as follows:

Flat shin bones—
    Heavy flats, 35 lbs. avg. per 100 pieces.
    Light flats, 28 to 35 lbs. per 100 pieces.
    Cull stock under 28 lbs.
Round shin bones—
    No. 1, 50 lbs. avg. per 100 pieces.
    No. 2, 40 lbs. avg. per 100 pieces.
    Culls, under 38 lbs. per 100 pieces.
Thigh bones—
    No. 1, 90 lbs. avg. per 100 pieces.
    No. 2, 70 lbs. avg. per 100 pieces.
    Culls, under 66 lbs. per 100 pieces.

The following is a test on 1,922 lbs. of round shin bones and 1,456 lbs. of flat shin bones. The yield was as follows:

No. 1 round shin bones	327 lbs. or	9.68%
No. 2 round shin bones	815	24.13
No. 1 flat shin bones	195	5.77
No. 2 flat shin bones	640	18.95
Cull rounds ....88 Cull flats 77	165	4.88
Total	2,142 lbs.	

No. 2 neatsfoot oil	303 lbs. or	8.97%
Tallow	44	1.30
Brown grease	45	1.33
Tankage, dry basis	50	1.48
Concentrated tankage from cooking water	88	2.60
Total tankage and oils	2,672 lbs.	
Shrinkage	706	20.91
Total	3,378 lbs.	100.00%

Horns.—Horns are saved and graded also, although there is not much market for horns at the present time (1921) except for horn tips of No. 1 quality, which in some plants are sawed off and saved.

The following is the usual grading of horns:
Steer horns No. 1, 100 lbs. avg. per 100 pieces.
Steer horns No. 2, range from 65 to 80 lbs., average 70 lbs. per 100 pieces.
Steer and cow horns No. 3, range 30 to 65 lbs. per 100 pieces.
Bull, crab and cull horns are graded No. 4.

Hoofs.—White and striped hoofs are also sometimes saved and are cared for in the bone room. No. 1 quality of whites generally weigh 22 lbs. per 100 pieces; No. 2 quality of whites range from 16 to 20 lbs. and must average 18 lbs. for 100 pieces. Striped hoofs should average about the same weight as number ones.

All hoofs and pig toes not utilized for manufacturing are calcined in a steam drier, which requires about 8 hours, and are then placed in a dry corner of the bone room, from where they may be shipped, or if the packer is manufacturing fertilizer they may be ground and used for this purpose. Dry steamed bone also may well be stored in the bone room, where it is dry, waiting the time that a sufficient quantity may be accumulated so that it may be ground and prepared for the market.

CATCH BASINS

Catch basins or fat traps are frequently sadly neglected, or are often not present in the packing plant to any extent whatever, except perhaps for one final basin. This is a serious mistake on the part of any packer.

As has been mentioned under the article on Inedible Tankhouse, it is important to care for each product at as near its source of production as possible, and to keep it from mixing with other products of lower quality, thereby reducing the grade and losing the packer money. For instance, adjacent to a gut-cleaning table there should be a catch basin to catch the water which drains from that table, thereby preventing the fat particles from running away with the water, going in with the general run of waste waters leaving the plant.

A type of basin should be selected which will perform as follows: First, slow down the rate of flow of the waste water to such an extent as to allow the fat sufficient time to rise to the surface; second, the basin should be made so that it can be cleaned readily and the manure and dirt at the bottom removed. The old type of narrow basins with under and overflows, which had very much the appearance of a mill race, are not efficient in producing the desired results.

One of the most important points, as stated, is to see that the water travels at a slow rate of speed. Catch basins should be skimmed constantly, and the material put into the tank house in as fresh a condition as possible. This will prevent high acid in the resulting greases and tallows, and will result in large savings thereby.

Suspended or settled solid matter in catch basins easily ferments and increases the percentage of free fatty acids in the collected fats, in addition to affecting the color or bleach of the rendered product. Certain sanitary catch basins are designed so as to carry the heavy solid matter out of the basin continually, besides producing a big saving of space.

HANDLING HOG HAIR

Hog hair is a by-product which, when handled properly, is a source of considerable revenue. The amount in dollars and cents which can be recovered per hog depends upon the condition of the hair market, the season of the year and the number of hogs killed. The average hog yields one-half to three-quarters of a pound of hair, and it can easily be figured out what the gross revenue will be by calculating the number of hogs killed daily.

Field curing or curing in the open air has been abandoned by practically all packers, as it has been found not alone unsanitary but also impractical, and the process today consists of curing, drying and curling.

The smaller packer, however, does not produce hair in large enough quantities to justify expensive methods, and usually bales his hair in a regular baling press after it has been washed and dried in a hair dryer.

The hair as it is received from the scraping machine should be placed in a vat of cold water. In determining the size of the vat, one should use a basis of 45 cubic feet of vat per 100 hogs. The hair should be left in this vat for one day and the water kept cool. However, if the vat is too large, the hair may be held over for another day.

After the hair has been thoroughly cooled a solution of 3 to 8 oz. of soda lye is added for every 100 hogs slaughtered. The strength of the solution depends a great deal upon the condition of the hair. The more scurf that has to be removed, the stronger the solution should be, and the cleaner the hair, the weaker the solution should be.

The hair lying in water is now agitated continuously; steam is turned on until the water has boiled. As soon as it has reached the boiling point, the steam is reduced so that the water will merely simmer, and this is kept up for four to 10 hours, or until such time as the scurf will easily slip off the hair; in other words, until such time as it becomes slimy. Do not use more caustic soda than is absolutely necessary, as it has an effect upon the gloss of the hair when it is added.

The hair is now passed from the cooking vat to a hair-picking machine, and while the machine is in operation, it is constantly being fed with boiling hot water and steam. Sometimes it is necessary to put the hair through the picking machine twice or three times before it is thoroughly cleaned.

Now the hair is ready for the dryer. There are several hair-drying machines on the market of various types, but in the smaller packinghouses rooms are usually provided which are equipped with either steam coils or have forced hot air circulation for the drying of the hair.

It usually requires from three to six hours to dry a batch of hair. It is very essential that all the machines be kept scrupulously clean, and especially the drying rooms, which should be cleaned after every operation, as otherwise the odor in this room will be very objectionable.

If it requires sterilization to drive the odor from this room, a hydrochloric acid treatment is used in order to deodorize it.

After the hair has been thoroughly dried it is usually pressed in the ordinary steel press and is ready for the market, unless the hair is to be dried further. However, a drying process is used in very few plants.

Tanking of Hog Hair

When the market for hog hair does not justify its treatment as described, it is usually mixed with tankage.

There are two ways of mixing hog hair with tankage. One is to cook it with the offal, and the other is to cook the offal and the hog hair separately, and then mix the hog hair with the tankage and let it go through the dryer.

It is not good practice to mix raw hog hair with the tankage, however. The second method, that of treating separately, is the one to use. By this method a more satisfactory product can be turned out, inasmuch as the average run of hog hair, when placed with good offal, will have a tendency to darken the grease. So, in case it is the desire to turn out a grease of the lightest possible coloring, it is advisable to cook the hog hair separately in a plain tank, and thereafter mix it with the tankage and put it through the dryer.

If advantageous to sell hog hair for the price of tankage, the operator will of course only secure an additional small percentage of ammonia units; whereas, if he handles the hog hair separately, it is very probable that he can secure a higher price for it.

There is no specific rule as to the amount of hog hair to mix with tankage, but the hair resulting from the average killing per day can be mixed with the day's killing offal.

COST OF AND RETURN ON BY-PRODUCTS

The detail to which any manufacturer may go in developing by-product manufacture depends entirely upon his local conditions. The various factors which enter into this are the volume available, the labor and the market.

Each packer should make a thorough investigation of each and every avenue along this line, and determine for himself by a series of complete tests whether or not the ramifications of his manufacture should be extended.

It has been shown that the percentage of total return which packers receive from by-products is much smaller than that generally conceived in the public mind. For instance, of the total return on the various classes of live stock, the percentage of money received from by-products ranges from .4 to 6.9. The following table will indicate the method of this calculation:

	% from Meat	% from By-Product	% from Hide or Pelt
Cattle	80.5	6.9	12.6
*Hog	99.6	.4	Sold with meat
*Sheep	69.4	6.1	24.5
Calf	94.6	5.4	Sold with carcass

*In this calculation lard is estimated as meat, while tallow is figured as by-product.

Chapter V—MISCELLANEOUS

SAUSAGE MANUFACTURE

Sausage is one of the most popular forms of meat food among a large percentage of the consuming public. Especially is this true among certain nationalities and in certain communities.

The art of sausage making has been developed to a high degree through long experience and careful tests. Good sausage can be made only from good materials, and the selection of the raw product is therefore very important. Plenty of cooler space, proper chilling, cleanly handling and close supervision, the proper weighing of both meats and spices, throughout the entire process are all-essential. It is especially important that all machinery be thoroughly cleaned and oiled after the completion of each day's work. Keeping these points in mind, the sausage manufacturer may utilize and find a market for hearts, livers, trimmings, cheek meat, etc., that otherwise would probably find their way to the tank house.

The seasoning of sausage is naturally one of the most important things to consider. The spices should be pure, properly ground, and well balanced. The formulas here given will afford sufficient idea of the use of spices in the various classes of sausage.

There are two general classes of sausage—fresh sausage and summer or dry sausage. Fresh sausage depends entirely upon the seasoning, cooking and smoking for its flavor and palatability. Summer sausage is subjected to a further process; namely, that of curing at specified temperatures and humidity in a curing room. Mold has a part in the curing of certain summer sausages and their flavoring, similar to the action of mold in the curing and flavoring of Camembert and Rocquefort cheese.

Sausage Yields.—Sausage yields are enhanced by the addition of moisture or cereal or both. Potato flour, bread crumbs, and special sausage flour are among the cereals used. Under the United States government regulations the use of cereal in sausage is limited to 2 per cent, and most states and local regulations conform to this rule. Packages must be branded accordingly, the percentage of cereal being stated on the label.

Some of the fundamental principles in making fresh sausage can never be disregarded without poor results. For instance, sausage makers figure on using not more than 30 per cent of product such as hearts, tripe, lips, and items of that kind, which are known to have poor binding qualities.

Boneless beef, especially boneless chucks and beef cheek meat, are the two products that have the greatest binding qualities from the beef.

Large pieces of lean trimmings from the pork, especially from the shoulder, and pork cheek meat have the best binding qualities of the pork products.

Formerly it was the prevailing custom in the sausage trade to bone out hot bulls for sausage meat, because this meat has a great absorption power for moisture, and because of the fact that the binding qualities of fresh beef are greater than those of cold beef. In fact, each day beef and pork trimmings are held before using in sausage reduces their binding qualities.

The Emulsion Method.—A substitute for the method of boning out hot bulls has been adopted in many factories in the shape of a specially prepared emulsion. The usual method is to put the meats through the Enterprise grinder and chop and add the amount of moisture that it will take up. It is the usual custom to use about 60 per cent of ground ice in with the meat at the time of grinding. The percentage of ice will vary, because cow meat will take up 60 per cent, shank meat will take up 70 per cent, and bull meat as high as 90 per cent. The usual method is to put the beef in the silent cutter for ten minutes, working the ice in gradually and adding a curing formula of about the following basis per thousand pounds of meat: 35 lbs. salt, 15 lbs. sugar, 30 ozs. saltpetre.

It is quite important to see that the emulsion is well spread in the pans, so as to avoid shortening. After twelve hours it will have the consistency of dough, and will show an increased yield over meats direct from the cutting block.

It is quite necessary, however, to watch the smoking of this sausage, which has to be done quite carefully, and avoid having the fire too hot. A temperature of about 120 degrees Fahr. is most satisfactory.

Moisture in Meats.—Following are tables showing the moisture capacity of various meats for sausage making:

Per Cent of Moisture

Pork Meats

Back bone trimmings	58.00
Belly trimmings	55.90
Neck bone trimmings	50.87
Shoulder trimmings	48.20
Ham trimmings	41.29
Back fat trimmings	38.84
Shoulder fat	11.26
Ham fat	10.44
Neck fat	6.71
Pork cheeks	57.69
Pork hearts	74.99
Skinned ham fat	8.89

Sheep Meats

Sheep hearts	71.55
Mutton cheeks	65.09

Beef Meats

Boneless chucks	70.76
Regular beef trimmings	68.13
Hanging beef tenderloins	67.76
Beef cheeks	66.50
Flank beef trimmings	54.00

SAUSAGE FORMULAS

Sausage Formulas.—There are few standard formulas used, except in the largest packing plants and sausage factories. Every superintendent or sausage maker as a rule has his own formulas or recipes for different kinds of sausage.

Much depends upon the kind of meat used, the quality, and strength of spices; and furthermore, sausage recipes must be in line with the demand of the trade in certain localities. What will sell well in one section or climate will not meet the taste or demands of another.

To cover the field of sausage recipes a volume in itself would be required. Each nationality has its own style of sausage or bologna. A selected list of formulas is here given which it is believed covers the field of sausage making in a representative way.

There are four kinds of pork trimmings recognized in general sausage practice: 1, Regular pork trimmings; 2, Fat-back pork trimmings; 3, Extra lean trimmings; 4, Mortadella meat (extra lean pieces of meat cut from hams, etc.). The packer terms No. 1 a "B" trimming; No. 2, "C" trimming; No. 3 an "A" trimming, and No. 4 a (diamond X) trimming.

Where "sweet pickle meats" are called for in formulas, they may include hearts, weasand meats, ox lips, giblets, etc.

For ready reference these formulas are divided into four groups: Fresh Sausage, Smoked Sausage, Cooked Sausage and Summer Sausage.

Fresh Sausage Formulas

Country Style Pork Sausage:

100	lb. regular pork trimmings	2½	lb. sugar
2½	lb. salt	⅝	oz. savory
2	oz. sage	7	oz. white pepper.

(For Southern trade, 4¼ oz. roasted Japanese peppers.)

Grind through ¼-inch plate of Enterprise grinder, put in mixer, add seasoning, and if under government inspection only 3 lbs. water to 100 lbs. meat is permitted. Mix for 2 min. and stuff in hog casings. Yield in this test, 103 lbs.

No. 1 Pork Sausage or Farm Sausage:

50	lb. meat from cala (pork shoulder) butts	2½	oz. sage
		¾	oz. savory
50	lb. ham fat	¼	oz. cardamom
30	oz. salt	1	oz. ginger
6	oz. white pepper	¼	oz. marjoram.

Chop under the rocker knives for 10 minutes, then throw on the spice and chop for ten minutes, constantly turning over the meat with a paddle. Usually stuffed in sheep or hog casings.

Fresh Sausage

50 lb. fresh reg. pork trimmings.   1 oz. mace.
25 lb. fresh tripe.   4 oz. sage.
25 lb. fresh pork or head beef meat.   7 lb. flour.
2 lb. 8 oz. salt.   25 lb. water.
8 oz. white pepper.
Stuff in No. 1, wide or stump hog casings.

Pork Sausage

    100 lb. fresh reg. pork trimmings.   8 oz. white pepper.
    2 lb. 10 oz. salt.   1 oz. mace.
    4 oz. sage.   10 lb. water.

    Stuff in No. 1 hog casings, and double link 3 in. long.

Fresh Link Sausage

    55 lb. fresh reg. pork trimmings.   4 oz. sage.
    25 lb. fresh tripe.   1 oz. mace.
    20 lb. No. 1 beef trimmings.   7 lb. flour.
    2 lb. 8 oz. salt.   25 lb. water.
    8 oz. white pepper.

    Stuff in stump or No. 1 hog casings and double link 3 in. long.

Bockwurst

    20 lb. fresh knuckle meats.
    30 lb. fresh reg. pork trimmings.
    1 lb. 6 oz. salt.   1 doz. eggs.
    5 oz. white pepper.   1 lb. flour.
    8 oz. chives or leek.   14 lb. water.

    Grind through $\frac{1}{8}$-in. plate of grinder, put in silent cutter, add spices, eggs, flour and water, and chop for 5 min. Stuff in medium or wide sheep casings and link 5 in. long.

Smoked Sausage Formulas

Good Frankfort Style Sausage

    30 lb. No. 1 beef trimmings.   2 lb. 8 oz. salt.
    10 lb. pork hearts   4 oz. onions.
    25 lb. pork cheeks   4 oz. mace.
    30 lb. pork trimmings   4 oz. white pepper.
    15 lb. tripe.   1 oz. red pepper.
    ———   3 oz. saltpetre.
    110 lbs.   4 oz. sugar.
        5 lb. flour.
        38 lb. water and ice.

    If meats cured ahead, figure salt, sugar and saltpetre accordingly. Stuff in extra wide sheep casings and link 4 in. wide.

Frankfurt Sausage, no cereal: *No. 1 Frankfurt Sausage*, no cereal:

58 lb. bologna beef   60 lb. No. 1 pork trimmings
30 lb. sweet pickle meats   40 lb. boned beef
30 lb. fat pork trimmings   24 oz. fine salt
32 lb. tripe   10 oz. white pepper
2 lb. salt   2 oz. coriander seed
8 oz. white pepper   2 oz. ground ginger
1½ oz. mace   Yield, 115 lbs.
2 pieces of garlic
Yield, 165 lb.

    May be stuffed in sheep, hog or narrow beef casings, smoked lightly, and immersed in boiling water for 30 minutes.

SAUSAGE FORMULAS

Medium Frankfort Style Sausage

- 40 lb. pork cheeks.
- 26 lb. pork trimmings.
- 26 lb. S. P. pork trimmings.
- 20 lb. hearts.
- 20 lb. tripe.
- 68 lb. fresh beef trimmings.
- 8 oz. white pepper.
- 2 oz. red pepper.
- 4 oz. mace.
- 10 lb. flour.
- 8 oz. sugar.
- 2 lb. salt.
- 6 oz. saltpetre.
- 50 lb. water.

If meats cured ahead, figure salt, sugar and saltpetre accordingly. Stuff in wide sheep casings and link 5 in. long.

Vienna Sausage:

- 55 lb. pork trims, 40 per cent fat
- 45 lb. tender beef trims
- 55 oz. salt
- 4 oz. sugar
- 2½ oz. saltpeter
- 5 oz. white pepper
- 3 oz. coriander seed
- 1 oz. mace
- 5 pieces garlic.

Add part of the salt and all of the saltpeter and sugar in a vessel or mixer when partly ground. Add balance of seasoning when ground. Stuff in sheep casings. Smoke ½ hour at 125 degrees Fahrenheit, then scald in hot water and hang to dry. Yield, 101 lb.

Smoked Link Sausage in Oil

- 100 lb. D. C. pork head meat.
- 150 lb. D. C. beef head meat.
- 100 lb. D. C. hearts or caps.
- 75 lb. S. P. pork trimmings.
- 75 lb. tripe.
- 25 lb. flour.
- 2 lb. 10 oz. white pepper.
- 5 oz. red pepper.
- 10 oz. mace.
- 10 oz. allspice.

Stuff in No. 1 stump or wide hog casings.

Frankfort Style Sausage in Oil

- 80 lb. D. C. pork cheek meat.
- 45 lb. D. C. hearts.
- 85 lb. D. C. beef cheeks.
- 45 lb. S. P. pork trimmings.
- 45 lb. tripe.
- 15 lb. flour.
- 1 lb. 8 oz. white pepper.
- 3 oz. red pepper.
- 6 oz. mace.
- 6 oz. coriander.

Use no water. Stuff in wide sheep casings and link 5 in. long.

Bologna Style Sausage, no cereal:

- 65 lb. bologna beef
- 30 lb. sweet pickle meats
- 20 lb. tripe
- 35 lb. fat pork trimmings
- 45 oz. fine salt
- 3 oz. saltpetre
- 15 oz. black pepper
- 9 oz. ground cloves
- 3 oz. coriander seed.

Stuffed in beef bungs or middles; yield 168 lb.

Detailed operating instructions are not given with each formula, it being taken for granted that the sausage-maker knows the fundamentals of his business. Besides, every sausage-maker has his own methods, and usually prefers to follow them.

Those who desire details not given concerning any formula may obtain them upon application to the Editor of THE NATIONAL PROVISIONER, Chicago.

Long, Large and Round Bologna Style Sausage

- 80 lb. beef cheek meat.
- 50 lb. beef tongue trimmings.
- 25 lb. tripe.
- 25 lb. hearts
- 20 lb. S. P. pork trimmings.
- 10 lb. ham fat.
- 12 lb. flour.
- 1 lb. 8 oz. black pepper.
- 5 oz. coriander.
- 4 oz. onions.
- 6 oz. allspice.
- 6 lb. salt.
- 8 oz. sugar.
- 6 oz. saltpetre.

If meats cured ahead, figure salt, sugar and saltpetre accordingly.

Knoblaugh Style Sausage
Bologna Style Sausage

- 30 lb. pork cheeks.
- 15 lb. pork trimmings.
- 25 lb. beef cheeks.
- 10 lb. weasand meat.
- 10 lb. S. P. pork trimmings.
- 10 lb. ham fat.
- 100 lbs.
- 7 lb. flour.
- 8 oz. white pepper.
- 2 oz. coriander.
- 2 oz. onions.
- 30 lb. water and ice.
- 3 lb. salt.
- 4 oz. sugar.
- 3 oz. saltpetre.

If meats cured ahead, figure salt, saltpetre and sugar accordingly. Stuff bologna in 1 lb. beef weasands. When making Knoblaugh, add 1½ oz. garlic and stuff in beef rounds.

Bologna Style Sausage in Bags
(Paraffined)

- 100 lb. beef cheeks.
- 30 lb. hearts.
- 50 lb. tongue trimmings.
- 20 lb. tripe.
- 10 lb. fat.
- 8 lb. flour.
- 3 oz. allspice.
- 5 oz. coriander.
- 1½ lb. black pepper.
- 6 lb. salt.
- 8 oz. sugar.
- 6 oz. saltpetre.

Use no water. If meats are cured ahead, figure salt, saltpetre and sugar accordingly. Stuff in cloth bags to hold about 6 lb., and hold in cooler over night; then cook for 4 hours at 150-155 degs. Shower with cold water for 3 min., then hang 1 hour or more to allow heat to escape. Then paraffine.

Bologna Style Sausage in Oil

- 210 lb. D. C. beef head meat.
- 90 lb. D. C. hearts or caps.
- 120 lb. D. C. ox lips.
- 120 lb. D. C. tongue trimmings.
- 60 lb. tripe.
- 30 lb. 12 oz. flour.
- 3 lb. 2 oz. black pepper.
- 1 lb. 12 oz. coriander.
- Allspice.

Veal Sausage

- 65 lb. pork cheeks.
- 25 lb. lean pork trimmings.
- 10 lb. fresh tripe.
- 6 lb. flour.
- 10 oz. white pepper.
- 2 oz. mace.
- 25 lb. water and ice.
- 3 lb. salt.
- 3 oz. saltpetre.
- 4 oz. sugar.

SAUSAGE FORMULAS

Polish Sausage:

30 lb. fat pork trimmings	1½ lb. salt
40 lb. sweet pickle meats	1½ oz. garlic
20 lb. tripe	1 oz. marjoram
8 oz. mixed pepper	3 lb. sugar.
10 lb. beef	

The tripe and sweet pickle meats are made fine in the silent cutter, while the pork trimmings and beef are put through a ⅜ in. plate. Mix, add cereal, stuff in hog casings 12 in. long and cook 30 minutes at 155°. Yield, 98 lb.

Polish Style Sausage

50 lb. beef cheeks.	1 lb. 8 oz. black pepper.
100 lb. beef trimmings.	15 lb. flour.
30 lb. pork hearts.	6 oz. garlic.
30 lb. tripe.	90 lb. water and ice.
45 lb. regular pork trimmings.	9 lb. salt.
45 lb. sweet pickle trimmings	6 oz. saltpetre.
6 oz. marjoram.	12 oz. sugar.

If meats cured ahead, figure salt, sugar and saltpetre accordingly. Stuff in wide stump or No. 1 hog casings and link 8 in. long.

Pressed Luncheon Specialty
  (*New England Pressed Ham*)

400 lb. dry cured extra lean pork trimmings.
80 lb. dry cured extra lean beef trimmings.

Grind beef trimmings through 7/64 in. plate of the grinder, then chop three minutes in silent cutter, adding 5 lb. of water to 100 lb. of meat.

Minced Luncheon Specialty
  (*In beef bladders*)

50 lb. lean pork trimmings.	50 lb. pork head meat.
100 lb. beef trimmings.	20 lb. water.

Use 3 oz. white pepper to 100 lb. meat. Cure meats ahead. Bladders to be selected to make weight from 5 to 8 lb.

Delicatessen Luncheon Specialty

25 lb. fresh lean pork head meat.	2 oz. sugar.
10 lb. fresh backfat trimmings.	1 oz. saltpetre.
10 lb. tripe.	1 lb. 6 oz. salt.
5 lb. hog livers.	½ doz. eggs.
2 lb. cracker meal.	1 gal. milk.
4 oz. white pepper.	1 lb. butter.
1 oz. mace.	6 oz. onions.

Chop fine in silent cutter and bake in oven in 6 lb. tins.

Southern Style Luncheon Specialty
  (*In cloth bags*)

170 lb. dry cured beef head meat.
400 lb. dry cured pork cheek meat.

Grind beef head meat through 7/64 in. plate of grinder, then chop fine in silent cutter, adding 5 lb. of water to 100 lb. meat. Grind pork cheeks through 1 in. plate.

Berlin Style Luncheon Specialty

    200 lb. dry cured lean pork trimmings.
    150 lb. dry cured beef trimmings.
    200 lb. dry cured pork cheek meat.

Grind beef trimmings through 7/64 in. plate of grinder, then chop 3 minutes in silent cutter, adding 5 lb. of water to 100 lb. meat. Grind pork cheeks through 1 in. plate.

Prepared Luncheon Specialty
(In cloth bags)

    200 lb. dry cured lean pork trimmings.
    150 lb. dry cured beef trimmings.
    200 lb. dry cured pork cheek meat.

Grind beef trimmings through 7/64 in. plate of grinder, then chop 3 minutes in silent cutter, adding 5 lb. of water to 100 lb. of meat. Grind pork cheek meat through 1-in. plate of Enterprise hasher.

Cooked Sausage Formulas

Liver Sausage

500 lb. fresh hog livers.	10 lb. onions.
100 lb. fresh hog rinds.	6 lb. white pepper.
350 lb. fresh tripe.	24 lb. salt.
70 lb. fat.	40 lb. flour.
250 lb. snouts.	3 lb. marjoram.
90 lb. jelly.	1 lb. thyme.

Liver Pudding:

25 lb. hog livers	10 lb. of cube fat
30 lb. fat pork trimmings	2½ lb. salt
10 lb. pig skins	8 oz. black pepper
20 lb. tripe	3 oz. marjoram
5 lb. hog giblets	1 lb. onions.

The livers should be scalded and the pig skins cooked; then put all the meat through a ⅛ in. comb on the grinder, after which put all in a mixer and add cube fat and spices, and as much moisture as is necessary to make stuffing easy. This pudding is usually stuffed in hog bungs, then cooked 50 minutes at 170°. Yield, 105 lb.

Head Cheese

250 lb. ox lips.	2 lb. white pepper.
150 lb. pickled pig ears.	15 lb. flour.
300 lb. pickled pork snouts.	6 oz. coriander.
100 lb. pickled pork rinds.	4 oz. thyme.
90 lb. jelly.	6 oz. caraway, if required.
6 lb. salt.	6 lb. onions, if required.

Head Cheese:

35 lb. pork ears
20 lb. pork lips and snouts
10 lb. pork skins
20 lb. beef lips

15 lb. beef tripe
8 oz. black pepper
3 oz. caraway.

The ears, pork lips and snouts should be cooked 2 hours, the pork skins 1 hour and the beef tripe 4 hours, if green; if prepared, no cooking needed. The snouts, beef and pork lips and beef tripe should be put through the head-cheese chopper twice, while the ears and skins should be put through three times. Then add the spice and mix well. Stuff in hog stomachs and cook 2 hours at 175° Fahr.

Suelze

65 lb. pickled pig snouts.
25 lb. D. C. or pickled pork cheeks.
10 lb. tripe.

12 lb. jelly.
6 lb. 45-grain vinegar.

Blood Pudding:

23 lb. beef blood
34 lb. cooked tongue trims
18 lb. tripe
25 lb. cube fat

8 oz. black pepper
2 lb. salt
3 oz. allspice
2 oz. marjoram

Cook the tripe and put through $\frac{1}{8}$ in. comb grinder, also cook the cube fat, put the cooked tongue trimmings through a $\frac{3}{4}$ in. comb grinder, then mix all the meats with the blood and stuff in beef hungs. Cook at 200° for 3 hours. Yield, 92 lb.

Thuringer Style Blood Pudding:

20 lbs. beef blood
15 lb. sweet pickle snout meat
15 lb. pig skins
25 lb. cube fat
25 lb. sweet pickle hog lips

1 lb. onions
3 oz. allspice
7 oz. black pepper
3½ oz. marjoram.

Cook cube fat and cut in cubes. All other meats should be put through head cheese grinder three times except skins, which should go through $\frac{1}{8}$ in. plate on grinder. Then mix all meat with the blood and spices and stuff in hog middles. Cook 2 hours at 200° Fahr. Yield, 91 lb.

Blood Head Cheese

50 lb. pickled pork ears.
50 lb. pickled pork snouts.
30 lb. pickled pork rinds.
50 lb. pickled hog lips.
35 lb. blood.

3 lb. onions.
4 lb. salt.
1 lb. black pepper.
4 oz. allspice.

Tongue Sausage

200 lb. S. P. hog tongue.
200 lb. S. P. ox lips.
200 lb. S. P. hog snouts.
75 lb. hog rinds.
60 lb. cut fat.

150 lb. blood.
8 lb. salt.
3 lb. black pepper.
5 oz. allspice.
3 oz. ground cloves.

Blood Sausage

- 50 lb. hog rind, pickled.
- 40 lb. pickled ham fat.
- 140 lb. pickled snouts.
- 75 lb. blood.
- 50 lb. pickled ox lips.
- 50 lb. fresh tripe.
- 5 lb. salt.
- 5 lb. flour.
- 1½ lb. black pepper.
- 4 oz. allspice.
- 2 oz. ground cloves.

Veal Loaf

- 75 lbs. boiled veal.
- 10 lbs. boiled ham.
- 20 lbs. bread crumbs.
- ½ lb. salt.
- ½ pt. onion juice.
- 6 doz. eggs.
- ¼ lb. black pepper.
- 2 oz. sage.
- 3 oz. cloves.
- 2 oz. allspice.

Chop the veal and ham fine, beat the eggs, and add the onion juice, then mix the entire contents together and put in an oven and bake until brown.

Mince Meat

- 800 lb. green apples.
- 100 lb. dried apples.
- 98 gallons sweet cider.
- 12 lb. citron.
- 175 lb. currants.
- 270 lb. seeded raisins.
- 550 lb. brown sugar.
- 10 lb. beef suet.
- 75 lb. cooked meat.
- 16 lb. mixed spices.
- 2½ lb. sodium benzoate.

Dry or Summer Sausage

Materials.—Summer sausage—or dry sausage, as it is more accurately designated—requires the very highest quality of meat products, and following are points with reference to meat qualities which should receive attention:

First, it is not safe to use pork or beef that has been cut more than 72 hours. The sooner you use the trimmings after cutting, the better the binding qualities. Therefore it is rather difficult to use shipped trimmings in manufacturing summer sausage.

Second, it is quite necessary to watch the proportion of fat pork to the proportion of beef, as too much fat pork will seriously affect the binding quality of the sausage.

The general gradings of the products going into summer sausage, from the standpoint of binding qualities, are as follows:

Beef.—No. 1. Mixed beef trimmings, shank meat, boneless chucks. No. 2. Weasand meat, hanging beef tenderloins, beef cheeks. No. 3. Number two beef cheeks, sheep hearts.

Pork.—No. 1. Extra lean large pieces pork trimmings; butts. No. 2. Regular pork trimmings, small pieces; head meat.

The following kinds of meat are not used for the best grades of summer sausage: Beef hearts, hog hearts, tripe and ox lips.

Smoking Summer Sausage.—In starting the fire in the smoke house, use as little wood as possible, say one stick of ash cord-wood, with only enough fire to keep the sawdust smoldering and not blazing. Keep adding sawdust until you have sufficient fire to scatter it over the entire bottom of the smoke house, keeping the sawdust ignited only from the coal of the wood with which you first started the fire, which generally lasts through the entire 24 or 48 hours. If your smoke houses are naturally cold, it may be necessary for you to keep more fire than has been mentioned, in order to keep to temperature up to 70°.

Smoking summer sausage requires the greatest possible care. If the temperature is allowed to get too high and remain any length of time, your sausage will sour. If the fire is kept too low and the smoke is too dense, the sausage will have a smoky ring, the same as if it had not been dried properly before being put into smoke.

Drying Summer Sausage.—After the sausage is smoked it is taken to the drying room, which should be held at a temperature of 46° to 53°, the proper temperature being 48°, if it can be obtained. The drying room must be fitted with steam coils running beneath the sausage and around the sides of the rooms, and must also be supplied with plenty of windows, for at all times the windows must be kept open a little to allow fresh air to enter, no matter how cold the temperature. If the weather is damp, the window nearest the top of the room should be opened a little. If the room is supplied with fans, they should be kept going and the windows closed, steam to be turned on to dry the atmosphere, providing the weather is not too warm and the room can be kept as low as 53°.

Summer sausage in hog bungs can stand more draught or air than summer sausage in beef casings; consequently, beef casings are hung nearer the center of the room, where they will get plenty of fresh air, but no draught.

By modern mechanical methods of air conditioning the sausage-maker can turn out a product of good color, relatively free from mold. Goods made under these conditions can be shipped direct from the dry room, and do not have to be washed. This apparatus is somewhat expensive, but gets the best results.

Summer sausage is usually sold in three different weights:

First. New sausage between 10 and 25 days after smoke.

Second. Medium dry, from 30 to 60 days.

Third. Dry, 90 days and over.

In case summer sausage is not sold in a reasonable length of time, it may be packed in boxes and carried in a temperature of 36° above zero.

Before shipping, however, it is advisable to take it out and wipe it well before repacking.

Summer Sausage Formulas.—Following are formulas for dry sausage which have been tried out with success:

Cervelat

    40 lb. beef chucks, closely trimmed.
    90 lb. pork trimmings, closely trimmed.
    20 lb. shoulder fat.
    5 lb. salt.
    2 oz. whole pepper.
    12 oz. white pepper.
    3 oz. saltpetre.

The beef should be ground through an Enterprise 7/64 in. plate, after which it is placed on a rocker with the fat and seasoning and rocked for about five minutes. Then the pork trimmings and the fat are added and the whole rocked from 25 to 30 minutes, according to the speed of the machine, which should be from 52 to 54 strokes a minute. After the meat has been rocked, it is taken to a cooler, where the temperature is not lower than 38° or higher than 40°, and spread upon benches for this purpose, about ten or twelve inches thick. There it is allowed to remain for three days, after which it is stuffed by hand machines into hog bungs or beef middles, as required.

The sausage is then taken to a room where there is plenty of air but no draughts and allowed to hang for two or three days, according to the weather, at a temperature of 50°. If the weather is damp, great care must be taken that the sausage does not slime, and it is sometimes necessary to keep the temperature up to 55°, in order to keep the room as free from dampness as possible. If the sausage commences to slime, there is great danger of its turning sour or becoming hollow in the center.

In this case it is necessary for the sausage to be put into smoke sooner than otherwise, as the smoke of course stops the slime. But if it is put into smoke too soon the sausage will have a dark ring, and this is almost as serious as if the sausage had turned sour. Again, if the heat is too severe on the sausage, or the temperature of the smoke houses is not watched and is allowed to get too high, the sausage will sour.

When the sausage is ready for smoke, under favorable circumstances, it should be hung in a smoke house where the temperature reaches 70° or 72°, at which point it must be kept through the entire process of smoking, which consumes about twenty-four hours for beef middles and 48 hours for hog bungs, according to the weather.

Holsteiner

50 lb. shank meat.
10 lb. fresh beef cheek meat.
30 lb. fresh pork trimmings.
10 lb. pickled pork trimmings.
30 lb. fresh pork cheek meat.
20 lb. shoulder fat.
Same seasoning as Cervelat.

Shank meat, beef cheek meat and pork meat are ground through the Enterprise 7/64 in. plate. Pickled pork trimmings are ground through the Enterprise 7/64 in. plate. Shoulder fat is cut into small pieces, but not shaved. Ground material and shoulder fat are put on the block with the seasoning, and chopped five or ten minutes, when the pork trimmings are added, the whole being chopped 15 or 20 minutes. This meat is very coarse, and after chopping should be mixed in a mixer for three or four minutes, when it is taken to the cooler and handled the same as Cervelat, after which it is stuffed by hand stuffers into large beef rounds.

It is then allowed to hang in the same room where green Cervelat is hung, and about the same length of time, the same precaution being taken with this as with Cervelat as to slime, etc. This sausage is to be handled very carefully in smoke, as too much heat wrinkles it, and the object of smoking is to impede drying and prevent this sliming, which is done mostly when the weather becomes warm.

A great deal of Holsteiner is allowed to dry naturally without smoking, and where you have plenty of room it is better for the sausage to dry naturally than to smoke it, especially in the winter time. However, in damp weather, and through the summer months, it is always advisable to smoke it a little.

This sausage can be made throughout the summer, providing drying rooms can be regulated so as to keep cool and dry.

B. C. Salami.—This sausage is identically the same as Holsteiner, handled the same in every respect, and stuffed in beef middles, 11 inches in length, which give it its name, Beef Casing Salami. The same care is necessary with this sausage as with Holsteiner from the time it leaves the block until it is ready for shipment.

Swedish Mettwurst

- 125 lb. fresh boneless chucks.
- 25 lb. fresh pork trimmings.
- 20 lb. shoulder fat.
- 4 lb. salt.
- 12 oz. white pepper.
- 6 oz. coriander.
- 3 oz. saltpetre.

Chopped a little finer than Holsteiner or B. C. Salami, but not as fine as Cervelat. Chucks are trimmed and salted on a bench over night with about 3 pounds of salt to 100 pounds of meat. Handled in other respects the same as Cervelat. Stuffed in beef middles; and instead of tying the end, use a short wooden skewer.

Immediately after the sausage is stuffed it is put into vats holding about 800 pounds of sausage, and covered with plain pickle, about 65° strength. Allow to remain in pickle three days, then hang up in same room with green sausage. Allow to hang until moderately dry, which will take from two to three days. Put into smoke and handle identically the same as Cervelat in beef middles.

D'Arles Summer Sausage

- 75 lb. lean beef trimmings.
- 75 lb. lean pork trimmings.
- 5 lb. salt.
- 2 oz. saltpetre.
- 6 oz. sugar.
- 6 oz. crushed white pepper.
- 4 oz. whole white pepper.
- 2 oz. garlic.

Run beef once through $\frac{1}{8}$ in. plate on grinder; then put through rocker with pork trimmings for four minutes, the rocker to make 54 revolutions per minute.

H. C. and B. C. Cervelat

- 65 lb. pork trimmings.
- 15 lb. pork cheek meat.
- 70 lb. beef trimmings.
- ———
- 150 lb.
- 5 lb. salt.
- 3 lb. flour.
- 1 lb. sugar.
- 4 oz. saltpetre.
- 12 oz. pepper.
- $53\frac{1}{3}\%$ pork product.
- $46\frac{2}{3}\%$ beef product.

Cappicola

- 100 lb. cured boneless pork ham butts.
- 1 lb. 8 oz. cayenne pepper.
- 2 oz. fennel seed.
- 100% pork product.

Milano
 105 lb. pork trimmings.
  60 lb. beef trimmings.
 ―――
 165 lb.

 6 lb. salt.
 1 lb. 4 oz. pure spices.
 1 oz. garlic.
 63.6% pork product.
 36.4% beef product.

Sorrento Salami
 148 lb. pork trimmings.
  17 lb. beef trimmings.
 ―――
 165 lb.

 5 lb. salt.
 1 lb. 4 oz. pure spices.
 80% pork product.
 20% beef product.

Pepperoni
 60 lb. pork trimmings.
 90 lb. beef trimmings.
 ―――
 150 lb.
  5 lb. salt.
 12 oz. sugar.

 8 oz. pepperoni.
 2 oz. pepper.
 ¼ oz. fennel seed.
 40% pork product.
 60% beef product.

Gothaer
 120 lb. pork trimmings.
  30 lb. beef trimmings.
 ―――
 150 lb.
  5 lb. salt.
  1 lb. flour.

 4 oz. saltpetre.
 1 lb. sugar.
 12 oz. pepper
 80% pork product.
 20% beef product.

Export Goteborg
 50 lb. pork trimmings.
 40 lb. pork cheeks.
 60 lb. beef trimmings.
 ―――
 150 lb.
  5 lb. salt.
  1 lb. sugar.

 4 oz. saltpetre.
 12 oz. pepper.
 1 oz. cardamom.
 1 oz. coriander.
 60% pork product.
 40% beef product.

Landjaeger
 60 lb. pork trimmings.
 90 lb. beef trimmings.
 ―――
 150 lb.
  5 lb. salt.
  2 oz. garlic.
  1 lb. sugar.

 4 oz. saltpetre.
 12 oz. pepper.
 3½ oz. cherry wine.
 3 oz. caraway seed.
 40% pork product.
 60% beef product.

Lyon
 140 lb. pork trimmings.
  10 lb. beef trimmings.
 ―――
 150 lb.
  5 lb. salt.
 12 oz. pepper.

 12 oz. sugar.
 4 oz. saltpetre.
 1 oz. garlic.
 ⅕ oz. allspice.
 93⅓% pork product.
 6⅔% beef product.

H. C. and B. C. Arles Frisses

- 110 lb. pork trimmings.
- 40 lb. beef trimmings.
- ―――
- 150 lb.
- 5 lb. salt.
- 12 oz. pepper.
- 12 oz. sugar.
- 4 oz. saltpetre.
- 2 oz. garlic.
- 1/5 oz. allspice.
- 73 1/3% pork product.
- 26 2/3% beef product.

Mortadella

- 125 lb. pork trimmings.
- 25 lb. veal or beef trimmings.
- ―――
- 150 lb.
- 5 lb. salt.
- 1 oz. saltpetre.
- 12 oz. sugar.
- 5 oz. pepper.
- 2 oz. garlic.
- 1 lb. curacao.
- 8 oz. alcohol.
- 8 oz. gelatine.
- 83 1/3% pork product.
- 16 2/3% beef product.

Farmer Holsteiner

- 50 lb. pork trimmings.
- 10 lb. pork hearts.
- 20 lb. pork cheeks.
- 70 lb. beef trimmings.
- ―――
- 150 lb.
- 5 lb. salt.
- 4 lb. flour.
- 1 lb. sugar.
- 4 oz. saltpetre.
- 12 oz. pepper.
- 1 oz. coriander.
- 53 1/3% pork product.
- 46 2/3% beef product.

Domestic Goteborg

- 50 lb. pork trimmings.
- 20 lb. pork cheeks.
- 10 lb. pork hearts.
- 70 lb. beef trimmings.
- ―――
- 150 lb.
- 5 lb. salt.
- 3 lb. flour.
- 1 lb. sugar.
- 4 oz. saltpetre.
- 12 oz. pepper.
- 1 oz. cardamom.
- 1 oz. coriander.
- 53 1/3% pork product.
- 46 2/3% beef product.

Wholesale Sausage Factory and Provision Plant

On Page 162 is shown a typical arrangement of a wholesale sausage manufacturing plant which also produces hams, bacon, boiled hams and certain other specialties. The layout covers fully the various departments in a conveniently arranged manner.

Raw products are received on the loading platform, from whence they are conveyed either into the cooling rooms or are lowered by means of an elevator into a basement for storing.

The product can conveniently be taken into the cutting and manufacturing room from either the basement or the refrigerators. The cured sausage meats are usually taken from the cooler, brought into the manufacturing room, put through the hasher, cutter and mixer and then into the stuffers.

The illustration shows an overhead track located between two stuffing tables, so that the link sausage from both tables may be easily placed on the carrier and from there conveyed into the smoke houses.

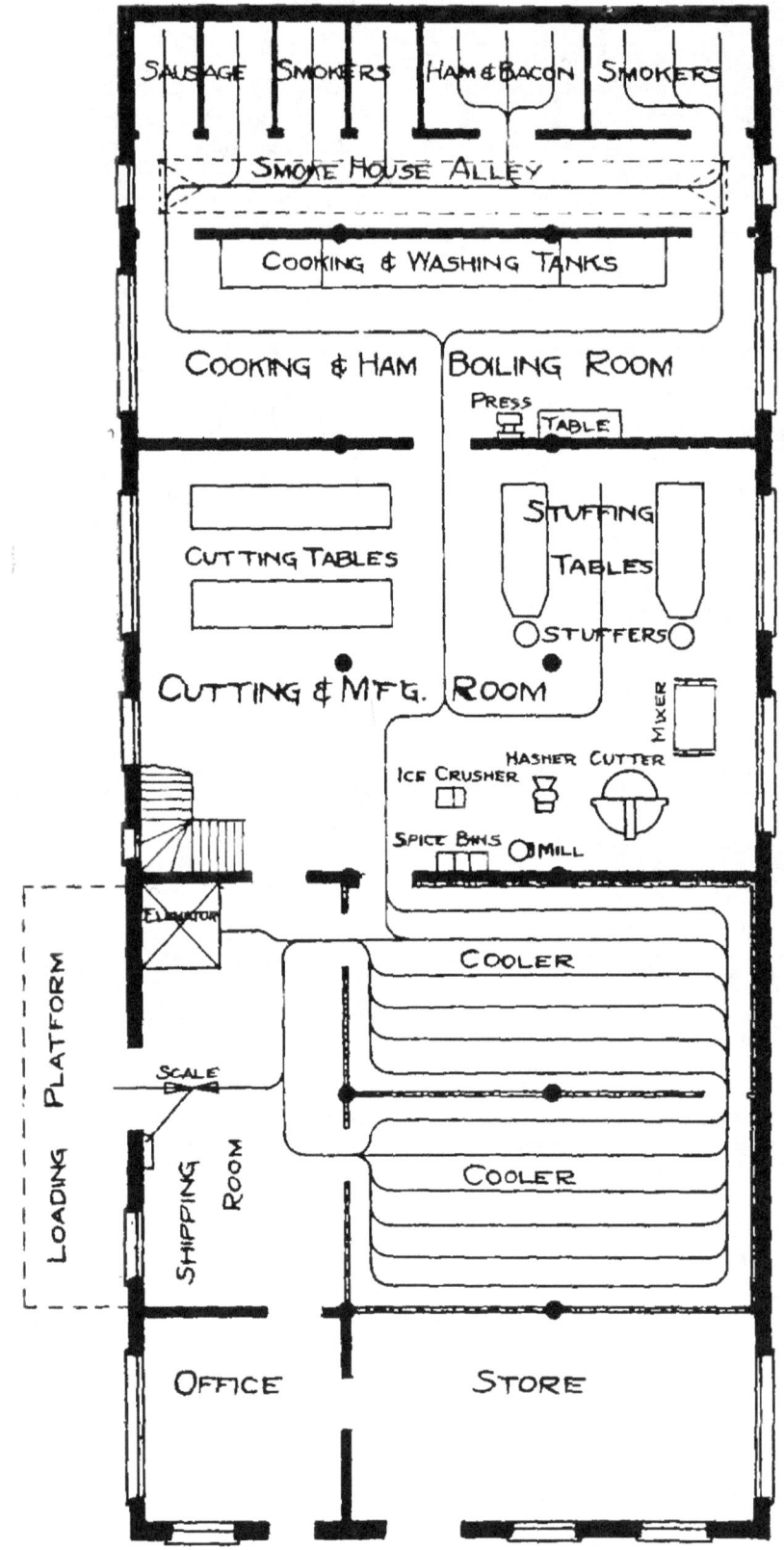

Layout of Wholesale Sausage Factory and Provision Plant

The six smoke houses are of different sizes, for sausage, hams and bacon, and the plan also shows a smoke house alley for the quick removal of any smoke when the smoke house door is open. This is a feature very often overlooked by builders of smoke houses, but leading architects try to provide for it in every modern plant now designed.

After the products leave the smoke houses they pass through the cooking and boiling room. This room is provided with vats for boiling of bologna and cooking of hams. A few boning tables also are shown, and the ham press; and the finished product can be removed from this point out onto the platform or into either one of the two coolers. Cutting tables for boning out meats, hams, etc., are also provided in the main cutting room.

The shipping room is under observation of the office at all times, and a little retail store is provided in the front.

The basement is used for the storing of raw products, and the rear department for smoke house firepits, while the part under the coolers, shipping, cutting and manufacturing rooms is for coolers and freezers.

MEAT CANNING

The object of canning meat is to preserve and protect it for future use. The principle employed is that of the destruction of all the microorganisms or bacteria and yeasts by means of heat after the product is placed in an air-tight receptacle, tin usually being employed. The sterilization process is accomplished by means of sufficient heat, by steam pressure, or two or three successive boilings in water are sometimes employed for certain classes of meat products. The latter process is not quite so sure as the former. The successive boilings are necessary, due to the fact that some organisms are spore producers, which means that they are very resistant in certain stages of their development and therefore it is necessary to heat them a second or third time so that this particular class of organisms may be killed when they are in the vegetative state.

Various machines have been developed to aid the canning process, such as the stuffing machines, vacuum machine, can-washing machinery, capping machinery, soldering machinery, etc. It will not be necessary here to enter into the various details relative to the operation of these different machines, as that is supplied by the manufacturer. It should be stated, however, that the best of machinery and good quality tin should always be used, for one of the greatest sources of loss in canning departments is that produced by "leaks," which means that both the value of the can and the product is practically destroyed. It is a dangerous practice to reprocess leaky cans, for there may be developed poisonous by-products from the result of bacterial action which may cause sickness or even death.

In processing in steam retorts a temperature of 250° Fahr. is generally used, sometimes 260°, the length of time depending upon the size of the tins to be processed and the product contained therein.

While the meat canning business has had its ups and downs, there will always be a consistent demand for canned meats. There are yet in both hemispheres great fields for development. The building of railroads, the opening up of new country, the felling of forests, mining, and many like occupations are of such character that meats preserved in this manner will be demanded.

Formulas.—Following are a number of formulas representative of various classes of canned meat products which will be of interest to those engaged in canning:

Ox-Tail Soup

125 gallons water.
½ gallon sherry wine.
1 quart of caromel.
1,000 lbs. ox tails.
10 lbs. oil.
30 lbs. chopped onions.
20 lbs. chopped turnips.
¾ lb. whole cloves.
¼ lb. ground black pepper.
¼ lb. bay leaves.
5 lbs. salt.

Bring the above to a boil and allow to simmer for 2 hours.
Process: ¼-lb. cans, 40 minutes, 250°; ½-lb. cans, 50 minutes, 250°.

Mock Turtle Soup

60 gallons water.
1 gallon sherry wine.
55 calves' heads.
25 lbs. calves' livers.
60 calves' hearts.
50 lemons.
1½ pints Worcestershire sauce
5 lbs. chopped carrots.
5 lbs. chopped onions.
5 lbs. chopped turnips.
½ lb. whole cloves.
¼ lb. sweet marjoram.
½ lb. white pepper.
2 lbs. green pot herbs.
10 lbs. butter or oil.
8 lbs. flour.
3 lbs. salt.

Cook for 3 hours. Process as for ox-tail soup.

Plain Soup Stock

1,000 lbs. of beef meat and bones.
125 gallons water.
25 lbs. chopped onions.
25 lbs. chopped carrots.
25 lbs. chopped turnips.
10 lbs. celery or celery seed.
1 lb. whole cloves.
½ lb. thyme.
½ lb. bay leaves.

The bones should be cracked to expose the marrow and after cold water is added bring to a boil and skim. Allow to simmer for six hours, then add the vegetables and spices and allow to simmer for two hours before adding the bay leaves. Add enough water to make 125 gallons; then strain and remove the grease.

Consomme

500 lbs. of lean beef.
500 lbs. veal knuckles.
125 gals. water.
25 lbs. chopped leeks.
25 lbs. chopped celery.
10 lbs. chopped carrots.
35 lbs. of oil or butter.
1 lb. pepper.
¼ lb. bay leaves.
¼ lb. thyme.

Bring to a boil and allow to simmer for six hours.

Potted Ham

10 lbs. ham meat.
¼ lb. mace.
½ lb. cloves.
1 oz. bay leaves.

Cover the ham with water then bring to a boil and allow to simmer for 4 hours. Grind the mixture to a paste and add ¾ lb. of white pepper, ¼ oz. cayenne pepper, 2 oz. powdered salt. Process the same as for beef.

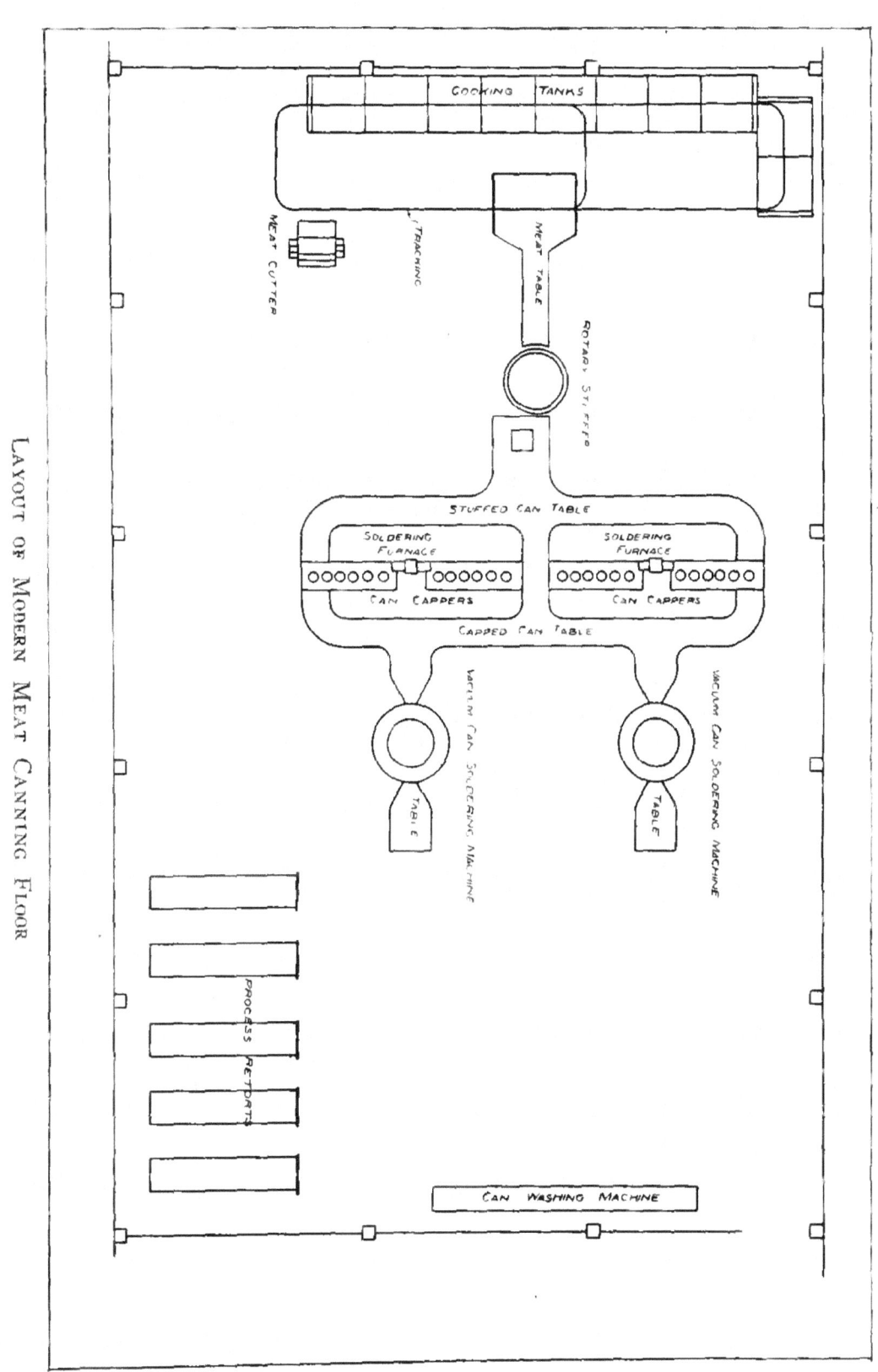

Layout of Modern Meat Canning Floor

Potted Tongue

75 lbs. salted tongue.	2 oz. powdered all-spice.
25 lbs. salt pork.	1 oz. coriander seed.
3 lbs. salt.	½ oz. bay leaves.
½ lb. ground black pepper.	1 pt. onion juice.
2 oz. powdered cloves.	

Soak the tongues for 12 hours in water, add fresh water, then bring them to a boil, allow to simmer for 4 hours, grind fine and mix. The process is the same as for Potted Beef.

Beef Tongues

These are soaked in water for 12 hours, then add fresh cold water and bring to a boil, let simmer for 2½ hours, then remove the skins and put in cans. Process: 1 lb. can large tongue 250° for 2¼ hours; 1 lb. can medium tongue, 250° for 1½ hours; 1 lb. can small tongue, 250° for 1 hour.

Pig Tongues are handled the same as beef tongues, except the process is as follows: 6 lb. cans, 250°, 2½ hours; 3 lb. cans, 250°, 2¼ hours; 2 lb. cans, 250°, 1½ hours; 1 lb. cans, 250°, 1 hour.

Sheep Tongues are handled the same as pig tongues.

Potted Beef

90 lbs. beef meat.	2 lbs. black pepper.
20 lbs. salt pork.	¼ lb. cayenne pepper.
3 lbs. salt.	1 oz. nutmeg.
1½ lbs. mixed spices.	

The beef should be put into boiling water and cooked till tender. Put the pork in cold water, bring to a boil and cook for 3 hours; then grind the beef and the pork to a paste and mix thoroughly with salt and spices. Process: 1 lb. cans, 250°, 1 hour; ½ lb. cans, 250°, ¾ hr.; ¼ lb. cans, 250°, 40 minutes.

Roast Beef:

Remove from the fire when about half roasted, then pack in tins, and cover with a little gravy, well spiced. Process: 3 lb. cans, 230° for 3 hours; 2 lb. cans, 230°, 2½ hours; 1 lb. can, 230°, 12 hours.

Corned Beef:

A word or two relative to the curing of corned beef will not be amiss here, as the process has undergone a radical change recently. It used to require from 12 to 15 days to prepare corned beef and get it into the tins. Now it can be accomplished in the same number of hours by using the following short cure method.

Depending upon the quality, the beef to be corned should be cooked for six hours, at 158° F. The same kettles used for cooking may be used for curing. After cooking, the soup should be drained off and 25° pickle containing the maximum amount of saltpeter or nitrate of soda used in any pickle should be added, the temperature of the same being 140° F. It is important to use 70% of second pickle and only 30% of new pickle. Little or no sugar in the formula is required.

Pickling will require six hours, during which time the product should be constantly agitated. The temperature should be down to 120° F. at the end of that period.

After curing, the beef is cut into pieces, put in cans and covered with juice composed of one-half bouillon in which the beef has been boiled and one-half bouillon from calves' feet. The latter may be flavored with laurel cloves and a little nutmeg. Processing should take place as follows: 3 lb. cans, 240°, 3¼ hours; 2 lb. cans, 240°, 2¾ hours; 1 lb. can, 240°, 1¼ hours.

Spices.—Depending upon the class of trade and other controlling factors, each canner has prepared various formulas and spicing methods. Following is a list of spices ordinarily used in meat canning:

Sage, marjoram, savory, rosemary, cloves, coriander, basilic coriander, cinnamon, bay leaves, mace, nutmeg, Hungarian paprika, yellow pepper, cayenne pepper, red pepper, white pepper, black pepper, thyme.

ANIMAL GLANDS

The packer killing any considerable number of animals may do well to calculate the profit to be obtained by the saving of various ductless and other glands, also hog stomach linings, all of which if properly saved and cared for may find ready sale to the manufacturer of biological products.

The following glands from cattle are frequently saved:

Thyroid gland. This gland lies close to the larynx, is of a dark color, somewhat resembling a leech. A pound may be recovered from 25 cattle, at a cost of about twenty cents. The extract from this gland is used by the medical fraternity in treating simple goiter and in supplying subjects deficient in thyroid substance.

Thymus gland. These glands are known to packing house men as neck sweetbreads. They lie close to the breast on the outside, at a point where the arteries flow from the body to the head. In calves and young cattle this gland sometimes weighs 2½ to 3 lbs. In cattle over 3 years old this gland is very small, and in six-year-old cattle it fades away to practically nothing. Thymus substance is used by physicians in treating rickets.

Suprarenal gland. Two of these glands are found in each animal, lying close to the kidney, being about the size of a lima bean. The suprarenalin substance is used as a styptic to stop blood flow, as a heart stimulant, and for cobra bites.

Parotid gland. These very small glands lie close to the ear; they are of a dark gray color and difficult for the inexperienced to discover. They are best removed from the head when the cheek meat is cut off. They are used by physicians in treating subjects deficient in saliva.

Pituitary gland. These are found at the base of the brain in sphenoid bone. This is a very important gland, its use in surgery for preventing shock and relieving gas pains, and also its use in cases of obstetrics having been thoroughly demonstrated by the medical profession.

Mammary gland. These are best removed from young heifers, not over two years old, and when manufactured for medicinal purposes are used in cases of menstrual disorders.

Pineal gland. The pineal gland is found on the surface of the brain and appears like a small white worm about three-quarters of an inch long. Immediately upon removal, which is done just before removing the brain from the skull, it should be placed in ice water. It is used in treating cases of mental backwardness.

Spleen. The spleen when saved is dessicated and is generally used for supplying iron to the system.

Ovaries. The ovaries of pregnant cows and sows are frequently saved; they are usually a light pink in color. The substance manufactured from them is used in cases of surgical and natural menopause, in menstrual troubles and ovarian disturbances.

The pancreas gland when saved is desiccated and used for predigesting foods like milk, starch, fat, etc.

The orchic gland when saved is used in cases of sexual neurasthenia.

Parathyroid glands. These are very small glands found at the root of the tongue. The substance manufactured from them is used in cases of paralysis agitants; also in cases of tetanus.

The thyroid, ovary and pancreas glands are frequently saved from hogs as well as from cattle, and in addition the stomach lining of hogs free from connective and fatty tissue is usually saved for the manufacture of pepsin, an important pharmaceutical preparation. In saving this lining, it should be washed gently in cold water, just enough to free it from dirt, and then packed in ice until used. Four or five linings are required to make a pound.

The glands usually saved from sheep are prostate, suprarenal, thyroid and thymus. The description and location of these is similar to those found in cattle.

Glands are not packed in salt of any kind, but are usually packed in open-head tierces. If they are used within thirty-six hours they are preserved by packing in ice and holding in a refrigerator temperature of about 36°. If they cannot be used within three days' time, then they are held in storage at 15° above zero. If held longer, or for export, they are packed in salt.

Gallstones.—Although they are not animal glands, gallstones may be properly referred to here. Good, yellowish unbroken gallstones are of high value and are saved for export to the Far East. They are found in the gall bladders of cattle and sometimes in the bladders of horses. The shape is round, oval or square. Gallstones should never be dried in sunlight, but carefully wrapped up in a thin rag and exposed to the heat of a stove or radiator, placed in a room with excellent ventilation. Unsufficiently dried gallstones will break or burst easier than well-dried stones. Broken stones are worth half the value of good, sound stones.

PACKING HOUSE CHEMISTRY

Packing house chemistry includes, among other things, the control by chemical analysis of an almost endless variety of materials, including not only the principal products and by-products, but also many, if not most, of the raw materials necessary in the conduct of the packing business.

It is impossible to give in the available space more than a few of the most commonly used and most necessary methods for the analysis of packing house byproducts and purchased materials. It will be necessary, therefore, to confine this article to the methods of analysis of such products as fertilizers, tallows, greases, etc., and of such materials as salt, saltpeter and sugar.

Sampling of Materials

The sampling of materials is of at least as much importance as the analysis of the sample. Proper sampling is a matter of good judgment, experience and unremitting care on the part of the sampler.

It is obvious that if a sample is not truly representative of the particular lot of material from which it was drawn, it makes no difference whether the analysis of that sample is correct or not. A handful of samples drawn at random from a pile cannot be expected to represent the pile. Samples must be drawn from all parts of a pile, bulk carload, bin, etc., and the coarser the material, the larger must be the sample. This is particularly true of heterogeneous materials, such as, for example, unground bone tankage, in which it is important to get a fair proportion of both bone and meat.

When materials are packed in bags, barrels, drums, etc., a representative number of packages must be sampled. When the lot to be sampled includes only a few packages, a sample should be taken from each package. When a carload, say of fifty, one hundred or more packages is to be sampled, at least five and preferably ten per cent of the number of packages should be sampled. Frequently it will be necessary to sample top, center and bottom of barrels, where a core cannot be obtained. Materials like salt, saltpeter and sugar, which do not require grinding, should be as carefully sampled as any other materials.

In the case of fats and oils, sampling is not very difficult, as suitable triers both for tierces and other packages, as well as for tank cars, have long been standard tools in the industry.

The carefully drawn sample must be quartered down in the usual manner on a clean surface, taking care to reduce larger pieces to a smaller size during the process of quartering. In general, the coarser the material, the larger the sample which should be delivered to the chemist.

Fats and oils also need careful attention with regard to thorough mixing and subdivision into smaller samples for analysis, particularly in the case of low grade fats containing considerable amounts of moisture and suspended impurities.

All samples which are not air-dry, that is, carry so much moisture that they will lose weight on exposure to the air, must be collected and delivered to the chemist in suitable air-tight containers. On such samples a double moisture determination must be made in many cases and only experience can tell in which cases the double moisture determination is not necessary. It is always necessary in the case of wet tankage, blood, bone, etc., frequently in the case of wet salt, and the procedure is described below.

Fertilizer Materials

Blood, Tankage, Bone, Etc.

Sampling.—In the case of wet samples a moisture determination is made on a comparatively large portion of the sample as received, say 500 or 1,000 grams. This may be dried on special shallow trays or on tin pail covers of suitable size, preferably at a temperature of 70° C. (158° F.) in a suitable oven with forced draft. The sample is air dried in this manner, so that it can be properly ground or otherwise reduced to the requisite fineness.

The amount of moisture lost in this drying is added to the moisture determined on the finely ground sample as described below.

Dry materials are ground, reduced by quartering and the final sample ground so as to pass a thirty or forty mesh sieve, depending on the material. The finer a sample can be ground for analysis, the more nearly uniform it will be and the easier it will be to obtain a satisfactory weighed portion for analysis.

Moisture.—In samples of shipments of dried blood and ground tankage, the moisture is determined both on a representative portion of the sample as received and also on a portion of the finely ground sample, prepared for analysis. Five to ten grams of each sample are weighed into suitable dishes, preferably shallow aluminum dishes, provided with well-fitting covers, so that the dried samples will not gain weight while on the balance. The samples are dried in a suitable oven at 105° C. (221° F.) to constant weight. Five hours is usually sufficient. On removing the samples from the drying oven, they should be covered, collected in a desiccator and weighed. The loss in weight is calculated to moisture.

Nitrogen (Ammonia).—The method used is essentially the Kjeldahl-Gunning-Arnold method, as described in Official Methods, A. O. A. C. (1920), p. 7.

Reagents.—Potassium Sulphate: The powdered potassium sulphate used for pharmaceutical purposes is satisfactory.

Sodium Hydroxide Solution: Dissolve the entire contents of a freshly opened ten-pound tin of granular caustic soda in an equal weight of water in an iron kettle. When completely dissolved, allow to stand over night in order to allow the carbonates to settle out. Next morning siphon off the clear liquid and then dilute the clear solution to a density of 1.38, or about 35 per cent, which usually takes about ten pounds of water. For routine use, the sodium thiosulphate used in the distillation of the ammonia may first be dissolved in a little water and added to the sodium hydroxide solution before dilution.

Congo Red Indicator: Dissolve 1 gram of congo red (Gruebler) in 10 cc. of alcohol and 90 cc. of water.

N/2 Hydrochloric Acid: Standardized preferably against pure potassium acid phthalate. Any other pure standard substance usually employed for this purpose may be used instead.

N/4 Potassium Hydroxide: This must be free from carbonates and standardized against standard N/2 hydrochloric acid.

Apparatus.—Care of the Still: In addition to running blanks whenever fresh lots of sulphuric acid, caustic soda and potassium sulphate are opened up, the still must be cleaned thoroughly by steaming out. Fill the 650 cc. Kjeldahl flasks about one-half full of tap water and connect as in the nitrogen determination. Turn off the flow of water on the condensers and allow the steam to pass from the receiving bulb with distilled water and permit the condensers to cool before turning on the water. In this way all-glass condensers may be used without danger of breakage. The rubber stoppers on the distilling bulb must be watched and replaced when necessary.

The most satisfactory distillation will be obtained, with least breakage of Kjeldahl flasks, by use of a fairly low flame, placed directly under the middle of the flask until the boiling point is reached, when it should be shifted to one side, thus producing a rolling motion in the solution. If the flame is placed to one side before the boiling point is reached, the strain on the glass at point of contact with the support will tend to break the flasks. It is always better to have a low flame close up than to have a high flame far below the flask.

Determination.—Spread the properly prepared sample on a sheet of glazed paper, mix thoroughly and weigh the proper factor weight on a nitrogen-free filter paper. An 11 cm. filter, such as S. & S. No. 597, Swedish 1F, Whatman's No. 1, or equivalent, answers this requirement. Care and judgment are required to get a representative sample on the balance pan. Roll up each weighed portion in its filter paper and drop it into a 650 cc. Kjeldahl flask, then add 5 grams of potassium sulphate and 0.5 gram of pure mercury, followed by 25 cc. concentrated sulphuric acid, and digest. Continue the digestion for not less than two hours after final color has been reached. Usually the entire digestion is complete in about three hours. Blank determinations must be made whenever a new lot of sodium hydroxide solution, potassium sulphate or of sulphuric acid are begun. Such blanks are determined on 2-gram portions of pure sugar, such as Crystal Domino sugar.

When the digestion is finished, cool, dilute with 250 cc. of water, add a little granulated zinc and then 2 grams of sodium thiosulphate with the necessary quantity, about 100 cc., of sodium hydroxide solution. Distill off not less than 200 nor more than 250 cc. into a 500 cc. Pyrex Erlenmeyer flask, containing N/2 hydrochloric acid. Allow the receiving bulb to drain for a few minutes after lowering the flask and wash the tip with distilled water before removing flask. Titrate with N/4 potassium hydroxide, using congo red indicator in the least possible quantity, so as to make the change in color as sharp as possible. About ten drops are usually required.

The quantities of N/2 hydrochloric acid to be drawn off are shown in the following table:

	Sample	N/2 Acid
Regular tankage	1.7034 grams	20 cc.
Blood and concentrated tankage	1.7034 grams	40 cc.
Bone	1.7034 grams	10 cc.

The number of cc. N/2 hydrochloric acid neutralized by the ammonia gives the percentages directly as follows:

Factor Weight Used	Each cc. N/2 Hydrochloric Acid Neutralized is Equivalent to:
1.7034	0.5% ammonia
0.8517	0.25% ammonia
1.4010	0.5% nitrogen
0.7005	0.25% nitrogen

For the determination of nitrogen or ammonia in other materials, such as sulphate of ammonia, nitrates, mixed fertilizers, etc., the Official Methods of the Association of Official Agricultural Chemists (1920) should be consulted.

The following factors are used for the conversion of nitrogen to ammonia and protein:

$$\text{Ammonia} = \text{Nitrogen} \times 1.2155$$
$$\text{Protein} = \text{Nitrogen} \times 6.25$$

Phosphoric Acid: (Volumetric Method)

This method is used for tankages, bone meals, acid phosphates and mixed fertilizers.

For purposes of settlement or in case of controversy with official control or other laboratories, the gravimetric method must be used. See Official Methods, A. O. A. C. (1920) p. 1.

Reagents.—Molybdate Solution: Make up the following two solutions:

I.	Molybdic Acid 85%	60 gm.
	Ammonium Hydroxide (sp. gr. 0.90)	75 cc.
	Distilled Water	150 cc.
II.	Nitric Acid (sp. gr. 1.42)	250 cc.
	Distilled Water	400 cc.

Pour solution I slowly and with constant stirring into solution II. Keep the mixture in a warm place over night, or until a sample deposits no yellow precipitate of ammonium phospho-molybdate when warmed to 40° C. Decant the clear solution as completely as possible and to every 950 cc add 50 cc of nitric acid (sp. gr. 1.42). If the solution is not perfectly clear and brilliant, it must be filtered before use.

Ammonium Nitrate Solution: A 5% solution of commercial nitrate, free from phosphate, in distilled water.

N/2 Potassium Hydroxide: To be standardized against standard N/2 hydrochloric acid prepared as described under Determination of Nitrogen.

N/4 Nitric Acid: To be standardized against standard N/2 hydrochloric acid prepared as described under Determination of Nitrogen.

Preparation of Solution.—The usual factor weight of sample is 3.7043 grams and this weight is used for all totals[1] except phosphate rock, bone meal, and acid phosphate.

Determination.—Digest in a 300 cc Kjeldahl flask with 30 cc nitric and 10 cc hydrochloric acid. Boil down to a volume of about 10 cc, add 10 cc of nitric acid and 50 cc of water and boil. When the sample contains sulphates, as in the case of acid phosphate or mixed fertilizers con-

---

[1] For phosphate rock, bone, or any other total running over 18%, use 3.0870 grams. For insolubles and acid phosphate totals use 1.8522 grams.

taining acid phosphate, add 10 to 15 cc of a saturated solution of barium nitrate. If 3.0870 grams have been weighed out, the digestion is made up to a volume of 500 cc; all others to 300 cc. Filter through a dry filter. For totals take an aliquot of 25 cc, diluting to 100 cc with distilled water, and for insolubles 100 cc.

Neutralize the solution with ammonium hydroxide (sp. gr. 0.90) from a burette, using nitric acid to neutralize any excess of ammonia, then add 10 grams of ammonium nitrate crystals. Heat in a water bath, held at 45° C for 15 minutes to allow the solution to come to the temperature of the bath. Precipitate the phosphoric acid by adding, with stirring, 20 to 45 cc of molybdate solution, depending upon the amount of P_2O_5 present. Precipitation must be made in sets of not more than four[2] allowing five minutes to elapse before precipitation of a second set. Digest for 15 minutes at 45° C, stirring at the end of five and ten minutes respectively.

Take out each set of four in the order of precipitation and filter at once[3] with suction. Wash three or more times by decantation[4] with ammonium nitrate solution, transferring the precipitate to the filter during the last washing. Then wash with distilled water, being very careful to wash every trace of acid from the beaker, funnel and precipitate. Always test the last washing for neutrality with litmus paper. Return the precipitate and paper pulp mat from the funnel to the beaker in which precipitation was made and rinse the funnel with a fine stream of water.

Dissolve the precipitate in N/2 potassium hydroxide, using from 5 to 10 cc excess, and titrate back with N/4 nitric acid, using phenolphthalein as indicator.

Calculation of Results

Amount weighed out cc	Volume of solution cc	Aliquot taken cc	1 cc N/2 alkali is equivalent to % P_2O_5	1 cc N/4 alkali is equivalent to % P_2O_5
3.7043	300	25	0.5	0.25
1.8522	300	25	1.0	0.5
1.8522	300	100	0.25	0.125
3.0870	500	25	1.0	0.5

Fat

Reagents.—Petroleum Ether: This is the light fraction of gasoline coming over below 75° C (167° F). A suitable gasoline may usually be obtained by purchasing gas machine gasoline (sometimes sold under the name pepsin naphtha). This grade of gasoline usually has a gravity of 80° to 88° Baumé, and yields on distillation about two-thirds of the desired low boiling fraction.[5]

---

[2] It is of the greatest importance that the excess of molybdate solution be filtered off immediately after removal from the precipitating bath, so as to avoid the danger of precipitating molybdic acid.

[3] Filtration of a second set of four, can, of course, be begun as soon as the first decantation on the previous set has been completed, so that three sets can be kept digesting simultaneously.

[4] The number of washings by decantation depends on the amount of precipitate. The larger this is, the more washings by decantation should be made.

[5] The residue may be used in the determination of soluble mineral matter from the standard analysis of greases and also for cleaning greasy glassware, etc.

Determination.—The residue from the determination of moisture may be used for the determination of fat in tankages, bone, etc. This dry residue is carefully wrapped in filter paper, from which the fat has been previously extracted and extracted with petroleum ether in any standard extraction apparatus for six to eight hours. The petroleum ether is then evaporated off over a steam bath in a current of dry filtered air and finally dried to constant weight in an air oven at 105° C (221° F) or in a vacuum oven, if one is available.

Fats

Greases, Tallows, Etc.

(A) Quick Methods of Testing for Factory Control

Moisture.—Weigh 25 grams of the well mixed sample into a three-inch casserole and heat on an asbestos board over a burner or on a hot plate, taking care that the temperature of the fat does not exceed 130° C (266° F.). During the heating rotate the casserole gently to avoid loss by spattering. The completion of the test is judged by the disappearance of bubbling or of foam or the absence of moisture on a cold watch glass, when held directly over the casserole. Avoid overheating the sample, as indicated by smoking or darkening. Cool in a desiccator and weigh. The dried residue is used for the determination of insoluble impurities, as described below.[6]

Insoluble Impurities.—Melt the dried residue from the moisture determination and add about an equal volume of gasoline tailings. Mix thoroughly, allow to settle and decant the clear solution as completely as possible. Add another portion of about 25 cc of gasoline tailings. At this point the analyst can, after some practice, estimate the amount of impurities by inspection in nearly all cases.

In exceptional cases, where the insoluble residue is of unusual appearance or in unusual bulk, the clear solution may be decanted off again, the residue washed once with redistilled petroleum ether, the residual solvent evaporated off on the steam bath, the residue dried on the hot plate or in the air oven, cooled and weighed.

(B) Standard Methods of Testing

Sampling.—The sample must be representative and at least three pounds in weight and taken in accordance with the Standard Methods for the Sampling of Commercial Fats and Oils.[7] It must be kept in an air-tight container in a dark cool place.

Soften the sample if necessary, by means of a gentle heat, taking care not to melt it. When sufficiently softened, mix the sample thoroughly by means of a mechanical egg beater or other equally effective mechanical mixer.

Moisture and Volatile Matter.—Weigh out five grams of the prepared

---

[6] Samples containing excessive amounts of moisture, which cannot be dried directly on the hot plate, are first dried in an air oven at 105° C, until most of the moisture has been driven off. The drying is then completed on the hot plate as described above.

[7] A copy of these methods may be obtained from the Committee on Analysis of Commercial Fats and Oils of the American Chemical Society, of which Mr. W. D. Richardson, Chief Chemist, Swift and Company, Chicago, is Chairman.

sample into a moisture dish and dry to constant weight in an air oven at 105° C (221° F.). If the sample is dried too long, it will gain weight.

If a vacuum oven is available, this determination is best carried out in such an oven instead of an air oven, thus avoiding the danger of a gain in weight, due to the absorption of oxygen, in the later stages of the drying. The loss in weight is calculated as moisture and volatile matter.

Insoluble Impurities.—Dissolve the residue from the moisture and volatile matter determination by heating it on the steam bath with 50 cc of kerosene.[8] Filter the solution through a Gooch crucible properly prepared with an asbestos mat or through a filter paper in a glass funnel.[9] Wash the insoluble matter five times with 10 cc portions of hot kerosene, and finally wash the residual kerosene out thoroughly with petroleum ether. Dry the crucible and contents to constant weight, as in the determination of moisture and volatile matter and report results as Insoluble Impurities.

Soluble Mineral Matter.—Place the combined kerosene filtrate and washings from the insoluble impurities determination in a platinum dish or large porcelain crucible. Place this in an ashless filter paper folded in the form of a cone, apex up. Light the apex of the cone, whereupon the bulk of the kerosene will burn quietly. Ash the residue in a muffle to constant weight, taking care that the decomposition of alkaline earth carbonates is complete, and report the result as Soluble Mineral Matter.

Free Fatty Acids.—The alcohol to be used is denatured alcohol, Formula 30, of such purity that it will give a definite, distinct end point with phenolphthalein.

Determination.—Weight 1 to 15 grams of the prepared sample into an Erlenmeyer flask, using the smaller quantity in the case of dark-colored, high acid fats. Add 50 to 100 cc hot, neutral alcohol, and titrate with $N/2$, $N/4$ or $N/10$ sodium hydroxide, depending on the fatty acid content, using phenolphthalein as indicator. Calculate to oleic acid. Each cc of $N/2$ sodium hydroxide is equivalent to 0.141 grams of oleic acid.

Unsaponifiable Matter

Extraction Cylinder.—The cylinder shall be glass-stoppered, graduated at 40 cc, 80 cc and 130 cc, and of the following dimensions: Diameter about 1⅜ in., height about 12 in.

Petroleum Ether.—Redistilled petroleum ether, boiling under 75° C, shall be used. A blank must be made by evaporating 250 cc with about 0.25 gram of stearine or other hard fat (previously brought to constant weight by heating) and drying as in the actual determination. The blank must not exceed a few milligrams.

Determination.—Weight 5 grams (± 0.20 gram) of the prepared sample into a 200 cc Erlenmeyer flask, add 30 cc of redistilled 95 per cent (approximately) ethyl alcohol and 5 cc of 50 per cent aqueous potassium hydroxide, and boil the mixture for one hour under a reflux condenser.

---

[8] Commercial kerosene mixed with an equal volume of gasoline tailings, as described under "Determination of Fat in Tankage," may be used for this work.
[9] If filter paper is used, it must be very carefully washed, especially around the rim, to remove the last traces of fat.

Transfer to the extraction cylinder and wash to the 40-cc mark with redistilled 95 per cent ethyl alcohol. Complete the transfer, first with warm, then with cold water, till the total volume amounts to 80 cc. Cool the cylinder and contents to room temperature and add 50 cc of petroleum ether. Shake vigorously[10] for one minute and allow to settle until both layers are clear, when the volume of the upper layer should be about 40 cc. Draw off the petroleum ether layer as closely as possible by means of a slender glass siphon into a separatory funnel of 500-cc capacity. Repeat extraction at least four more times, using 50 cc of petroleum ether each time. More extractions than five are necessary where the unsaponifiable matter runs high, say over 5 per cent, and also in some cases where it is lower than 5 per cent, but is extracted with difficulty. Wash the combined extracts in a separatory funnel three times with 25 cc portions of 10 per cent alcohol, shaking vigorously each time. Transfer the petroleum ether extract to a wide-mouth flask or beaker, and evaporate the petroleum ether on a steam bath in an air current. Dry as in the determination for "Moisture and Volatile Matter." Any blank must be deducted from the weight before calculating unsaponifiable matter. Test the final residue for solubility in 50 cc petroleum ether at room temperature. Filter and wash free from the insoluble residue, if any, evaporate and dry in the same manner as before.

Total Impurities.—The sum of the moisture and volatile matter, soluble mineral matter, insoluble impurities and unsaponifiable matter is reported as total impurities.

Titer

Standard Thermometer.—The thermometer is graduated at zero and in tenths of degrees from 10° C to 65° C, with one auxiliary reservoir at the upper end and another between the zero mark and the 10° mark. The cavity in the capillary tube, between the zero mark and the 10° mark is at least 1 cm. below the 10° mark, the 10° mark is about 3 or 4 cm. above the bulb, the length of the thermometer being about 37 cm. over all. The thermometer has been annealed for 75 hours at 450° C and the bulb is of Jena normal 16-in. glass, or its equivalent, moderately thin, so that the thermometer will be quick-acting. The bulb is about 3 cm. long and 6 mm. in diameter. The stem of the thermometer is 6 mm. in diameter and made of the best thermometer tubing, with scale etched on the stem, the graduation is clear-cut and distinct, but quite fine. The thermometer may be certified by the U. S. Bureau of Standards.

Glycerol Caustic Solution.—Dissolve 250 grams potassium hydroxide in 1000 cc dynamite glycerin with the aid of heat.

Determination.—Heat 75 cc of the glycerol-caustic solution to 150° C and add 50 grams of the melted fat. Stir the mixture well and continue heating until the melt is homogeneous, at no time allowing the temperature to exceed 150° C. Allow to cool somewhat and carefully add 50 cc 30 per cent sulfuric acid. Now add hot water and heat until the fatty acids separate out perfectly clear. Draw off the acid water and wash the fatty

---

[10] It is necessary to shake thoroughly and vigorously in order to secure accurate results. The two phases must be brought into the most intimate contact possible, otherwise low and disagreeing results may be obtained.

acids with hot water until free from mineral acid, then filter and heat to 130° C as rapidly as possible while stirring. Transfer the fatty acids, when cooled somewhat, to a 1-in. by 4-in. tube, placed in a 16-ounce salt-mouth bottle of clear glass, fitted with a cork that is perforated so as to hold the tube rigidly when in position. Suspend the titer thermometer so that it can be used as a stirrer and stir the fatty acids slowly (about 100 revolutions per minute) until the mercury remains stationary for thirty seconds. Allow the thermometer to hang quietly with the bulb in the center of the tube and report the highest point to which the mercury rises as the titer of the fatty acids. The titer should be made at about 20° C for all fats having a titer above 30° C and 10° C below the titer for all other fats. Any convenient means may be used for obtaining the temperature or 10° below the titer of the various fats. The best means for this purpose is a chill room; second, an artificially chilled small chamber with glass window; third, immersion of the salt-mouth bottle in water or other liquid of the desired temperature.

Color.—The color of tallows and greases is determined by matching against standard yellow and red Lovibond tintometer glasses, preferably in a Wesson type of Lovibond tintometer. This tintometer is furnished with suitable glass tubes and the lighter colored fats are read through a column $5\frac{1}{4}$ inches high, while darker colored fats are read through a $1\frac{1}{4}$-inch column. Melted fats are read in a uniform temperature of not less than 130° nor more than 150° F.

Bleach

Lard.—Weigh 150 grams of the melted sample into a porcelain casserole and heat to 160° F, stirring constantly with a thermometer. Now add 0.75 grams (0.5%) of standard English fullers' earth[11] and keep at 160° F for ten minutes, stirring constantly. At the end of ten minutes, filter through a fluted filter into a 4-ounce oil bottle, returning the filtrate to the filter, until a clear, brilliant filtrate is obtained. The oil bottle must be filled to the shoulder ($5\frac{1}{4}$-inch). Read the color in the cottonseed oil tintometer against standard Lovibond glasses at a temperature of not less than 130° nor more than 150° F.

Tallows.—Heat the sample to 180° F and add 6 grams (4%) of standard English fullers' earth; otherwise proceed exactly as in the case of lard, except that the height of column of the oil is to be made to conform to the specifications in the method for color.

Greases.—Heat the sample to 220° F and add 6 grams (4%) of standard English fullers' earth, otherwise proceed exactly as in the case of tallows.

Fullers' Earth.—Bleach two sets of well-settled prime steam lard and No. 1 tallow, one set with the earth to be examined and one set with standard English fullers' earth in the manner as described under the respective

---

[11] A supply of standard English fullers' earth may be obtained from Mr. Thomas G. Caldwell, Law & Company, Wilmington, N. C., Secretary, American Oil Chemists Association.

For methods for testing fullers' earth on cottonseed oil and therefore also for bleaching cottonseed oil, the Official Methods of the American Oil Chemists Society should be obtained from the same source as the standard earth.

fats and oils above. Report results by comparison with standard earth. The sample to be examined must first be passed through a 100-mesh sieve before it is tested.

Salt

Preparation of Sample.—Reduce all samples to pass a 60-mesh sieve before analyzing, except in the case of very wet salts. In such cases, a moisture determination is to be made on a representative portion of the sample as received. The remainder of the sample is rapidly dried, preferably on the hot plate, ground to pass a 60-mesh sieve and transferred to a tightly stoppered container for analysis as described below.

Moisture.—Dry ten grams of the salt as received at 105° C for 5 hours. (The moisture in rock salt is usually under 0.1% and if the sample is normal in appearance, moisture need not be determined.) In the case of the very wet salts mentioned above, as large a sample as conveniently possible, but not to exceed 100 grams, is taken for the determination of moisture.

Insoluble Impurities.—Dissolve 25 grams of the sample of salt in 250 cc of distilled water, making a 10% solution and filter through a dry filter. In most cases, the soluble impurities can be readily estimated by inspection of the solution before filtration and of the filter paper after filtering. In cases where there is more insoluble matter than usual, the solution is filtered through a tared Gooch crucible and the residue is dried and weighed in the usual manner.

Lime and Magnesia.—Reagents:

Standard Calcium Chloride Solution: Dissolve 1.79 grams of pure calcium carbonate in 5 cc of concentrated hydrochloric acid and make up to one liter with distilled water; 5 cc of this standard solution, when used according to the following method, represents 0.1% CaO and MgO.

Ammonium Hydroxide: A 10% solution of ammonia (187 cc of ammonia, sp. gr. 0.90, diluted to 500 cc with distilled water).

Ammonium Carbonate: A 10% solution of ammonium carbonate.

Determination.—Transfer 50 cc of the 10% salt solution to a 250 cc graduated cylinder, add 10 cc each of the 10% ammonia and ammonium carbonate solutions, dilute with water to 150 cc and mix. Take a known volume of the standard calcium solution, transfer it to another 250 cc cylinder and add the same amounts of ammonia and ammonium carbonate, as above. Dilute to the same volume as that of the sample and mix. Compare the degree of cloudiness and observe whether it requires more or less of the standard solution to equal the turbidity of the sample.

Example: If it requires 25 cc of the standard solution to produce the proper degree of cloudiness, the salt contains 0.5% CaO and Mgo. If 40 cc are required, it contains 0.8%, etc.

Sulphates.—Reagents:

Standard Sodium Sulphate Solution: Dissolve 4.05 grams of crystallized sodium sulphate in 1000 cc of distilled water. Each 5 cc of this solution used in the following method represents 0.1% SO_3 in 50 cc of the 10% solution of the sample.

Barium Chloride: A 10% solution of barium chloride.

Hydrochloric Acid: A 10% solution of hydrochloric acid (dilute 100 cc of 1.20 specific gravity acid to about 400 cc with distilled water).

Determination.—Take another 50 cc of the 10% solution of the sample, transfer to a 250 cc cylinder, add 10 cc each of the hydrochloric acid and barium chloride solutions, dilute to 150 cc and mix. Take a known volume of the standard sulphate solution, transfer to a 250 cc cylinder and add 10 cc each of the reagents as above. Dilute to 150 cc, mix and compare the cloudiness of this solution with that produced in the sample.

Example: If it requires 25 cc of the standard sulphate solution to equal the degree of cloudiness in the sample, then the salt contains 0.5% SO_3. If 30 cc are required, then it contains 0.6% SO_3.

Borax.—To 15 cc of the 10% solution of the salt in a small beaker, add 1 cc of concentrated hydrochloric acid. At the same time, prepare a blank, and also two standards containing one part in 100,000 and one part in 1,000,000 of boric acid. Suspend a strip of turmeric paper in each of these solutions. The relative red coloration which the different solutions produce on the turmeric paper indicates at the same time the presence and amount of borax in the salt.

Nitrate of Soda

Report moisture, insoluble impurities, lime (CaO) and magnesia (MgO) and sulphates. If the salt contains borax, it must be rejected.

Sampling.—The sample is made up by mixing equal portions of the individual samples drawn from each barrel.

Analysis.—Total Impurities: The total impurities in nitrate of soda include the determinations made on salt and, in addition, the determination of sodium chloride (common salt).

Reagents: N/10 Silver Nitrate Solution: One liter contains 16.989 grams silver nitrate.

N/10 Ammonium Thiocyanate: Contains approximately 8 grams of ammonium thiocyanate per liter and must be standardized against standard N/10 silver nitrate solution.

Ferric Alum Indicator: A saturated solution of ferric alum.

Nitric Acid: Free from lower oxides of nitrogen, secured by diluting the usual pure acid with about ¼ part of water and boiling until perfectly colorless.

Determination.—Transfer an aliquot of the solution used for the determination of impurities representing 2.5 grams of the original sample to a 200 cc casserole and determine chlorine by titration as follows: Add an excess of N/10 silver nitrate solution, heat to boiling, allow to settle, filter into another porcelain casserole, wash and titrate back the excess of silver nitrate in the filtrate with N/10 ammonium sulphocyanid solution, using ferric alum as indicator. The titration is best accomplished in a dark place by gas light.

One cc N/10 silver nitrate = 0.00585 gm. sodium chloride.

The difference between 100 and the percentage of total impurities (total refraction) gives with a fair degree of accuracy the purity of the sodium nitrate. This may be checked if desired by determining the nitrogen in the sample, according to the official ferrous sulphate-zinc-soda method as follows:

Place 0.5 gram of the sample in a 600-700 cc flask, add 200 cc of distilled water, 5 grams of powdered zinc, 1-2 grams of crystallized ferrous sulphate and 50 cc of sodium hydroxide solution (36° Beaumé). Distill, collect in the usual way in N/2 hydrochloric acid and titrate.

Sodium Nitrate = Nitrogen x 6.0678.

Borax.—Sodium nitrate intended for curing of meats in inspected establishments is always tested for freedom from borax. This test is carried out in the laboratories of the U. S. Bureau of Animal Industry. This method of testing is as follows:

Sampling.—Draw samples of two tablespoonsful from each barrel in the shipment. This is best done by boring an inch hole approximately 8 inches below the top and drawing the sample with a short trier. Place the sample in a clean, dry glass container, numbering each sample and the barrel from which it was drawn with a serial number, so that the barrel may be identified if positive tests for borax are obtained.

Reagents.—Turmeric Paper: Dissolve 0.05 grams of curcumin in 100 cc of Squibbs absolute alcohol and immerse in this solution one strip at a time heavy white filter paper of coarse texture and highly absorbent character (such as S. & S. No. 598, size 23 in., extra heavy) cut in 2-inch strips. Dry these strips in a dark place from one-half to one hour. Cut into strips $\frac{3}{8}$ in. by 2 in. and preserve in brown tightly stoppered bottles.

Standard Boric Acid Solutions: Make up sodium nitrate solutions (100 grams of sodium nitrate per liter of 10% hydrochloric acid) containing ten, twenty, thirty and fifty milligrams of boric acid per liter and preserve in glass stoppered bottles.

Borax Test.—Place 1 gram (approximately 15½ grains or ¼ level teaspoonful) of the well mixed sample in a test tube (½ in. by 4 in.), add 10 cc of 10% hydrochloric acid, and a small pinch of urea (about the size of a match head). Shake well and then, using a pair of clean tweezers, dip a strip of turmeric paper in the solution. Place the moistened paper on a glass plate and dry slowly, away from direct light, at a temperature as near 125° F. as possible. The testing must be carried on in a room which is free from ammonia fumes, and in all cases the test papers must be examined as soon as dry, as they fade rapidly.

If the percentage of boric acid does not exceed 0.05%, it can be determined by comparing the color of the turmeric paper with standard papers, prepared as follows:

Place 10 cc of each of the standard boric acid solutions in test tubes (½ in. by 4 in.). Add a small pinch of urea to each and finish exactly as in testing the saltpetre.

If the sample contains 0.05% or over of boric acid, the borax should be determined quantitatively. See Official Methods A. O. A. C. (1920) p. 123.

Sugar

Sampling.—The sampling of sugars, particularly of second sugars and raw sugars, must be carefully done, so as to be sure that the samples do not dry out before they are analyzed. After drawing samples from a representative number of sacks or barrels and collecting such samples in an air-tight container, such as a glass fruit jar with a well-fitting rubber ring, the composite sample is quickly and thoroughly mixed and the moisture determination made as promptly as possible.

Analysis.—Moisture: Five to ten grams are dried in a suitable moisture dish at 105° F. to constant weight. Usually five hours is sufficient. The loss in weight is calculated to moisture.

Insoluble Impurities.—Raw sugars sometimes contain insoluble matter, which, while not amounting to a great deal, when expressed in terms of percentage, can nevertheless become very noticeable and objectionable in actual practice. Ordinarily the insoluble impurities are determined by dissolving a convenient weight, say five or ten grams, of the sugar in 50 to 100 cc of water, filtering through a tared filter, washing thoroughly and drying. The percentage gain in weight of the filter is reported as insoluble impurities. The filter must be carefully dried and weighed in the same manner before filtration as after filtration of the sample.

In case of sugars of questionable purity, such as referred to above, a considerable quantity of a heavy solution, say 250 or 500 cc of a solution of about the same strength as made up in the curing department, will, after allowing to settle in a tall cylinder, give a very fair idea of the volume and character of the objectionable sediment.

Reducing Sugar.—Reducing sugar may be determined, if desired, in accordance with the Official Methods of A. O. A. C. (1920) p. 77.

White granulated sugars contain very little, if any, reducing sugar, but the percentage of reducing sugar in second and raw sugars usually serves as an index of their quality.

Physical Examination.—The physical examination of sugars should include a report on their odor and flavor. Sugars which have a moldy or fermented rum-like odor should be rejected. Sugars which are very damp should be stored so that they can be used up ahead of drier sugars, as the latter will keep better in storage.

PACKINGHOUSE COST AND ACCOUNTING METHODS

By J. H. Bliss, of Swift & Company, Chairman, Committee on Standardized Accounting, Institute of American Meat Packers.

The cost finding and accounting practices in the packing industry have been often misunderstood simply because they apparently are so different from practices in other industries. Much confusion and some criticism in the past has resulted from a lack of understanding of the principles underlying packinghouse cost finding and accounting procedure. The differences are not differences of principles, for the practices in this industry are based on sound accounting and business principles. The differences in the accounting are the results of the differences in the character of the operations performed.

The Standardized Cost Accounting Committee of the Institute of American Meat Packers has been busy for some time working out suggestions and recommendations as to the handling of packinghouse cost and accounting procedure. Their efforts are largely confined to the specific problems arising in the industry.

The following article was prepared for the purpose of explaining some of the peculiarities in packinghouse accounting procedure and contrasting these practices with those prevailing in other industries. In this article an effort is made to contrast and explain the cost finding and accounting procedure prevailing in the packing industry with that of other industries, without entering into a thorough or detailed discussion of either.

Three Types of Cost Finding

In a general way, and from the point of view of cost finding, the various types of manufacturing operations may be divided into three classes, each requiring a particular method of figuring or cost finding. These may be stated as follows:

a—Costs for ordinary manufacturing operations
b—Costs for operations producing major products and by-products
c—Costs for operations producing joint products.

It is the character of the operations that makes differences in methods of cost figuring necessary and determines the type of cost figuring to be applied to any specific kind of business. It is fundamental that the method of cost figuring should fit the particular operations to which it is applied.

In the consideration of any accounting or cost finding method, the first information to be developed is—what are the operations and transactions to be recorded. Then the cost finding and accounting procedure must be handled in a method adapted to these particular operations and transactions.

In the packing industry are found many and various operations. All three types of operations mentioned above are represented in the various branches of the business. Therefore in the cost finding and accounting for that industry each of these various types of costs are utilized. To each kind of operation or transaction is applied the most appropriate cost finding and accounting method.

Costs for Ordinary Manufacturing Operations

Ordinary manufacturing operations, as found in the great majority of industrial enterprises, consist of the putting together or building up of materials, labor, and expenses, into a finished product. This is the most common type of manufacturing operations, and generally thought of when cost figuring is mentioned. The steps in such cost finding would be as follows:

1—A known amount of material of a determinable cost is used.
2—A definite amount of labor of ascertainable cost is expended.
3—An average amount of overhead expense is absorbed.
4—The whole is the cost of the finished product.

This represents the steps in cost figuring for the ordinary type of manufacturing operations and is commonly known as cost finding.

This method of cost figuring is applicable to all business showing ordinary manufacturing processes such as:

>   The manufacture of Automobiles.
>   The manufacture of Soap.
>   The manufacture of Foundry Products.
>   The manufacture of Furniture.
>   Etc.

and in fact the manufacture of anything the operations of which are the putting together and building up of materials, labor, and expenses into a finished product.

Costs for Operations Producing Major Products and By-Products

A varied and somewhat different type of operations are those which involve the taking apart or breaking down of some materials of known value into several or many parts, of which one is the major product and the others by-products. All of these products being derived from the same material they are in the nature of joint products, that is, all produced out of something of known cost by the same operations. It is impossible to determine the cost of *each product separately*, although *the cost of all may be figured*.

The cost of the major product from such operations is usually computed in the following manner:

1—Starting with the known cost of the materials used.
2—Add the costs and expenses incurred.
3—Making the total outlay.
4—Deduct therefrom the *value* of the by-products produced.
5—The balance is the cost of the major or prime product.

It should be noted that by this method of figuring, the cost of the prime product is arrived at by deducting or crediting to the total outlay the full value of the by-products produced. The costs of such by-products are not ascertainable and the cost for the major product would not be ascertainable except by deducting from the total outlay the full value of all of the by-products produced.

The problem of costs for major products and by-products is found in several industries, including such as:

>   The Petroleum Industry.
>   The production of many farm products.
>   The production of gas, coke, etc., by gas plants.
>   The cattle business in the packing industry.

This method of cost figuring was recognized by the United States Tariff Commission in figuring the cost of wool, mutton being the by-product, and also in figuring the cost of mutton, wool being the by-product.

It is essential in figuring the cost of major products that the full value of by-products be credited. By full value of by-products is meant the marketable value of such by-products in their first commercial stage, less any costs and expenses to be incurred in processing them to such stage, and in marketing them. This gives the full value of the by-products in their present form at the time of production, and is the valuation to be used in cost figuring. If an arbitrary or different valuation were used, the *cost of* the major product would not be determined correctly.

By this method of figuring the results of the entire operations, the profit or loss appears upon the disposition of the major product. The by-product having been figured at their full value, should not show either profit or loss when disposed of, except as market conditions change or costs of processing and marketing have been over- or under-estimated.

This is the only known method of figuring costs when major products and by-products are produced out of the same operations, and is customarily applied to such businesses.

Costs for Operations Producing Joint Products

The third type of operations are those which are the taking apart or disintegration of something of known cost into several or many parts, *none* of which may be termed major products. They are simply joint products, that is, many products resulting from the same operations performed on a material of known cost. Being joint products it is impossible to figure the cost of *each separately*, though the cost of the *whole group might be readily computed*. For such operations it is almost always impracticable to apply an average cost to all of the various products, for the reason that usually some are high-grade and of relatively high value, while others are of medium or lower grades with relatively lower values, and an average cost would therefor be entirely misleading.

The practice in cost finding for such operations producing joint products is therefore about as follows:

1—Figure the value of all of the products derived from the given operations.
2—Figure the cost of materials going into such operations.
3—Figure the amount of expenses incurred.
4—The total of the materials and expenses, making the total outlay.
5—Compare the total cost expended to the total value of all products derived, arriving at the margin of profit or loss between the total costs and total value of products made.

This method of figuring emphasizes three points:

> The total cost or outlay.
> The total value of the products made.
> The margin of profit or loss on such operations.

In some cases the values of these various products may then be adjusted by spreading the profit or loss shown on the entire production,—that is, the value of each of the products may be reduced by a portion of the profit or increased by a portion of the loss, so that the total value of the products as a group is adjusted to the total of the outlay. The effect of this procedure is simply to spread the total cost over the various products made, on the basis of the relative market values of the various products.

In other businesses this adjustment of spreading the profit or loss over the products, is not made, especially if such products are merchantable commodities and may be marketed in their present stage or passed on to other departments of the business for further manufacturing, in which processes their identity is entirely lost. The reason for this will be made more clear later.

The problem of joint costs is found in several industries, for instance:

The Sorting and Grading of Tobacco.—Several grades of different values are sorted out of a given amount of tobacco purchased at a certain price. The costs of the various grades sorted out are figured by spreading the total outlay over the various grades on the basis of the relative market values of such quantities of each grade.

The Glue Business.—Various grades of glue are produced out of the same operations. The total cost of the processing is spread over these different grades made, on the basis of the relative market values of such quantities of each grade produced.

The Cottonseed Oil Industry.—In the cottonseed oil industry, oil meal, hulls, linter, etc., are produced by the crushing of cotton seed. Each product being a merchantable commodity in the form produced, the practice in this industry is to compare the cost of seed crushed to the total value of the product produced, arriving at the gross margin realized on such operations. The net results are then determined by deducting the expenses from this gross margin.

The Hog Business of the Packing Industry.—An example of this will be shown later.

These are not costs as ordinarily thought of, in fact, the usual methods of cost figuring cannot be applied to operations like these producing joint products. The cost figuring and accounting has to follow and fit the operations.

The differences between ordinary costs and costs of joint products may be illustrated by the building up of an automobile and the taking apart of a secondhand automobile. In the building up of an automobile, costs may be figured in the customary manner. Known quantities of materials, parts, labor, and overhead, are expended, and the totals thereof make the total cost of the finished automobile. On the other hand assume the purchase of a secondhand automobile for a certain sum, and proceed to dismantle it. The problem of joint costs would be like attempting to determine the cost of the engine or any part of it, or the cost of a fender, or a tire, or any other part from an automobile, when the only facts known are, first, the total cost of the machine as bought, and, secondly, the expenses incurred in dismantling. Obviously the costs of any of these parts cannot be determined. The cost of all of them as a group may be figured, but not the cost of any one item. It is a matter of joint products.

Costs of Ordinary Manufacturing Operations

We are all quite familiar with the cost figuring procedure for ordinary manufacturing operations, and little time need be spent in elaborating on that. It is simply a putting together of materials of known costs, a determinable amount of labor, and an average amount of expenses, into a finished product.

This method of cost figuring is used in the packing industry wherever it is appropriate. It is applicable to such operations as the:

> Manufacture of Sausage.
> Manufacture of Soap.
> Manufacture of Oleomargarine.
> Manufacture of Boxes.
> Manufacture of Cooperage and Other Supplies.
> Manufacture of Commercial Fertilizer.
> Manufacture of Canned Meats.
> Etc.

Cattle Business As An Example of Operations Producing Major Products and By-Products

The cattle business in the packing industry is a good example of that class of manufacturing operations producing major products and by-products. Dressed beef is the major product from this operation. By-products includes hides, oleo oil, stearine, tallow, tongues, other edible small products, bones, horns, tankage, fertilizer, etc. The dressed carcass of beef is the largest single product made from cattle by these operations and represents about 50-60% of the live weight. This is the prime product produced in the slaughtering and packing operations in the cattle business.

Beef is customarily marketed at wholesale in the form of dressed carcasses, either as quarters, or sides, or whole carcasses. When the beef carcass is received by the retailer, it is cut up into various parts, such as roasts, steaks, etc. The retailer faces a problem in joint costs. He knows what the whole carcass costs him, knows what the various cuts that he makes from it are worth. He knows that the market value of these various cuts varies widely, and he is therefore unable to apply or spread the total cost over them on any average basis. These retail cuts are simply joint products for which individual costs cannot be computed.

Since the beef is customarily marketed in the form of dressed carcasses by packing houses, it is the one major product resulting from the cattle operations. Following is an example of cost figuring for a representative lot of cattle:

Example of Cost-Figuring for Lot of Cattle

1. Live cost, 46 head, weight 52,390 lbs.; averaging 1,138; native steers, at $7.00..$3,667.30
2. Expenses, killing, dressing, chilling (estimated for month).. 118.03
3. Allowance, condemnations, trimmings, etc. (average)......... 17.77

   Total outlay on lot.....................................$3,803.10

4. Less credit for hides (cured values less expenses curing and marketing) ....$ 387.06
5. Less credit for fats (value of products less expenses)....... 116.30

6.	Less credit for other by-products (value of products less expenses preparing)	125.73
	Total by-products credits	$ 629.09
7.	Balance—plant cost of carcasses—in cooler	$3,174.01
8.	Dressed carcasses weighed 29,615 lbs., yield 56.53% of beef.	
9.	Average dressed cost per cwt	$ 10.72
10.	Add selling costs and expenses	1.87
11.	Total cost of lot, per cwt	$ 12.99

It will be noted that this example emphasizes clearly the effect of the yield of beef on prices, creating the spread between the live price and the cost of the beef produced. The total live weight was 52,390 and the weight of the dressed beef only 29,615. This is a yield of 56.53% of beef out of the live weight. The live cost was $7 per cwt., which means $7 for each hundred pounds of live weight. Considering that the yield of beef is only 56.53% would have the effect of almost doubling this price as applied to the dressed beef produced. This is the most important single factor in accounting for the spread between the cost of live cattle and the cost of dressed beef, and is probably one of the most confusing factors in the public mind.

Note also that the value of the by-products is considerably in excess of all of the expenses, so that the dressed beef might be sold at a good margin over cost and still be sold for less money than was paid out for the live animal.

Hide Credits.—The credit for hides is arrived at in the following manner:

From the current market price for this particular grade of cured hides, is deducted sufficient to allow for the shrinkage and expenses in curing and marketing. This gives the present value of the green hides per cwt. This price is applied to the actual weight of hides produced by each lot.

Fat Credits.—The credit for raw fats is likewise figured backwards. Starting with the market value of the yield of oil and stearine from fats, the expenses of rendering and marketing are deducted and an allowance is made for the yield of such oils and stearine. This gives the basis for valuing the fats produced from each lot in their present raw state.

The credit for other by-products, which includes everything else produced out of the animal, are all computed in the same manner. That is, from the values of the finished products at current markets are deducted the expenses of preparing and marketing, which gives the present value of such by-products in their raw state.

Value of By-Products.—The values of by-products should be determined on the basis of current market prices, for that is the only known basis. It is impossible to predict what the market for these by-products may be 30, 60 or 90 days hence, when they are disposed of, and to use other than the present market in figuring their values would be mere guess work.

When the current market price is used, any loss or gain realized upon

the ultimate disposition of the by-products will be made up of two factors:
1—Loss or gain due to changes in market values of by-products between date of slaughter and date of sale.
2—Loss or gain due to adjustments in expenses and yields.

As the by-products are figured at their full value, based upon the present market, it follows that the results of these transactions will show as a profit or loss in the disposition of the beef.

Adjustment of Results.—These results shown on the disposition of the beef, being the profit or loss between the cost as figured and the price sold for, are subject to periodical adjustment as follows:

a—Loss or gain on realization of by-products.
b—Difference in expenses between estimated and actual.

Applying these adjustments gives the final results, profit or loss, on the cattle business as a whole.

Cattle are customarily bought in the central live stock markets in lots ranging from one or a few head to several carloads. Usually the lot as bought is the basis for cost figuring. In some cases purchases of small numbers may be combined into one lot for slaughtering and cost figuring purposes, especially if they are of the same kind and grade.

Almost all of the beef products are sold fresh and have to be marketed within a very limited time after slaughter. The markets for beef and for beef products are changing from day to day. It is necessary, therefore, that the costs of these lots killed daily should be figured within a very few hours after the operations take place. This necessitates estimating expenses and yields of various products on the basis of averages. Hence the results in the beef business within a period, for a day or a week may be subject to some minor adjustments, but when the business for a period is considered and the adjustments for expenses and by-product values are applied, these become the final figures for the cattle business.

Dressed beef is customarily inventoried at cost, for costs are figured and are available. By-products, however, are customarily inventoried on the basis of market values, for no other values are or can be determined.

Hog Business As An Example of Operations Producing Joint Products

The hog business in the packing industry is a good example of operations producing joint products. Pork products are customarily marketed in the form of cuts, loins, hams, bellies, shoulders, etc. Very little pork is marketed in the form of a dressed carcass, as in the case of dressed beef.

Another feature of importance to be noted is that part of these pork products may be marketed fresh, as is usually the case with loins, butts, etc., and other parts will be cured and smoked and finally marketed in the form of bacon, smoked hams, etc.

The hog business differs from the cattle business essentially in that there is no one major or prime product. It is purely a case of joint products, or many products made by the same operations, out of certain raw materials of a known cost.

Figuring a Test on Hogs.—The following is an example of how a test on a lot of hogs would be figured. It should be noted that it follows quite well the procedure previously mentioned in this section. The yield column shows the pounds of each kind of product produced out of each

100 pounds of live weight on the average for the lot. Applying the current price to this amount of product, as extended, gives the value of each product out of 100 pounds of live weight. For example: This lot produced on the average, 13½ pounds of fresh hams for each 100 pounds of live weight. These hams were worth 13 cents per pound. Hence the hams produced were worth $1.76 per 100 pounds of live weight. So with each other product. The complete test follows:

Test on a lot of hogs:

Lot of Hogs of Average Live Weight of 250 Pounds

Products	Average Weights	% Yield Out of Live Wt.	Current Market Price	Extension
Fresh Hams	16/18	13½%	13c	$1.76
Fresh Shoulders	12/15	10%	10½c	1.05
Fresh Bellies	14/16	12%	11¼c	1.35
Fat Backs	8/10	7%	7c	.49
Pork Loins	8/10	10%	14½c	1.45
Spare Ribs	....	1%	11c	.11
Prime Steam Lard	....	14½%	8c	1.16
Trimmings	....	2%	7½c	.15
Miscellaneous	....	3%	4c	.12
Yield and gross value		73%		$7.64
Expenses (per cwt., alive)				.62
Hogs, per cwt.				$7.02
Hogs cost				6.70
Profit per cwt., alive				$ .32

This is the method used in figuring a test on a lot of hogs and provides the information needed by the management in following the markets, and judging whether or not it is profitable to buy at prevailing prices. Note that it emphasizes the following points:

 a—What all of the products are worth at the present market, reduced to a live weight basis.
 b—The expenses of operation.
 c—The cost or present market value of live hogs.

Obviously the difference between the value of the products and the total costs would represent the profit or loss between the markets at these figures. This is the method that a man doing a hog business would use in figuring whether or not his purchases were profitable, and in judging whether or not it would be advisable to extend operations.

Fresh Pork Department.—These figures also indicate the basis upon which the fresh pork or killing and cutting department is usually established in most packing houses. Such a department would draw together the following figures:

 As charges to the department—
  The cost of the hogs purchased.
  All expenses incurred.

As credits to the department—
Value of all products sold or transferred from it.

These factors together with the inventories at the beginning and end of the period, make up the fresh pork or killing and dressing departmental accounts, indicating the profit or loss for a period on current hog business operations.

The foregoing statement emphasizes as well that there is no one major product in the hog business. Note that no one item produced is as much as 15% of the total live weight of the hog. Also note the wide range in the value of the products realized, from the highest, pork loins at 14½c per pound, to trimmings at 7½c per pound.

Inventories of such hog products are necessarily valued on a market basis. There are no actual costs for the various products. As these products move throughout the various processes and manufacturing operations, the identity of specific items is lost, so that the only available basis for valuing inventories is the current market for such products.

Departmentization of the Business

Departmentization of the business and of the accounts is one of the features of the present-day packing organizations. Departmentization is necessary because of the large variety of operations and activities which any sizeable modern plant conducts. Any plant may have a variety of operations peculiar to itself, and will naturally have to be departmentized on a plan particularly adapted to its physical layout and operations.

Some plants handle only a cattle, calf, and sheep business. Others may handle only a hog business. Some will handle all classes of live stock. Some plants do not process by-products to any extent, but dispose of them raw or in the lesser stages of manufacture, while others process their by-products to various degrees. Some plants do not engage in allied industries,

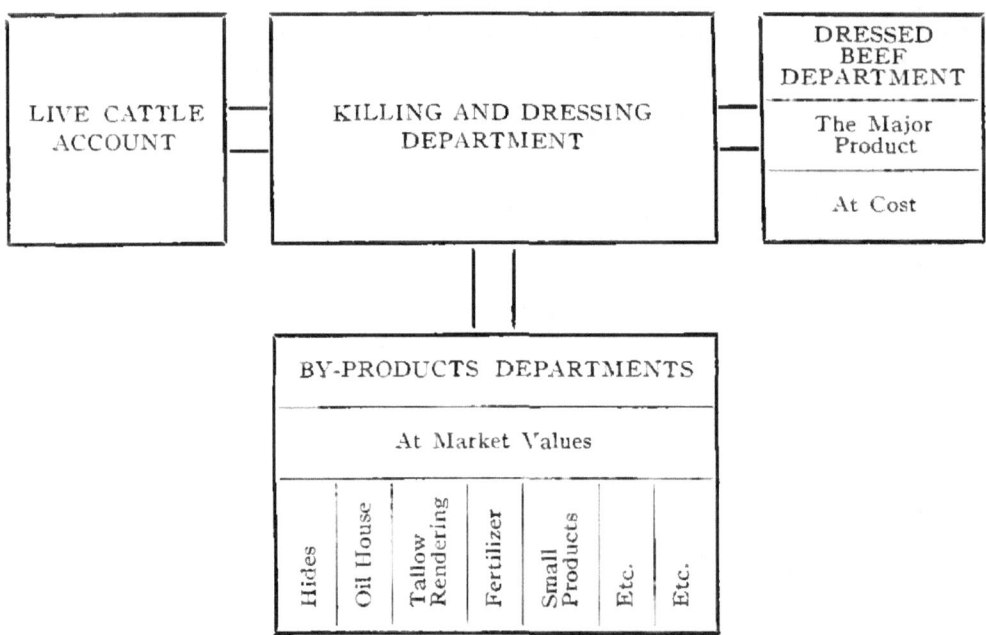

A—Plan for Departmentizing Cattle Business

while other organizations may have allied business, such as glue, soap, fertilizer, etc. The handling of the business and accounting therefor requires careful and logical departmentization, appropriate to the particular plant and its activities.

Departmentizing the Cattle Business.—In the cattle business the departmentization should fit the operations and support the cost figuring and accounting procedure. Chart numbered "A" indicates briefly a general plan for departmentization of the cattle business. This will of course be varied somewhat to meet local conditions, but the general plan will usually be found about the same in various establishments.

Departmentizing the Hog Business.—The departmentization of the hog business will naturally be somewhat different because the operations and processes differ materially. One of the outstanding features of the hog business is that the operations are a succession of processes. The dressed carcasses are customarily cut up into the form of cuts—hams, shoulders, loins, bellies, etc. Some of these may be sold fresh, some may be put into cure, some may be carried in storage from a period of surplus production to a period of scarcity. As the product comes from cure, some of it may be sold in that stage, and some of it processed further, into the form of smoked hams, bacon, etc. Each one of these various processes

    Killing, Dressing and Cutting,
    Storage,
    Curing,
    Smoking,
    Etc.,

are in fact distinct competitive industrial businesses, and the departmentization has to recognize these peculiar features.

For the hog business, the departmentization will be something along the lines indicated in the following chart:

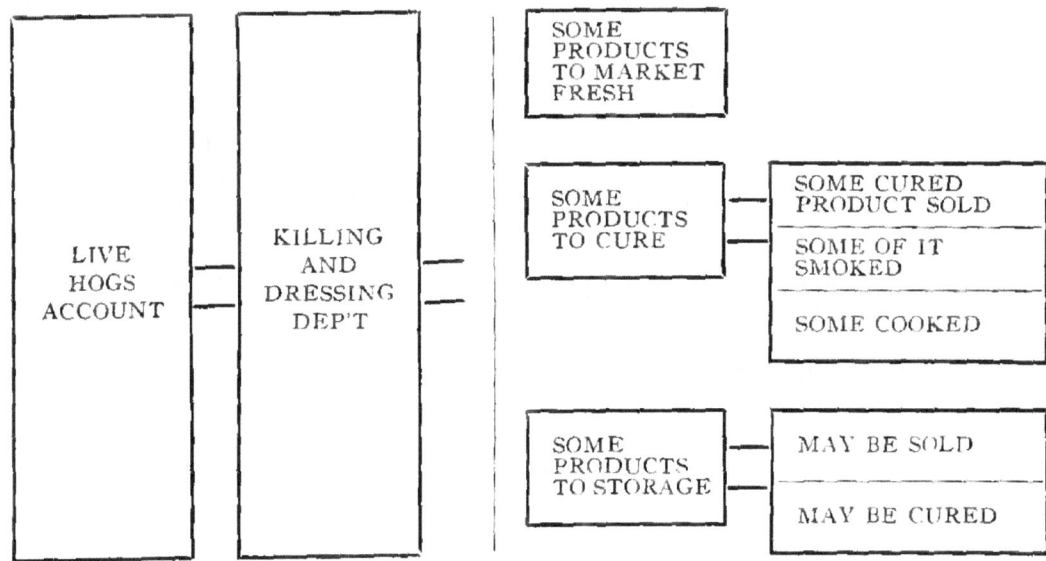

B—Plan for Departmentizing Hog Business

Departments Should Be Handled On Commercial Basis

The present-day packing organization is essentially a group or aggregation of separate industrial activities or competitive businesses. Each department or activity meets the competition of other firms engaged in that particular line of business. There are those who only have a slaughtering business, selling all products fresh. They meet the competition of all slaughterers, including as well, local and farm slaughterers. Some organizations process by-products into further manufactured stages, and in these activities meet the competition of concerns handling packinghouse products, such as rendering concerns, hide dealers, etc.

Some concerns cure and smoke pork products and in this operation meet the competition of firms who do only this class of business. Some plants make sausage and must meet the competition of all sausage makers, whether they do slaughtering and curing business or not. So that the larger organizations, engaging in many of these operations and processes, meet competition of other firms at each of these various stages of manufacturing or processing.

Accounting for Each Department.—It is necessary, therefore, that each of these separate competitive businesses within the organization be handled, operated, and accounted for, on an independent commercial basis, the same as its competition is operated and accounted for. That is the only way these larger groups of businesses can be handled or operated. That is the way the organization is built up and the way the business is manned and operated. It is simply an aggregation of separate competitive businesses.

This condition means accounting for each department on a commercial basis, charging it for its goods bought at their full value, whether bought from another part of the organization or from the outside, charging it all its factory and general expenses incurred, and selling its product at the full market value, whether it is to the outside or to some other part of the organization for further manufacture.

"Opportunity Costs."—The delivery and receipt of these inter-organization sales or transfers at their full value is known as transfers at "opportunity costs." "Opportunity Costs" are well recognized in economics, but little considered among accountants. It means simply that any product is worth to a department of an organization what the preceding department had the opportunity to get for it if placed in the outside market.

In this industry, at the close of each process or manufacturing operation, the management has the opportunity or option of either selling the product on the market in that stage or transferring it to some other department for further processing and manufacturing, the product to be ultimately sold in some other form.

For example: Oleo oil is a merchantable commodity, and may be sold as such or may be transferred to an oleo factory for use in making oleomargarine. Fresh hams and bellies may be sold on the market as such or transferred to a curing department to be cured. Cured hams and bellies may be sold as such as they are merchantable commodities, or may be transferred to the smoke house and manufactured into smoked hams and bacon.

So at each stage in the many processes there is the option of selling the product in its then present form, or passing it on for further manufacturing operations, to be finally sold in a more processed form. This is one of the peculiar features of the industry. The product in its various stages of manufacture is almost always a finished merchantable commodity and may be disposed of in that form. "Opportunity Cost," therefore, is the cost to a department, of products received from some preceding operation and represents the value the preceding department could get for such products if they were placed on the market in that form.

The markets for various products are the factors which determine the disposition of products in any of these stages. If the market for the product at any stage is more than could be realized if it was manufactured into some other product, then it is more profitable to sell it in the first stage. If the markets for further manufactured products indicate that such processing would yield a profit over what would be realized by immediate sale, then it is more profitable to carry the processing operations further.

The fluctuations of the markets for all of these various products is the factor that stimulates or contracts the processing operations and adjusts the supply of each kind of the various products to the demand for such.

Each one of these various departments handled on a commercial basis receives a final accounting, and a profit or loss is realized in each one. This is entirely in order, for as each succeeding operation is performed on the product, labor and expenses are encountered, investment in buildings, machinery, equipment, inventories, etc., is used. The additional manufacturing operations incur additional costs, use additional investments, and should be expected to yield a price sufficient not only to cover the expenses, but to yield a return on the investment used.

The results of the hog business in total are composed of the results of each one of these various departments or processes taken together. For operating purposes, and the conduct of the business, it is necessary to know how each department stands on a commercial, competitive basis.

Unrealized Loss or Gain.—Inventories of these various products in the process of manufacture will naturally include an element of unrealized loss or gain. To eliminate this would be impossible, for the reason that these various products are unidentifiable in any of these manufacturing processes. A vat of hams, for instance, may include hams from several lots of hogs bought at different times and at varying prices—a pile of Dry Salt Cuts in the cellar would naturally include cuts made from many lots of hogs bought at different times and at different prices. It is impossible to follow the products and identify them as they pass through the various operations. Hence it is impossible to allocate any departmental losses or gains. It should be noted, however, that the margin, profit or loss, in the packinghouse operations averages a very small percentage of the total value of the products handled, and that as inventories are taken on the same basis at the beginning and end of a period, the element of anticipated profit or unrealized loss in inventories is quite negligible indeed.

Influence of Markets on Packinghouse Operations and Accounting

Packinghouse operations are conducted between two very sensitive and constantly fluctuating markets—the live stock market on the one hand where the live animals are purchased, and the product market on the other hand where beef, pork and other products are sold. Each of these markets is constantly fluctuating, changing daily or almost hourly, in response to conditions within itself. The general movements of these markets, of course, have to be relative, for live stock is worth what its products will bring in the product market less a margin for the services of manufacturing and distributing. Any individual market, however, is subject primarily to the conditions of supply and demand within itself, for the reason that these are perishable products and have to move into consumptive channels at whatever price will induce the trade to absorb the supply at any time.

Beef, for instance, is not sold in any market for a fixed or figured list price, but for what that product will bring in the market in which it is offered for sale. Retail dealers come to the wholesale houses and inspect the beef offered for sale and make bids. They customarily shop around and take advantage of the best bargains offered. Each individual sale to a retailer is a trade in itself, resulting from bids and askings. This product being perishable, it must be sold within a few hours after it is received at packinghouse wholesale branch. Necessarily, therefore, it moves at the best prices obtainable under the local conditions of supply and demand.

Live cattle, on the other hand, are customarily marketed by farmers in the established open live stock markets. They are consigned to and offered for sale to packinghouse buyers by commission merchants who act as agents for the producers. The prices for live animals are affected by the prices at which products are selling and the demand for them, and the quantity of cattle offered on the market.

Fewer cattle coming to market means less beef to be placed on the product market. Less beef at any time in a product market to supply the demand there, tends to raise the selling price. More cattle coming to the live stock markets means more beef on the product market. A large supply of beef on the product market with steady demand can only be moved at lower prices. And it must be moved.

So it is the fluctuation in prices for products and animals that stimulates or contracts the amounts coming to such markets. The trends of prices in the product markets reflect back to the livestock market. Price changes in the live stock market stimulate or hold back shipments by farmers. Naturally they attempt to place their stock on the most advantageous market. When prices are good it stimulates shipments coming in. When a surplus appears and prices are depressed, it naturally holds back shipments.

Working of Supply and Demand.—Fluctuation in prices determined in open competitive markets is the only agency which adjusts the production of goods to the demand for them. It is the desires of the consuming public, and the prices they are willing to pay for goods and services, that leads to the expenditure of human effort necessary to produce such goods. The forces controlling prices in markets between which the packing industry is conducted, do not differ fundamentally from the forces controlling markets for any competitive industry. The significant feature is the sensitiveness of these markets for live stock and packinghouse products, which is due to the perishable character of the products that necessitates prompt movement into channels of consumption.

In this connection reference should be made to an article appearing in the Journal of Political Economy, Volume XXIX Number 8, of October, 1921, entitled "Unit Costs as a Guiding Factor in Buying Operations," by Mr. George E. Putnam. This article emphasizes in a very clear and able manner the fact that unit costs do not *determine* selling prices. It is the demand for goods that leads to the supply, and the present and prospective consumers' prices that lead to producers' prices.

The fact that the markets for live stock and packinghouse products are unusually sensitive means that those controlling such packinghouse operations must be promptly and well supplied with market information and cost statistics. They must have at hand at all times the most recent information as to the markets and costs of live animals; they must know what the products are selling for, wherever such products are marketed; and must have up-to-the-moment and reliable information as to costs. These conditions impose on the accounting forces of packinghouse organizations more exacting requirements than commonly met in other industries. In addition there are the peculiar cost and accounting problems arising out of the character of the operations, which have to be considered.

In all, the packing industry probably places greater responsibility on its accounting and statistical forces than most industries, and the information and statistics prepared must be dependable and supplied promptly. Old cost figures and statistics may afford consolation, but do little good for the current business.

LOCATION OF PACKING PLANTS

One of the most important elements making for success in the packing business is proper selection of the location for a new plant. This is a fundamental matter, for unless great care is taken in choosing a site it is difficult to build a successful business.

The greater number of packing plants in the United States at the present time are located in the corn belt states, Illinois, Indiana, Iowa and Missouri. But in recent years there has been a scattering of packing plants, and now there are, besides the plants in the large centers, many plants throughout the entire United States which supply local regional markets.

Factors in Location.—The successful plants which have been established have taken into consideration in selecting a location the following factors:

> Accessibility to live stock markets.
> Economical distribution of the finished packinghouse products.
> Railroad facilities.
> Labor supply.
> Cost of power.
> An abundance of water.
> Drainage and sewerage facilities.

Financing.—In addition to these physical factors, the adequate financing necessary to the success of a packing plant should be borne in mind. Not only is this so because of the funds needed for live stock purchases from day to day, but for carrying of stocks of some packinghouse products for months. Modern methods, machinery and equipment are costly, and competition is keen. This is especially true in handling perishable products, subject to quick deterioration if not properly handled, and if poorly refrigerated.

The facts of quick turnover of stocks to prevent loss, and keen competition, have tended to make the packing business one of small profit margin on its sales. By not realizing this, or understanding the fundamental factors in picking the location of a plant, a considerable number of failures have resulted, especially from the operations of professional promoters and misguided enthusiasts.

Packinghouse Promotion.—In various sections of the country projects have been launched to establish packing plants. After being constructed and put in operation, many have had to shut down because of bad location and lack of working capital, or inexpert management.

These examples only go to show that the meat packing business must rely on small profit margins, and that it requires not only time, but technical skill and infinite labor, to build a good business.

Packers' Profits.—Old-established packing businesses rely on this small margin as much as new concerns, and this includes all revenue from by-product and even from private car lines. One business for the ten-year period, 1911 to 1920, had a net profit per dollar of sales of 2.07 per cent, and a net profit on net worth (capital and surplus) of 10.18 per cent.

These figures show moderate fluctuations before the world war, an increase of profits while business was being done on a rising scale of prices, and the decline of profits since the armistice, with the price movement downward. The big year in the business was 1917, when the prices of livestock and products were enhanced by gains upon stocks on hand. These gains were all lost, however, in the three years 1918, 1919 and 1920. The average profit for the last five years was less than for the five years 1911-1915.

Consideration of these facts and factors only adds to the force of the advice concerning the selection of location for a new packing enterprise.

CONSTRUCTION OF PACKING PLANTS

The change due to economic factors, such as increased freight rates and change in location of livestock supplies, has brought about the erection of packing plants in localities in the United States which no one would have considered twenty-five years ago. The new plants which have been constructed, due to the previously-mentioned economic factors, have been laid out and constructed very carefully by highly-trained packinghouse engineers and architects.

In designing these plants a great deal of attention has been given to construction of buildings to meet the conditions for a given locality. There is no general rule for the laying out of a packing plant, because in certain plants a great percentage of hog products may go into curing cellars; whereas, in other plants conditions may be the reverse.

There are today packing plants of the smaller type which slaughter hogs, and they are out of the cooler and sold the next day. In other localities little packing is done during the summer, and the biggest business is done during the winter months; a plant in such a section would have to be constructed accordingly.

It is not the intention here to describe building materials, equipment or construction details, which belong to the sphere of the trained packing-house architect and engineer, but typical plant layouts are given to show the general arrangement of modern packing plants.

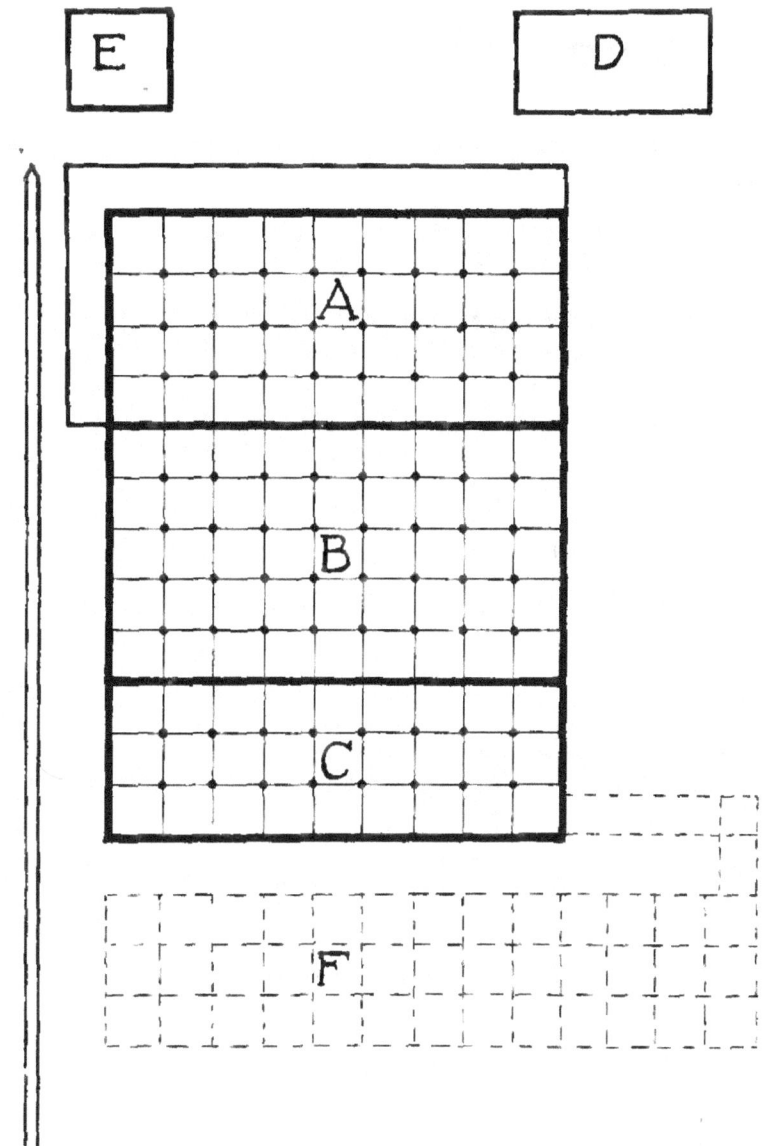

A—Plant With Capacity of 3,000 to 5,000 Hogs Per Week

Illustration A shows a plant of a capacity of from 3,000 to 5,000 hogs per week, located in a western state. A is the cooler building, where products are loaded on the cars and trucks. B is the manufacturing building, containing the killing and offal floors. C is lower than the other parts of the building, and contains waiting pens on top, and tankhouse and power equipmet. The livestock pens, F, are in the rear. E is the office, and D is the garage.

This type of construction is considered very modern, and this particular plant, in its generalities, has been copied by many packers.

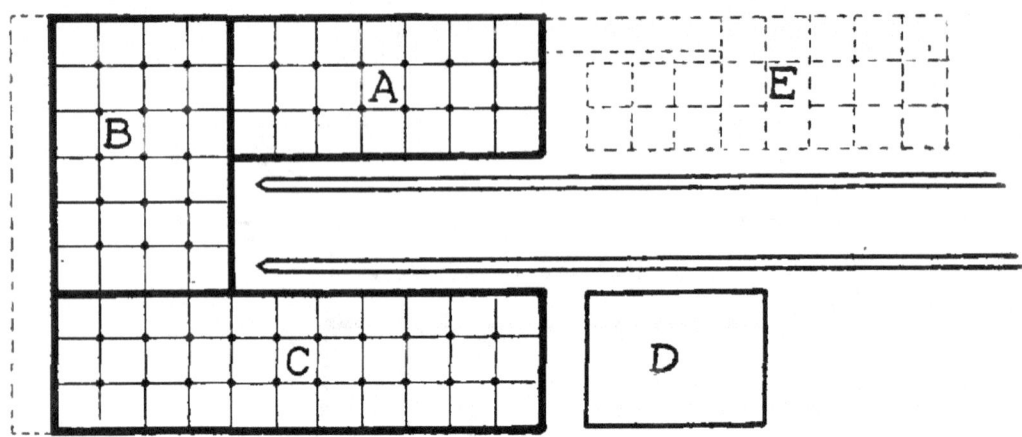

B—Another Type of Plant With Interior Court

Illustration B shows a packing plant arranged with a large court. A is the killing building, also containing the offal and tankhouse. B is the manufacturing. C is the cooler. D is the garage, and the pens are marked E.

A plant of this type has the advantage of an abundance of daylight for the killing floors and the manufacturing building.

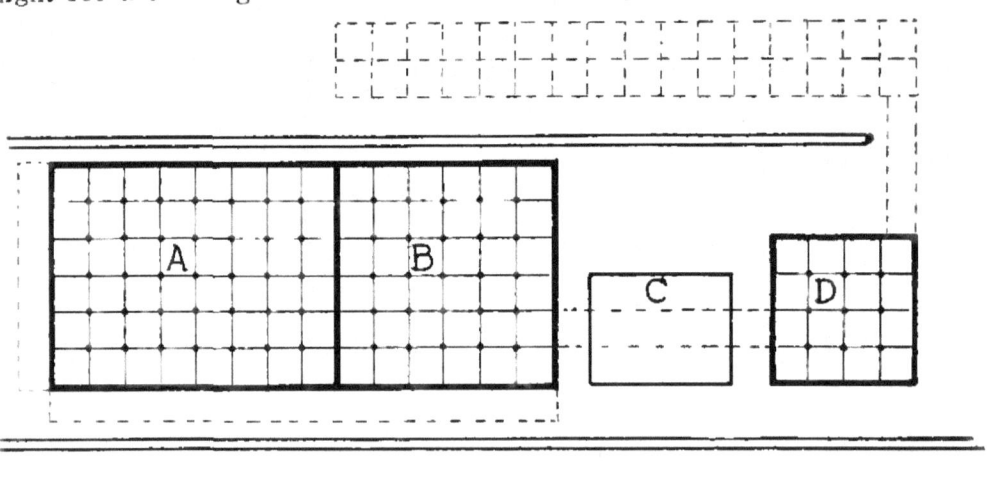

C—Large Plant With Separate Pen House and Power House

Illustration C illustrates another modern type of layout, a large packing plant with a capacity of from 6,000 to 15,000 hogs per week. A is the cooler building. B is the manufacturing building and tankhouse.

C is the powerhouse, and D is the live stock pens, from which the livestock go over a bridge into building B. Building E is the garage.

Illustration D shows a small or medium-sized packing plant with a capacity of 1,000 to 2,000 hogs per week. This building is laid out along the same lines as the one shown under illustration A, except that it is on a smaller scale, and the office is in the building.

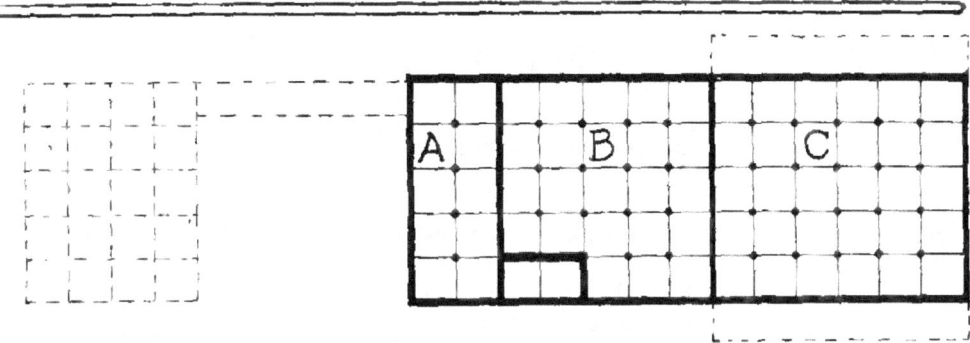

D—Medium-Sized Plant, Compactly Laid Out

Illustration E contains a slaughtering plant adapted for either cattle or hogs, and where there is no manufacturing done. It is located in an eastern city, and the products are quickly disposed of. The pens are very conveniently located between the railroad tracks, and E illustrates the holding pens, where the animals are led into building B by way of an outside chute. This is the slaughtering floor. Building A represents the coolers. D is the tankhouse. F is the powerhouse. C contains the elevators used to elevate the by-products up to the tank-charging floor.

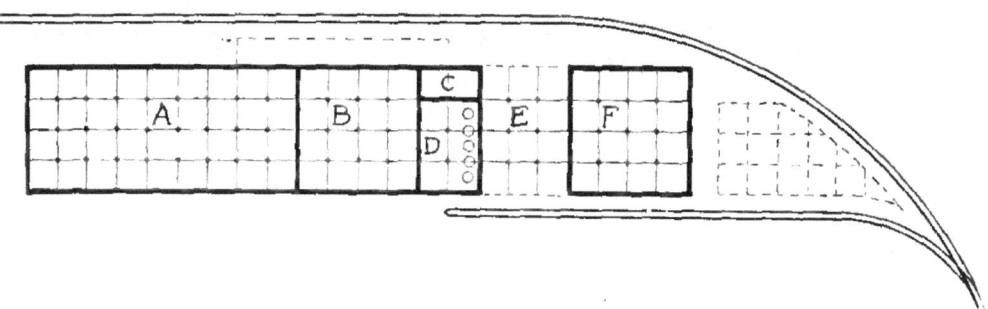

E—Killing Plant, Where No Manufacturing Is Done

It can readily be seen from the foregoing illustrations that the design of a packing plant must be judged by the conditions it has to meet.

Great progress has been made during recent years in sanitation, due to modern methods and to the regulations of the United States Department of Agriculture, which have helped a great deal to bring packinghouse reputation to a high standard so that American packinghouse practice is now considered the most sanitary and up-to-date in the world.

PACKINGHOUSE REFRIGERATION

No attempt will be made here to enter into all the technical details of mechanical refrigeration, but to cover the subject in a general way as it applies to modern packinghouse practice. Great progress has been made in recent years in the science of refrigeration, and it has been indispensable to the growth and development of the packinghouse industry.

Methods.—There are four principal systems of distributing refrigeration now in use in the modern packinghouse.

        a The Direct Expansion System
        b The Brine Circulating System
        c The Brine Spray System
        d The Curtain (Brine) System

In addition some few packinghouses use some of these systems applied slightly differently. A brine spray system is also used in which the spray is located in a pipe or funnel placed on the ceiling, and which has eliminating bunkers or coil loft, saving considerable height in the building.

Another new system recently installed in a large Eastern plant may be termed a "water cooling system." Instead of spraying brine on the cooler

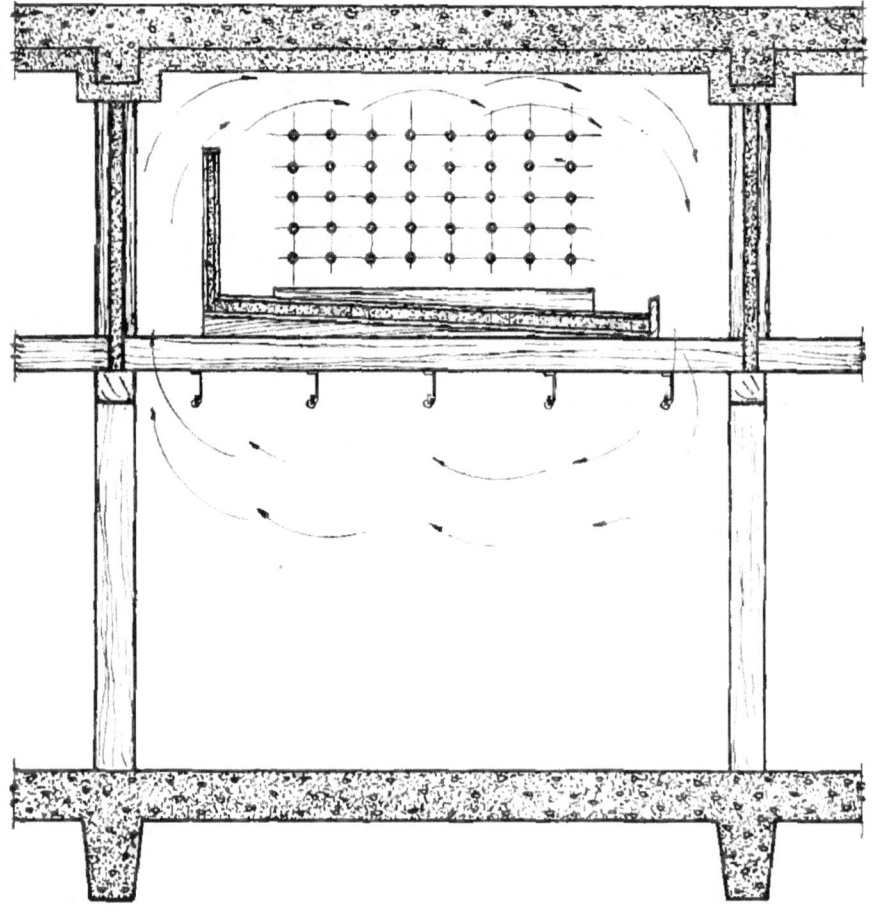

A—Showing Arrangement of Coils, Either Direct Expansion or Brine, on a Single Bunker

decks, water which has been cooled in Beaudalot coolers is used. Needless to say, such a system must be watched very closely to keep the water from freezing.

The indirect cooling system is rarely found in packing plants in the United States, but is used extensively in Europe and in tropical countries. It is the same method used in the United States in candy and chocolate factories. The coils are located in a well-insulated chamber, from which the cold air is taken and forced through ducts in the various rooms to be cooled. This allows a control to be established over the volume of cold air and the humidity.

A number of packers have installed forced air circulating systems in connection with air-conditioning equipment, insuring them perfect temperature and moisture control in rooms where this is actually required. If continued progress is made in scientific packinghouse refrigeration, this method will find many users in this field.

The Direct Expansion System.—This is the most common method in use today in American packinghouses. The ammonia coils are placed

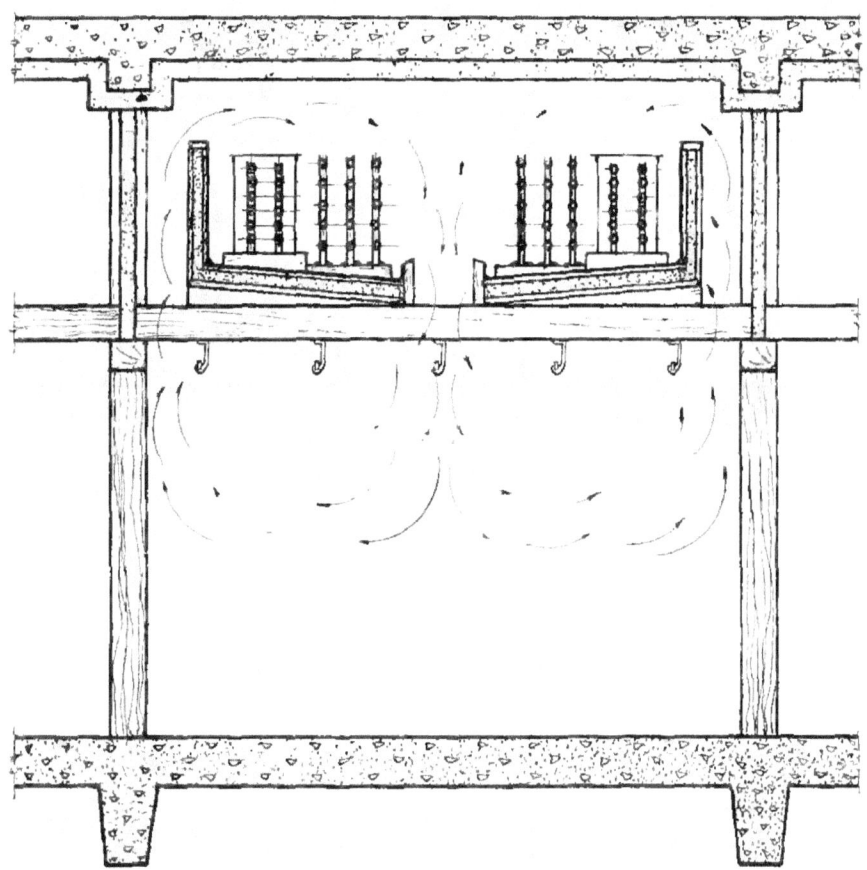

B—Direct Expansion System, With Hold Over Brine Tanks, Used in Small Packing Plants, Showing Also Double Bunker for Rapid Circulation

directly in the rooms to be cooled, either in the coil lofts or on the sides or ceilings. Depending upon the amount of pipe in the rooms, the various sets can be controlled by various expansion valves.

The Brine Circulating System.—Instead of extra-heavy direct expansion coils being used, the coolers are equipped with brine pipes. Cold chloride of calcium or common salt brine is pumped through these coils. This brine is cooled either in a large brine tank equipped with direct expansion coils, or the brine is pumped through additional brine coolers of various types.

The Brine Spray System.—Cold brine is produced by the same method as with the brine circulating system, except that instead of coils, brine spray nozzles are placed on the water-proofed bunker decks (see illustration C). The spraying brine falls on to the deck and gravitates or is pumped back to the brine tank.

The Curtain System.—This system is still in use in some large packing plants, and is in a way similar to a brine spray system, except that the brine trickles down on curtains made of burlap or canvas instead of being sprayed.

Irrespective of the various advantages and disadvantages of any particular system, it is advisable to engage competent refrigerating engineers

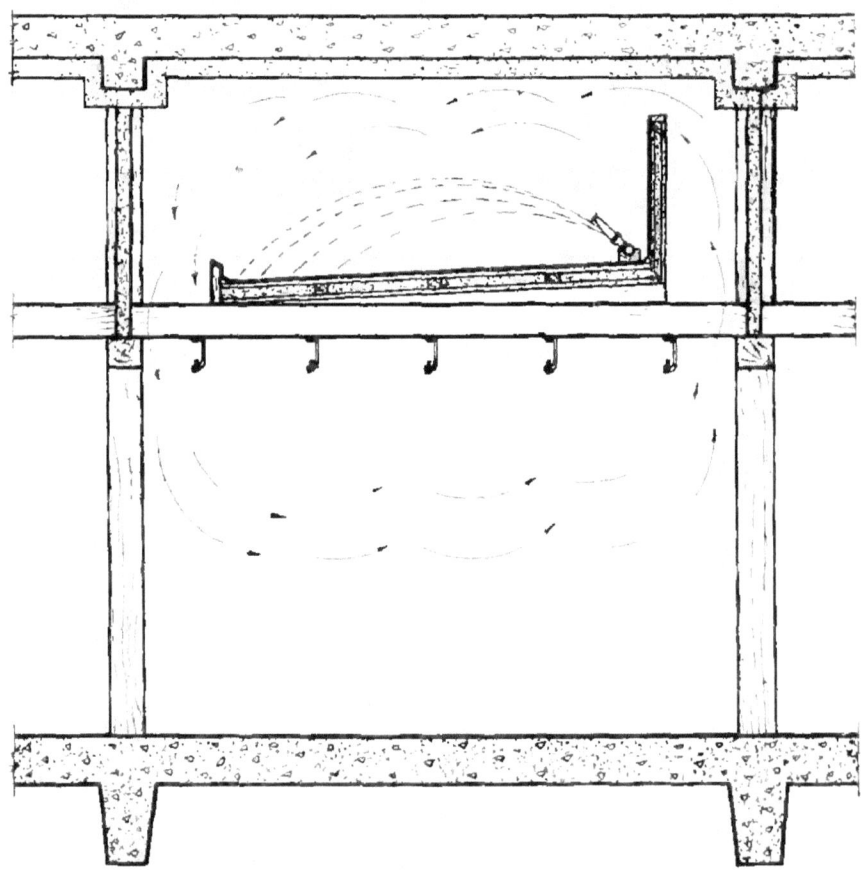

C—Brine Spray System

for designing or laying out a refrigerating plant, as much depends upon the installation of a refrigerating system in the packing plant. Power, water, temperatures and insulation have to be taken into consideration, and it is a true saying that poor refrigeration will make a packer poorer.

Air Circulation in Coolers

Of equal importance to a proper refrigerating plant is the circulation of cold air in the coolers. As cold air is heavier than warm air, coils should be arranged in such a manner that this principle can be best applied.

The overhead bunker or coil loft method has proved very satisfactory, and is principally used in the packinghouse field. The deck or coil loft is well insulated to prevent the condensation of warm air under the coil loft. Another very important point is that the coils should not be placed too close either to the ceiling or bunker floor, as the frost may go through the ceiling or bunker floor if the coils are not kept properly defrosted. Illustration "A" shows a direct expansion system and single pan coil loft, with the arrow indicating the flow of air.

In smaller packing plant coolers, and where rapid circulation is required,

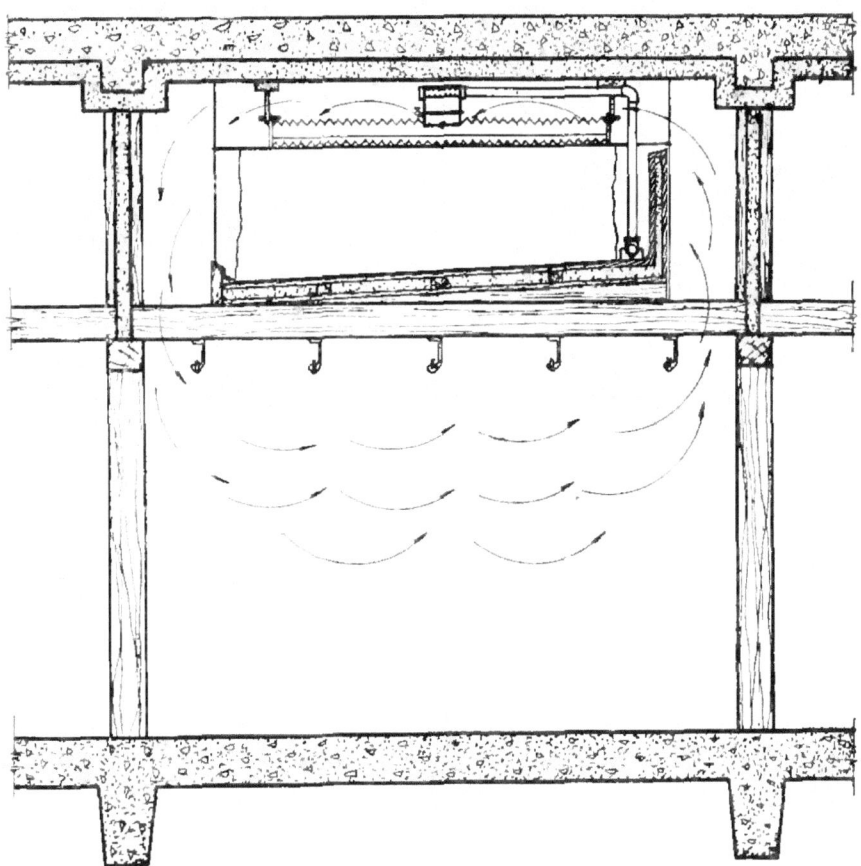

D—Curtain System

as in the chill room, the double coil loft is used, as is shown in illustration "B." Besides the direct expansion coils this drawing also shows brine hold-over tanks, which are used to assist in maintaining the temperature when the machine is shut down during the night in the smaller plants.

A brine spray system is shown in illustration "C," and illustration "D" shows the application of the brine curtain method.

Proper cold air circulation will prevent the accumulation of excessive dampness and moisture, as well as extreme dryness, which may cause excessive shrinkage. The colder the air is when entering the cooler, or the colder the medium used for chilling, the greater the amount of moisture that will be condensed, and therefore the drier the air will be in the cooler and the greater the shrinkage.

In other words, the greater the spread between the incoming refrigerating medium and the outgoing medium, the greater will be the shrinkage; and vice versa, the nearer the temperature of the incoming medium to the outgoing medium, the less the shrinkage will be. As stated, however, these are problems which should be decided only after consulting an efficient refrigerating engineer regarding the matter.

Systems of Making Refrigeration

The two principal refrigerating systems now in use should be mentioned—the Compression and the Absorption System. The first-named can be found in the majority of plants, and is a system familiar to most packers. The absorption system is not found in so many packing plants, due no doubt to the unfamiliarity of the packer with this system, which in many cases can be most advantageously installed in conjunction with a Compression System and operated at a very low cost.

Unbiased, competent refrigerating engineers should always be consulted when a new refrigerating plant is required, as much of the success of the packing plant depends upon economical and reliable refrigeration.

Chapter VI—VEGETABLE OILS

VEGETABLE OIL REFINING

As many of the large packing plants now place on the market lard substitutes, margarin and cooking oils, the vegetable oil industry has become of considerable interest and importance to meat packers. This discussion of the refining of vegetable oils, compound and margarin manufacture is written from a practical standpoint, as a guide to superintendents or operating engineers.

There are no standardized operating instructions for conducting a refinery, as the locality where a plant may be situated, or where it markets its product, determines the character of the product to a large extent. Again, practice changes with the superintendent operating the plant, who usually applies his own ideas.

There is much secrecy concerning practices in this particular field, and information concerning refining processes has been guarded very closely. These so-called secrets, however, are more or less common knowledge today in the refining industry.

In the following pages there is given a detailed description of the various operating processes as practiced commercially.

Refining

As the crude oils received at the refinery contain many impurities, such as meal, moisture, coloring, mucilaginous and nitrogenous colloidal matter, and free fatty acids, these must be removed for storing and stabilizing the oil, and also to improve the flavor and color for edible use. It is for these reasons that crude oil should never be put in storage.

Refining Processes.—The refining or purifying of vegetable oils is accomplished by several processes. The most used and practical method, however, is the caustic alkali treatment.

Caustic soda neutralizes the free fatty acids which coagulate and separate from the oil, and carries with it other impurities and coloring matter, and produces a neutral or "free from fatty acid," light-colored oil.

Mechanical methods of separating the clear oil are the centrifugal method, the direct filtering method, etc. These, however, require additional special agents. The most practical method is that of gravity, by the batch process, which can be assisted by such agents as salt, brine, silicate of soda, and others, which are restricted by patents.

Equipment.—The equipment required for the standard commercial refinery consists of the following units:

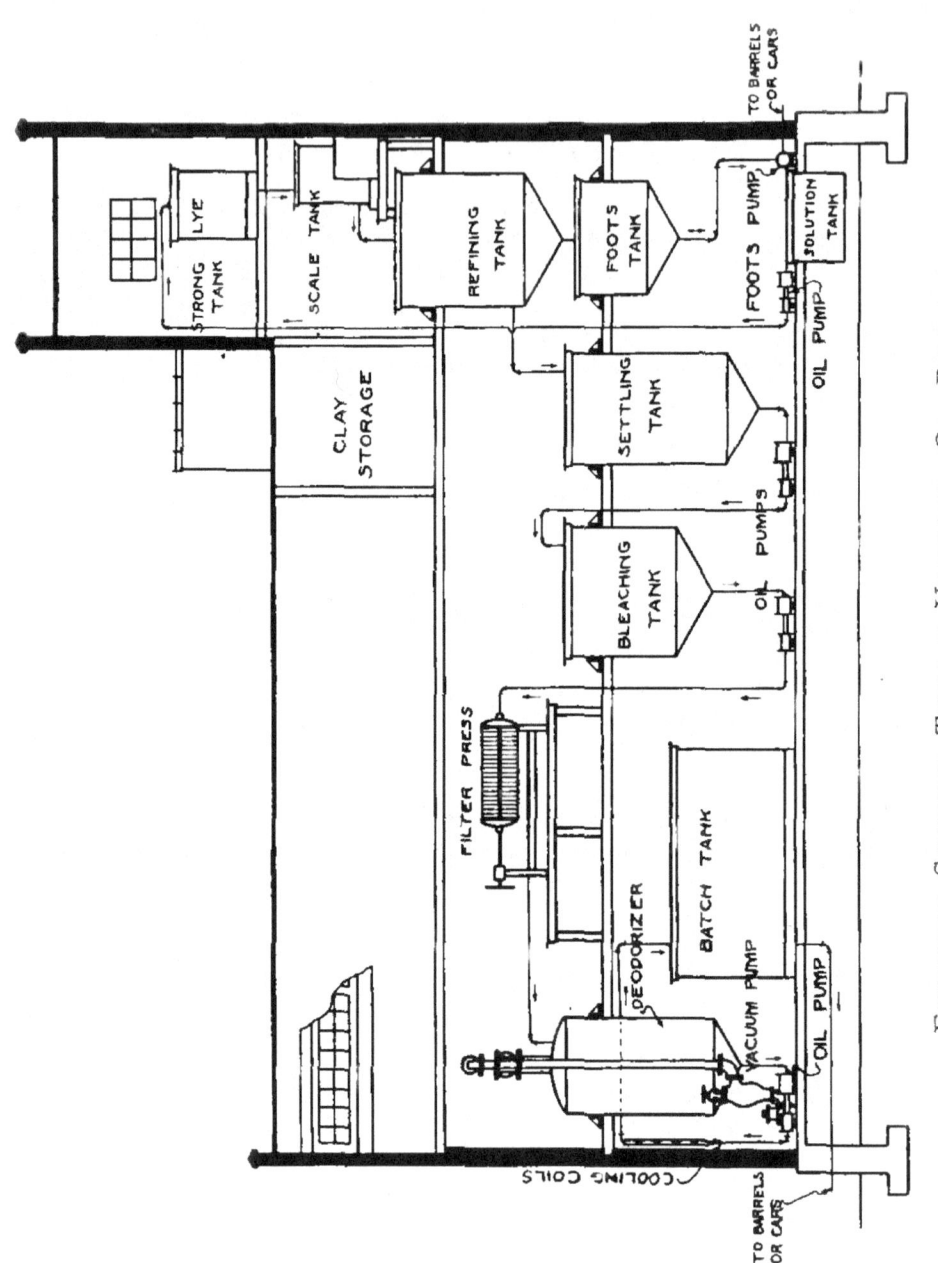

Elevation Showing Typical Vegetable Oil Refinery

a. Caustic dumping tank
b. Lye pump
c. Lye mixing tank
d. Lye scale tank and scale
e. Crude oil pump
f. Refining tank and motive power
g. Settling tank
h. Soap stock tank
i. Soap stock pump
j. Bleach tanks and motive power
k. Filter press pump
l. Filter presses
m. Batch tanks (raw filtered oil)
n. Compressed air outfit
o. Deodorizers (vacuum type)
p. Vacuum units
q. Finished oil pump
r. Cooling coils
s. Batch tanks (finished oil)
t. Superheaters
u. Filling and storage tanks

The refining process is as follows: *Making Caustic Solution.*—A caustic soda solution is usually prepared the evening before. After the caustic soda drums have been cut open they are rolled to the dumping tank to admit the water. If the empty drums are to be used for shipping soap, etc., only the seal is knocked off. The caustic soda in the tank is now covered with water and left to dissolve, and the liquor should test 30 to 40 degrees Beaumé.

The free fatty acid content of the crude oil should be determined by analyzing a representative sample. From this test the amount of caustic soda to be used can be determined.

Several hand refiners are made, using various lyes, and the most favorable results will be the guide for the refining kettle. The lye will vary in amounts from 3 to 15% or more, and strengths of 25° or more down to 10° Beaumé, or less, depending on the variety of oil and its quality.

Pumping to Tanks.—While the crude oil is pumped to the refining tank, the strong caustic liquor is pumped to the lye mixing tank and brought down to the gravity Beaumé desired. The strong liquor is some 40 to 50° strength and is diluted with water and by agitation to the requirements, and run to the lye scale tank in sufficient quantity ready for immediate use.

Experienced operators allow the oil to stand quietly after pumping into the refinery to allow any air and foam to settle, as otherwise it might increase the soap stock in the process.

Agitation.—Agitation is now started on the crude oil, and it is usually brought up to a temperature around 85° Fahrenheit. Cocoanut oil requires higher temperatures, and individual refiners have their personal preferences.

Agitation is now increased and the required quantity of lye added to the oil. The oil is closely watched for the "break," or grain. This break or graining of the oil is more or less similar to the "coming" of butter in churning cream. In fact, the entire operating of a refinery is quite similar. The time of agitation depends upon the "break" of the oil. At this point the oil is heated up to 105° to 110° F., and sometimes higher—depending entirely upon the type of the refining equipment and the kind of oil to be refined.

This graining can be more easily observed by dipping small samples from the mass frequently, and observing its action and progress. As

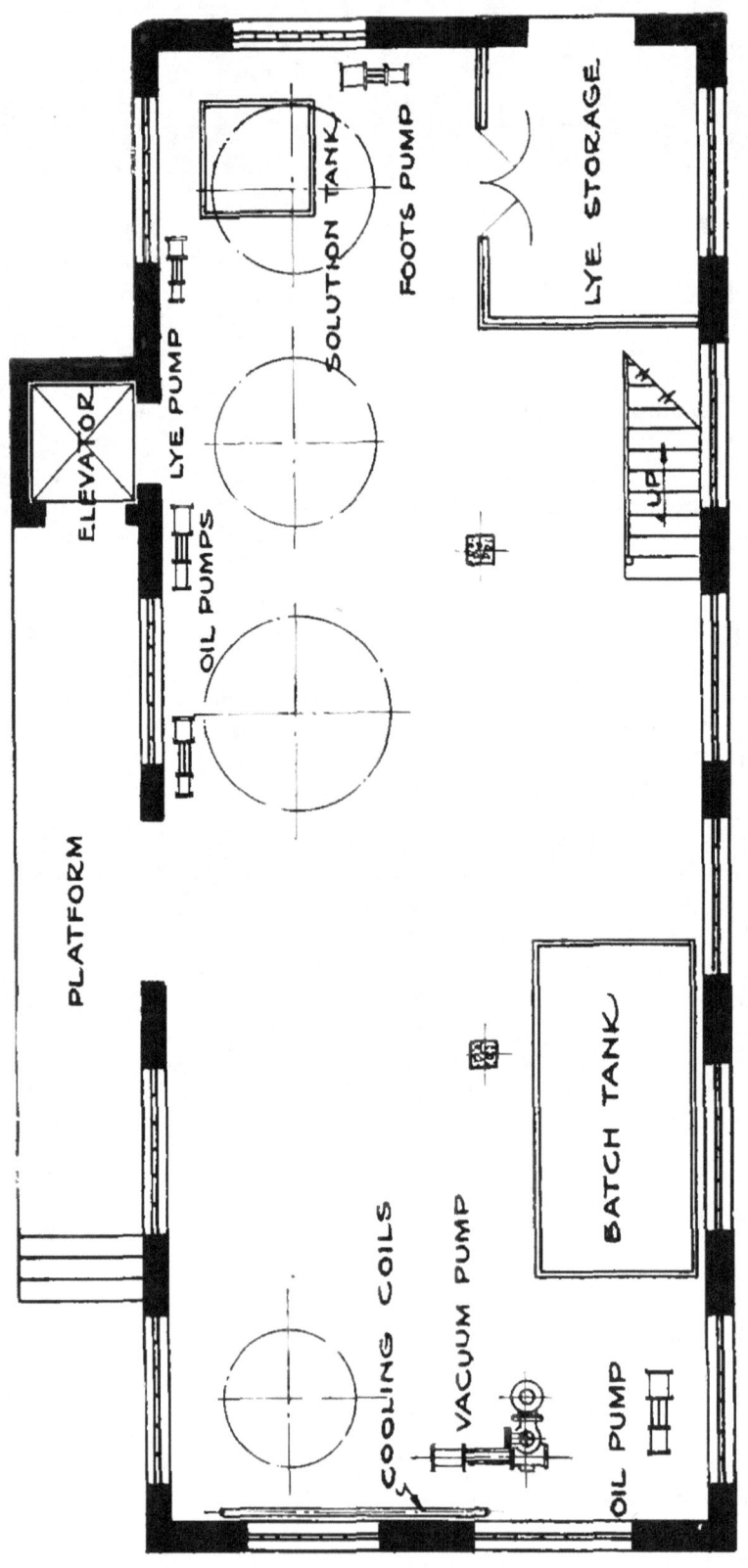

First-Floor Plan of Vegetable Oil Refinery

the temperature increases the small specks will increase in size and have a tendency to soften or run together and settle out.

As the soap stock grains become softer, the speed of the agitator is reduced to avoid breaking them apart. In other words, the aim is to gather or collect the soap stock into flakes of increased size, and separate the mass from the refined oil.

Refining operations usually require from forty-five minutes up to one hour or more, depending upon the oil used. Cocoanut oils, or similar oils, must be handled rapidly, while imported soya bean oils usually require considerably more time. Egyptian and South Sea Island cold pressed oils also refine slowly and with difficulty

Grades of Oils.—There is a happy medium as to the quality of the oil that is refinable. With too low a grade, an excess of free fatty acids and settlings, requiring a greater percentage of caustic soda to neutralize, produce an excess of soap stock; and the remaining oil, being the smaller part, is lost in the mass, and as the refiner says, the kettle "goes to soap."

The maximum limit for cottonseed oil is in the neighborhood of 25 to 30 per cent free fatty acid. Cocoanut oils refine with 15 to 25%. The majority of vegetable oils may be refined similarly, with the exception of castor oil, which is very difficult to separate from its emulsified mass.

Settling.—The refined oil is usually permitted to stand and settle until the following morning, at which time the refined oil is syphoned off through the special connection to the settling tank, care being used to draw off every last bit from the layer of semi-solid soap stock. This soap stock with some oils is quite liquid, and this final separation is somewhat difficult. With other oils, soap stock is quite stiff and hard, also its condition depends on the strength of caustic and quantity used.

Soap Stock.—After the yellow oil is drawn off, the remaining soap stock in the refining tank is flowed out the bottom outlet to a soap stock receiving kettle, from where it is pumped to temporary storage or to soap-boiling kettles, or filled into barrels by the small refiners. The refiner, now empty, is washed and made ready for a second batch.

This by-product soap stock will not store permanently, particularly the cotton oil soap stock, owing to the fact that it ferments quite rapidly. Therefore, by the larger refiners it is given a further treatment, in a soap-boiling kettle with strong caustic; that is, it is "killed," and after boiling up and salting out can be pumped to permanent storage as "killed" foots. Or, as is often done with the cottonseed foots, it may be washed up further with successive caustic treatment and made into a "settled" or "pitched" soap, or "boiled down" quality.

The brown or blackish cottonseed soap stock can be made into a golden yellow hard soap, much valued for wool scouring and filling of other laundry grades. This "washed up" soap stock is a base for most of the washing powders, to which soda ash is added and further finished in special manner and method to a flaky or grainy washing powder.

Filtering the Refined Oil.—"Yellow" cottonseed oil has more or less fine particles of soap stock and moisture, which all must be removed before the oil is of merchantable or usable quality.

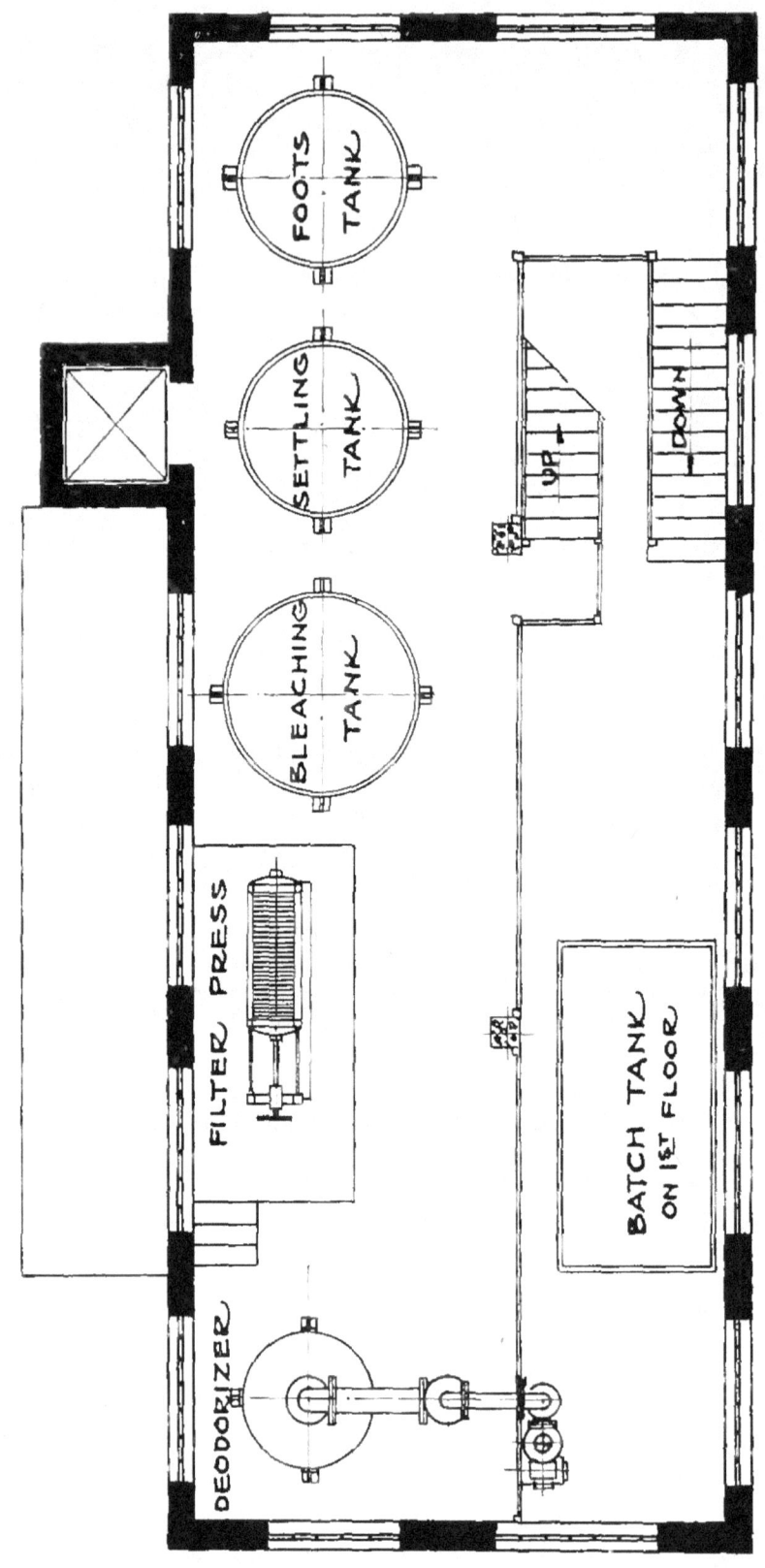

Second-Floor Plan of Vegetable Oil Refinery

Some refiners wash the oil with hot water, with or without salt, in order to dissolve out all traces of this colloidal soap and excessive alkali.

The oil is pumped back to the refiner, or into an agitating tank, or an agitator may be supplied to the settling tank. This is often connected to a vacuum system, and while this may improve the oil in this respect, there appear to be other more simple and as efficient methods. Furthermore, washing produces an excess loss due to emulsions formed, even though it is centrifuged or recovered in the acidulating tank, and some claim it tends to produce a rancidity in the finished oil.

A practical and rapid method of clarifying refined oil is by filtration, using a small amount of filtering medium—fullers' earth, filter cell, etc. The refined oil is gravitated or pumped from the settler to the bleaching kettle, brought up to about 140° to 160° F., rapidly agitated, and about one-quarter of one per cent of an unbleachable fullers' earth or other filtering medium added.

After some fifteen or twenty minutes the oil is started through the filter press with the pump. The first oil from the press will be slightly clouded with earth that passes through the cloths, and must be returned through proper lines to the bleaching kettle. The cloths soon become coated with the earth, and when the oil flows perfectly clear and brilliant it is directed to the receiving or holding tanks for further disposition.

The color and flavor determine the grade or quality. This color can be controlled, to a degree, by the amount of fullers' earth used, due consideration being given to the earthy flavor produced by excess of same For cottonseed oil to be of prime color, it must be no darker than 35 yellow and 7.6 red on the Lovibond color scale.

Bleaching

If the refined oil is a bleachable oil, and a "white" or bleached oil is required, it may be treated directly from the settler; preferably, however, after filtering in a similar manner to the above process, differing in details.

In bleaching, the oil is brought up to temperature of 215° to 230° F., and two to three per cent of bleachable fullers' earth is added, depending on the oil as determined in the laboratory test, and on the color desired. Some oils requiring very much more, and the usable limit depends on market values of earth and oil, more or less. The time required in the process is not less than one-half hour before directing the oil to the holding tank. Agitation is continued until the oil is all pumped from the bleaching kettle.

An addition of "bleaching black" of about three-tenths per cent with the fullers' earth assists the bleaching very materially on practically all vegetable oils.

The yellow oil must be dry for best results, and should be free from any soap stock, as moisture retards the effect of the earth and the latter may be re-dissolved in the hot oil and will darken the oil, and with cocoanut oil it will settle out in the finished oil as colloidal soap.

Some refiners bleach their oils under vacuum, which has its advantages, especially in drying the oil.

In pumping the oil through the filter presses the pressure must not be

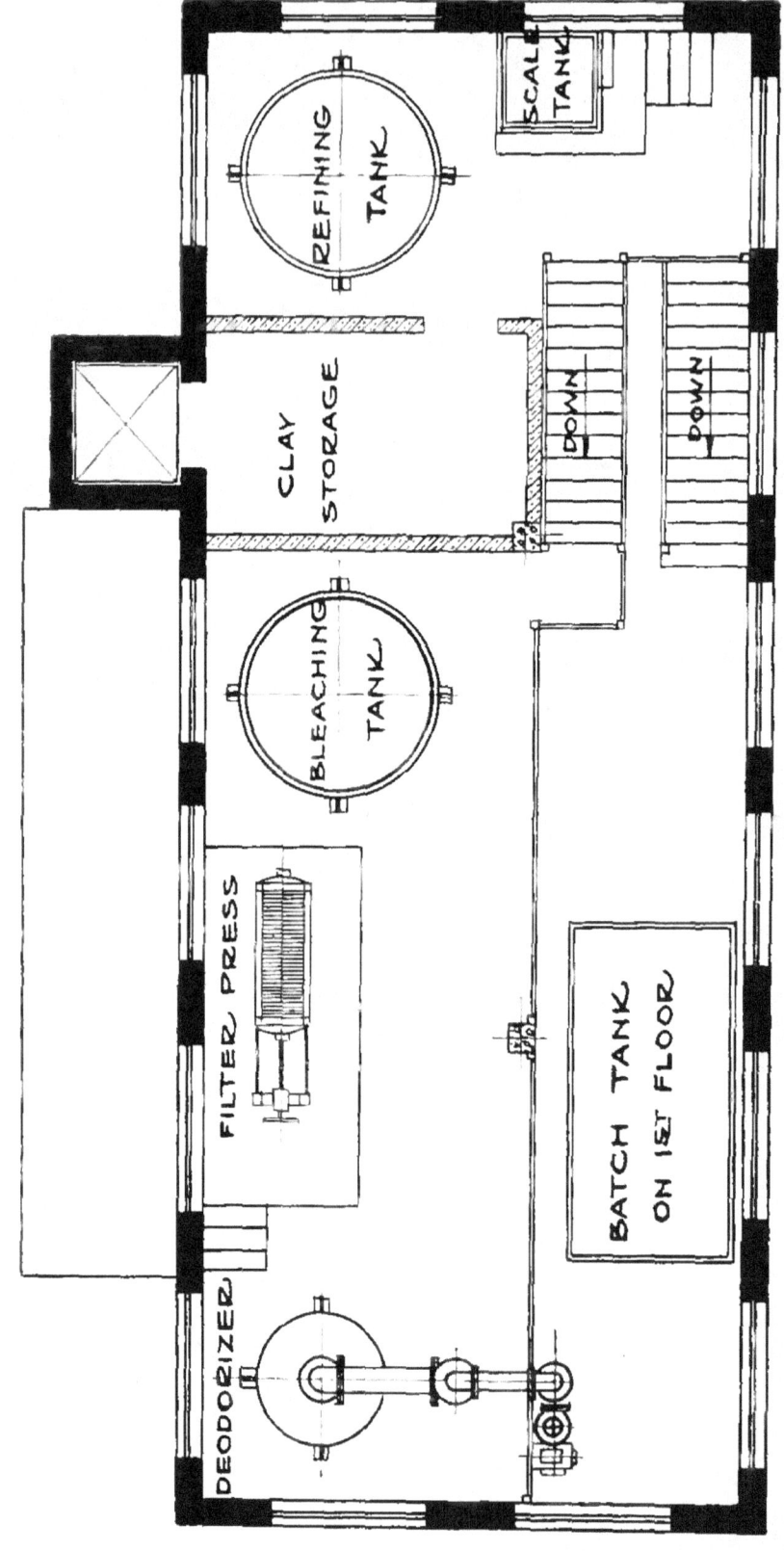

Third and Second-Floor Plan of Vegetable Oil Refinery

allowed to rise too high on the cloths, as they will break and leak earth. This can be observed on the pressure gauge; depending on the strength and age of the cloths, as a rule, it should not exceed 50 to 60 pounds at the maximum. A flow at 20 to 30 pounds is to be recommended for speed and safety.

The color limit of a Prime White Cottonseed oil is 20 yellow and 2.5 red (Lovibond color scale). However, this highly earth-treated or bleached oil is not of prime flavor and is not merchantable in this form, having a very strong or "earth" flavor, and this flavor must be removed by deodorization before it can be used as a commercial grade, or as an edible product.

Deodorizing

On the proper deodorization of edible oils and fats depends absolutely the quality and competitive commercial value of their finished products. A product of the highest quality can be improperly deodorized and spoiled in the process, and thus produce a strictly "off" quality, reduced in commercial value.

The very best and most modern equipment is not a full guarantee of the best-finished product, as much depends on the operation. On the other hand, oils and fats of low grade and off quality can be greatly improved, and much of them finished as a prime grade. However, other things being equal, the better the raw product, the better the finished quality.

The deodorizing of oils and fats by the modern methods removes not only the foreign odors and flavors, but much of the characteristic nature of the products, and in many cases the various types of oils cannot be distinguished one from another when well deodorized.

Methods.—Deodorizing of oils is possible in various types of equipment and under many modifications of details. The process in general is the treatment at elevated temperatures with open steam, being both a washing and volatilization process.

Modern methods are usually in vacuum (or exclusion of air), and superheated (open) steam is usually used, or the oil may be superheated directly. The vacuum may be effected by various means, such as vacuum pumps, surface condensers, atmospheric condensers with necessary water leg, etc.

Vacuum System.—The operation with the vacuum system, in general, is as follows: Vacuum should be held on the tank, operating or idle. Draw or flow the oil to tank, filling it about two-thirds full. Closed coil should be below surface of oil. The height of the tank should be 1½ or 2 times the diameter, and the most economical and flexible quantity per batch is upwards of two to three thousand gallons.

The oil is brought up to 220° F. with closed coils; open steam lines are drained, and steam turned on tank slowly (to avoid jumping the oil) through the open steam spider. The open steam is superheated up to about 550° to 600° F. As the temperature rises the closed coils are shut off and the temperature is allowed to increase with the open steam *only* up to 300° to 340° F.

With vacuum the adjustments should be gauged so that the vacuum maintains fully and as efficiently as possible, with the open steam feeding to the tank.

Treatment is continued four to six hours or more, depending on the relative efficiency of the installation, type and quality of the oil, and results required. Occasionally a sample of the oil is drawn and its odor noticed; also taste on cooling and compare with the original oil and that of a properly finished sample under similar conditions. Experience will acquaint the operator or chemist as to finishing point or time of treatment.

When finished, all steam connections are shut off tightly (vacuum continued) and the hot oil pumped through the cooling coils to remove the extreme heat down to 100° to 140° F. *before* permitting air to contact with the oil. This is important.

The deodorized oil is now suitable for compound (if a bleached oil), butterine, cooking, frying, etc., being very mild, bland, and practically tasteless and odorless.

Operating Points.—Animal products will not stand as high a heat in deodorizing as vegetable oils and fats, and the higher the melting point as a general rule, the lower the temperature that can be safely used on all products.

The oil recovered in the drip tank and baffle box of the vacuum system is more or less rancid, and should be returned to "off" qualities or possibly to the crude refiner.

The deodorizing tank should be kept perfectly clean and free from air if of the vacuum type. Periodical inspection and cleaning should be given.

To clean the tank it should be boiled up with (full charge) water and caustic (one-half barrel or so), and rinsed thoroughly with clear water. Dirty tanks produce poor oil and may darken the color.

COMPOUND LARD

Vegetable or Lard Compound is of a variety of mixtures and formulas; some contain hog lard, but mostly they are composed of large percentages of deodorized vegetable oils, with beef tallow, oleo stearine or hydrogenated vegetable oil added as hardening or stiffening medium.

The formulas are governed mainly by market prices of commodities, and at times as many as five or six different oils or fats are used in one product.

Sufficient hardening substances must be used to produce a finished product of a melting point above the temperatures into which it is shipped and consumed, and at the same time it must not be too stiff or brittle. The old rule of chemical titre is no longer a guide, since the commercial use of such oils as corn and soya bean oil have become possible, as shown on the following titre table:

	Titre
Prime Steam Lard	36-37 degrees C.
Lard Stearine	40-44 degrees C.
Tallow	42-44 degrees C.
Oleo Stearine	49-51 degrees C.
Corn Oil	17-20 degrees C.
Soya Oil	22-25 degrees C.
Cottonseed Oil	30-34 degrees C.
Hydrogenated Cottonseed Oil	50-60 degrees C.

Some compounds have been successfully marketed with a titre of only 29 degrees C. The usual method in manufacturing is to weigh out the given formula, drop to formula batch tanks and mixing by aid of heat, air or pumps in tanks, and held at 130° to 150° F.

Methods.—From the batch tanks the mixture is flowed or pumped through cooling coils or over a *water* roll, and brought on to the main refrigerated lard roll at 95° to 105° F., from which it will drop to the picker pan at a temperature of 45° to 60° F., depending on the operator, formula, season, etc.

The picker pan or trough, with the revolving beater shaft, mixes the semi-solid compound into a white smooth mass, and it should be pumped through strainers to the package without undue delay. Following this rapid beating in this form the semi-stiff compound immediately "sets" in the package to a firm semi-hard product. Too much beating is to be avoided.

The manner of finishing compounds is a great aid in producing a product that will "stand up," also assisting to produce a better color on the finished package. Some packers cool their mixture to the semi-stiff degree in an upright agitating tank; others have a cylindrical horizontal drum with internal agitators, and cooled on the outside by water or brine. Both of these types produce a product more or less grainy in texture and not of the whitest color. Different formulas work differently; the harder the product the greater the capacity and the easier the operation.

Some refiners do not use the pre-cooling coil. However, the method outlined is to be recommended, both for quality of finished product and economy of operation. It is also recommended that all the products used for hardening, both vegetable and animal, be deodorized, either separately or in the mixed formulae.

Formulas.—Compound formulas vary with season, location, temperatures and market values. Following are illustrations:

Cottonseed oil........60% to 50%	Cottonseed oil....... 50%
Edible tallow........40% to 50%	Soya oil............ 30%
Cottonseed oil......80% to 85%	Cocoanut oil........ 7%
Oleo stearine........20% to 15%	Hydrogenated oil.... 13%
Cottonseed oil......80% to 90%	Peanut oil 43%
Hydrogenated oil.....20% to 10%	Soya oil............ 45%
	Hydrogenated oil.... 12%
Cottonseed oil......75% to 78%	Cottonseed oil....... 30%
Edible tallow........15% to 10%	Peanut oil.......... 30%
Oleo stearine........10% to 12%	Soya oil............ 27%
Cottonseed oil....... 75%	Hydrogenated oil.... 13%
Cocoanut oil........ 8%	
Oleo stearine........ 17%	Cottonseed oil....... 70%
Cottonseed oil....... 45%	Edible tallow....... 15%
Soya oil............ 30%	Oleo stearine........ 10%
Cocoanut oil......... 7%	Hog lard.......... 5%
Oleo stearine........ 18%	

Peanut oil can be used instead of cottonseed oil, or corn oil can be substituted for soya oil. At times market values prohibit the use of any oils or fats other than cottonseed oils and hydrogenated fats. Then again, this situation may be reversed.

When the product is purely vegetable in its formula no government inspection is required in its manufacture, but when animal products enter into its makeup its manufacture comes under the meat inspection laws and supervision of the U. S. Department of Agriculture.

Equipment.—The principal items for a practical compound plant are:

a. Fat melting tank
b. Oil stock tanks
c. Scale tank and scale
d. Formula batch tanks
e. Circulating pump
f. Cooling coil
g. Lard roll and motor
h. Refrigerating outfit
i. Draw-off pump
j. Filling line

WINTER OIL

The average Cottonseed Oil is composed in the main of two fats or glycerides known as olein, of low melting point, and stearine, of higher melting point. The usual Summer Yellow grade in cool temperatures, around 50° F., will separate out more or less of this solid stearine fat of high melting point. In other words, the oil is fully liquid only in summer; hence the Summer Yellow grade name. This characteristic, therefore, prevents its being used satisfactorily as a salad oil, or in salad dressings which are subjected to refrigeration, and the removal of this stearine becomes necessary for certain uses and trade requirements.

This can be done practically and commercially by chilling the Summer Oil in a special manner, and filtering out the separated stearine from the clear oil, which oil is then known as Winter or Salad Oil, since it remains liquid at medium winter temperatures.

Some refiners guide their process and operation to produce a high melting point stearine, others aim for low melting point stearine and high test oil. Naturally the former is the slower process and produces less stearine than the latter. The proportion of stearine and winter oil varies, therefore, from as low as 7% up to possibly 20% to 25% stearine, balance being the winter oil. This is governed by the stock used, as oils from different sections and of different seasons vary in stearine content.

Equipment.—The principal units for this process are:

a. Refrigeration units
b. Chill room
c. Special chilling tanks
d. Pressure tanks or equivalent
e. Large filter presses
f. Cold compressed air supply
g. Warming coils
h. Receiving tank, etc.

The entire operation is conducted in a chill-room, and the chilling tanks should be located above all receiving tanks and filter presses, as gravity or pressure replaces all conveying pumps.

Method.—The general method is as follows:

Summer oil (yellow or white), thoroughly dry, is delivered to the graining tanks in the chill room at a temperature of 65° to 75° F. One set of tanks is filled each day, as it requires 3 to 4 days or more per cycle.

The chill-room is maintained at a temperature of 38° to 50° F. The oil is allowed to cool down slowly, with occasional gentle agitation either by hand, mechanical or cooled air. Some refiners use no agitation, which produces a high-grade oil, but requires much longer time.

Gradually stearine will be noticed to separate after 2 to 3 days or so. Other things being equal, the slower the graining the better grade of winter oil. The finishing point of the graining must be determined by hand filtration tests, guided by experience, which will be during the second or third day, or even later.

Filtration.—Along about this period a certain point is reached, not very pronounced at times, at about 48° to 50° F. in cooling, at which point the temperature of the oil remains more or less constant, then rising slightly until all heat or crystallization is removed. At this point the oil is allowed to flow automatically to the filter presses or into the pressure tank by suction, the upper clearer portion being withdrawn first, and followed with the lower portion containing the bulk of stearine.

The oil is sent through the filter press, using either gravity or air pressure, the clear winter oil filtering through, leaving the crystallized stearine on the filter cloths in the press. The first clouded oil must be returned to unfiltered oil.

When the chill tank or batch is exhausted, pressure is gradually raised as the filtering slackens, due to the cloths becoming more thickly coated with stearine. Pressure is allowed to increase up to 30 or 40 lbs., depending on the oil and other conditions, types of cloths, etc. Filter cloths of various types and quality are used, for instance, No. 10 duck, twill cloth, chain weave, also heavy double muslin.

The pressure is continued for some little time on the press, in order to drain all of the winter oil, and all air used must be of the same temperature as the oil.

The filtered winter oil is delivered from the trough through an oil seal outside of the chill room into batch tanks. The oil being cold, will condense and absorb moisture. Therefore, reasonable heat is kept on the batch tank to bring the oil up to outside atmospheric temperature. The convenient arrangement for avoiding this condensation is to conduct the fresh-cooled filtered oil through a double-pipe coil directly from the press, whereby it becomes heated in passing to storage, before contacting with the warmer air.

Recovering the Stearine.—The filter press is duly opened and the stearine scraped from the cloths and barreled for further disposition, or sent to a receiving tank, according to requirements. The convenient arrangement in this connection is the use of a steam-heated type of press, through which warm air can be forced, and the stearine gradually melted from the press, out of the reverse flow direction, thus removing it and cleaning the press without disassembling. A further method is to pump warm winter oil through the press and meet the stearine.

If neither warm air or warm oil is used, the cloths must be removed and melted up for rapid work when they filter sluggishly. Stearine as recovered can be used for various purposes, principally lard compound or butter substitutes, or it can be mixed in small proportions with Summer Yellow oil up to 10 or 15 per cent for cooking oils, especially during the summer months.

The above practically completes one cycle of the operation. The following day the contents of the second tank will be filtered, and so on. The exact details of the general operation vary somewhat, depending on the type of oil, quality, season, etc.

The entire operation must be conducted in the chill room, and all compressed air used must be pre-cooled. It will happen often that the oil will stand this cold test beyond the official limit of five hours, which merely indicates that it is of a higher grade. To be full standard it is required to remain clear during and through the five-hour period only.

Deodorizing.—Winter oils are customarily marketed in a deodorized form. Some refiners deodorize before pressing. However, it is of considerable advantage to press the raw oil and deodorize the finished winter oil, as this removes all flavor of press cloths, etc.

However, great care must be observed in that no Summer oil from the general plant becomes mixed with the winter oil; as in this case it will cause a "weak" test oil and counteract the wintering operation. Considerable trouble is experienced in refineries through this feature, also because of traces of dissolved moisture in the finished oil, which causes the oil to become cloudy in the cold test, preventing its passing the official requirements.

In some localities salad oil is used as a cooking oil, which is thoroughly feasible, but somewhat expensive when compared with the summer grades of cooking oil.

HYDROGENATION OF OILS AND FATS

While hydrogenating of oils and fats is in a way an industry of its own, and not usually practiced by the smaller refiners or packers, reference to it should be made here, as quite a few packing plants use the hydrogenating process.

This is the process whereby the liquid or fat is transformed into one of very high melting point, upwards of 140° F. (60° C.), and it can be applied to almost all the fat products—animal, fish and vegetable—to varying degrees, depending on the chemical structure of the fat.

Chemically, it is the process of changing various unsaturated glycerides, principally oleins ($C_{18} H_{34} O_2$) to the saturated glycerides of the stearine series ($C_{18} H_{36} O_2$), by adding two atoms of hydrogen (H) to same. This is affected by aid of certain metallic catalyzers, nickel being the most practical.

This process requires a supply of pure hydrogen, which is produced by several methods, usually by the electrolytic method, and requires quite an investment for the hydrogen generating cells.

The working equipment of a hydrogenating plant is not extensive, but as the process is controlled by many patent rights, its universal use becomes more or less limited. The main unit is the converter or hardening tank, which is of various types and forms, and all more or less covered by patents.

The treatment is carried on at an elevated temperature of 320° to 390° F., for three to six hours or more, depending on purity of oils, activity of catalyzer, degree of hardness desired, etc. From the converter, after cooling slightly, it is filtered to separate the intermixed catalyzer, which is returned to the converter for successive batches. The hardened product is solidified either in open tanks or over cooling cylinders—some spray the hot oil into cooling towers—and can be shipped in slack barrels or usual bags.

Cocoanut oil will not harden to any appreciable extent, while the usual seed oils, also animal and fish oils, will harden to melting points of 125° to 150° F., or higher with certain oils.

This process produces a manufactured product which will substitute for and even excel the natural and necessarily limited supply of animal fats for lard compound, margarin, soaps, etc. The product is fully as nutritious as a food as the natural hard fats, and equally as digestible as those of equal melting points.

While the main condition to be desired in a hydrogenated fat is its melting point, these are judged and guided relatively by the iodine number or titre of the total fatty acids. The methods for determining these characteristics can be obtained of the Bureau of Chemistry, U. S. Department of Agriculture, and are more or less difficult and technical for the layman, requiring the skill of a trained chemist.

MANUFACTURE OF MARGARIN

Oleomargarin is made from animal and vegetable fats, churned with milk. The principal oils used are oleo oil, neutral lard, peanut oil, cottonseed oil, and cocoanut oil. The usual specifications for the animal oils are found in previous chapters of this Encyclopedia.

The vegetable oils must each come up to the particular requirement necessary, depending upon the kind and nature of the oil. In general it should be said that these oils should be free from moisture and all material not fat, also they should be devoid of rancidity and acidity. The importance of having pure high quality oils to start with cannot be over-stated.

The milk used for making "starter" should also be given very careful attention. It is best to get milk that is produced in sanitary surroundings, free from all objectionable odors, and low in numbers of bacteria. It should be shipped in a cold condition, below 40° F. being best. Immediately upon being received at the factory the milk should be pasteurized, either by the flash or holding method, the latter being preferred. In the "flash" process the milk is heated to 170° F. for thirty or sixty seconds. In the "holding" process the milk is maintained at a temperature of 145° F. for thirty minutes.

After being pasteurized, ripening is the next step, and it is at this point that great skill is required in order to develop the proper flavor and aroma which is later to be imparted to the margarin upon being emulsified with the oil. The ripening should not be carried to the point where the milk will curdle; so that many oleomargarin manufacturers ripen only to .4 of 1 per cent actual acidity, and never over .5 of 1 per cent. Much work has been done in developing the various pure cultures of lactic acid producing organisms, the by-product of this development being esters and alcohols which provide proper flavor and aroma.

The milk room should be separate, and should contain only the pasteurizing vats and starter cans. Pasteurization by the batch method is best carried out at 145° for half to three-quarters of an hour, whereupon it is immediately cooled to 68° F., then inoculated with ½ of 1 per cent of the pure culture and then allowed to ripen to .4 of 1 per cent acidity.

Churning.—One of the most important steps in the manufacture of margarin is the churning, the temperature of the mixture in the churn depending upon the formula used. In general, such a temperature should be used as will create a perfect emulsion of the milk and oils; 80° to 85° is often employed for white goods or even a somewhat colder temperature, while for natural color goods a little higher temperature is usually employed, so as not to destroy the color.

Two general methods of crystallizing the emulsion are used, one the vat method and second the sluice method. The advantage of the second is that the crystallized margarin gravitates directly into trucks, whereas in the vat method it must be removed by manual or mechanical labor.

Ripening.—After crystallizing, some manufacturers have allowed the margarin to ripen further, in order to develop flavor, the time required for this being about twelve hours at a temperature of 70°. This, however, depends upon the grade of goods. Other manufacturers immediately put their goods onto the workers, salt them, and print and package them, allowing the flavor to develop during transit.

In general, it may be said that the fresher the goods gets to the buying public the better. The working of the goods has an important bearing on the consistency and body of the same, European methods having been developed much further along this line than the average American methods. The latter are being rapidly improved, however.

A good oleomargarin should contain about 2½ per cent of salt, 13½ per cent of moisture and ½ of 1 per cent of casein, and it should also have a smooth, uniform velvety body, with no trace of visible moisture. The flavor of the product is the most important thing to consider, a clean, acid buttermilk flavor being most desirable.

Formulas.—The following formulas indicate in a general way the mixtures of oils which give desirable results in producing the various classes of margarin:

High grade white goods
>   No. 1 oleo oil, 800 lbs.
>   Neutral lard, 700 lbs.
>   No. 1 peanut oil, 600 lbs.
>   Ripened milk, 70 gallons.
>   Salt, 200 lbs.

Second grade white goods
>   No. 2 oleo oil, 700 lbs.
>   Neutral lard, 700 lbs.
>   White cottonseed oil, 650 lbs.
>   Ripened milk, 60 gallons.
>   Salt, 200 lbs.

High grade natural color goods
>   Yellow oleo oil, 1,200 lbs.
>   Neutral lard, 300 lbs.
>   Yellow cotton oil, 450 lbs.

Butter, 650 lbs.
Ripened milk, 60 gallons.
Salt, 200 lbs.

Vegetable Margarins

Purely vegetable oleomargarin came into prominence at the time of the World War, due to the scarcity of animal fats and the high price of butter and animal oleomargarin. The cocoanut butters consist largely of cocoanut oil and some peanut oil, churned with milk. As the melting point of cocoanut oil is about 76° F., it is necessary to use a hydrogenated vegetable product during the summer season in order to make goods that will stand up well. Special treatment and temperatures are of course required to make a satisfactory product of this nature, each manufacturer varying his formula, depending upon the locality and the trade.

It should be stated that in all of the formulas given there is a certain loss of salt in the working, so that the finished goods will contain only about 2½ or 3 per cent.

Taxation

Most oleomargarins sold in the United States are free of artificial coloring, due to the fact that the law imposes a tax of 10 cents per pound on oleomargarin artificially colored, whereas the tax on uncolored goods is only ¼ cent per pound. The margarin manufacturer of today is confronted with certain requirements, not only by the United States Government, but by nearly every state, many of these requirements in the various states being different in character. Manufacturers are required to pay a federal tax of $600, wholesale dealers in colored oleomargarin are required to pay $480, wholesale dealers in uncolored goods pay $200, retail dealers in colored goods $48, and retail dealers in uncolored goods $6. Manufacturers must also be bonded for $5,000.

The use of oleomargarin as a spread for bread is growing. It is a safe food, the fats going into it coming from animals which have been U. S. inspected and passed, and the milk entering therein being subjected to pasteurizing temperatures which destroy disease producing bacteria. It has also been conclusively shown that animal oleomargarins contain the much-talked of vitamines necessary for child growth and development.

PART TWO

Statistics

FOREWORD

While no hard and fast sub-divisions have been made, this Statistical Section can be said to fall naturally into four parts. These, in the order in which they are presented, are as follows:

1. Charts and maps (livestock and meats statistics).
2. Tables (special compilations of livestock and meat production and consumption figures).
3. Traffic charts and tables of rates on cattle, beef, packinghouse products, and icing charges.
4. Trade Term Definitions (in use both in domestic and foreign commerce).

Many times packers want to refer at short notice to figures showing production and consumption of beef, pork, mutton and lamb, stocks on hand for a period of years, range of cattle, hog and sheep prices for years back, trend of exports of meat products, etc.

Another kind of information where figures are of great aid is in the field of traffic. It is of importance to be able to refer to railroad rates on livestock and compare them with what they have been for the last fifty years. Such comparison shows whether the rates at any particular time are fair and in which way they should be changed. Of more concern is the question of packinghouse products rates. Reference to the whole subject in a chart gives the packer a chance to refresh his memory and size up the rate situation from his own point of view. The same is true of icing charges.

A third field where a source of constant reference is necessary is that of trade term definitions. There are many troubles that arise out of ignorance of the right interpretation of a trade term under which sales and shipment should be made. The question of what the seller must do on the one hand and what the buyer must do on the other hand has to be absolutely clear. And just what this means in the packing business requires careful and accurate knowledge.

This Statistical Section has been prepared to give the reader the information just outlined by means of charts, maps, tables and otherwise in as brief space as possible in order to make the figures and facts clear and of ready reference.

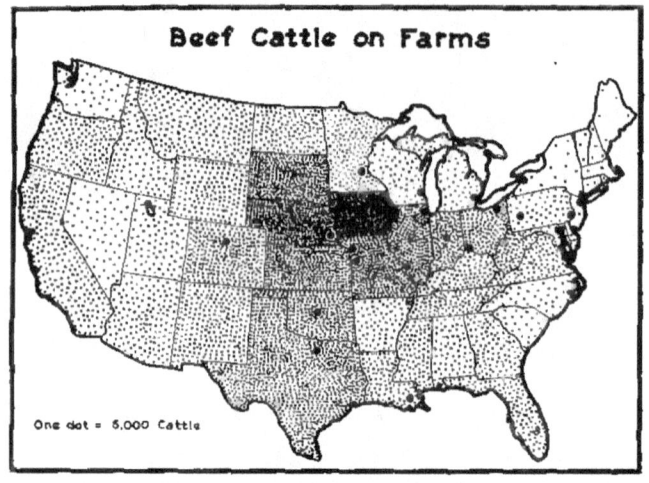

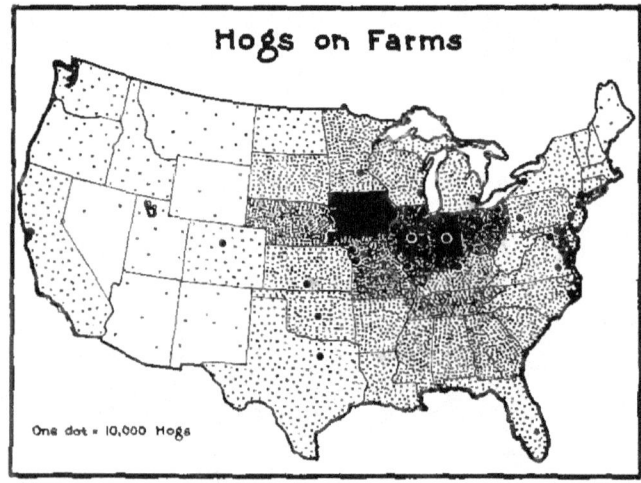

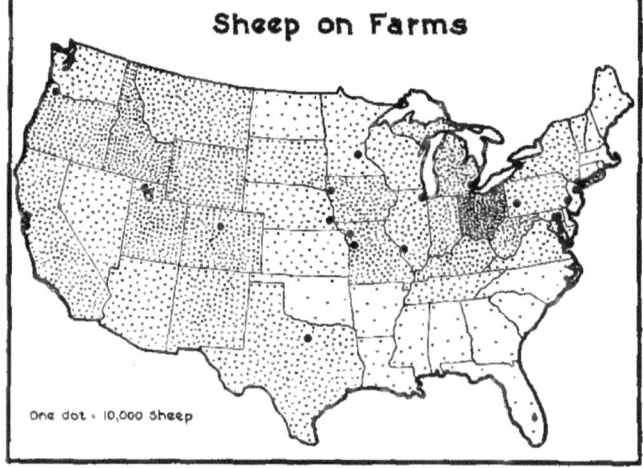

Charts by Commercial Research Dept., Swift & Co.

Sources of U. S. Meat Supply

Number on Farms January 1, 1920. Data from U. S. Census Bureau. Large dots show the location of Principal Slaughtering Centers.

STATISTICS

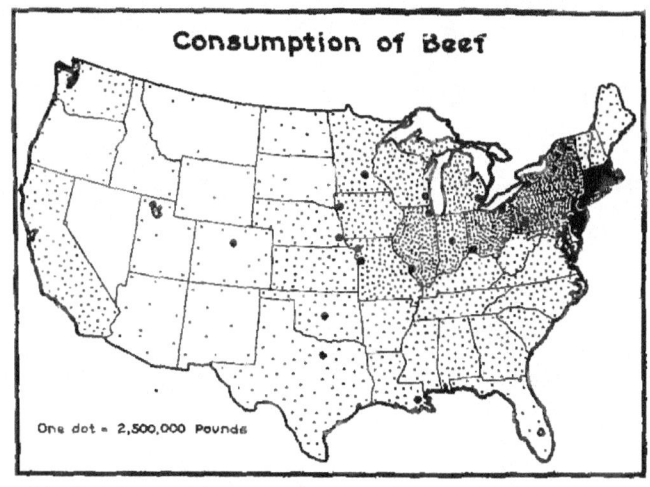

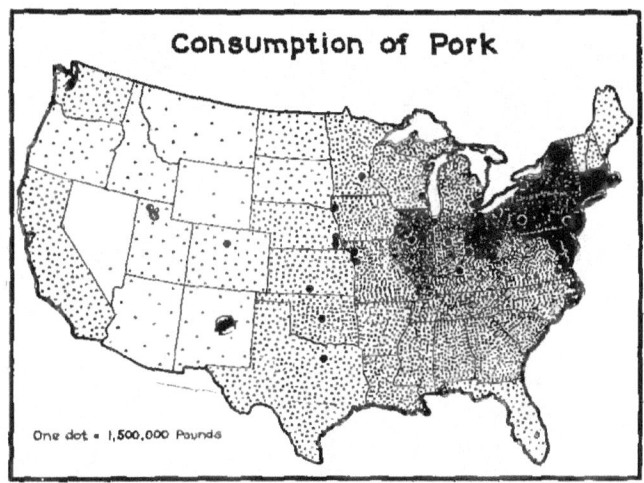

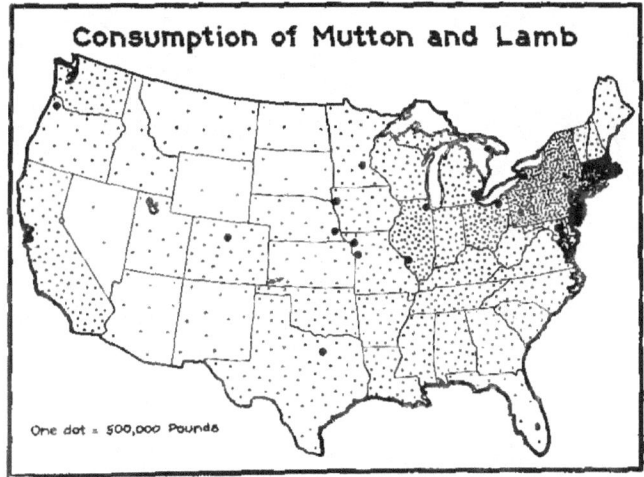

Charts by Commercial Research Dept., Swift & Co.

Areas of U. S. Meat Consumption

Based on per capita consumption as estimated by U. S. Bureau of Crop Estimates for 1919, and on the 1920 population.

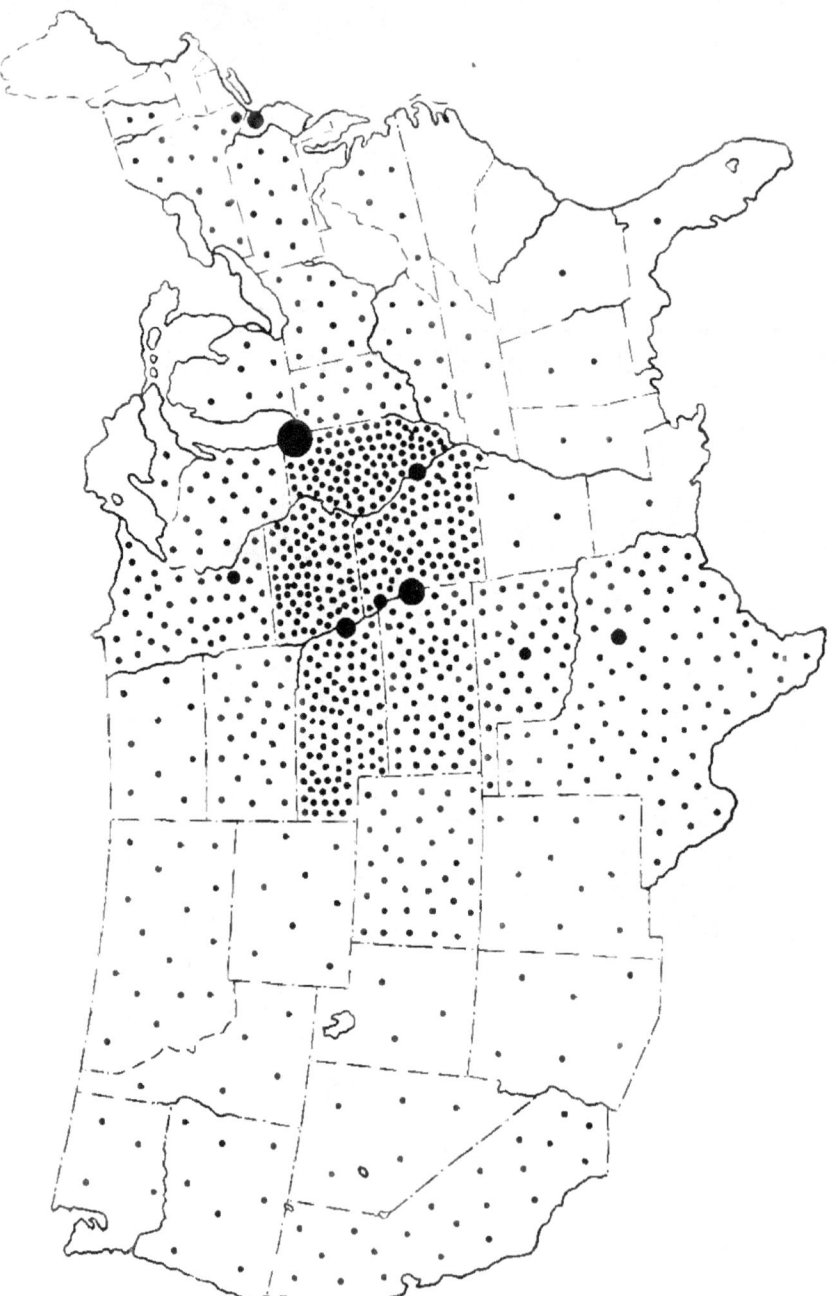

Cattle Loading Points and Slaughtering Centers

This map of the United States shows the cattle loadings by States during the year of 1918, and also the principal cattle slaughtering centers. Each dot represents 1,000 carloads. Statistics from U. S. Bureau of Markets and Crop Estimates.

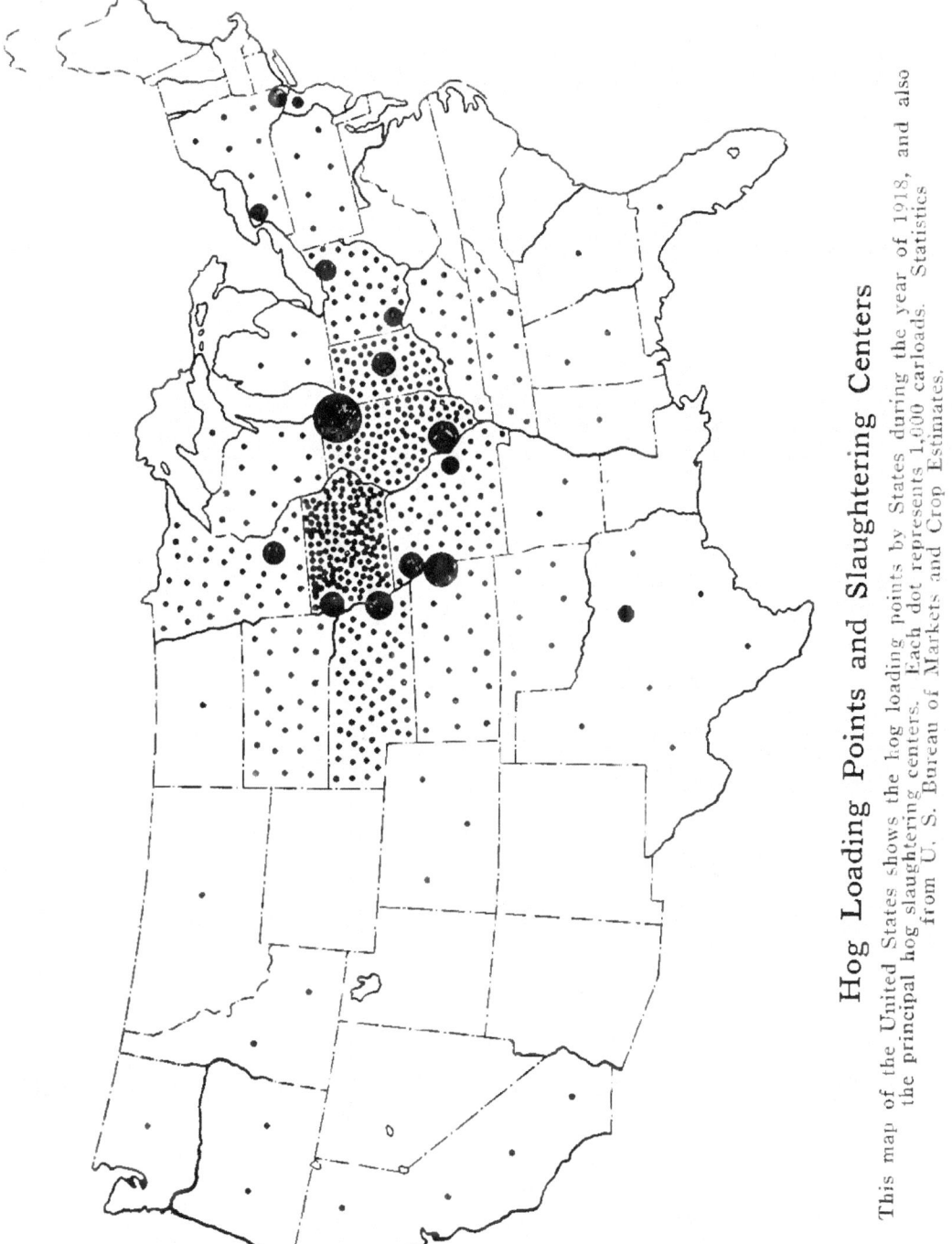

Hog Loading Points and Slaughtering Centers

This map of the United States shows the hog loading points by States during the year of 1918, and also the principal hog slaughtering centers. Each dot represents 1,000 carloads. Statistics from U. S. Bureau of Markets and Crop Estimates.

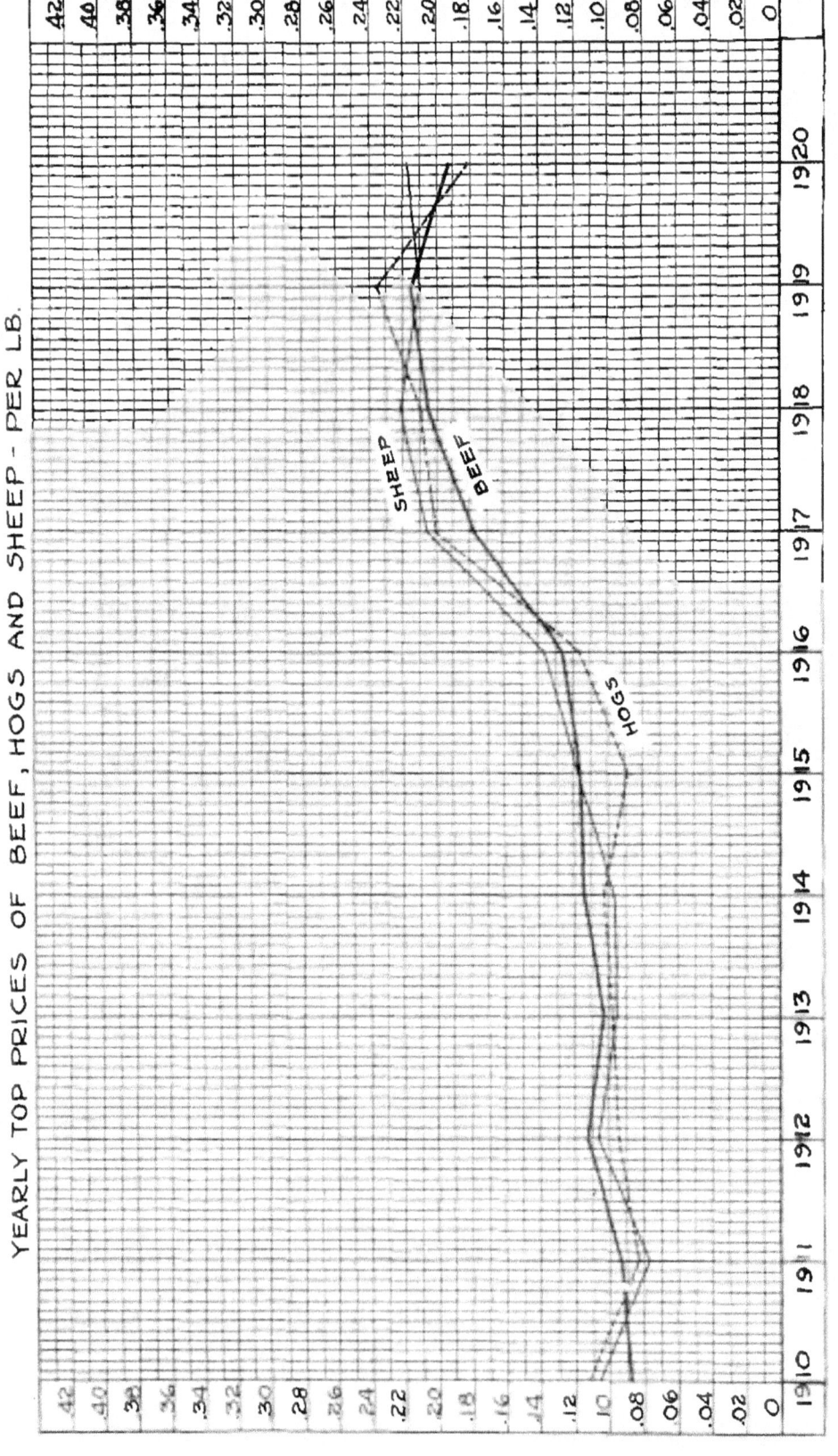

Chicago prices. "Sheep" includes lambs.

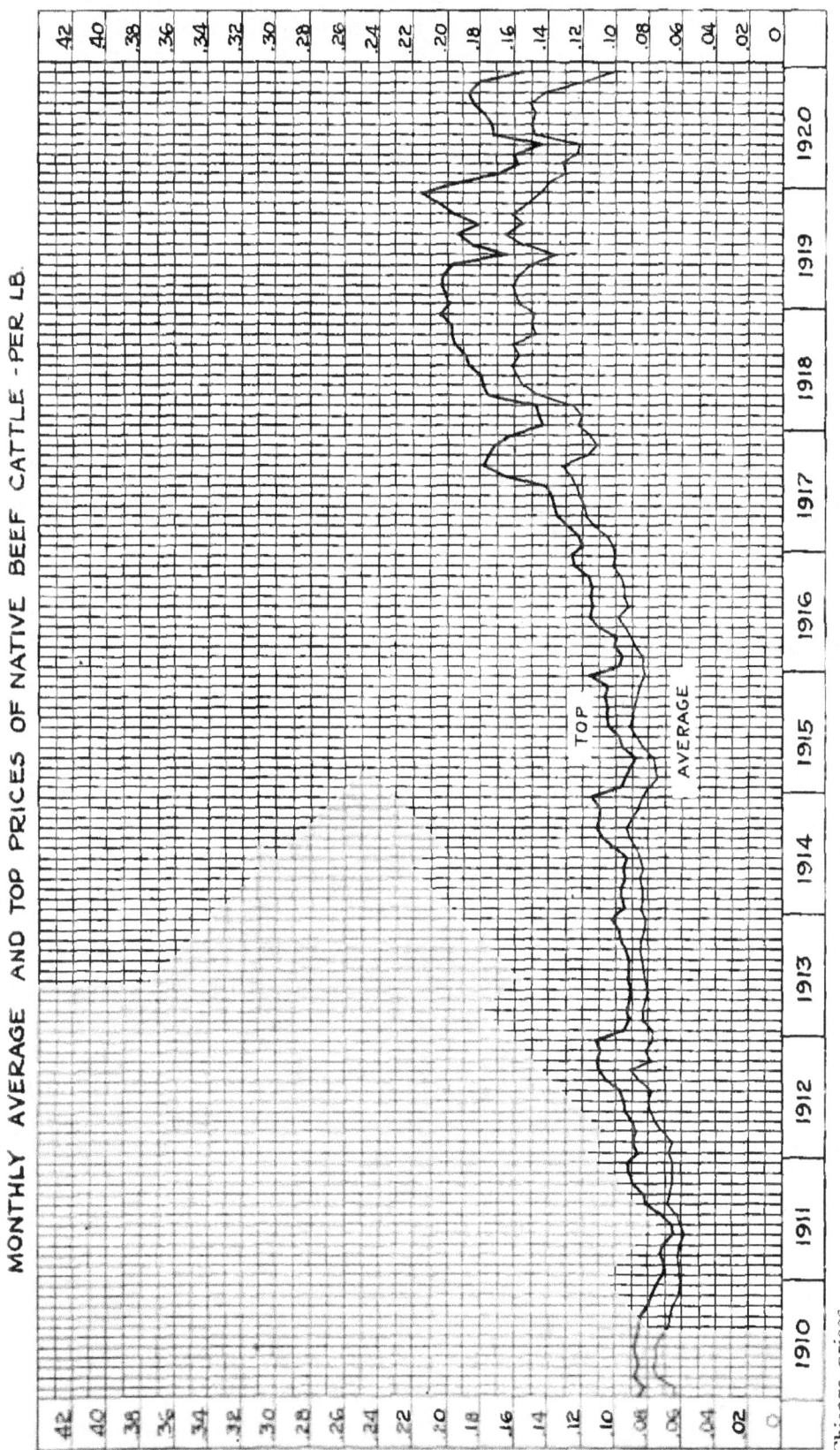

Chicago prices.

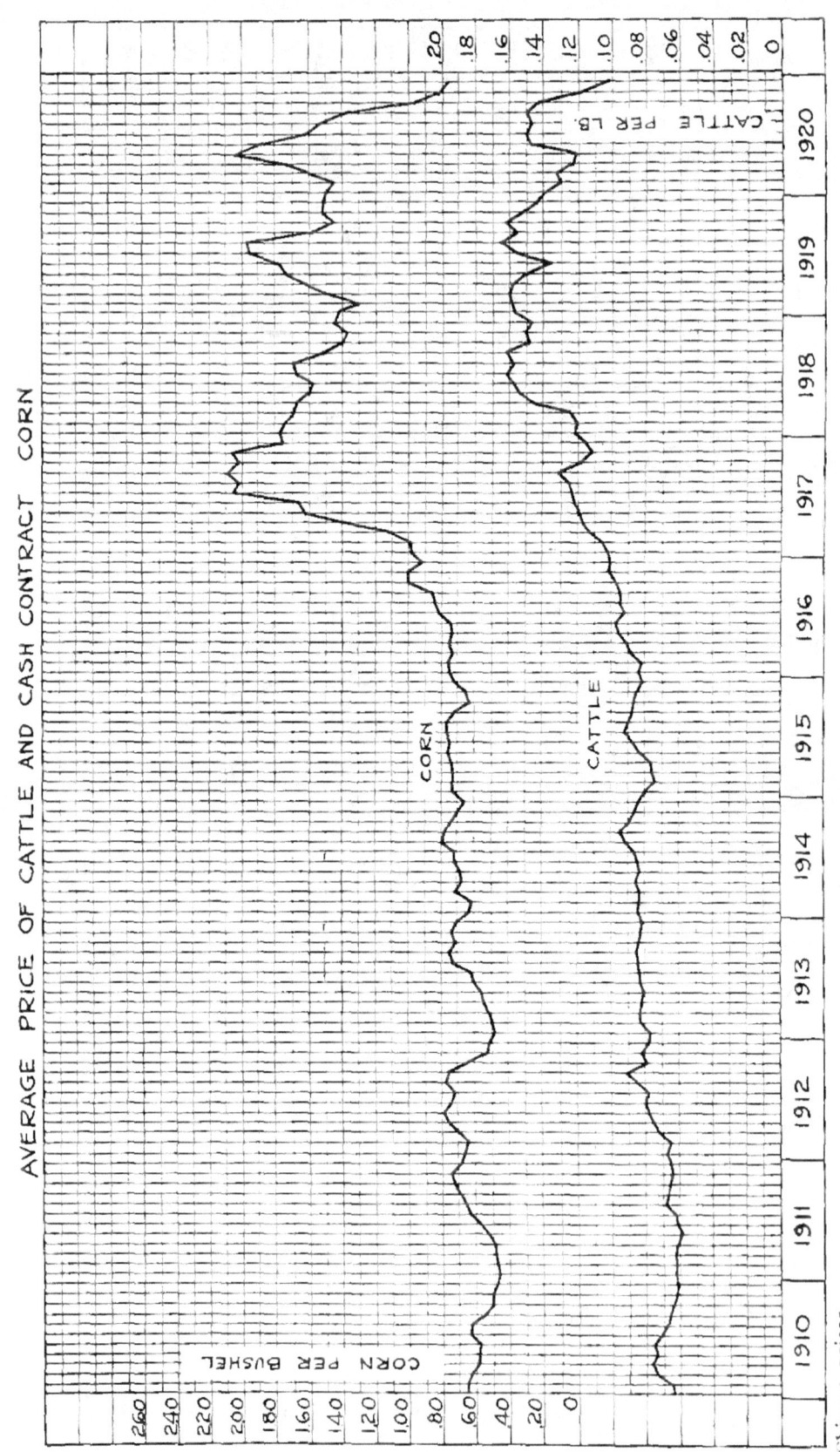
Chicago prices.

HIDE, TALLOW AND OLEO OIL PRICES

Packer Hides, Heavy Native Steers, Chicago

Average monthly prices, per 100 pounds

Month	1909	1910	1911	1912	1913	1914	1915	1916	1917	1918	1919	1920	1921
Jan	$15.80	$17.33	$13.00	$16.03	$18.69	$17.75	$23.27	$23.20	$32.50	$32.00	$28.00	$40.10	$17.20
Feb	15.13	15.35	13.00	15.69	18.12	18.21	22.97	22.84	31.00	29.25	28.25	40.00	14.56
Mar	14.41	14.40	12.88	15.55	17.40	18.19	21.00	22.40	30.10	26.05	27.55	37.00	13.06
Apr	14.53	15.17	13.08	15.88	16.97	18.12	19.28	23.20	30.50	26.62	30.94	35.88	10.60
May	16.30	15.56	13.87	16.97	17.07	18.43	21.60	25.50	31.56	31.00	36.60	36.00	12.37
June	16.88	17.50	15.63	17.45	17.66	18.84	23.56	26.75	32.10	32.60	40.80	36.00	14.00
July	17.18	16.10	16.30	17.75	18.12	19.68	26.42	26.70	31.75	33.00	50.10	30.70	13.60
Aug	16.88	16.19	15.75	18.90	18.70	20.70	27.09	26.09	32.50	30.00	52.70	28.12	14.00
Sept	17.22	15.94	16.20	19.50	19.47	21.37	26.19	26.35	32.60	30.00	46.38	28.50	14.00
Oct	17.90	15.77	16.34	19.53	19.87	21.25	26.35	28.31	34.19	30.00	48.25	26.20	14.90
Nov	17.88	14.87	16.50	19.93	19.65	21.94	25.90	32.38	35.00	29.00	47.20	22.25	15.75
Dec	17.93	13.72	16.41	19.41	18.42	22.90	24.77	33.40	34.70	29.00	40.38	20.00	16.50
Average	16.50	15.66	14.91	17.72	18.35	19.78	24.03	26.43	32.37	29.88	39.76	31.73	14.21

Tallow, Packer's Prime, Loose, Chicago

Average monthly prices, cents per pound

Month	1909	1910	1911	1912	1913	1914	1915	1916	1917	1918	1919	1920	1921
Jan	.0648	.0738	.0756	.0649	.0674	.0719	.0670	.0908	.1156	.1763	.1095	.1813	.0675
Feb	.0647	.0731	.0731	.0639	.0684	.0716	.0691	.0917	.1213	.1741	.1009	.1603	.0650
Mar	.0630	.0760	.0688	.0650	.0681	.0706	.0689	.0965	.1289	.1748	.1025	.1608	.0575
Apr	.0616	.0775	.0616	.07	.0698	.0694	.0683	.1050	.1489	.1747	.13	.1563	.0559
May	.0619	.0756	.0621	.0706	.0669	.0681	.0667	.1071	.1806	.1736	.1488	.1409	.06
June	.0620	.0734	.0608	.0684	.0666	.0666	.0639	.0991	.1814	.1719	.16	.1198	.0543
July	.0608	.0759	.0609	.0673	.0688	.0656	.0638	.0903	.1681	.1795	.2058	.1231	.0569
Aug	.0606	.0775	.0651	.0695	.0741	.0672	.0638	.0848	.1684	.1875	.1888	.1278	.0666
Sept	.0639	.0816	.0719	.07	.0763	.0696	.0644	.0969	.1733	.1947	.1703	.1348	.0738
Oct	.0681	.0823	.0689	.0705	.0745	.0659	.0775	.1048	.1715	.20	.1873	.1156	.0725
Nov	.0738	.0815	.0691	.0688	.0747	.0666	.0813	.1161	.1753	.1916	.1722	.0869	.0683
Dec	.0738	.0767	.0647	.0681	.0731	.0666	.0879	.1180	.1781	.1494	.1643	.0658	.0656
Yearly	.0649	.0771	.0669	.0681	.0707	.0683	.0703	.1000	.1593	.1790	.1534	.1311	.0637

Oleo Oil, Extra, Chicago

Average monthly prices, cents per pound

Month	1909	1910	1911	1912	1913	1914	1915	1916	1917	1918	1919	1920	1921
Jan	.1294	.1455	.0981	.1205	.1333	.0969	.1425	.1242	.2070	.2225	.3125	.2906	.1325
Feb	.1225	.1294	.0950	.1228	.1163	.0969	.1469	.1266	.1994	.2375	.3019	.2788	.1288
Mar	.1165	.1308	.0839	.1313	.1213	.0950	.1458	.1251	.1988	.2519	.2825	.2575	.1238
Apr	.1359	.1344	.0784	.1300	.1158	.0935	.1325	.1328	.2113	.2475	.2953	.2381	.1081
May	.1359	.1250	.0878	.1300	.1069	.0991	.1150	.1405	.2403	.2395	.3206	.2100	.1022
June	.1344	.1133	.0825	.1248	.1088	.0975	.1048	.1325	.2325	.2419	.3291	.1945	.0953
July	.1247	.1091	.0853	.1182	.1129	.0940	.0984	.1325	.2106	.2570	.3330	.1800	.0944
Aug	.1178	.1115	.0934	.1316	.1164	.1055	.0963	.1365	.1985	.2600	.3056	.1681	.1115
Sept	.1196	.1206	.1184	.1356	.1156	.1343	.1018	.1413	.2250	.2625	.3025	.1975	.1291
Oct	.1263	.1144	.1144	.1466	.1049	.1281	.1225	.1450	.2395	.2780	.2825	.2044	.1253
Nov	.1375	.1031	.1178	.1363	.1006	.1397	.1309	.1770	.2219	.2863	.3075	.1775	.1143
Dec	.1485	.0992	.1100	.1350	.0956	.1403	.1258	.2038	.2159	.3019	.3050	.1598	.1013
Yearly	.1291	.1197	.0971	.1302	.1124	.1101	.1219	.1432	.2167	.2572	.3065	.2131	.1139

232 THE PACKERS' ENCYCLOPEDIA

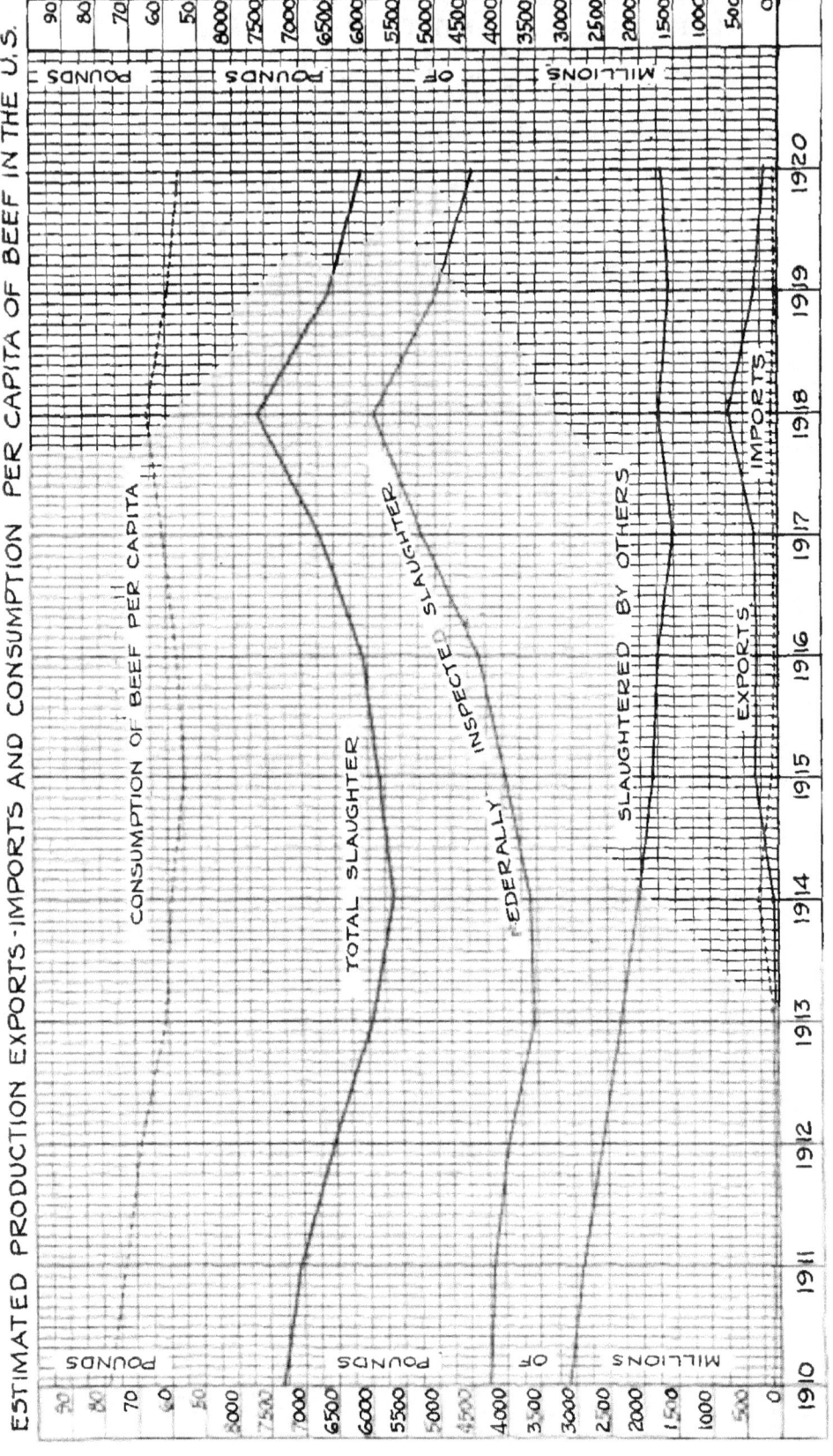

The Chart contrasts beef consumption and beef slaughters, and compares inspected and non-inspected slaughters, for a period from 1910 to 1920. The tables opposite give the figures from 1907 to 1921.

Estimated Annual Production, Exports, Imports, and Consumption Per Capita of Beef in the United States

Calendar Year	Slaughter			Exports (Domestic)	Imports (Less Re-exports)	Consumption	
	Total	Federally Inspected	Other			Total*	Per Capita
	Million Pounds	Million Pounds	Million Pounds	Million Pounds	Million Pounds	Million Pounds	Pounds
1907	7,320	4,336	2,984	352	....	6,968	79.7
1908	6,676	3,955	2,721	228	....	6,448	72.4
1909	7,071	4,189	2,882	163	....	6,908	76.2
1910	7,323	4,240	3,083	110	....	7,213	78.1
1911	7,036	4,137	2,899	92	....	6,944	73.9
1912	6,509	3,938	2,571	56	....	6,453	67.5
1913	5,913	3,595	2,318	47	38	5,904	60.8
1914	5,639	3,601	2,038	95	260	5,804	58.9
1915	5,816	3,979	1,837	399	109	5,526	55.6
1916	6,118	4,362	1,756	287	19	5,850	58.1
1917	6,686	5,169	1,517	376	28	6,338	62.0
1918	7,320	5,638	1,682	728	111	6,703	64.7
1919	6,283	4,774	1,509	314	35	6,004	57.2
1920	6,463	4,578	1,885	164	45	6,500	61.1
1921	6,194	4,113	2,081	52	27	6,227	57.7

*Includes differences between quantities in storage at beginning and end of year.

Since the passing of the big western ranges, cattle raising and beef production suffered a steady decline until the advent of the World War. The totals show that the war greatly stimulated production until the end of 1918. Since that year there has been rapid falling off, the 1921 total being over a billion pounds less than that of 1918.

Exports of beef at one time formed a large and important branch of the foreign trade of the United States, but by 1913 they had largely disappeared and foreign beef began to come in. The exports were large from 1915 to 1918 solely because of the war needs and have since fallen to a pre-war basis.

Imports of meat previous to 1913 were so small that they were not enumerated separately in the commerce reports. Imports of beef in 1914, however, were quite considerable. At this period the sources of cheap beef in the Southern Hemisphere, especially Argentina, had developed enormously and they had, in fact, supplanted the United States in the overseas trade with Europe.

Consumption of beef, as seen in the table, was at a low point in 1915, having decreased 24 pounds per head of the population during the preceding 8 years from 1907. Since the war it has receded 7 pounds per head.

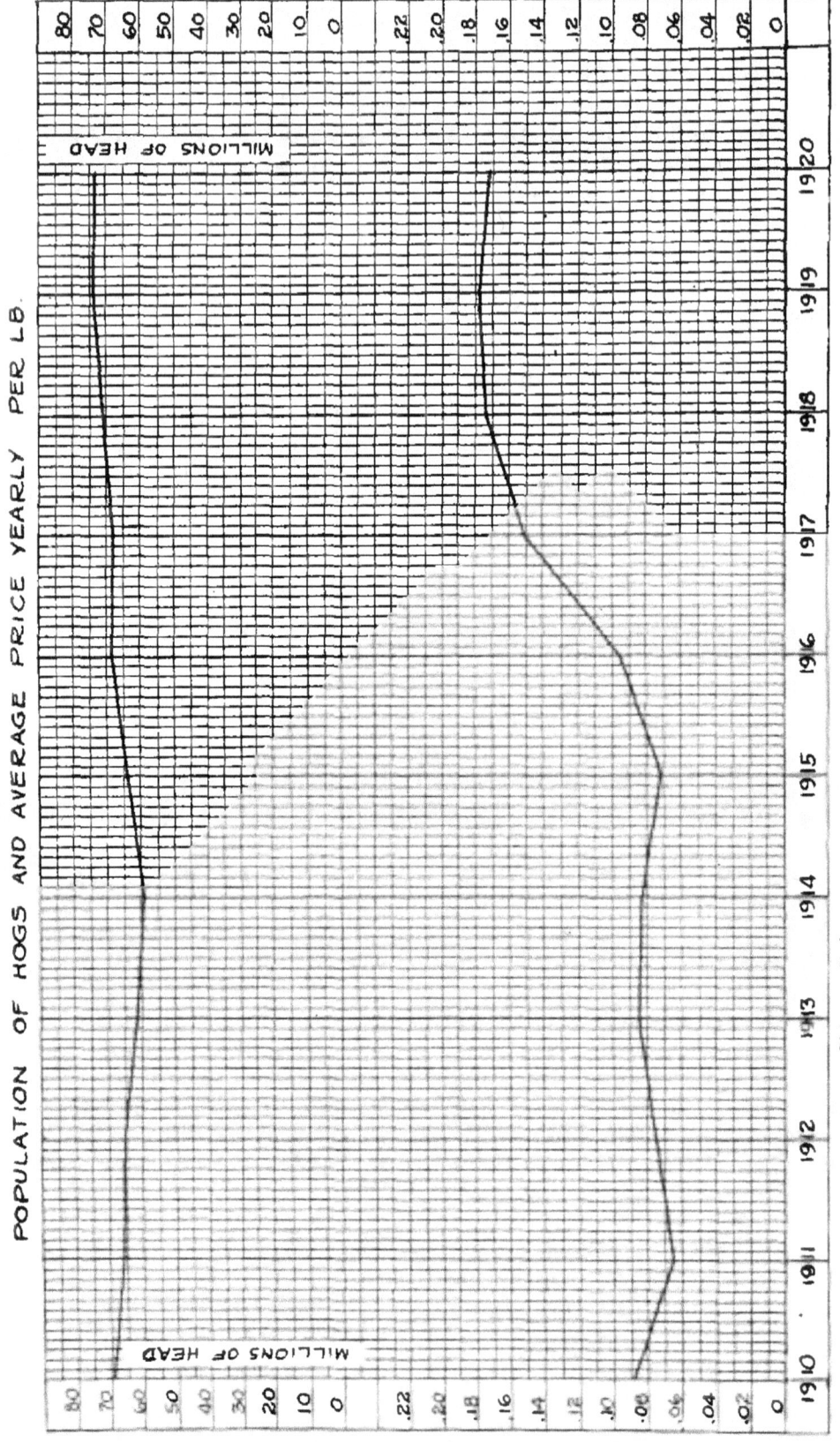

U. S. Dept. of Agriculture Census. Chicago Prices.

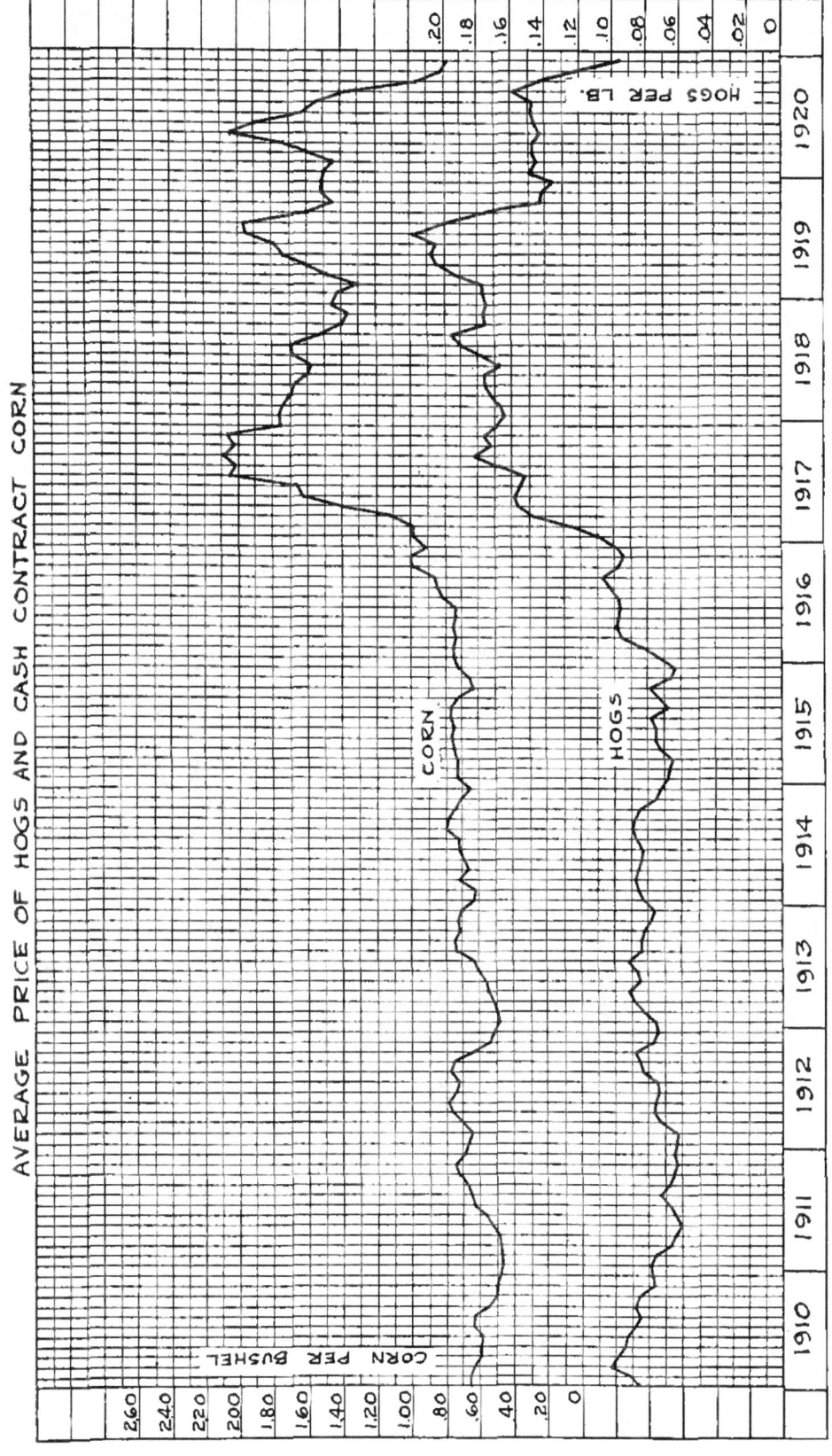

Chicago prices.

236 THE PACKERS' ENCYCLOPEDIA

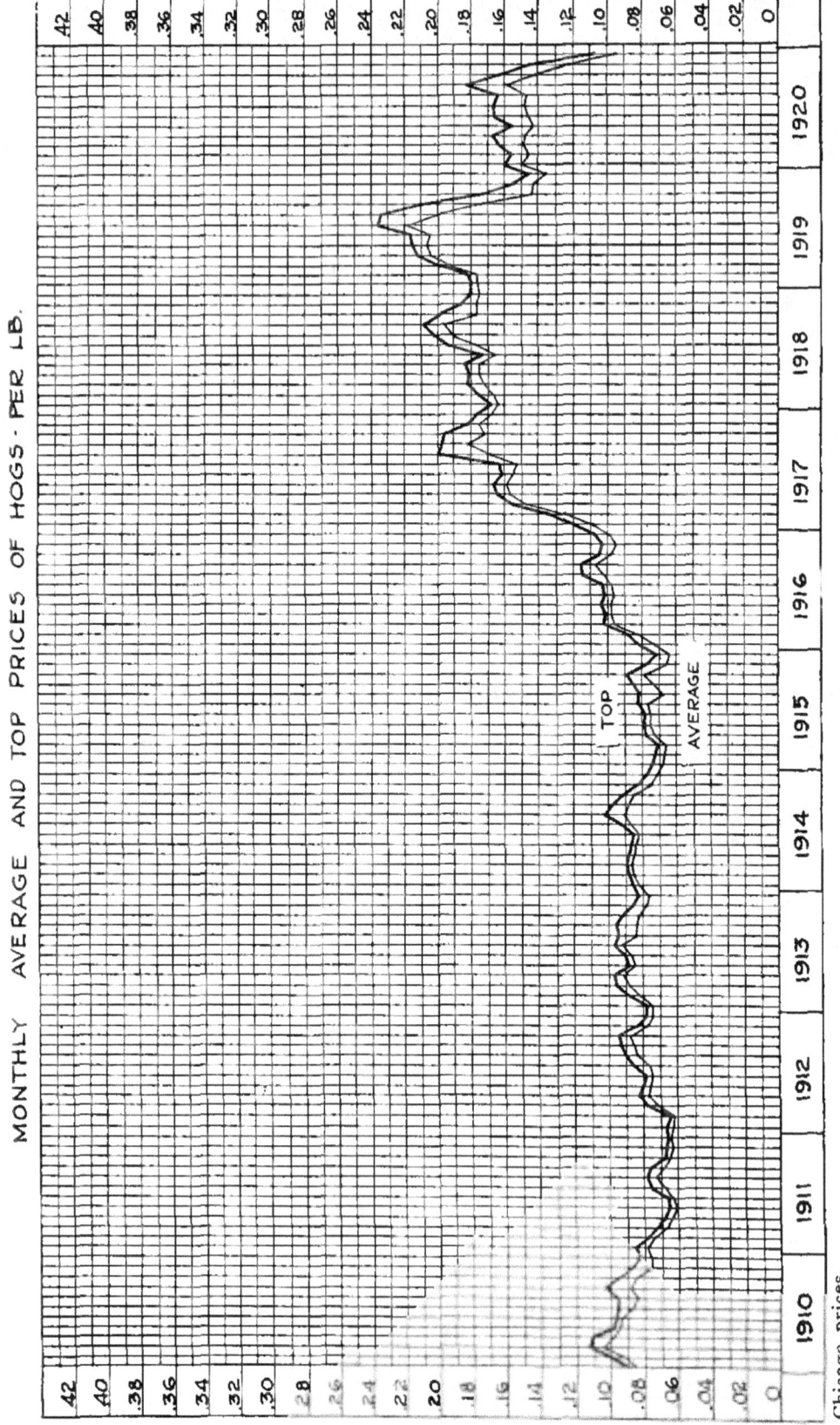

Chicago prices.

SEASONAL TURNING POINTS IN CHICAGO HOG MARKET PRICES, 1901 TO 1921

(Prices per 100 lbs.)

Year	Highest Weekly Average in Spring Rise		Advance From Previous Winter Low		Lowest Weekly Average in Early Summer Decline		Decline From Spring High		Highest Weekly Average in Late Summer or Fall Rise		Advance From Early Summer Low		Lowest Weekly Average in Winter Decline		Decline From Summer or Fall High	
	Week Ending	Price	Amount	Per Cent	Week Ending	Price	Amount	Per Cent	Week Ending	Price	Amount	Per Cent	Week Ending	Price	Amount	Per Cent
1901	April 6	$6.05			May 11	$5.73	$0.32	5.3	Sept. 28	$6.79	$1.06	18.5	Nov. 16	$5.67	$1.12	16.5
1902									July 26	7.79			Dec. 6	6.08	1.71	22.0
1903	Mar. 28	7.47	$1.39	22.9	Aug. 1	5.24	2.23	29.9	Sept. 26	5.87	.63	12.0	Nov. 28	4.28	1.59	27.1
1904	Mar. 12	5.45	1.17	27.3	May 28	4.59	.86	15.8	Oct. 1	5.94	1.35	29.4	Dec. 10	4.46	1.48	24.9
1905	April 15	5.57	1.11	24.9	June 3	5.31	.26	4.7	Aug. 19	6.09	.78	14.7	Nov. 25	4.80	1.29	21.2
1906	April 21	6.65	1.85	38.5	May 26	6.39	.26	3.9	July 14	6.79	.40	6.3	Dec. 1	6.09	.70	10.3
1907	Feb. 16	7.10	1.01	16.6	Aug. 24	5.79	1.31	18.5	Oct. 12	6.46	.67	11.6	Feb. 15[1]	4.31	2.15	33.3
1908	April 4	6.08	1.77	41.1	May 30	5.41	.67	11.0	Sept. 19	7.08	1.67	30.9	Dec. 19	5.52	1.56	22.0
1909									Sept. 18	8.23			Oct. 16	7.64	.59	7.2
1910	April 2	10.88	3.24	42.4	Aug. 6	8.03	2.85	26.2	Sept. 10	9.12	1.09	13.6	Nov. 26	7.01	2.11	23.1
1911	Jan. 7	8.07	1.06	15.1	May 6	5.89	2.18	27.0	Aug. 12	7.43	1.54	26.1	Dec. 23	6.07	1.36	18.3
1912	April 20	7.96	1.89	31.1	July 6	7.40	.56	7.0	Oct. 12	9.07	1.67	22.6	Dec. 21	7.18	1.89	20.8
1913	Mar. 29	9.20	2.02	28.1	May 10	8.40	.80	8.7	July 19 / July 26	9.15	.75	8.9	Nov. 29 / Dec. 6	7.65	1.50	16.4
1914	April 11	8.80	1.15	15.0	June 13	8.10	.70	8.0	Aug. 15	9.40	1.30	16.0	Feb. 20[2] / Feb. 27[2]	6.65	2.75	29.3
1915	June 26	7.70	1.05	15.8	Aug. 21	6.60	1.10	14.3	Oct. 16	8.50	1.90	28.8	Dec. 25	6.35	2.15	25.3
1916	May 20	10.00	3.65	57.5	June 10	9.40	.60	6.0	Sept. 23	10.85	1.45	15.4	Nov. 25	9.50	1.35	12.4
1917	May 19	16.15	6.65	70.0	July 21	14.95	1.20	7.4	Sept. 29	18.90	3.95	26.4	Oct. 27	15.55	3.35	17.7
1918	May 11	17.80	2.25	14.5	June 15	16.50	1.30	7.3	Sept. 21 / Oct. 6	20.15	3.65	22.1	Oct. 26	16.25	3.90	19.4
1919	May 10	20.90	4.65	28.6	June 7	20.20	.70	3.3	July 26	22.20	2.00	9.9	Dec. 13	12.80	9.40	42.3
1920	April 10	15.40	2.60	20.3	May 22	13.95	1.45	9.4	Sept. 25	16.70	2.75	19.7	Dec. 18	9.10	7.60	45.5
1921	Mar. 12	10.34	1.24	13.6	June 4	7.85	2.49	24.1	July 30	10.35	2.50	31.8	Nov. 19	6.73	3.62	35.0

[1] 1908. [2] 1915.

Compiled by U. S. Bureau of Markets and Crop Estimates.

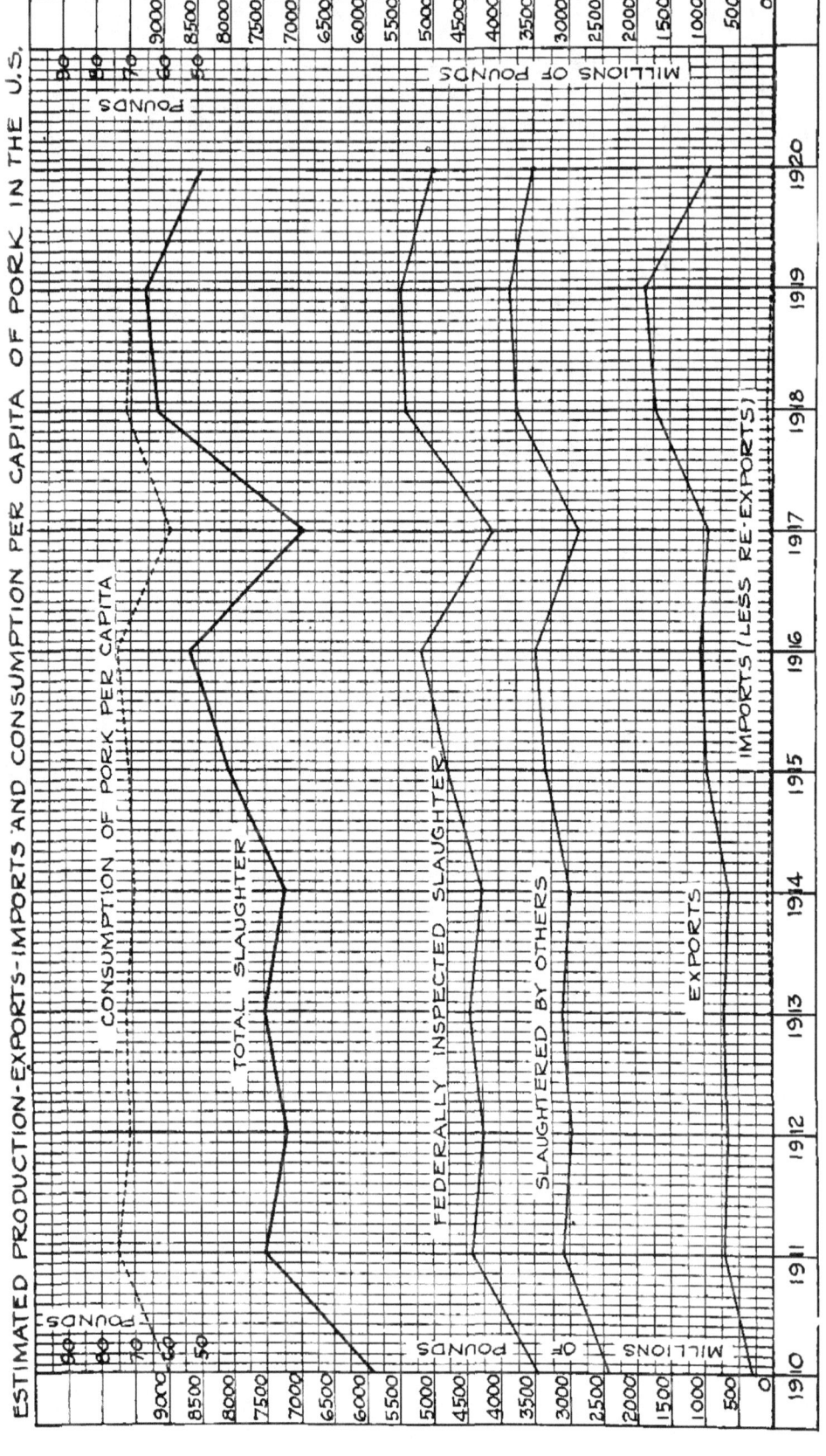

Estimated Annual Production, Exports, Imports, and Consumption of Pork in the United States

Calendar Year	Slaughter			Exports (Domestic)	Imports (Less Re-exports)	Consumption	
	Total	Federally Inspected	Other			Total*	Per Capita
	Million Pounds	Million Pounds	Million Pounds	Million Pounds	Million Pounds	Million Pounds	Pounds
1907	7,491	4,420	3,071	1,014	....	6,477	74.1
1908	8,226	4,853	3,373	619	....	7,607	85.4
1909	6,690	3,946	2,744	472	....	6,218	68.6
1910	5,881	3,470	2,411	313	....	5,568	60.3
1911	7,511	4,481	3,080	456	....	7,055	75.1
1912	7,189	4,242	2,947	440	....	6,749	70.6
1913	7,492	4,420	3,072	456	3	7,039	72.5
1914	7,228	4,264	2,964	377	38	6,889	69.9
1915	8,050	4,749	3,301	906	7	7,151	72.0
1916	8,634	5,196	3,448	1,011	2	7,625	75.7
1917	6,901	4,071	2,830	943	10	5,968	58.4
1918	8,854	5,551	3,303	1,724	97	7,227	69.8
1919	8,933	5,584	3,349	1,897	11	7,047	67.1
1920	8,193	5,133	3,060	925	6	7,338	68.9
1921	8,475	5,351	3,124	748	1	7,851	72.8

Estimated Annual Production, Exports, Imports, and Consumption of Lard in the United States

Calendar Year	Slaughter			Exports (Domestic)	Imports (Less Re-exports)	Consumption	
	Total	Federally Inspected	Other			Total*	Per Capita
	Million Pounds	Million Pounds	Million Pounds	Million Pounds	Million Pounds	Million Pounds	Pounds
1907	1,693	993	690	589	....	1,094	12.5
1908	1,834	1,094	760	582	....	1,272	14.3
1909	1,506	888	618	458	....	1,048	11.6
1910	1,344	793	551	379	....	965	10.5
1911	1,717	1,013	704	605	....	1,112	11.8
1912	1,643	969	674	553	....	1,090	11.4
1913	1,713	1,011	702	575	....	1,138	11.7
1914	1,652	975	677	460	....	1,192	12.1
1915	1,840	1,086	754	487	....	1,353	13.6
1916	1,973	1,164	809	454	....	1,519	15.1
1917	1,577	930	647	383	....	1,194	11.7
1918	2,015	1,263	752	555	....	1,460	14.1
1919	2,089	1,327	762	784	....	1,305	12.4
1920	2,022	1,326	696	636	....	1,390	13.1
1921	2,095	1,384	711	893	....	1,214	11.3

*Includes differences between quantities in storage at beginning and end of year.

The production of lard does not necessarily follow that of other pork products in relation to the number of hogs slaughtered in any given period of time. Certain conditions, as the plentifulness and cheapness of corn, determine whether the feeder shall market his hogs in fat or lean condition, and other economic factors determine whether the packer shall produce more or less lard than usual in the slaughtering process.

240 THE PACKERS' ENCYCLOPEDIA

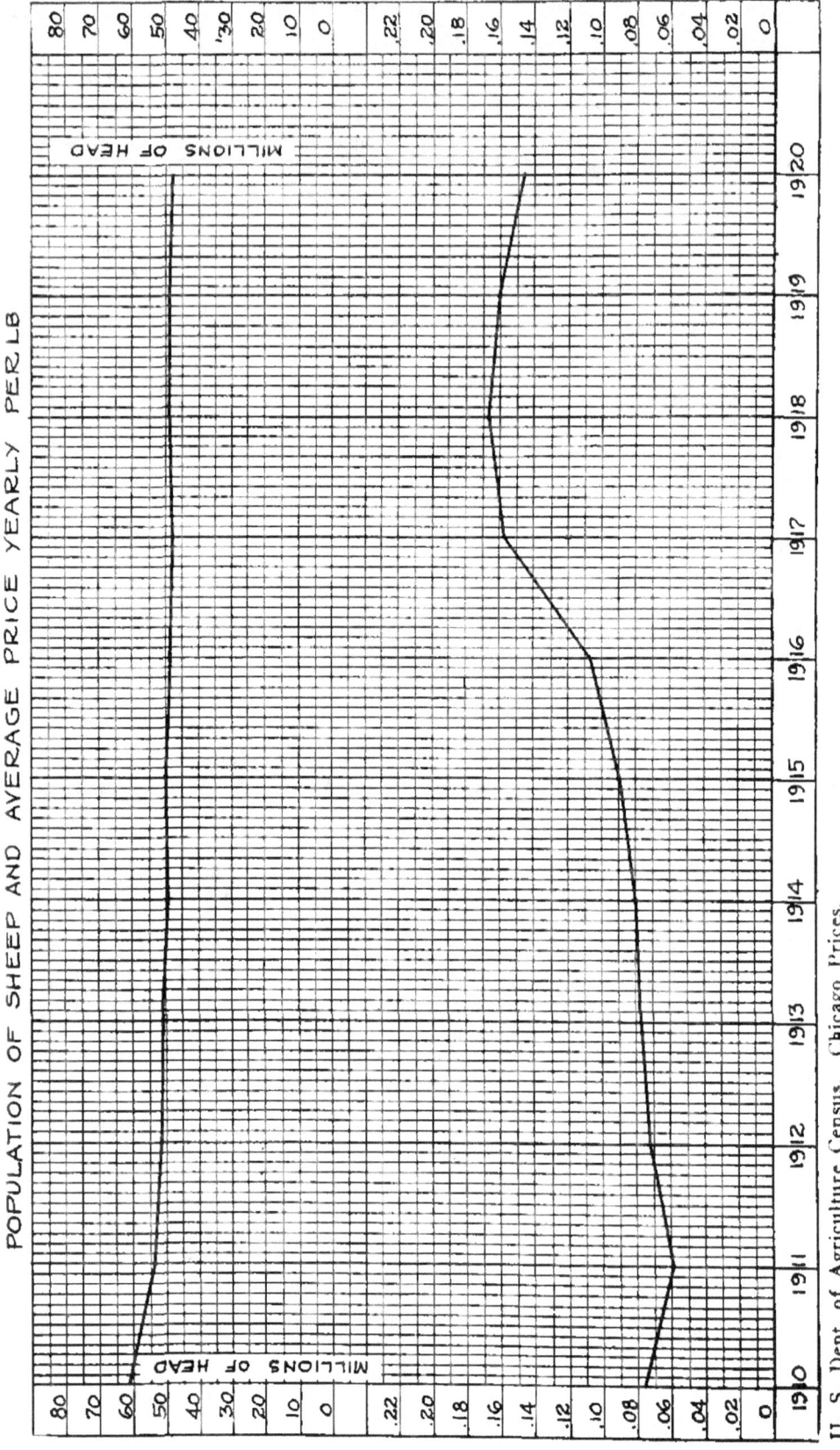

U. S. Dept. of Agriculture Census. Chicago Prices.

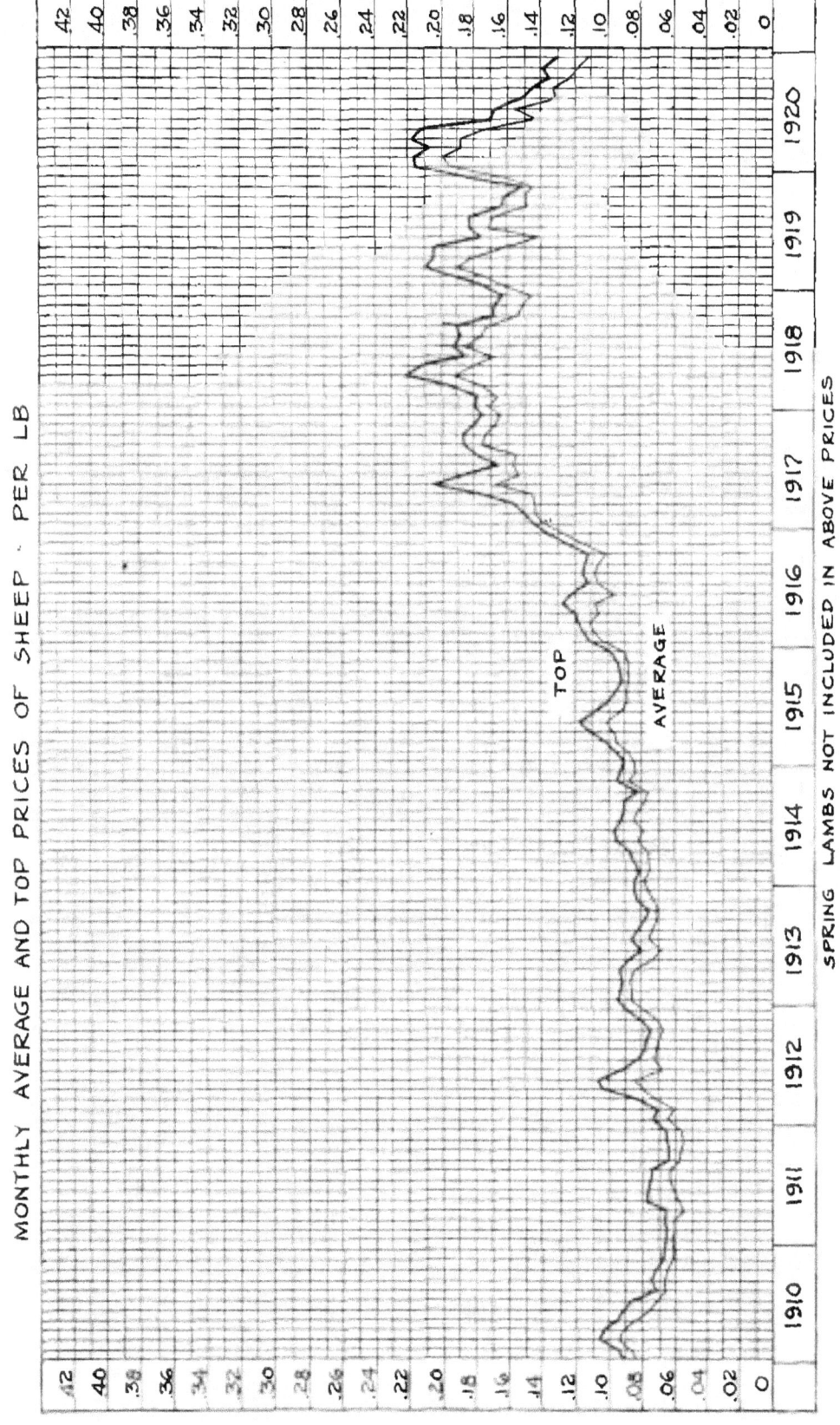

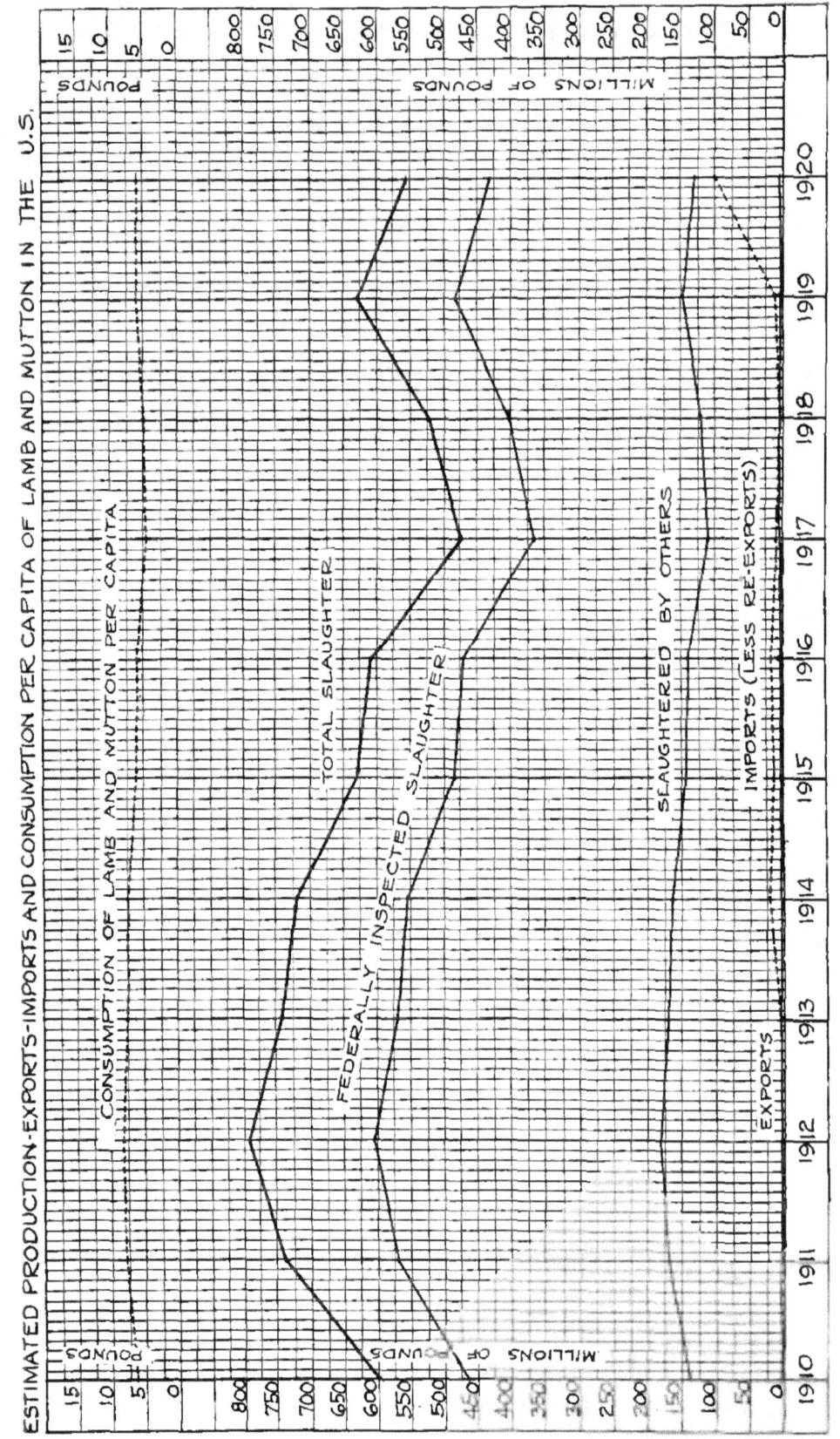

Estimated Annual Production, Exports, Imports, and Consumption of Mutton and Lamb in the United States

Calendar Year	Slaughter			Exports (Domestic)	Imports (Less Re-exports)	Consumption	
	Total	Federally Inspected	Other			Total*	Per Capita
	Million Pounds	Million Pounds	Million Pounds	Million Pounds	Million Pounds	Million Pounds	Pounds
1907	559	431	128	1	....	558	6.4
1908	555	428	127	1	....	554	6.2
1909	604	466	138	2	....	602	6.6
1910	600	463	137	2	....	598	6.5
1911	738	569	169	3	....	735	7.8
1912	788	608	180	5	....	783	8.2
1913	738	569	169	5	1	734	7.5
1914	720	555	165	4	20	736	7.5
1915	626	482	144	4	12	634	6.4
1916	612	472	140	5	16	623	6.2
1917	473	364	109	3	6	476	4.7
1918	489	381	108	2	1	488	4.7
1919	602	470	132	3	8	607	5.8
1920	538	423	115	4	61	537	5.0
1921	626	494	132	20	9	677	6.3

*Includes differences between quantities in storage at beginning and end of year.

The production and consumption of mutton and lamb is small in comparison with beef and pork. It averages about one-tenth of beef and one-twelfth of pork. The table shows the production to have been greatest from 1911 to 1914, in each of which years it exceeded 700 million pounds. The year of lowest production was 1917, when the yield was only 473 million pounds. Since 1917, the trend has been generally upward.

The proportion of federally inspected slaughter is greater with sheep and lambs than with any other class of livestock. Nearly four-fifths of the total mutton and lamb produced is inspected in establishments having Government supervision.

Normally there is very little foreign trade in mutton or lamb, but 1920 saw a new departure in heavy imports of Australasian product. These were not readily marketed and a large proportion was reexported.

The table shows the per capita consumption ranging between 8.2 pounds (highest) in 1912, and 4.7 pounds (lowest) in 1917 and 1918. There has been a steady rise in the last three years.

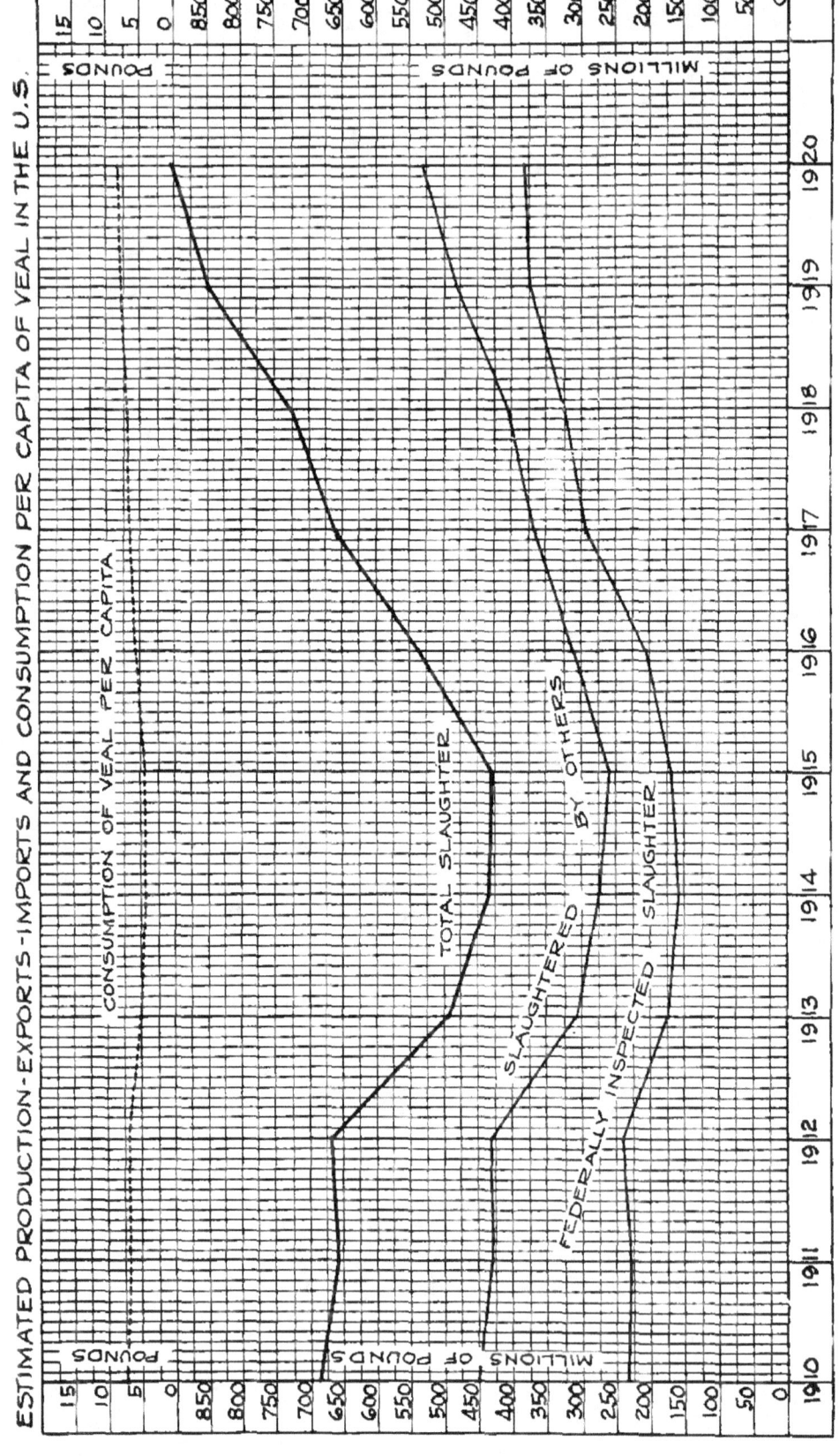

Estimated Annual Production, Exports, Imports, and Consumption of Veal in the United States

Calendar Year	Slaughter			Exports (Domestic)	Imports (Less Re-exports)	Consumption	
	Total	Federally Inspected	Other			Total	Per Capita
	Million Pounds	Million Pounds	Million Pounds	Million Pounds	Million Pounds	Million Pounds	Pounds
1907	626	210	416	....	....	626	7.1
1908	605	203	402	....	....	605	6.8
1909	684	230	454	....	....	684	7.5
1910	687	235	452	....	....	687	7.4
1911	657	229	428	....	....	657	7.0
1912	668	239	429	....	....	668	7.0
1913	488	176	312	....	....	488	5.0
1914	433	158	275	....	5	438	4.4
1915	428	168	260	....	1	429	4.3
1916	536	220	316	....	1	537	5.3
1917	662	296	366	....	1	663	6.5
1918	791	352	439	....	1	792	7.6
1919	860	378	482	....	5	865	8.2
1920	936	402	534	....	8	944	8.9
1921	864	367	497	....	4	868	8.0

Veal production as a rule follows that of beef. The unusually large slaughter in recent years, however, contrasts rather curiously with the considerable decline in cattle slaughter in the same period. It is accounted for partly by the droughty conditions in the West which induced heavy marketings of young stock during 1919, and the relatively higher prices for calf products since that time.

Country slaughter of veal is proportionately much larger than for any other class of animals. The federally inspected slaughter of calves in 1909 was about one-third of the total slaughter, and although it is increasing it is estimated to be still well below one-half of the total.

The consumption of veal practically corresponds to the production as there are no exports recorded and the imports are insignificant. The per capita consumption for the whole period has averaged close to 7 pounds per annum. It was lowest in 1915 (4.3 pounds) and highest in 1920 (8.9 pounds).

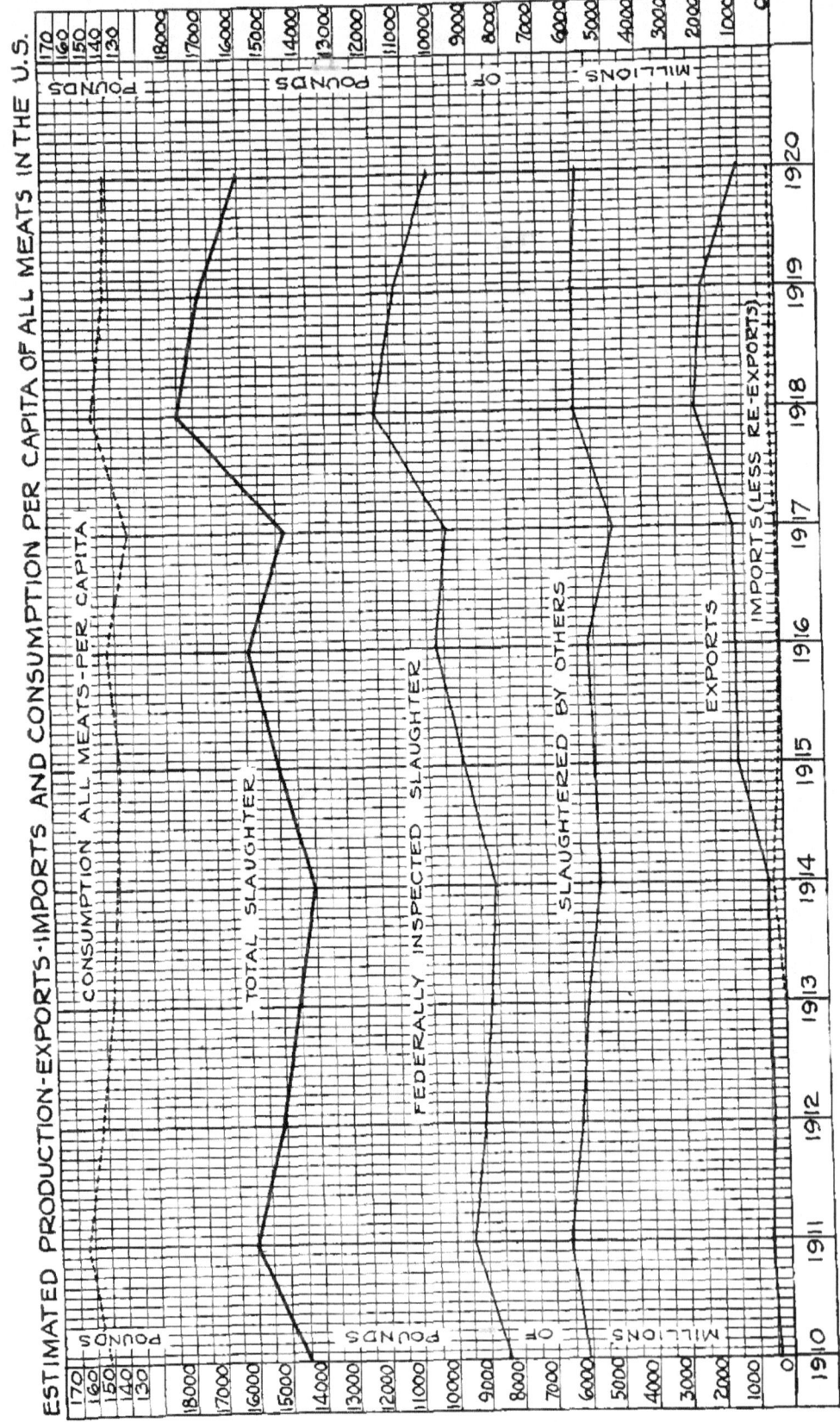

Estimated Annual Production, Exports, Imports, and Consumption of All Meats** (Excluding Lard) in the United States

Calendar Year	Slaughter			Exports (Domestic)	Imports (Less Re-exports)	Consumption	
	Total	Federally Inspected	Other			Total*	Per Capita
	Million Pounds	Million Pounds	Million Pounds	Million Pounds	Million Pounds	Million Pounds	Pounds
1907	16,003	9,399	6,604	1,367	....	14,636	167.4
1908	16,067	9,441	6,626	848	....	15,219	170.9
1909	15,060	8,835	6,225	637	....	14,423	159.0
1910	14,502	8,412	6,090	412	....	14,090	152.5
1911	15,946	9,368	6,578	534	....	15,412	163.9
1912	15,162	9,030	6,132	486	....	14,676	153.5
1913	14,640	8,763	5,877	507	41	14,174	145.9
1914	14,039	8,585	5,454	475	323	13,887	140.9
1915	14,937	9,384	5,553	1,309	129	13,757	138.5
1916	15,922	10,248	5,674	1,304	38	14,656	145.5
1917	14,740	9,906	4,834	1,322	44	13,462	131.8
1918	17,469	11,927	5,542	2,454	210	15,225	146.9
1919	16,687	11,209	5,478	2,214	59	14,532	138.4
1920	16,135	10,538	5,597	1,093	120	15,324	144.0
1921	16,160	10,325	5,835	820	41	15,624	144.8

*Includes differences between quantities in storage at beginning and end of year.
**Includes small quantity of goat meat not given separately.

The figures in the table above are merely the addition of the various meats in the previous tables plus a small quantity of goat meat. The latter however, furnishes only about one-tenth of a pound per capita of the total meat consumption in the country.

It may be seen from the table that the banner year in meat production was 1918, when about 17½ billion pounds were produced. About two-thirds of this meat was examined and certified as fit for human food by Federal inspectors. One-third, or 5½ billion pounds, was subject to State or local inspection, or no inspection at all, and practically all of this was slaughtered and consumed within State boundaries. During the last three years there has been little change in the production totals, but the heavy exports in 1919 brought down the consumption in that year about 6 pounds per head of the population.

LIVE STOCK POPULATION IN THE UNITED STATES

Number of cattle, sheep and hogs in the United States on January 1 of each year from 1900 to 1922, according to the estimates of the U. S. Department of Agriculture are as follows:

	Milch Cows	Other Cattle	Total Cattle	Total Sheep	Total Hogs
1900	16,292,360	27,610,054	43,902,414	41,883,065	37,079,356
1901	16,833,657	45,500,213	62,333,870	59,756,718	56,982,142
1902	16,696,802	44,727,797	61,424,599	62,039,091	48,698,890
1903	17,105,227	44,659,206	61,764,433	63,964,876	46,922,624
1904	17,419,817	43,629,498	61,049,315	51,630,144	47,009,367
1905	17,572,464	43,669,443	61,241,907	45,170,423	47,320,511
1906	19,793,866	47,067,656	66,861,522	50,631,619	52,102,847
1907	20,968,000	51,566,000	72,534,000	53,240,000	54,794,000
1908	21,194,000	50,073,000	71,267,000	54,631,000	56,084,000
1909	21,720,009	49,379,000	71,099,000	56,084,000	54,147,000
1910	21,801,000	47,279,000	69,080,000	57,216,000	47,782,000
1911	20,823,000	39,679,000	60,502,000	53,633,000	65,620,000
1912	20,699,000	37,260,000	57,959,000	52,362,000	65,410,000
1913	20,497,000	36,030,000	56,527,000	51,482,000	61,178,000
1914	20,737,000	35,855,000	56,592,000	49,719,000	58,933,000
1915	21,262,000	37,067,000	58,329,000	49,956,000	64,618,000
1916	22,108,000	39,812,000	61,920,000	48,625,000	67,766,000
1917	22,894,000	41,689,000	64,583,000	47,616,000	67,503,000
1918	23,310,000	44,112,000	67,422,000	48,603,000	70,978,000
1919	23,475,000	45,085,000	68,560,000	48,866,000	74,584,000
*1920	23,772,000	43,392,000	67,164,000	39,025,000	59,344,000
*1921	23,594,000	41,993,000	65,587,000	37,452,000	56,097,000
*1922	24,028,000	41,324,000	65,352,000	36,048,000	56,996,000

*These figures are revised estimates of the U. S. Department of Agriculture, published February 15, 1922.

Live Stock Top and Average Prices at Chicago

Yearly top and average prices of native beef cattle, hogs and fat lambs at Chicago, as compiled by the Chicago Drovers Journal:

	Beef Steers		Hogs		Lambs	
	Top	Average	Top	Average	Top	Average
1920	$19.25	$13.30	$18.25	$14.15	$21.75	$14.60
1919	21.50	15.50	23.60	17.85	21.00	16.00
1918	20.50	14.65	20.95	17.45	22.10	16.60
1917	17.90	11.60	20.00	15.10	20.60	15.60
1916	12.60	9.50	11.60	9.60	13.60	10.75
1915	11.60	8.40	8.95	7.10	11.85	9.00
1914	11.40	8.65	10.20	8.30	9.60	8.00
1913	10.25	8.25	9.70	8.35	9.50	7.70
1912	11.25	7.75	9.42	7.55	10.60	7.20
1911	9.35	6.40	8.30	6.70	7.85	5.95
1910	8.85	6.80	11.20	8.90	10.60	7.55

MEAT ANIMALS SLAUGHTERED IN THE UNITED STATES

Number of animals slaughtered annually under Federal inspection, and estimated total number slaughtered (including farm) in United States:

Calendar Year		Cattle	Calves	Sheep and Lambs	Goats	Hogs
1907	U. S. Inspected..	7,633,365	2,024,387	10,252,070	56,750	32,885,377
	Total............	13,469,900	6,026,800	13,300,600	161,000	55,737,900
1908	U. S. Inspected..	7,279,260	1,958,273	10,304,666	42,981	38,643,101
	Total............	12,845,000	5,829,900	13,368,800	121,900	65,496,800
1909	U. S. Inspected..	7,713,807	2,189,017	11,350,349	100,659	31,394,896
	Total (census)...	13,611,422	6,515,976	14,724,699	285,553	53,219,568
1910	U. S. Inspected..	7,807,600	2,238,587	11,408,020	100,379	26,003,463
	Total............	13,540,600	6,552,600	14,800,200	284,800	44,073,500
1911	U. S. Inspected..	7,619,096	2,183,533	14,020,446	38,891	34,232,955
	Total............	12,958,100	6,264,500	18,189,500	110,300	58,022,000
1912	U. S. Inspected..	7,252,378	2,277,946	14,979,265	72,894	33,052,727
	Total............	11,979,000	6,348,000	19,433,400	206,800	56,022,000
1913	U. S. Inspected..	6,978,361	1,902,414	14,405,759	75,655	34,198,585
	Total............	11,477,600	5,284,500	18,689,400	214,600	57,973,500
1914	U. S. Inspected..	6,756,737	1,696,962	14,229,343	175,906	32,531,840
	Total............	11,004,500	4,661,400	18,460,500	499,000	55,148,100
1915	U. S. Inspected..	7,153,395	1,818,702	12,211,765	153,346	38,381,228
	Total............	10,822,100	4,639,500	15,843,000	435,000	65,064,000
1916	U. S. Inspected..	8,310,458	2,367,303	11,941,366	198,909	43,088,708
	Total............	12,026,700	5,773,900	15,492,200	564,300	73,035,600
1917	U. S. Inspected..	10,350,052	3,142,721	9,344,994	165,660	33,909,704
	Total............	13,723,900	7,030,700	12,123,800	470,000	57,488,800
1918	U. S. Inspected..	11,828,549	3,456,393	10,319,877	137,725	41,214,250
	Total............	15,750,449	7,767,193	13,230,577	390,125	65,732,450
1919	U. S. Inspected.	10,089,984	3,969,019	12,691,117	87,380	41,311,830
	Total............	13,635,084	9,041,019	16,264,817	247,480	66,680,330
1920	U. S. Inspected..	8,608,691	4,058,370	10,982,180	42,477	38,018,684
	Total............	13,242,691	8,822,370	14,079,680	120,477	60,635,884
1921	U. S. Inspected..	7,608,280	3,807,568	13,004,905	12,133	38,982,356
	Total............	12,271,280	8,654,568	16,673,005	34,433	62,172,856

U. S. inspection of horses at slaughter was commenced in September, 1919, the number so inspected to date being 1919, 433; 1920, 894; 1921, 2,562. **A large proportion of this horseflesh is exported.**

MEAT PACKING IN THE UNITED STATES

Number and cost of animals slaughtered, and quantities and values of principal products manufactured in the United States during the year 1919, compiled from returns made by 1305 establishments, according to the U. S. Census for 1919, with comparisons for 1914, are as follows:

Materials		1919	1914
Total cost		$3,774,901,000	$1,441,663,000
Animals slaughtered	Cost	$3,055,495,000	$1,199,642,000
Beeves	Number	10,818,000	7,149,000
	Cost	$1,055,319,000	$490,108,000
Calves	Number	4,395,000	2,019,000
	Cost	$95,720,000	$27,623,000
Sheep, lambs, goats and kids	Number	13,523,000	15,952,000
	Cost	$146,965,000	$84,813,000
Hogs	Number	44,519,000	34,442,000
	Cost	$1,757,491,000	$597,098,000
All other materials	Cost	$719,406,000	$242,021,000

Products		1919	1914
Total value		$4,246,290,000	$1,651,965,000
Fresh meat:			
Beef	Pounds	4,932,284,000	3,658,334,000
	Value	$846,806,000	$421,297,000
Veal	Pounds	422,928,000	194,699,000
	Value	$83,884,000	$26,299,000
Mutton, lamb, goat and kid	Pounds	501,201,000	629,233,000
	Value	$120,451,000	$74,676,000
Pork	Pounds	2,112,243,000	1,877,099,000
	Value	$532,075,000	$226,535,000
Edible offal and all other fresh meat	Pounds	516,983,000	296,667,000
	Value	$59,832,000	$20,576,000
Cured meat:			
Beef, pickled and other cured	Pounds	129,960,000	91,572,000
	Value	$28,360,000	$14,395,000
Pork, pickled and other cured	Pounds	4,145,232,000	2,929,310,000
	Value	$1,217,420,000	$393,605,000
Canned goods	Pounds	305,943,000	160,799,000
	Value	$96,904,000	$26,418,000
Sausage:			
Canned	Pounds	161,002,000	74,004,000
	Value	$27,985,000	$9,845,000
All other	Pounds	629,701,000	435,147,000
	Value	$145,601,000	$58,350,000
Lard	Pounds	1,372,550,000	1,119,189,000
	Value	$415,817,000	$120,414,000
Lard compounds and substitutes	Pounds	521,122,000	396,398,000
	Value	$123,724,000	$33,037,000
Oleo oil	Gallons	20,339,000	16,502,000
	Value	$30,953,000	$11,926,000
Other oils	Gallons	6,721,000	6,715,000
	Value	$9,153,000	$4,010,000
Tallow and oleo stock	Pounds	242,084,000	209,614,000
	Value	$36,536,000	$13,733,000
Oleomargarine	Pounds	123,639,000	60,388,000
	Value	$36,778,000	$8,819,000
Hides and pelts:			
Cattle hides	Number	10,818,000	7,159,000
	Value	$185,020,000	$69,959,000
Calf	Number	3,353,000	1,464,000
	Value	$24,797,000	$3,513,000
Sheep, lamb, goat and kid	Number	12,244,000	15,917,000
	Value	$33,780,000	$13,624,000
Fertilizers and fertilizer material	Tons	391,000	294,000
	Value	$18,315,000	$8,737,000
All other products*	Value	$172,099,000	$92,197,000

*Includes value of ammonia, butter, butter reworked, condensed milk, glue, glycerine, hog hair, ice, sausage casings, scrapple, soap, wool, etc., and amount received for slaughtering and refrigeration for others.

EXPORTS OF MEAT PRODUCTS 1910-1921

Beef

	Fresh Beef	Canned Beef	Pickled & Other Cured	Tallow	Oleo Oil	Total Beef & Beef Products
	Million Lbs.	Million Lbs.	Million Lbs.	Million Lbs.	Million Lbs.	Million Lbs
1910	56	12	35	16	105	224
1911	29	11	42	46	163	291
1912	9	9	29	29	94	170
1913	7	4	25	28	101	165
1914	31	31	34	10	85	181
1915	263	70	43	27	109	511
1916	182	54	37	15	84	372
1917	216	66	68	8	33	391
1918	514	141	44	4	69	772
1919	174	54	43	39	76	386
1920	90	24	26	21	74	235
1921	10	6	25	14	128	186

Pork

	Fresh Pork	Canned Pork	Pickled Pork	Cured Hams & Shoulders	Bacon	Lard	Neutral Lard	Total Pork & Pork Products
	Million Lbs.	Million Lbs.	Million Lbs.	Million Lbs.	Million Lbs.	Million Lbs.	Million Lbs.	Million Lbs.
1910	1	4	42	131	128	369	10	685
1911	2	5	51	189	198	553	53	1,051
1912	3	5	54	176	192	495	58	983
1913	3	4	54	172	213	536	39	1,021
1914	1	3	37	142	184	438	22	827
1915	24	8	59	267	524	451	35	1,368
1916	55	7	55	287	593	427	27	1,451
1917	49	6	39	243	578	373	10	1,298
1918	12	5	37	537	1,105	539	6	2,251
1919	27	6	34	597	1,190	761	23	2,638
1920	38	2	39	185	637	612	23	1,536
1921	56	1	33	232	415	869	24	1,630

Other Meat Products

	Mutton	Sausage (Canned and All Other)	Sausage Casings	Total Value Meat Products
	Lbs.	Lbs.	Lbs.	
1910	1,997,000	*	35,467,821	$ 127,303,473
1911	2,574,000	*	40,640,686	155,864,543
1912	5,076,000	3,746,960	28,828,200	144,421,752
1913	4,789,000	7,529,212	28,776,526	157,486,469
1914	3,847,000	5,092,437	25,433,526	137,737,493
1915	4,231,000	13,345,189	28,514,453	259,064,321
1916	5,258,000	11,998,350	7,492,608	279,198,960
1917	2,862,000	18,053,301	7,815,814	369,539,310
1918	1,631,000	12,378,956	4,194,748	845,260,801
1919	3,009,000	12,087,621	25,477,028	1,014,165,889
1920	3,575,409	17,667,381	25,238,187	463,256,812
1921	7,515,438	8,998,222	31,521,187	298,212,479

*Quantities not available.

PROVISION PRICES AT CHICAGO

Monthly Range Cash Prices Mess Pork

	1910	1911	1912	1913	1914	1915	1916	1917	1918	1919	1920	1921
January	$20.25	$19.75	$15.00	$17.50	$20.25	$17.95	$18.62	$28.00	$47.00	$42.00	$39.50	$24.00
	22.50	21.00	16.00	19.25	22.25	19.15	20.45	31.75	48.50	47.50	42.00	25.50
February	21.50	19.25	15.00	19.10	21.25	16.87	20.00	28.25	47.50	40.75	37.00	23.50
	24.75	21.50	15.87	20.37	22.25	19.50	21.00	32.12	50.50	45.00	40.00	25.00
March	24.75	16.00	15.25	19.87	20.75	16.75	20.75	32.00	48.50	43.50	37.00	22.00
	26.75	19.50	17.12	20.87	21.75	17.75	23.25	35.25	50.50	47.25	39.00	24.00
April	20.65	15.50	16.87	19.50	19.35	16.62	22.87	35.00	49.35	46.50	36.00	18.00
	25.75	16.50	19.62	20.62	21.25	17.62	24.25	39.00	49.25	54.00	38.00	21.50
May	21.75	14.75	18.00	19.62	19.25	17.45	22.50	37.00	39.50	52.12	34.00	15.75
	23.50	17.50	19.50	20.72	20.25	18.00	24.50	39.55	46.25	56.00	37.00	17.25
June	21.25	14.87	18.37	20.50	19.80	16.60	20.75	37.75	40.35	51.50	33.00	18.50
	24.25	16.00	19.00	21.25	21.35	18.00	25.75	40.10	43.80	56.00	34.75	19.25
July	24.00	15.62	17.37	21.12	21.65	13.12	25.37	40.00	43.60	51.00	25.50	19.25
	27.00	17.50	18.62	22.75	23.50	16.80	27.50	41.15	46.00	55.75	32.62	19.50
August	21.50	16.12	17.32	21.25	18.45	13.25	25.87	40.00	43.00	41.50	24.50	19.50
	24.00	18.12	18.25	22.62	24.50	14.05	28.25	44.00	45.30	54.00	26.42	20.50
September	18.25	14.75	16.50	21.50	18.50	12.00	28.00	42.50	30.50	35.50	22.90	20.00
	21.75	15.87	17.87	22.75	21.25	13.50	29.00	46.25	43.10	45.00	27.00	22.00
October	17.25	14.50	16.12	20.50	16.50	13.12	28.00	41.50	33.50	35.00	22.50	19.50
	18.75	16.00	17.75	22.00	18.50	15.25	29.50	46.50	39.00	43.00	26.00	21.00
November	17.00	15.62	16.25	20.25	16.75	14.25	27.75	42.50	35.00	42.00	23.75	19.00
	18.00	16.50	18.00	21.50	17.50	16.50	29.50	52.00	48.00	43.00	25.00	19.50
December	17.00	14.75	16.00	20.25	16.50	16.25	29.00	47.00	46.25	39.50	23.50	17.50
	20.00	16.50	18.00	21.75	17.25	17.50	29.50	52.00	49.50	43.50	25.50	19.00
Year	17.00	14.50	15.00	17.50	16.50	12.00	18.62	28.00	33.50	40.75	22.50	17.50
	27.00	21.50	19.50	22.75	24.50	19.50	29.50	52.00	50.50	56.00	42.00	25.50
Average	21.75	16.85	17.20	20.73	20.01	16.14	25.08	39.63	44.51	46.28	31.33	20.43

Monthly Range Cash Prices Short Rib Sides

	1910	1911	1912	1913	1914	1915	1916	1917	1918	1919	1920	1921
January	$11.10	$ 9.87	$ 7.50	$ 9.00	$10.25	$ 9.12	$ 9.75	$13.25	$23.00	$21.25	$18.25	$10.75
	11.12	10.75	8.62	10.50	11.62	10.37	10.87	15.62	24.62	25.50	20.00	12.75
February	11.37	8.87	7.62	9.62	10.75	8.87	10.00	14.75	23.62	21.00	17.25	10.50
	12.87	10.50	8.50	10.87	11.50	10.12	11.37	17.12	25.75	25.50	19.50	12.25
March	12.37	8.00	7.62	10.12	10.62	8.87	10.75	16.50	23.87	24.00	17.25	10.50
	13.95	9.50	9.75	11.50	11.50	9.62	12.37	18.50	25.00	27.25	19.50	12.25
April	11.62	7.37	9.00	10.75	10.37	9.00	11.62	18.00	22.05	26.50	17.00	8.25
	13.62	8.62	10.25	11.75	11.25	10.25	12.87	20.75	24.37	28.87	19.00	11.00
May	12.12	7.25	9.62	11.12	10.62	9.62	12.12	19.75	20.65	28.20	17.00	9.17
	12.00	8.12	10.50	12.75	11.50	10.62	13.00	21.25	23.85	30.00	28.20	9.95
June	12.37	7.37	9.75	11.50	10.87	9.50	11.87	20.30	21.35	27.00	17.00	9.00
	13.62	8.50	10.62	12.25	12.00	10.50	14.00	22.00	23.50	30.00	18.62	10.87½
July	11.37	7.62	9.87	11.25	11.37	8.87	13.12	21.00	22.90	27.00	15.25	9.75
	11.37	8.62	10.62	12.37	10.37	14.00	22.10	25.00	29.37	18.00	11.50
August	11.00	7.75	10.12	10.25	11.25	7.87	13.12	21.60	23.87	21.50	14.00	8.75
	12.62	9.37	11.12	12.25	13.00	9.37	14.75	24.12	25.02	28.00	16.50	11.25
September	10.75	8.00	10.12	10.25	11.00	7.50	14.00	23.25	22.70	18.00	14.50	7.00
	12.25	9.25	11.12	12.00	12.75	9.25	15.00	27.12	24.60	23.50	18.62	10.25
October	10.00	7.75	10.00	10.12	9.75	8.50	13.50	25.50	20.00	17.75	14.00	5.50
	11.75	8.87	11.25	11.75	11.62	11.00	14.87	28.50	23.00	19.50	19.00	8.00
November	9.00	7.62	10.12	10.00	9.25	9.00	13.62	25.50	22.25	18.50	12.50	5.50
	10.87	8.75	11.12	11.62	11.00	11.00	15.00	28.25	27.25	20.50	16.75	7.75
December	9.00	7.62	9.00	10.25	9.12	9.25	12.75	23.00	23.50	17.00	10.75	7.00
	10.75	8.75	10.75	11.62	10.50	10.50	14.25	28.00	27.50	20.25	14.50	8.50
Year	9.00	7.25	7.50	9.00	9.12	7.50	9.75	13.25	20.00	17.75	10.25	5.50
	13.95	10.75	11.25	12.75	13.08	11.00	15.00	28.50	27.50	29.37	20.00	12.75
Average	11.60	8.55	9.79	11.08	11.08	9.57	12.87	21.49	23.64	23.99	16.80	9.49

Monthly Range of Prices of Cash Lard

	1910	1911	1912	1913	1914	1915	1916	1917	1918	1919	1920	1921
January	$11.70	$9.82	$9.05	$9.47	$10.62	$10.40	$9.87	$15.10	$23.50	$22.47	$22.50	$12.62
	11.90	10.67	9.45	10.30	11.17	11.05	10.92	16.40	25.40	23.77	24.45	13.30
February	11.97	9.00	8.65	10.12	10.30	9.65	9.75	15.72	25.40	22.05	19.75	11.32
	13.20	9.87	9.15	10.72	10.82	11.27	10.30	18.45	26.62	26.25	21.85	12.57
March	13.35	8.25	8.90	10.40	10.30	9.65	10.30	18.45	25.32	25.50	19.62	11.00
	14.65	9.15	9.80	11.25	10.70	10.27	11.42	20.30	26.47	28.60	21.32	12.05
April	12.00	7.70	9.72	10.65	9.82	9.70	11.20	20.27	24.55	28.67	18.75	9.35
	14.00	8.17	11.00	11.27	10.50	10.12	12.77	21.97	25.90	32.87	20.05	10.75
May	12.30	7.92	10.37	10.87	9.60	9.42	12.50	21.05	23.67	32.65	19.75	9.22
	12.80	8.20	10.90	11.25	10.10	10.30	13.15	22.77	25.62	34.70	21.10	9.75
June	11.85	8.00	10.62	10.92	9.72	9.10	12.22	21.70	23.87	33.50	20.00	9.22½
	12.47	8.30	10.92	11.15	10.10	9.75	13.20	21.67	25.35	35.85	20.55	10.50
July	11.55	8.12	10.32	11.15	9.60	7.55	12.57	20.15	25.50	33.72	17.80	10.50
	12.27	8.60	10.80	11.87	10.27	9.37	13.32	21.20	26.82	35.25	20.20	12.20
August	11.50	8.52	10.40	10.90	8.60	7.67	12.55	20.90	26.40	27.20	17.90	10.22½
	12.12	9.40	11.20	11.57	10.25	8.07	14.25	23.62	26.90	33.65	18.90	12.12½
September	11.95	8.97	10.82	10.95	9.10	7.85	14.15	23.37	26.65	23.95	18.35	9.65
	12.85	9.57	11.22	11.35	10.20	8.35	14.77	25.07	27.12	28.10	20.85	11.95
October	12.55	8.72	10.70	10.30	9.42	8.20	14.47	22.75	25.00	26.75	18.50	8.80
	13.10	9.10	11.97	10.97	10.75	9.65	17.00	25.00	27.12	29.57	20.75	10.05
November	9.70	8.97	10.65	10.47	10.05	8.70	16.50	25.07	26.25	23.87	18.45	8.50
	11.87	9.22	11.45	10.97	11.60	9.27	17.45	28.20	27.30	27.00	20.00	9.70
December	9.77	8.80	9.67	10.50	9.50	9.15	15.15	23.65	23.50	22.12	12.62	8.50
	10.92	9.17	11.10	10.75	10.50	9.82	16.85	25.75	26.62	24.00	17.00	8.80
Year	9.70	7.70	8.65	10.12	8.60	8.07	9.75	15.10	23.50	22.05	12.62	8.50
	13.10	10.67	11.97	11.87	11.60	11.27	17.45	28.20	27.30	35.85	24.45	13.30
Average	12.15	8.90	10.38	10.97	10.15	9.45	13.16	21.53	25.70	28.41	17.15	10.52

Balance of Trade in Vegetable Oils

Vegetable oil importation and exportation, as reported by the U. S. Department of Commerce, were as follows:

Year	Importation Lbs.	Exportation Lbs.
1912	282,062,000	382,940,000
1914	279,301,000	235,932,000
1916	430,181,000	208,301,000
1917	619,299,000	148,858,200
1918	867,377,000	128,894,010
1919	857,188,655	361,482,750
1920	632,209,143	272,810,610

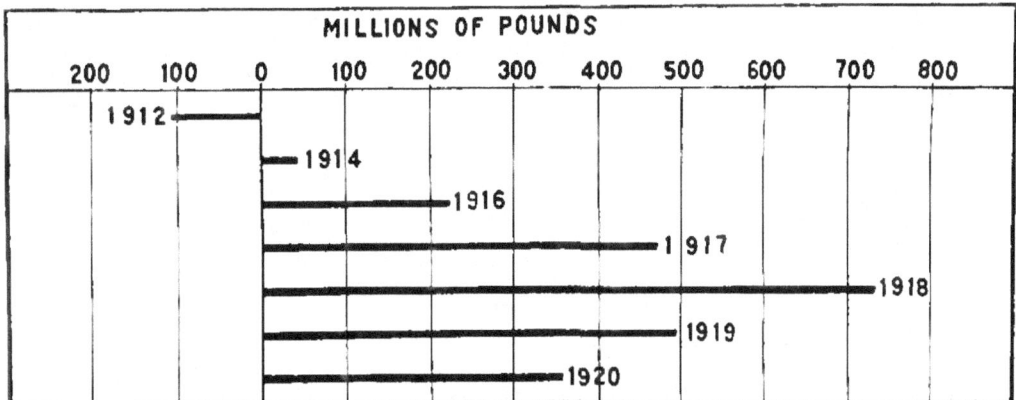

The above chart illustrates the balance of trade in vegetable oils. The line for 1912, extending to the left, illustrates in millions of pounds the amount by which exports from the United States exceed imports. Since 1912 imports have been larger than the exports, and the balance of trade increased steadily up to and including 1918. Since 1918 there has been a steady decline. The heavy lines extending to the right show the relative amounts of these import balances.

CANADIAN MEAT INDUSTRY STATISTICS

Summary and Comparison of the Yearly Reports of the Dominion Statistician.

	1919	1920
Total Plants	82	86
Alberta	7	7
British Columbia	6	6
Manitoba	7	8
Saskatchewan	2	3
Ontario	28	29
Quebec	15	16
New Brunswick	6	9
Nova Scotia	2	2
Prince Edward Island	9	6
Capital Invested	$93,363,791	$84,288,306

Employees	Male	Female	Male	Female
	11,770	1,452	10,837	1,141

	1919	1920
Salaries and Wages, etc.	$16,302,388	$16,691,471
Cost of Fuel	$1,033,913	$1,273,444
Animals Slaughtered		
Cattle	694,394	626,668
Sheep	523,998	624,436
Hogs	2,296,252	1,986,744
Calves	199,621	223,305
Dressed Weights	778,479,174 lbs.	749,916,403 lbs.
Cost Value	$175,133,821	$170,916,888
Sales Value	$233,936,913	$240,544,618

PRODUCTS	Lbs. 1919		Lbs. 1920	
Beef, fresh	317,467,956	$57,581,831	297,297,935	$55,239,777
Mutton	22,583,283	5,068,615	46,941,632	10,297,988
Pork	53,343,432	14,776,888	75,686,123	21,669,071
Veal	20,309,526	3,698,402	22,571,511	4,592,955
Other Meats, fresh	1,167,030	328,994	10,563,374	2,095,773
Beef, salted or cured	12,105,182	2,203,225	8,977,202	1,679,524
Pork, salted	90,379,798	27,659,867	45,170,076	13,276,170
Hams	76,302,075	25,928,198	44,392,002	15,801,386
Shoulders	17,603,271	4,931,363	32,799,091	10,131,900
Bacon and Sides	84,294,846	31,212,706	96,128,042	36,772,497
Sausage	27,151,818	6,712,217	28,547,527	6,353,748
Canned Goods	18,397,335	7,649,013	6,396,305	1,591,447
Lard	41,894,907	12,623,235	54,451,386	14,950,621
Tallow	14,219,539	2,055,109	14,051,223	2,031,904
Oleo and other oils, gals.	1,791,438	752,689	2,631,048	1,720,777
Oleomargarine	10,084,377	2,655,181	10,565,055	3,673,072
Stearine	173,167	45,914	9,371,518	2,347,516
Fertilizer, tons				
Tankage "	19,769	893,225	12,171	607,358
Bone "	9,836	590,008	5,699	480,864
Complete "	3,506	405,505	7,370	573,565
Glue	252,778	45,286	29,379	8,042
Gelatine	5,625,403	1,758,709		
Hides	31,239,916	9,649,129	32,546,932	10,561,070
Hides, No.	581,012			
Sheep Skins, No.	436,713	1,225,243	607,937	1,270,488
Calf Skins, No.	169,318	1,263,379	121,685	445,445
Wool	16,044	9,513	450	67
Hair	802,396	39,248	2,357,802	174,440
All other Products		12,174,177		13,748,422
Repairs, etc.				
Total Selling Value		$233,936,913		$240,544,618

LIVESTOCK PRODUCTION IN CANADA

	Milch Cows	Other Cattle	Hogs	Sheep
1914	2,673,286	3,363,531	3,434,261	2,258,045
1915	2,666,846	3,399,155	3,111,900	2,038,662
1916	2,833,433	3,760,718	3,474,840	2,022,941
1917	3,202,283	4,718,657	3,619,382	2,369,358
1918	3,538,600	6,507,267	4,289,682	3,052,748
1919	3,547,437	6,536,574	4,040,070	3,421,958
1920	3,504,692	6,067,504	3,516,678	3,720,783
1921	3,736,832	6,469,373	3,904,895	3,675,860

MEAT CONSUMPTION IN CANADA

	1910 Lbs. per capita	1920 Lbs. per capita
Beef and Veal	61	59.39
Pork and Lard	67	61.74
Mutton and Lamb	9	10.5
Total	137	131.6

LIVESTOCK SLAUGHTERED UNDER INSPECTION

Fiscal Years	Cattle	Hogs	Sheep
1909	298,241	1,532,796	191,792
1910	384,789	1,261,496	257,049
1911	411,308	1,452,237	329,017
1912	408,401	1,852,997	376,437
1913	450,390	1,607,741	455,647
1914	531,994	1,799,060	499,280
1915	530,425	2,598,338	447,173
1916	542,154	2,363,693	403,147
1917	648,859	2,245,511	416,575
1918	739,085	2,126,682	336,897
1919	887,773	2,334,354	397,961
*1920	831,715	1,785,235	662,763
†1920	965,394	2,171,650	601,170
†1921	570,702	1,686,059	245,770
*1921	497,457	1,636,389	206,929

In 1921 (fiscal year) there were 256,790 calves and 436,910 lambs slaughtered. In 1921 (calendar year) 217,845 calves and 440,922 lambs were slaughtered.

*Calendar year   †Fiscal year

CANADIAN MEAT EXPORTS*

(000's omitted.)

	1915 Lbs.	1916 Lbs.	1917 Lbs.	1918 Lbs.	1919 Lbs.	1920 Lbs.	1921 Lbs.
Bacon	76,801	144,918	207,213	199,957	120,622	223,643	98,234
Hams	17,958	8,732	4,403	7,875	4,066	(Included with Bacon)	
Pork	21,288	13,142	13,987	7,909	37,317	6,682	3,126
Beef	18,828	47,422	45,546	86,565	127,809	110,048	53,502
Mutton	1,064	99	167	855	1,933	6,140	6,405
Canned Meats	9,882	11,031	6,676	13,422	14,140	2,813	437
Other Meats	4,403	3,939	3,762	7,016	5,895	....	....
Total	150,227	229,287	281,756	323,602	311,786	364,970	161,704
Lard	2,689	24	1,405	1,955	2,640	7,622	2,096

*These figures are for fiscal years

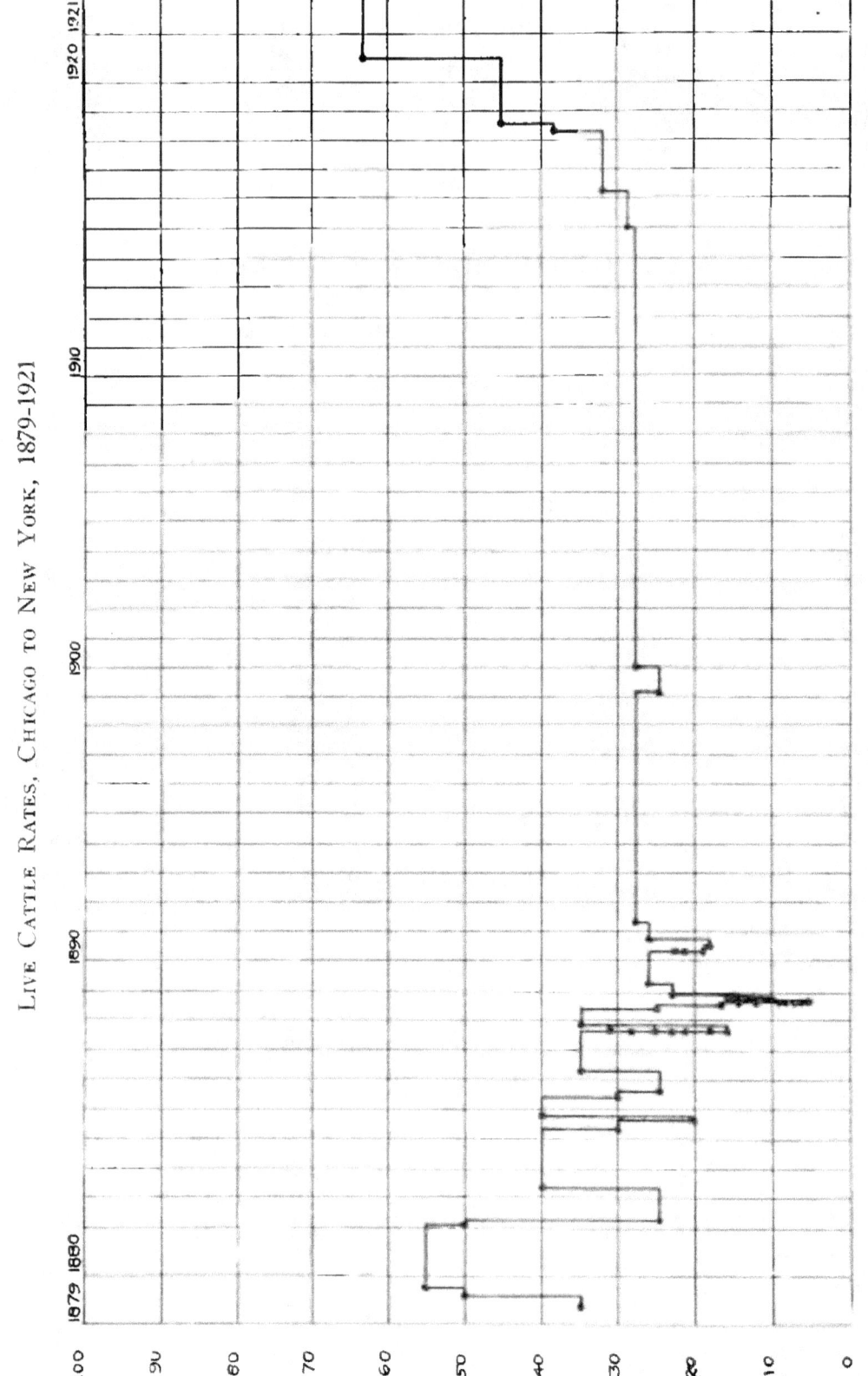

Live Cattle Rates, Chicago to New York, 1879-1921

Chart by Traffic Committee, Institute of American Meat Packers.

RAILROAD RATES OF INTEREST TO MEAT PACKERS

Foreword

It has been estimated that 85 per cent of the meat traffic of the United States moves into the Official Classification Territory, which is generally described as that territory lying east of the Illinois-Indiana state line, north of the Ohio River and north of the line of the Norfolk & Western Railway through Virginia.

The rates between Chicago and New York have always been the "yard stick" upon which all rates to, from, and between points in that territory are based. Hence, the showing made here is with respect to the basic rates between Chicago and New York.

A question of great importance to Western killers is the matter of relation between the live stock rates and the rates on the finished product. Obviously, the packer killing in the West is in a better condition to compete with Eastern killers when live stock rates are high and meat rates low. When the reverse is true (as at the time this book was printed) the Eastern packer has a strong advantage over his Western competitor.

From the time when shipments of meat first moved from Western packers to Eastern markets the question of rate relationshp was constantly agitated, until 1883 when a special committee was appointed by the railroads to agree upon a fixed relationship. After a series of conferences they recommended that the relative rates on live stock and beef should be as 40 to 77.

Basis of Live Stock and Beef Rates

No attempt was made to put this basis into effect, and the following year a new committee on arbitration was appointed, consisting of the following: T. M. Cooley, Chairman; G. F. Swift, Western packers; S. W. Allerton, Eastern packers. The three members were unable to agree upon a basis, and Judge Cooley finally rendered his decision, which has since become famous as the "Cooley award." This made the relationship as 40 to 70.

The rates were readjusted substantially upon this basis, so that for many years (except during rate wars) the live cattle rate from Chicago to New York and Boston stood at 28 cents, while the dressed beef rate was 45 cents.

It will be of interest to note how these rates varied over a long period of time.

Live cattle rates, Chicago to New York, 1879 to 1922, prevailed as follows:

Live Cattle Rate Changes

	Cents		Cents
June 9, 1879	35	December 8, 1884	40
August 4, 1879	50	May 3, 1885	30
November 5, 1879	55	July 1, 1885	25
March 14, 1881	50	March 1, 1886	35
May 9, 1881	25	November 21, 1887	31½
April 17, 1882	40	November 23, 1887	28½
May 5, 1884	30	November 24, 1887	25½
September 1, 1884	20	November 25, 1887	23

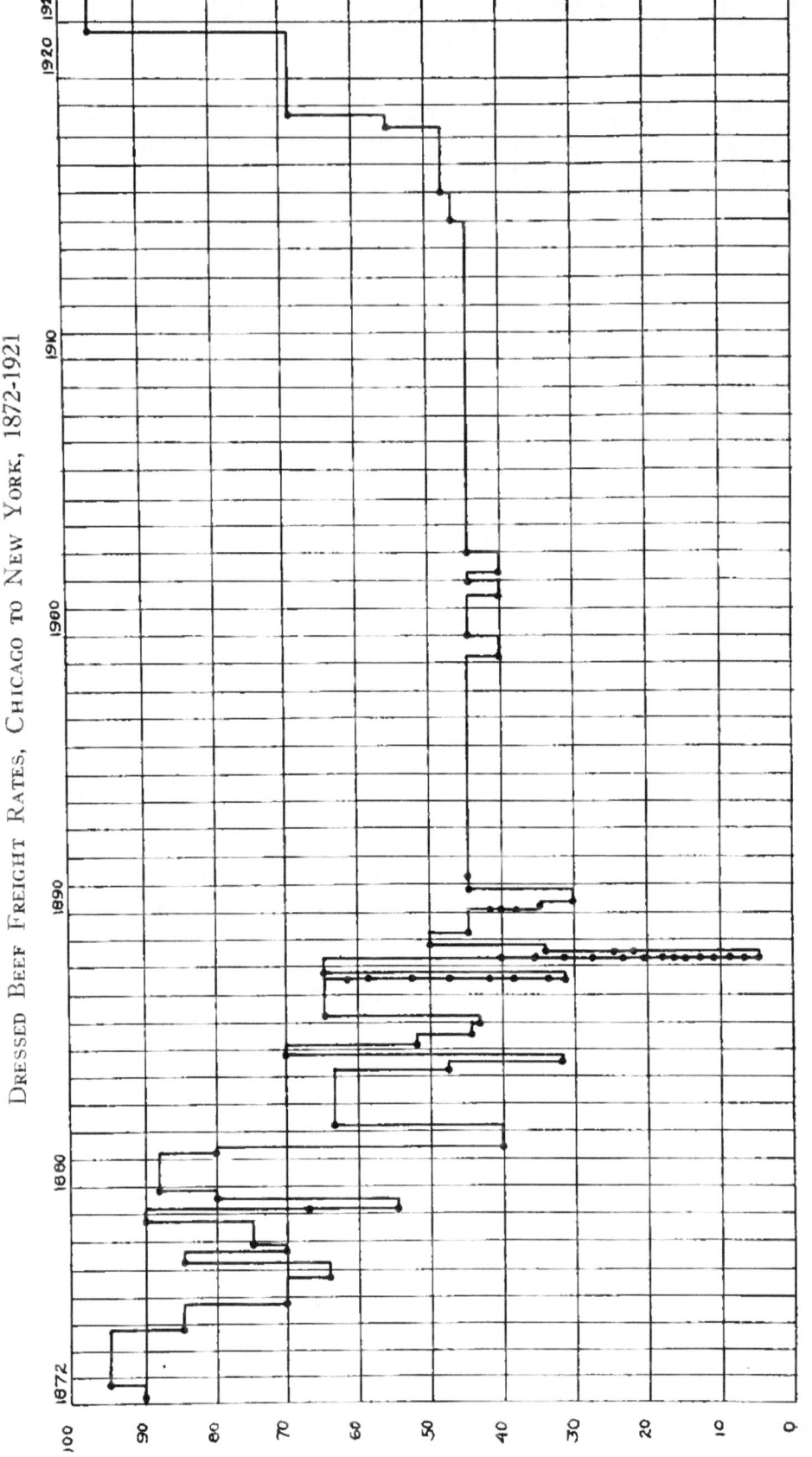

Chart by Traffic Committee, Institute of American Meat Packers.

	Cents		Cents
November 26, 1887	20½	September 24, 1888	15
November 28, 1887	18½	December 17, 1888	22½
November 29, 1887	16½	May 1, 1889	26
December 26, 1887	35	June 16, 1890	22½
May 14, 1888	25	June 26, 1890	21
June 18, 1888	16½	June 30, 1890	19½
July 2, 1888	14½	July 3, 1890	18
July 3, 1888	12½	November 24, 1890	26
July 5, 1888	11	April 20, 1891	28
July 6, 1888	9½	February 1, 1899	25
July 7, 1888	8½	January 1, 1900	28
July 9, 1888	7½	January 15, 1915	29.4
July 10, 1888	6½	January 13, 1916	33
July 11, 1888	5½	March 25, 1918	38
August 20, 1888	14½	June 25, 1918	45
August 25, 1888	10	August 26, 1920	63

Dressed Beef Rate Changes

Dressed beef, carload rates, from Chicago to New York, 1872 to 1922, prevailed as follows:

	Cents		Cents
March 26, 1872	90	June 26, 1888	35
November 20, 1872	95	June 29, 1888	30½
December 1, 1874	85	July 2, 1888	26½
November 2, 1875	70	July 3, 1888	23
November 18, 1876	65	July 5, 1888	20
April 30, 1877	85	July 6, 1888	17½
September 4, 1877	70	July 7, 1888	15½
October 22, 1877	75	July 9, 1888	13½
September 2, 1878	90	July 10, 1888	12
June 9, 1879	67½	July 11, 1888	10½
June 26, 1879	56	July 12, 1888	9
August 4, 1879	80	July 13, 1888	8
August 25, 1879	80	July 14, 1888	7
October 13, 1879	80	August 3, 1888	22½
November 10, 1879	88	August 20, 1888	25
March 14, 1881	80	September 24, 1888	35
May 9, 1881	40	December 17, 1888	50
April 17, 1882	64	May 1, 1889	45
May 5, 1884	48	June 16, 1890	42
September 1, 1884	32	June 20, 1890	39
December 8, 1884	70	June 26, 1890	36
May 3, 1885	52½	June 30, 1890	33
July 1, 1885	43½	July 3, 1890	30
January 1, 1886	43	November 24, 1890	45
March 1, 1886	65	February 1, 1899	40
November 21, 1887	58½	January 1, 1900	45
November 23, 1887	52½	July 29, 1901	40
November 24, 1887	47	January 1, 1902	45
November 25, 1887	42½	April 1, 1902	40
November 26, 1887	38½	January 1, 1903	45
November 28, 1887	34½	January 15, 1915	47.3
November 29, 1887	31	January 13, 1916	47½
December 26, 1887	65	March 25, 1918	55
June 25, 1888	40	June 25, 1918	69
		August 26, 1920	96½

Packing House Product Rates

The term "packinghouse products" as used in traffic work is ordinarily understood to include cured meats, lard, stearine, tallow, and canned meats. Customarily one rate applies on all products coming under this description. In the Eastern territory, however, an exception is noted, for here we find that bulk cured meats, carload, are rated fourth-class, while cured meats packed are fifth-class.

Prior to 1880 there were but four classes of freight on traffic from Chicago to New York. A fifth-class rate was made effective March 8, 1880, and twelve classes were established April 6, 1885. The twelve classes continued in effect until April 1, 1887, when the Official Classification became effective, this having the effect of establishing six classes only.

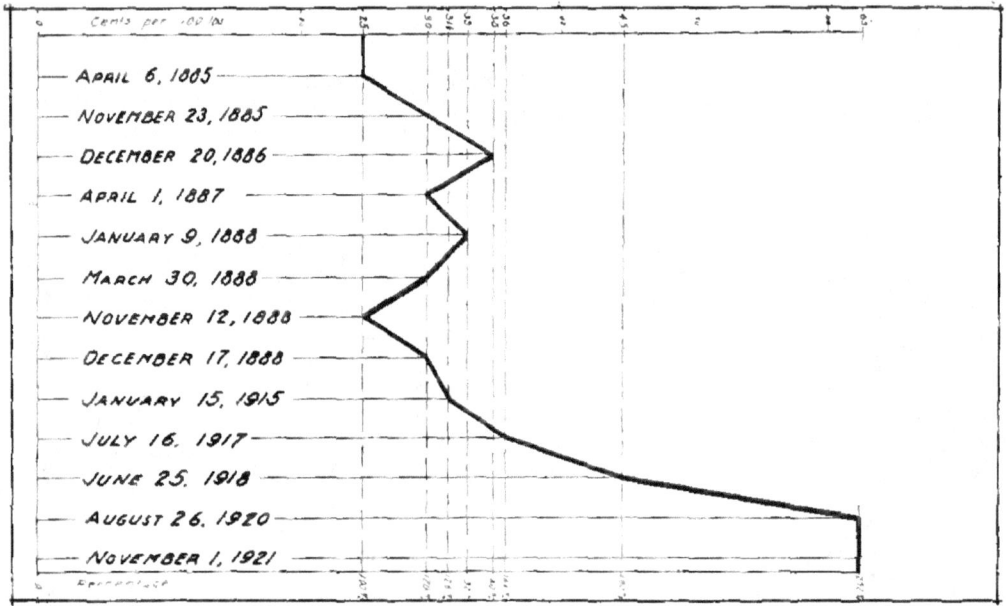

The chart here shown indicates the trend of freight rates on packinghouse products between the basing points, Chicago and New York, from 1880 to 1921.

History of Transit Icing Charges

The method of icing meat cars in transit is familiar to the trade. The history of transit icing charges in Official Classification territory will be of interest. A summary is here presented:

Prior to December 1, 1916, the charge was $2.50 per ton, including salt and labor.

1916-1918

On various dates between December 1 and December 15, 1916, applicable east of Buffalo-Pittsburgh territory individual lines increased the initial icing charge from $2.50 per ton (including salt and labor) to $4.00 per ton (including salt and labor), subject to a minimum charge of $4.00. No change in transit icing charge.

1919-1921

Under Freight Rate Authority 3384 and supplement thereto, carriers increased, applicable in C. F. A. and E. T. line territory, both the initial and transit icing charge to $4.00 per ton for ice, plus 75 cents per cwt. for salt, subject to a minimum charge of 1,000 pounds ice and 100 pounds salt. Owing to the varied interpretations placed upon this freight rate authority the increase became effective on various dates via the different lines, ranging from January 22, 1919, to March 15, 1919.

This basis is in effect at the present time (1922).

Average cost of icing cars on a basis of 4 tons of ice and 960 pounds of salt used in initial icing, and 2,800 pounds ice and 336 pounds salt used in an average of two re-icings in transit, based on the various charges shown above:

$2.50 per ton (including salt and labor).....................$13.50 per car
 4.00 per ton (including salt and labor)..................... 19.50 per car
 4.00 per ton for ice, plus 75c per cwt. for salt.............. 29.20 per car

---

Growth of Margarin Production

Butter substitutes, as reported by the U. S. Department of Agriculture, were produced as follows:

Year	Pounds	Year	Pounds
1909	110,000,000	1918	355,536,000
1914	143,900,000	1919	371,317,000
1916	202,444,000	1920	370,730,000
1917	290,902,000		

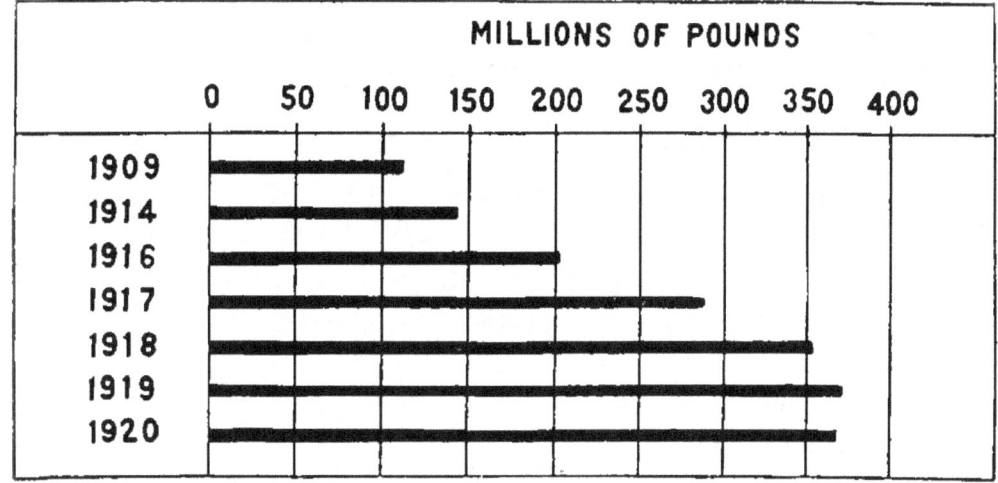

The accompanying chart shows the growth and development of the margarin business during the decade 1909-1920. The output of oleomargarine in the United States practically trebled in the ten years. The scale of quantities (millions of pounds) shown at the top omits 000,000.

DOMESTIC TRADE TERM DEFINITIONS

Foreword

Owing to the general demand for some definite and well established interpretations of trade terms ordinarily used in domestic commerce the Institute of American Meat Packers has prepared this work for distribution to its members and various trade associations.

The Institute of American Meat Packers is an association of meat packers located throughout the United States and Canada, and is organized to secure co-operation among and to lawfully further the interests of the industry. Members endeavor to promote and foster domestic and foreign trade in American meat products and to create fair trading conditions. Neither the Institute nor its various committees have anything to do with sales or the making of prices; and its members, acting independently, are in active competition with each other.

The following interpretations are respectfully referred to the members with the following recommendations:

1. That sales and shipments be made subject to these interpretations.
2. That reference to them be made on sales contracts and in preliminary correspondence relative to prospective sales.
3. That the trade terms indicated be used to the exclusion of other forms whenever practicable.

Definition of Domestic Trade Terms

ARTICLE NO. 1

Contracts of purchase or sale by or between members of the Institute of American Meat Packers shall be interpreted according to the following rules.

No rule herein, however, shall prejudice any specific agreement when such agreement is clearly set forth in a contract between buyer and seller, but such rules shall be used to interpret such contract as to terms on which the contract is silent or not sufficiently clear.

ARTICLE NO. 2

"F. A. S. (Named Port)"

Under this quotation:

A. Seller must

(1) transport goods to seaboard.

(2) store goods in warehouse or on wharf, if necessary, but in any case at risk and expense of buyer; also re-ice cars while being held at seaboard awaiting steamer, but at buyer's expense.

(3) place goods alongside vessel, either in a lighter or on the wharf, or, in the event buyer fails or refuses to furnish vessel upon arrival of goods at seaboard, store in warehouse or hold in cars at buyer's expense.

(4) be responsible for loss and/or damage to goods until they have been delivered alongside ship or on wharf or stored in warehouse.

B. Buyer must

(1) be responsible for loss and/or damage thereafter, and for insurance and/or demurrage and car service.

(2) handle all subsequent movement of the goods.

(3) furnish vessel ready and able to receive goods on their arrival at seaboard.

(4) pay cost of hoisting goods into vessel where weight of goods is too great for ship's tackle.

ARTICLE NO. 3

"F. O. B. (Named Point)"

Under this quotation:

A. Seller must

(1) Furnish a railroad or privately owned car, which shall be:

(a) Suitable for transportation of the product sold;

(b) Properly cleaned and equipped to receive the product;

(c) Initially iced and/or pre-cooled in accordance with buyer's instructions.

(d) In apparent good order and condition to protect the goods under reasonable and normal handling by the carrier.

(2) Place goods on or in cars.

(3) Secure railroad bill of lading;

(4) Be responsible for loss and/or damage to goods until they have been placed on or in cars and a bill of lading secured from carrier.

B. Buyer must:

(1) Furnish seller with shipping permit if any required;

(2) Furnish the seller with full and complete instructions for shipping, routing, inspecting, weighing, icing and reicing, or specifically authorize the seller to act for him in these matters, in which case the seller assumes no responsibility other than to use his (the seller's best judgment;

(3) Pay initial icing and reicing and transportation charges (including taxes, if any) to point of destination;

(4) Assume all shrinkage, expense, loss, and/or damage to goods after seller has complied with his obligations as set forth above.

ARTICLE NO. 4

"Chicago Freight Equalized" or "Chicago Freight"

Under this quotation:

A. A seller must:

(1) Furnish cars as specified in his obligation under article (3) above.

(2) Place goods on or in cars;

(3) Secure railroad bill of lading;

(4) Pay or allow any costs of freight or transportation by cheapest suitable all-rail route from shipping point to destination beyond Chicago in excess of such costs had shipment been made from Chicago to same destination. In case such costs to destination are less than if shipped from Chicago, seller shall be entitled to add such difference to the invoice;

(5) Make any necessary freight contract in accordance with buyers' shipping and routing instructions;

(6) Be responsible for loss of, and/or damage to goods until they have been delivered to the carrier and bill of lading secured, seller not being responsible for delivery of goods at Chicago or destination.

B. Buyer must:

(1) Furnish seller with shipping permit if any required;

(2) Furnish the seller with full and complete instructions for shipping, inspecting, weighing, routing, icing and reicing, or specifically authorize the seller to act for him in these matters, in which case the seller assumes no responsibility other than to use his (the seller's) best judgment;

(3) Assume all expense, loss of, and/or damage to goods after seller has fulfilled his obligations as above.

(4) Pay for initial icing and reicing.

ARTICLE NO. 5

"(Named Point) Freight Equalized"

Under this quotation:

When some other point than Chicago is named in a Freight Equalized contract, the respective duties and obligations of buyer and seller shall be the same as in Article (4) above, and the name of such other point shall be substituted for the word "Chicago" wherever "Chicago" appears in Article (4).

ARTICLE NO. 6

"C. A. F. (or Cafe) (Named Destination)"

Under this quotation:

A. Seller must:

(1) Furnish cars as specified in his obligations under Article (3) above;

(2) Place goods on or in cars;

(3) Secure any necessary freight contract or shipping permit;

(4) Secure railroad bill of lading, same to be endorsed "Lighterage Free" if so requested by Buyer, *provided* this can be done without additional expense to seller;

(5) Pay (or allow) freight and cost of icing and reicing, between point of shipment and destination;

(6) Be responsible for any loss of or damage to goods until they have been placed on or in cars and bill of lading secured from carrier—seller not being responsible for delivery of goods at destination.

B. Buyer must:

(1) Assume any expense, shrinkage, loss of, and/or damage to goods after seller has fulfilled his obligations as above.

ARTICLE NO. 7

"Delivered (Named Destination)"

Under this quotation:

A. Seller must:

(1) Furnish cars as specified in his obligations under Article (3) above;

(2) Place goods on or in cars;

(3) Secure any necessary freight contract or shipping permit;

(4) Secure railroad bill of lading;

(5) Pay (or allow) freight and cost of icing and reicing, between point of shipment and destination;

(6) Assume all responsibility for expense, shrinkage, loss of, and/or damage to goods until they are delivered by carrier at destination named.

B. Buyer must:

(1) Assume any expense, shrinkage, loss of, and/or damage to goods after arrival of goods at destination named.

ARTICLE NO. 8

"In Store"

Under this quotation:

A. Seller must:

(1) Furnish buyer with a Warehouse Receipt or Delivery order showing where stored and with suitable endorsement or instruction to warehouseman to surrender goods to buyer.

(2) Pay (or allow) all storage, insurance, or other charges up to date of sale to buyer;

(3) Be responsible for any loss of, and/or damage to the goods up to the time of delivery of warehouse receipt or delivery order, but not thereafter;

(4) Refund to buyer any shortage shown by the official certificate of inspection and/or weight as provided for in Section (2) below.

B. Buyer must:

(1) Accept such Warehouse Receipt or Delivery Order as actual transfer of title to property from seller to buyer, and assume any expense, shrinkage, loss of, and/or damage to goods thereafter;

(2) Notify the seller within two days after receipt of such Warehouse Receipt, Delivery Order, or document transferring title, of his desire to have such goods officially inspected and/or officially weighed. His failure to notify the seller of such desire within the time specified above shall constitute an acceptance of the goods and a termination of seller's liability for their condition and/or weight.

ARTICLE NO. 9

Terms of Payment

(A) Regardless of which of the above forms of sale govern the transaction, in all cases buyer must promptly make payment or honor draft according to its terms upon presentation, and arrange to properly handle goods after seller has fulfilled the conditions assumed by him according to the terms of the sale.

(B) In making payment for goods sold under terms specified in Articles numbered 3, 4 and 5 buyer must pay in funds current in the city from which goods are shipped, or if by seller's draft on buyer then buyer must pay with exchange and collection charges, so that seller may receive net the amount of the invoice.

(C) In making payment for goods sold under terms specified in Articles numbered 2, 6 and 7 buyer must pay in funds current in city of destination; seller to bear the expense of exchange and collection charges on his draft, so that buyer shall only be compelled to pay the net amount of the invoice.

(D) In making payment for goods sold "In Store," as specified in Article numbered 8, buyer must pay in funds current in the city in which goods are stored, immediately upon acceptance of Warehouse receipt or delivery order.

ARTICLE NO. 10

Time of Shipment

(A) When a time in which shipment is to be made is named, it shall (unless otherwise specifically provided) be understood to be at seller's option within that period.

(B) In all cases the date of the bill of lading shall be conclusive and exclusive proof of the date of shipment.

(C) When goods are sold for shipment at buyer's option, the buyer is privileged to demand shipment at any time during the period specified in the contract, procided he gives the seller a reasonable time in which to make the necessary arrangements to comply with his shipping instructions.

(D) When "immediate" shipment is specified, it must be made within the three succeeding business days after day of sale.

(E) When "quick" shipment is specified, it must be made within the five succeeding business days after day of sale.

(F) When "prompt" shipment is specified, it must be made within the ten succeeding calendar days after day of sale.

(G) When "shipment within a week" is specified, it must be made within the seven succeeding calendar days after day of sale.

(H) When shipment is specified as within ten, fifteen, twenty, or other named number of days, then it must be made within that number of calendar days succeeding the day of sale.

(I) If shipment during the first, or last, week of any month, then it must be within the first, or last, seven calendar days of that month respectively. In the same way the second week would be the 8th to the 14th days (both inclusive) of that month; the third week, the 15th to the 21st (both inclusive) of that month; and the fourth week the 22d to the 28th (both inclusive) of that month.

(J) If sale is made for weekly shipments during a specified month, then the first shipment must be made during the 1st to the 7th of that month; the second shipment during the 8th to the 14th of that month; the third shipment during the 15th to the 21st of that month, and the fourth shipment during the 22d to the 28th of that month.

(K) In no case shall seller be penalized on account of failure or delay of buyer to furnish shipping instructions.

(L) If shipment within the time named in contract is prevented by strikes, fire, flood, blockade, embargo, war, revolution, or epidemic, the time for shipment shall be extended for a period not exceeding ten days, provided the seller has notified the buyer in writing before the expiration of the contract period of his inability to make shipment as contracted for and stated his reasons therefor. If shipment cannot be made within such ten day extension then the contract shall be cancelled, without penalty to either Buyer or Seller.

ARTICLE NO. 11

Agreement for Delivery Within Time Limits

Sales made for "Delivery to a named point," at a specified date or within a specified time, should be either made on a C. A. F. or F. O. B basis, and thus be governed by the rules covering transactions of that kind, or else it should be considered a special transaction on which full details must be set forth in the contract.

ARTICLE NO. 12
Quality of Goods

Unless otherwise specifically provided, all sales are to be considered as standard to quality, condition, weights, drainage, saltage and/or packing. Such standard shall conform to the Rules and Regulations of the Board of Trade of Chicago.

ARTICLE NO. 13
Official Inspection and/or Weighing

(A) If official inspection is furnished at the request of either buyer or seller, the expense of same shall be borne by the party making the request.

(B) If official weighing of the goods is done at the request of either buyer or seller, the expense of same shall be borne equally by buyer and seller.

(C) In all cases, the official certificate of weight shall be exclusive and conclusive proof of the weight of the goods.

ARTICLE NO. 14
Method of Shipment

Unless otherwise specifically provided for all shipments are understood to be via all rail routes.

ARTICLE NO. 15
Explanation of Abbreviations

F. O. B. .... Free on board
F. A. S. .... Free along side
C. A. F. or Cafe ....Cost and freight

DEFINITIONS OF EXPORT QUOTATIONS

These are, in their order, the normal situations on which an exporter, manufacturer or shipper may desire to quote prices. It is understood that unless a particular railroad is specified, the property will be delivered to the carrier most conveniently located to the shipper. If the buyer, for the purpose of delivery, or in order to obtain lower transportation charges, desires that the goods be delivered to a carrier further removed from the shipper and entailing a greater cost than delivery to the carrier most favorably situated, the carrier to which the buyer desires delivery of the goods should be named in the quotation.

The term "cars or lighters," as used herein, is intended to include river, lake or coastwise ships, canal boats, barges, or other means of transportation, when so specified in the quotation.

---

1. *When the price quoted applies only at inland shipping point and the seller merely undertakes to load the goods on or in cars or lighters furnished by the railroad company serving the industry, or most conveniently located to the industry, without other designation as to routing, the proper term is:*

"F. O. B. (Named Point)"

Under this quotation:

A. Seller must

(1) place goods on or in cars or lighters

(2) secure railroad bill of lading

(3) be responsible for loss and/or damage until goods have been placed in or on cars or lighters at forwarding point, and clean bill of lading has been furnished by the railroad company

B. Buyer must

(1) be responsible for loss and/or damage and/or any necessary expense incurred thereafter

(2) pay initial icing and all transportation charges, including taxes, if any

(3) handle all subsequent movement of the goods

---

2. When the seller quotes a price including transportation charges to the port of exportation without assuming responsibility for the goods after obtaining a clean bill of lading at a point of origin, the proper term is:

"F. O. B. (Named Point) Freight Prepaid to (Named Point on the Seaboard)"

Under this quotation:

A. Seller must

(1) place goods on or in cars or lighters

(2) secure railroad bill of lading

(3) pay freight to named port, including tax, if any

(4) be responsible for loss and/or damage until goods have been placed in or on cars or lighters at forwarding point, and clean bill of lading has been furnished by the railroad company

B. Buyer must

(1) be responsible for loss and/or damage incurred thereafter

(2) handle all subsequent movement of the goods

(3) unload goods from cars

(4) transport goods to vessels

(5) pay initial and all icing, demurrage and/or storage charges

(6) arrange for storage in warehouse or on wharf where necessary

---

3. Where the seller wishes to quote a price, from which the buyer may deduct the cost of transportation to a given point on the seaboard, without the seller assuming responsibility for the goods after obtaining a clean bill of lading at point of origin, the proper term is:

"F. O. B. (Named Point) Freight Allowed to (Named Point on the Seaboard)"

Under this quotation:

A. Seller must

(1) place goods on or in cars or lighters

(2) secure railroad bill of lading

(3) be responsible for loss and/or damage until goods have been placed in or on cars or lighters at forwarding point, and clean bill of lading has been furnished by the railroad company

B. Buyer must
(1) be responsible for loss and/or damage incurred thereafter
(2) pay all transportation charges (buyer is then entitled to deduct from the amount of the invoice the freight paid from primary point to named port)
(3) handle all subsequent movement of the goods
(4) unload goods from cars
(5) transport goods to vessels
(6) pay initial and all icing, demurrage and/or storage charges
(7) arrange for storage in warehouse or on wharf where necessary.

---

4. *The seller may desire to quote a price covering the transportation of the goods to seaboard, assuming responsibility for loss and/or damage up to that point. In this case, the proper term is:*

"F. O. B. Cars (Named Point on Seaboard)"

Under this quotation:
A. Seller must
(1) place goods on or in cars
(2) secure railroad bill of lading
(3) pay all freight charges from forwarding point to port on seaboard
(4) be responsible for loss and/or damage until goods have arrived in or on cars at the named port
B. Buyer must
(1) be responsible for loss and/or damage incurred thereafter
(2) unload goods from cars
(3) handle all subsequent movement of the goods
(4) transport goods to vessel
(5) pay all demurrage, reicing at seaboard and/or storage charges
(6) arrange for storage in warehouse or on wharf where necessary

---

5. *It may be that the goods, on which a price is quoted covering the transportation of the goods to the seaboard, constitute less than a carload lot. In this case, the proper term is:*

"F. O. B. Cars (Named Port) L. C. L."

Under this quotation:
A. Seller must
(1) deliver goods to the initial carrier
(2) secure railroad bill of lading
(3) pay all freight charges from forwarding point to port on seaboard
(4) be responsible for loss and/or damage until goods have arrived on cars at the named port
B. Buyer must
(1) be responsible for loss and/or damage incurred thereafter
(2) handle all subsequent movement of the goods
(3) accept goods from the carrier
(4) transport goods to vessel
(5) pay all storage and other charges
(6) arrange for storage in warehouse or on wharf where necessary.

6. *Seller may quote a price which will include the expense of transportation of the goods by rail to the seaboard, including lighterage. In this case, the proper term is:*

"F. O. B. Cars (Named Port) Lighterage Free"

Under this quotation:

A. Seller must

(1) place goods on or in cars

(2) secure railroad bill of lading

(3) pay all transportation charges to, including lighterage at, the port named

(4) be responsible for loss and/or damage until goods have arrived on cars at the named port

B. Buyer must

(1) be responsible for loss and/or damage incurred thereafter

(2) handle all subsequent movement of the goods

(3) take out the insurance necessary to the safety of the goods after arrival on the cars

(4) pay the cost of hoisting goods into vessel where weight of goods is too great for ship's tackle

(5) pay all demurrage and other charges, except lighterage charges

---

7. *The seller may desire to quote a price covering delivery of the goods alongside overseas vessel and within reach of its loading tackle. In this case, the proper term is:*

"F. A. S. Vessel (Named Port)"

Under this quotation:

A. Seller must

(1) transport goods to seaboard

(2) store goods in warehouse or on wharf, if necessary, but in any case at risk and expense of buyer; also reice cars while being held at seaboard awaiting steamer, but at buyer's expense

(3) place goods alongside vessel, either in a lighter or on the wharf, or, in the event buyer fails or refuses to furnish vessel upon arrival of goods at seaboard, store in warehouse at buyer's expense

(4) be responsible for loss and/or damage until goods have been delivered alongside ship or on wharf or stored in warehouse

B. Buyer must

(1) be responsible for loss and/or damage thereafter, and for insurance

(2) handle all subsequent movement of the goods

(3) furnish vessel ready and able to receive goods on their arrival at seaboard

(4) pay cost of hoisting goods into vessel where weight of goods is too great for ship's tackle

8. *The seller may desire to quote a price covering all expenses up to and including delivery of the goods upon the overseas vessel at a named port. In this case, the proper term is:*

"F. O. B. Vessel (Named Port)"

Under this quotation:
A. Seller must:
(1) meet all charges incurred in placing goods actually on board the vessel
(2) provide the usual dock or ship's receipt
(3) be responsible for all loss and/or damage until goods have been placed on board the vessel
B. Buyer must
(1) be responsible for loss and/or damage thereafter
(2) handle all subsequent movement of the goods

---

9. *The seller may be ready to go farther than the delivery of his goods upon the overseas vessel and be willing to pay transportation to a foreign point of delivery. In this case, the proper term is:*

"C. & F. (Named Foreign Port)

Under this quotation
A. Seller must
(1) make freight contract and pay or allow transportation charges sufficient to carry goods to agreed destination
(2) deliver to buyer or his agent clean bills of lading to the agreed destination
(3) be responsible for loss and/or damage until goods have been delivered to the carrier and clean bill of lading obtained (seller is not responsible for delivery of goods at destination)
B. Buyer must
(1) be responsible for loss and/or damage thereafter and must take out all necessary insurance
(2) handle all subsequent movement of the goods
(3) take delivery and pay costs of discharge, lighterage and landing at foreign port of destination in accordance with bill of lading clauses
(4) pay foreign taxes, customs, duties and wharfage charges, if any

---

10. *The seller may desire to quote a price covering the cost of the goods, the marine insurance on the goods, and all transportation charges to the foreign point of delivery. In this case, the proper term is:*

"C. I. F. (Named Foreign Port)"

Under this quotation
A. Seller must
(1) make freight contract and pay or allow freight charges sufficient to carry goods to agreed destination
(2) take out and pay for necessary marine insurance
(3) deliver to buyer or his agent clean bills of lading to the agreed destination, and insurance policy and/or negotiable insurance certificate
(4) be responsible for loss and/or damage until goods have been

delivered to the carrier, and clean bill of lading obtained and covered by insurance policy and/or negotiable insurance certificate (seller is not responsible for the delivery of goods at destination, nor for payment by the underwriters of insurance claims)

(5) provide war risk insurance at buyer's expense on his request or when considered necessary by the seller

B. Buyer must

(1) be responsible for loss and/or damage thereafter, and must make all claims to which he may be entitled under the insurance directly on the underwriters

(2) take delivery and pay costs of discharge, lighterage and landing at foreign port of destination in accordance with bill of lading clauses

(3) pay foreign taxes, customs duties and wharfage charges, if any.

EXPLANATION OF ABBREVIATIONS

F. O. B.	Free on board
F. A. S.	Free along side
C. & F.	Cost and freight
C. I. F.	Cost, insurance and freight
L. C. L.	Less than carload lot

GENERAL INFORMATION

In reaching the conclusions set forth in this statement the Committees considered the fact that there are in more or less common use by manufacturers in different parts of the United States numerous variations of these abbreviations, practically all of which are employed to convey meanings substantially synonymous with those here defined.

For instance, there are manufacturers who quote "F. O. B. Cars," "F. O. B. Works," "F. O. B. Mill" or "F. O. B. Factory," meaning that the seller and buyer have the same responsibilities as those set forth in section 1.

The Committees considered all those variations and determined to recommend the use of "F. O. B. (named point)," as "F. O. B. Detroit," "F. O. B. Pittsburgh," etc. Of the considerable number of these abbreviations which are used in the United States, the Committees felt that the form "F. O. B. (named point)" is most widely used and understood, and therefore should be adopted as the standard of practice.

The chief purpose of these interpretations is to simplify and standardize American practice, and to that end it urges members to cease the use of synonymous abbreviations, and quote habitually in the terms here recommended.

It is further suggested that it is understood that all consular fees for legalizing invoices, stamping bills of lading or other documents required by the laws of the countries of destination, are payable by the buyer and are not included in the seller's prices.

PART THREE

Trade Directory

FOREWORD

No attempt at a comprehensive trade directory of the packinghouse and allied industries has been made up to this time. In gathering the data contained in this Directory Section every effort has been made to obtain accurate and reasonably representative information. Data listed is that provided by the concerns themselves. It should be remembered that changes constantly are taking place, and that this directory information is of the date of publication of this edition.

Lists here given include:

Packers and Slaughterers (United States, Canada, South America, South Africa, Australia and New Zealand).

Wholesale Meat Dealers and Provisioners (those who do not slaughter), and Wholesale Sausage Makers.

Renderers (including hide and tallow dealers).

Vegetable Oil Refiners.

Margarin Manufacturers.

Packinghouse and Oil Brokers.

Livestock Order Buyers.

The Packers' and Slaughterers' section contains the names of slaughtering establishments, or what is commonly defined as packing establishments. Butchers killing only for their own consumption in small quantities have not been listed as packers, but primarily those who do a wholesale business and also slaughter and pack for others. Wholesale meat and provision dealers, wholesale sausage makers, and other dealers in allied products are listed in a subsequent section.

It must be borne in mind that the greater part of the information has been furnished voluntarily by officials of the various packing companies. Information blanks were mailed to every packer in the United States, and those who are not fully listed have not complied with the request for information. As this is the first and only directory of its kind, a few packers seem to have been reluctant to give information.

Capacities given are weekly killing capacities. Some packers have furnished information which seems to be the actual killing per week, leading to the conclusion that at the time of the report they were not killing to capacity.

PACKERS AND SLAUGHTERERS

UNITED STATES

ALABAMA

ANDALUSIA

Swift & Company. (See Chicago.)

ANNISTON

Anniston Packing Co.

BIRMINGHAM

Alabama Packing Co.—Capital, $50,000. President, A. Goldberg; Vice-President, I. Lefschultz; Secretary, J. Goldberg; Treasurer, M. Goldberg.

Birmingham Packing Co.—23rd Ave. and 24th St.—Railroad, Birmingham Belt. A corporation. Capital, $100,000; issued, $100,000. President, C. H. Ungerman; Vice-President, J. P. Phillips; Secretary-Treasurer, F. N. Phillips; General Manager, E. G. Bruce. Employes, 150. Codes—Cross. Cattle, 600; hogs, 1,500. Sausage—Fresh and smoked. By-Products—Dried blood, tankage and hog feed. Government inspection. Refrigeration—150 and 125-ton De La Vergnes; direct expansion; 75 tons ice daily. Boiler capacity, 600 H. P. Trade Marks—Hams, "Golden Rod"; bacon, "Golden Rod"; lard, "Carnation."

BREWTON

Brewton Packing Company—President, J. E. Finlay; Vice-President and General Manager, A. B. McPhaul; Secretary-Treasurer, W. Y. Lovelace.

GADSDEN

Jones Abattoir Co.

HUNTSVILLE

Municipal Abattoir.

MOBILE

Haas-Davis Packing Co.—Railroad, Mobile & Ohio. A corporation. President, Wm. O. Haas; Vice-President, A. D. Davis; Secretary-Treasurer, F. E. Haas; General Manager, Leo Barmann; General Superintendent, Jos. McCallister. Employes, 120. Codes—Cross. Cattle, 500; hogs, 1,000; sheep, 50. Sausage—Fresh, smoked. By-Products—Tankage, hog and chicken feed. Refrigeration—35-ton York, spray and direct expansion. Boiler capacity, 300 H. P. Trade Marks—Hams, "Diamond D"; bacon, "Diamond D" and "Hi-Grade"; lard, "Diamond D"; compound, "Diamond D." Retail Markets—One, Mobile, Ala.

MONTGOMERY

Union Slaughter House Co.—Railroad, Seaboard Air Line. A corporation. Capital, $6,000; issued, $6,000. Stockholders, 2. President, Samuel Sabel; Secretary and Treasurer, Marx Sabel. Employes, 12. By-Products—Dried blood, tankage and inedible tallow.

SELMA

Selma Slaughter House Co.

Figures on cattle, hogs and sheep indicate AVERAGE WEEKLY KILLING CAPACITY.

ARIZONA

BISBEE

R. H. Mason Slaughter House.
Hogan & Eyzware.

CAMP VERDE

Len Young.

GLOBE

Bonne's Market—Broad St. Individual ownership. Employes, 20. Cattle, 100; hogs, 50; sheep, 50. All kinds of sausage. Refrigeration—Two 10-ton and one 5-ton Baker; brine circulating.

HUMBOLDT

Arctic Ice & Meat Co.—Also branch at Mayer.

JEROME

Jerome Meat Co.—Also branch at Clarksdale.

KIRKLAND

L. J. Haselfeld.

PHOENIX

Arizona Packing Co.—Tempe Road. Railroad, Arizona Eastern. A corporation. Capital, $2,000,000; issued, $1,662,900. Stockholders, 124. President, E. A. Tovrea; Vice-President, S. J. Tribolet; Secretary and Treasurer, W. M. Smith; General Manager, E. A. Tovrea; Manager of Department Heads, E. A. Tovrea; Comptroller, G. C. Baker; Live Stock Buyer, J. T. Hughes. Employes, 250. Codes—Cross, Western Union. Cattle, 1,800; hogs, 6,000; sheep, 6,000. Sausage—Fresh and smoked. By-Products—Dried blood, tankage, hog and chicken feed. Refrigeration—170-ton Vilter; brine and direct expansion; 8 tons ice daily. Boiler capacity, 300 H. P. Trade Marks—"Aripaco"; hams, "Cactus"; bacon, "Cactus" and "Apache"; lard, "Cactus" and "Apache"; compound, "Desert Bloom." Branch Houses—Bisbee, Douglas, Nogales and Clifton, Ariz. Retail Markets—Phoenix, Ray and Superior, Ariz.

Tribolet Packing Company.

TUCSON

Young Bros.—6th and Brachman Sts.

ARKANSAS

JONESBORO

Jonesboro Ice Co.—108 S. Main St. Railroad, St. Louis Southwestern. A corporation. Capital, $150,000. Stockholders, 24. President, J. F. Mason; Vice-President, Thos. Burress; Secretary-Treasurer, Gordan Matthews; General Manager, Thos. Burress; General Superintendent, Thos. W. Burress. Employes, 8. Cattle, 60; hogs, 75; sheep, 25. Sausage—Fresh, smoked and summer. By-Products—Chicken feed. Boiler capacity, 150 H. P.

LITTLE ROCK

Little Rock Packing Co.—Foot of E. 4th St. Railroad, Missouri Pacific. A corporation. Capital, $50,000. Stockholders, 6. President, Otto Finkbeiner; General Manager, Otto Finkbeiner. Employes, 45. Cattle, 100; hogs, 300 to 400. Sausage—Fresh, smoked and summer. By-Products—Tankage. City inspection. Refrigeration—20-ton Frick; direct expansion. Boiler capacity, 125 H. P.

Figures on cattle, hogs and sheep indicate AVERAGE WEEKLY KILLING CAPACITY.

CALIFORNIA

ANAHEIM

Anaheim Beef Co.—A corporation. President, John Ruether; Vice-President, Secondo Guasti, Jr.; Secretary and Treasurer, Alexander Huch. Fourteen branch retail stores.

BAKERSFIELD

Kern County Land Company—Packers, and large cattle ranch owners.

EL DORADO

Farmers' Exchange Packing Co.

EMERYVILLE

Alden & Agnew.

EUREKA

Russ Market Co.—Corner 3rd and G Sts. Railroad, Northwestern Pacific. A corporation. Capital, $200,000; issued, $200,000. Stockholders, 30. President, Wm. N. Russ; Vice-President, Mrs. F. G. Williams; Secretary, J. H. G. Weaver; General Manager, L. Carlson; General Superintendent, Wm. N. Russ; Slaughterhouse Foreman, C. Smith; Ranch Foreman, W. O. Perry. Employes, 45. Cattle, 40; hogs, 40; sheep, 70. Sausage—Fresh, smoked and summer. By-Products—Tankage, hog and chicken feed. City inspection. Refrigeration—By the Eureka Ice Company. Retail Markets—4 in Eureka. One wholesale market in Eureka.

FORT BRAGG

Noyo Land & Cattle Company.

IMPERIAL

Pacific Land & Cattle Co.—Railroad Ave. Railroad, Southern Pacific. A corporation. Capital, $250,000; issued, $201,300. Stockholders, 15. President, G. A. Lathrop; Vice-President, T. J. Green; Secretary, C. W. Allison; Treasurer, C. W. Allison; Auditor, H. S. Capron; General Manager, G. A. Lathrop; General Superintendent, C. W. Allison. Employes, 30. Cattle, 100; hogs, 100; sheep, 75. Makes sausage. By-Products—Commercial fertilizer. Refrigeration—25-ton Vilter; direct expansion. Boiler capacity, 80 H. P. Retail markets—3 in Brawley, 3 in Calexico, 2 in Calipatria, 1 in Holtville, 1 in El Centro and 1 in Westmoreland, Cal.

ISLETON

Isleton Meat Market—Transportation, Southern Pacific steamer. Individual ownership. Owner and General Manager, W. S. Hartin. Employes, 3. Cattle, 10; hogs, 20; sheep, 20. Sausage—Fresh, smoked.

LOS ANGELES

California Dressed Beef Co.—3820 Santa Fe Ave. Railroad, Santa Fe. A corporation. Capital, $500,000. Directors, R. L. Bliss, T. S. Reynolds, J. W. Ruwe. Employes, 150. Cattle, 1,200; hogs, 2,500; sheep, 2,500. Sausage—Fresh and smoked. By-Products—Fertilizer. Refrigeration—York; brine circulating system. Boiler capacity, 250 H. P.

California Packing Co.—900 Macy St.

Figures on cattle, hogs and sheep indicate **AVERAGE WEEKLY KILLING CAPACITY.**

LOS ANGELES

Hauser Packing Co.—2300 E. Ninth and Mateo Sts. Railroads, Atchison, Topeka & Santa Fe and Union Pacific. A corporation. Capital, $500,000; issued, $500,000. President, Julius Hauser; Vice-President, E. C. Hauser; Secretary, H. J. Hauser; Assistant Secretary, F. M. Hauser; Treasurer, L. A. Hauser; General Manager, H. J. Hauser; General Superintendent, Mr. Brown; Sales Manager, G. L. Shivel. Employes, 400. Codes—A. B. C. and Western Union; Code word, "Resuah." Cattle, 3,750; hogs, 6,000; sheep, 1,200. Sausage—Fresh, smoked and summer. By-Products—Dried blood, tankage, hog and chicken feed, "Big Gun Tankage," digester tankage and ground bone for poultry and fertilizer. Government inspection. Refrigeration—Two 75-ton Vilters; direct expansion. Boiler capacity, 800 H. P. Trade Marks—Hams, "Hauser Pride" and "Angelus"; bacon, "Hauser Pride" and "Angelus"; lard, "Angelus" and "Hauser Pride"; compound, "Fryola," "Hauser's Lard Substitute" and "Violet Brand Shortening." Also exports.

H. F. Lewis Packing Company—E. Vernon Ave. and Salt Lake Tracks. Railroad, Salt Lake. A corporation. Capital, $250,000; issued, $100,000. Stockholders, 3. President and Treasurer, H. F. Lewis; Vice-President, F. M. Lewis; General Manager, H. F. Lewis; General Superintendent, Geo. E. Sailor. Employes, 60. Cattle, 600; hogs, 800; sheep, 1,000. Sausage—Fresh, smoked and summer. By-Products—Dried blood, tankage, hog and chicken feed and fertilizers. Refrigeration—50 and 10-ton Vilter; brine circulating. Boiler capacity, 150 H. P. Trade Marks—"Crescent."

Los Angeles Packing Co.—Vernon & Boyle Aves. Railroad, Salt Lake. A corporation. Capital, $25,000; issued, $25,000. Stockholders, 4. President, D. Danny; Vice-President, Frank Murphy; Secretary and Treasurer, A. Miller; General Manager, Max Goldring. Employes, 25. Cattle, 200; hogs, 600; sheep, 600. Sausage—Fresh and smoked. By-Products—Tankage and hog and chicken feed. Refrigeration—10-ton York; air blast. Boiler capacity, 40 H. P.

Newmarket Co.—Vernon. Railroad, Santa Fe. President, R. H. Jeffries; Vice-President, J. C. Link. Cattle, 1,500; hogs, 2,000; sheep, 2,000. Sausage—Fresh. By-Products—Dried blood, tankage, hog and chicken feed.

Standard Packing Co.—Vernon and Santa Fe Aves. Railroad, Santa Fe. A corporation. Capital, $10,000; issued, $10,000. Stockholders, 4. President, Vice-President, Secretary and Treasurer, T. P. Breslin. Employes, 55. Cattle, 650; hogs, 1,200; sheep, 1,800. Sausage—Fresh and smoked. By-Products—Dried blood, tankage, hog and chicken feed. Refrigeration—15 and 25-ton York; direct expansion. Boiler capacity, 160 H. P. Trade Mark—"Poinsettia."

Woodward Bennett Co.—Box 112, Station H. A corporation. Capital, $350,000; issued, $350,000. Stockholders, 3. President, J. A. Woodward; Vice-President, G. J. Woodward; Secretary and Treasurer, E. J. Bennett; General Manager, J. A. Woodward; General Superintendent, Claud Safstrom. Employes, 125. Cattle, 1,700; hogs, 2,000; sheep, 8,000. Sausage—Fresh. City inspection. Refrigeration—30-ton York. Trade Marks—"W. & B.," and "Blue Ribbon."

Wilson & Co.—1000 Lyon St. (See Chicago.)

Figures on cattle, hogs and sheep indicate AVERAGE WEEKLY KILLING CAPACITY.

MARICOPA

West Side Market—George Fiester, W. W. Bley, Props.

NEEDLES

Frank C. Soto—Railroad, Santa Fe. Ownership, individual. Manager, S. R. Cord. Cattle, 50; hogs, 20; sheep, 20.

OAKLAND

Golden West Meat & Packing Co.—Railroad, Southern Pacific. A corporation. Capital, $200,000; issued, $200,000. Stockholders, 4. President, John Lacoste; Vice-President, Al. Lacoste; Secretary and treasurer, Geo. J. Dupuy; General Manager, A. W. Lugg. Employes, 30. Calves, 350; hogs, 1,200; sheep and lambs, 5,500. Municipal inspection. Refrigeration—7½-ton Cyclops; direct expansion.

Grayson-Owen Packing Company—6481 Bay St. A corporation. Capital, $300,000. President, P. B. Lynch; Vice-President, J. G. Westphal; Secretary and Treasurer, A. E. Corder.

Oakland Meat & Packing Co.—65th and Bay Sts. Railroad, Southern Pacific. A corporation. Capital, $500,000; issued, $500,000. Stockholders, 6. President, C. J. Hooper; Vice-President, O. L. Watson; Secretary, B. C. Chew; Treasurer, O. L. Watson; General Manager, O. L. Watson. Employes, 60. Cattle, 500; hogs, 350; sheep, 1,000. Sausage—Fresh, smoked. By-Products—Dried blood, tankage, hog and chicken feed, tallow, oleo oil, stearine. Refrigeration—30-ton Cyclops; brine circulating. Boiler capacity, 250 H. P. Trade Marks—Hams, "Shasta"; bacon, "Shasta"; compound, "Shasta."

Steinbeck Co.—1625-29 Broadway. Railroad, Southern Pacific. A corporation. Capital, $100,000; issued, $100,000, fully paid. Stockholders, 3. President, J. F. Reynolds; Vice-President, H. C. W. Steinbeck; Secretary, C. L. Culbert; Slaughter House Superintendent, F. A. Cook; Jobbing and Retail, A. L. Theiss; Chief of Office Staff, Arthur Theiss; Credit Manager, F. Marolf. Employes, 49. Cattle, 300; hogs, 500; sheep, 1,000. Sausage—Fresh, smoked and summer. By-Products—Tankage, hog and chicken feed, neatsfoot oil. Refrigeration—8-ton Cyclops, brine circulating, and 15-ton Vulcan, direct expansion. Buys steam from adjoining plant. Trade Mark—"Crystal Brand." Retail Market—1625-29 Broadway, Oakland, Cal. Slaughter House—West Berkeley, Cal.

Oakland Abattoir Co.
Lewis & McDermott.
Louis Schaffer & Sons.
U. M. Slater, Inc.

OROVILLE

Johnson & Openshaw.

POMONA

San Antonio Meat Co.—Railroad, Salt Lake. A corporation. Capital, $75,000; issued, $75,000. Stockholders, 6. President, M. L. Sparks; Vice-President, E. S. Ware; Secretary, A. H. Peck; Treasurer, A. H. Peck; General Manager, A. H. Peck. Employes, 60. Cattle, 200; hogs, 200; sheep, 200. Sausage—Fresh. By-Products—Tankage and digester tankage. Government inspection. Refrigeration—Two 10, one 3 and two 2-ton Parker, and one 5-ton Stevens; direct expansion. Boiler capacity, 50 H. P. Retail Markets—Palace Market, Pomona; Home Market, Cucamonga; Glendora Market, Glendora; Central Mar-

---

Figures on cattle, hogs and sheep indicate AVERAGE WEEKLY KILLING CAPACITY.

ket, Ontario; City Market, Azusa; Azusa Market, Azusa; Cash Market, Corona; San Dimas Market, San Dimas; LaVerne Market, LaVerne.

SACRAMENTO

C. Swanston & Son (Inc.)—Mail address, P. O. Box 323, Sacramento; office, abattoir and stock yards at Swanston. Railroads, Southern Pacific, Western Pacific and Sacramento Northern. A corporation. Capital, $500,000; issued, $500,000. Stockholders, 3. President, George Swanston; Vice-President, Robert Swanston; Secretary, E. A. Murray; General Manager, George Swanston; General Superintendent, Robert Swanston; F. H. Carroll, Manager San Francisco Branch; M. D. Gallagher, Sales Manager, Sacramento. Employes, 100. Codes—Swanston. Cattle, 600; hogs, 1,000; sheep, 1,000. By-Products—Dried blood, tankage, hog and chicken feed, edible and inedible tallow, neatsfoot oil, hides, pelts and lard. Government inspection. Refrigeration—85-ton Vulcan, brine or direct expansion; 7 tons ice daily. Boiler capacity, 180 H. P. Trade Marks—Ham, "Poppy" and "Capital"; boiled ham, "Poppy"; bacon, "Poppy" and "Capital"; picnics, "Capital." Branches—San Francisco, Cal., 8th St. near Townsend. Also exports.

Mohr & York.

SAN BERNARDINO

Oehl Packing Company—South E St. Sup., Martin Green.

SAN DIEGO

Chas. S. Hardy—Cor. 6th and G Sts. Railroad, Old San Diego. Individual ownership. Superintendent, J. J. Donnelly. Employes, 325. Cattle, 500; hogs, 1,000; sheep, 1,500. Sausage—Fresh and smoked. By-Products—Dried blood, tankage, hog and chicken feed and pulled wool. Government inspection. Refrigeration—40, 30 and two 10-ton Frick; direct expansion and brine spray; 5 tons ice daily. Boiler capacity, 500 H. P. Trade Marks—"Bay City," "Hardy's Best" and "Cotoc." Retail Markets—San Diego, La Mosa, El Cajolin and Coronado, Cal.

SAN FRANCISCO

Buttgenbath, Joseph, & Co., Inc.—336 4th St., San Francisco County. A corporation. Capital, $25,000; issued, $16,200. President, Mrs. Wilhemine Buttgenbath; Secretary and Treasurer, Hugo Buttgenbath.

Miller & Lux Incorporated—1114 Merchants Exchange. Railroads, Southern Pacific, Santa Fe., and Union Pacific. A corporation. Capital, $12,000,000; issued, $12,000,000. Stockholders, 5. President, J. Leroy Nickel; Vice-President, J. Leroy Nickel, Jr.; Secretary, D. Brown; General Manager, J. Leroy Nickel. Employes, 300. Codes—All; code word, "Millerlux." Cattle, 1,000; hogs, 5,000; sheep, 15,000. Sausage—Fresh. By-Products—Dried blood, tankage and hog and chicken feed. Refrigeration—150-ton Vilter and 100-ton Larsen-Baker; direct expansion; 4 tons ice daily. Boiler capacity, 1,350 H. P. Trade Mark—"1-1-1" and "M. & L."

H. Moffatt Co.—1 Montgomery St. A corporation. Capital, $50,000; issued, $50,000. Stockholders, 5. President, Wm. H. Moffatt; Vice-President, Mrs. A. M. Sharp; Secretary and Treasurer, Geo. W. Andrews; General Manager, Geo. W. Andrews. Employes, 30. Cattle, 250; sheep, 1,200. Refrigeration—One 40-ton Cyclops.

Figures on cattle, hogs and sheep indicate AVERAGE WEEKLY KILLING CAPACITY.

SAN FRANCISCO

Roth Blum Packing Co.—1490 Fairfax Ave. Railroads, Southern Pacific and Western Pacific & Santa Fe. Partnership. General Manager, Lester L. Roth; General Superintendent, Otto Birbe. Employes, 50. Codes—Yopp. Cattle, 650; hogs, 2,900; sheep, 10,000. Refrigeration—15-ton Cyclops; direct expansion. Boiler capacity, 300 H. P. Trade Mark—"Our Choice."

Western Meat Co.—6th and Townsend Aves. Railroads, Southern Pacific, Santa Fe, Western Pacific and Northwestern Pacific. A corporation. Capital, $2,500,000; issued, $2,500,000. President, C. J. Hooper; Vice-President, R. H. Benedict; Secretary, L. W. Snow; Treasurer, R. H. Benedict; General Manager, C. J. Hooper; General Superintendent, J. O. Snyder. Employes, 750. Codes—Bentleys, W. U., A. B. C.; code word, "Steadfast." Cattle, 900; hogs, 3,000 to 4,000; sheep, 10,000. Sausage—Fresh, smoked. By-Products—Dried blood, tankage, hog and chicken feed, fertilizer, glue, etc. Government inspection. Refrigeration—One 100-ton and one 200-ton Wolf and one 250-ton Ball at main plant; one 50-ton Vulcan at San Francisco branch; one 10-ton Vulcan at Oakland; one 10-ton Vulcan at Sacramento; one 5-ton Vulcan at San Jose; one Automatic at Fresno and one Automatic at Stockton; all direct expansion; 25 tons ice daily. Boiler capacity, 2,020 H. P. Trade Marks—Hams, "Mayrose"; bacon, "Mayrose"; lard, "Golden Gate"; compound, "Arrow" and "Califene." Branches—San Francisco, Sacramento, Stockton, Oakland, Fresno, and San Jose, Cal. Also exports.

SOUTH SAN FRANCISCO

South San Francisco Packing & Provision Co.—407 Front St. A corporation. Capital, $200,000; issued, $200,000. Stockholders, 7. President, L. Nonnenmann; Vice-President, Geo. A. Zimmermann; Secretary, H. Heinsohn; General Manager, L. Nonnenmann. Employes, 40. Hogs, 1,800. Sausage—Fresh and smoked. By-Products—Tankage. Refrigeration—Two 30-ton Cyclops; direct expansion.

Virden Packing Co.—P. O. Box 128. Railroad, So. San Francisco Belt. A corporation. Capital, $10,000,000; issued, $5,000,000. Stockholders, 4,000. President, Charles E. Virden; Vice-Presidents, J. C. Good, James T. Doyle, A. W. Virden; Treasurer, H. G. Brown; Secretary, F. F. Atkinson; General Auditor and Assistant Treasurer, George A. Ticoulet; Assistant Secretaries, C. E. Holloway and T. O'Leary; General Manager canning operations, F. E. Laney; Advertising Director, Peter B. Newman; Purchasing Agent and Traffic Manager, B. A. Wise; Provision Manager, John Tiedemann; Office Manager, L. B. Shippey. Employes, 300. Codes—Cross, Roy and Saunders; code word, cable, "Campfire." Cattle, 1,000; hogs, 4,000; sheep, 2,000. Sausage—Fresh, smoked and summer. By-Products—Tankage, hog and chicken feed, and others. Government inspection. Refrigeration—Two 40-ton Carbondales; brine; 10 tons ice daily. Boiler capacity, 900 H. P. Trade Marks—"Camp Fire," "Magnolia," "Thrift." Branches—1300 Bryant St., San Francisco; Sacramento, Los Angeles and Maryville, Cal. Also exports.

Figures on cattle, hogs and sheep indicate AVERAGE WEEKLY KILLING CAPACITY.

SAN FRANCISCO

McIntyre Packing Co.—807 Montgomery St.
Roth, Winter & Walsh—1271 Mission St.
Wm. Taafe Co.—5th and Third Aves.
J. G. Johnson—208 Macdonough Bldg., 333 Kearny St.
Henry Levy & Co.—1 Montgomery St.
Sumski & Harband—1510 Evans St.

SAN PEDRO

Nielson & Kittle Canning Co., Ltd.—Box B. Fish canners. Railroad, Los Angeles & Salt Lake. A corporation. Capital, $1,000,000; issued, $704,000. Stockholders, 20. President, C. O. Nielson; Vice-President, Adolph Larsen; Secretary-Treasurer, T. E. Madsen. Employes, 250. Codes—W. U., Bentleys, J. K. Armsbys, and A. B. C.—5th; code word, "Nilkit." National Canners inspection. Trade Marks—"Norseman," "Regent," "Nekco" and "N. & K." Also exports.

SANTA BARBARA

F. N. Gehl Packing Company.—Railroad, Southern Pacific. A corporation. Capital, $50,000; issued, $50,000. President, F. N. Gehl; Vice-President, Mrs. F. N. Gehl; Secretary and Treasurer, M. A. Burdick; General Manager, F. N. Gehl. Employes, 20. Cattle, 130; hogs, 150; sheep, 400. Sausage—Fresh, smoked and summer. Retail Markets—817 State St. and 605 State St., Santa Barbara, Cal.

Santa Barbara Packing Co.—636 State St. Railroad, Southern Pacific. A corporation. Capital, $100,000; issued, $100,000. Stockholders, 10. President, A. L. Hobson; Vice-President, E. C. Gammill; Secretary and Treasurer, E. B. Olsen; General Manager, E. C. Gammill. Employes, 37. Cattle, 45; hogs, 50; sheep, 100. Sausage—Fresh and smoked. Refrigeration—Two York and two Cyclops. Retail Markets—Four in Santa Barbara. Abattoir and stock yards at Goleta, Cal.

SANTA PAULA

Santa Paula Packing Company.
J. W. Barker.

SANGER

Sanger Meat Market.

SEBASTOPOL

Sebastopol Meat Co.

SUSANVILLE

Sierra Packing Co.

SUISUN

Sacovalley Meat Canning & Provision Co.—Railroads, Southern Pacific, Western Pacific & Suisun Bay. Corporation. Capital, $500,000. Stockholders, 200. President, B. F. Rush; Vice-Presidents, C. W. R. Thelen, Arth Hibborn; Treasurer and General Manager, C. W. R. Thelen; Secretary, H. Bird. Employes, 140. Cattle, 300; hogs, 300; sheep, 500. Sausage—Fresh, smoked. By-Products—Dried blood, tankage, chicken feed, inedible tallow, bone grease, neatsfoot oil. Trade Marks—"Sacovalley," "Thelen."

Figures on cattle, hogs and sheep indicate AVERAGE WEEKLY KILLING CAPACITY.

COLORADO

ALAMOSA

Creek Packing Co.—Railroad, Denver & Rio Grande. Individual ownership. President and owner, Walter Creek. Employes, 15. Cattle, 50; hogs, 100; sheep, 25. Sausage—Fresh, smoked and summer. Refrigeration—Ice. Boiler capacity, 40 H. P. Retail Markets—Alamosa, Creede, Del Norte and La Veta.

COLORADO SPRINGS

Smith's Packing Co.—118 S. Cascade St. Railroad, Denver & Rio Grande. A corporation. Capital, $200,000; issued, $90,030. Stockholders, 3. President, C. Louis Smith; Vice-President, A. E. Smith; Secretary, Carlos L. Smith; Treasurer, C. Louis Smith; General Manager, C. Louis Smith. Employes, 25. Cattle, 125; hogs, 200; sheep, 500. Sausage—Fresh, smoked and summer. By-Products—Tankage. Refrigeration—Two 20-ton Ideals; direct expansion. Boiler capacity, 100 H. P.

DENVER

Armour & Company. (See Chicago.)

Burkhardt Packing & Provision Company—1515 Lawrence St. A corporation. Capital, $150,000; issued, $71,500. President and Treasurer, Jacob Burkhardt; Vice-President, W. C. Burkhardt; Secretary, Chas. J. Burkhardt; Asst. Treasurer, Chas. C. Dinkelaker.

Coffin Packing & Provision Co.—Union Stock Yards. Railroad, Chicago, Burlington & Quincy. A corporation. Capital, $250,000; issued, $250,000. Stockholders, 4. President, W. N. W. Blayney; Vice-President, J. P. Murphy; Secretary, H. F. Blayney; Treasurer, J. P. Murphy; General Manager, J. P. Murphy. Employes, 150. Codes—Cross and Robinson. Cattle, 800; hogs, 2,500; sheep, 500. Sausage—Fresh and smoked. By-Products—Dried blood, tankage, hog and chicken feed, and commercial fertilizers. Refrigeration—25 and 50-ton Linde; direct expansion. Boiler capacity, 600 H. P. Branches—Denver, Pueblo, Glenwood Springs and Grand Junction, Colo.

Geuting-Nuckolls Packing Co.—1612-1614 Market St.

Liberty Packing & Provision Company.—58th and York Sts. A corporation. Capital, $400,000; issued, $150,000. Stockholders, 15. President, M. A. Hayutin; Vice-President, Max Milstein; Secretary, Morris Hayutin; Treasurer and General Manager, Louis Heller. Employes, 25. Cattle, 350; sheep, 1,000. Refrigeration—20-ton York; direct expansion. Boiler capacity, 120 H. P.

Midwest Packing Company.

Mountain States Packing Co.—4800 Washington St. Railroad, Colorado and Southern. A corporation. Capital, $6,000,000; issued, $200,000. Stockholders, 120. President, Chas. F. Kamrath; Vice-President, Fred Klink; Secretary, I. A. Wood; Treasurer, Allison Stocker; General Manager, Chas. F. Kamrath. Employes, 50. Codes—Cross, 3rd edition, Reo and Maurice Pincoffs. Cattle, 750; hogs, 1,500; sheep, 500. By-Products—Dried blood, tankage, tallow, grease, hides, hog hair, glue stock, etc. Refrigeration—30-ton York and 15-ton Althoff; direct expansion. Boiler capacity, 300 H. P.

Sargent, Edward W.—1730 Market St. Capital, $50,000. Individual ownership.

Swift & Company. (See Chicago.)

Figures on cattle, hogs and sheep indicate AVERAGE WEEKLY KILLING CAPACITY.

GRAND JUNCTION

Grover Packing Co.—Railroad, Denver & Rio Grande. A corporation. Capital, $100,000; issued, $70,000. Stockholders, 8. President and General Manager, Fred F. Grover. Employes, 35. Cattle, 200; hogs, 600; sheep, 100. Sausage—Fresh, smoked and summer. By-Products—Dried blood, tankage, chicken feed. Refrigeration—35-ton Sterling; direct expansion. Boiler capacity, 120 H. P. Trade Marks—"Royal" and "Home" hams, bacon, lard; "Eagle" compound.

Rettig Packing Co.—P. O. Box 103. Railroad, Denver & Rio Grande Western. Corporation. Capital, $100,000. General Manager, Adam Rettig. Employes, 25. Cattle, 400; hogs; 600; sheep, 100. Sausage—Fresh, smoked and summer. By-Products—Dried blood and tankage. Refrigeration—12-ton Althoff; direct expansion. Boiler capacity, 70 H. P. Trade Mark—"Mount Garfield."

MONTE VISTA

Busch-Nelson Meat Co.

The San Luis Packing Company.

PUEBLO

Nuckolls Packing Co.—Santa Fe Ave. and Spring St. Railroad, Denver & Rio Grande. A corporation. Capital, $1,000,000; issued, $631,000. Stockholders, 3. President, G. H. Nuckolls; Vice-President, Ezra Nuckolls; Secretary, M. C. Crum; Treasurer, Ezra Nuckolls. Employes, 190. Codes—Cross and Yopp. Cattle, 750; hogs, 3,000; sheep, 500. Sausage—Fresh and smoked. By-Products—Dried blood, tankage and hog feed. Government inspection. Refrigeration—150 and 25-ton York, spray and direct expansion, and 100-ton Vilter, direct expansion; 25 tons ice daily. Boiler capacity, 700 H. P. Trade Marks—Hams and bacon, "Peerless"; lard, "Open Kettle" and "Pikes Peak"; compound, "Little Chief." Branches—Salt Lake City, Utah; Trinidad, Durango, Montrose, Telluride, Colorado Springs, and Denver, Colo., and Santa Fe, N. Mex. Also exports.

ROCKY FORD

B. F. Stauffer—308 S. Main St. Individual ownership. General Manager, B. F. Stauffer; General Superintendent, H. D. Wine. Cattle, 25; hogs, 100; sheep, 10. Sausage—Fresh and smoked. By-Products—Tallow. Refrigeration—6-ton Althoff; brine circulating. Retail Markets—One at Rocky Ford and one at Swink, Colo

STERLING

Sterling Packing Co.—Samuel P. Rosenbaum and John L. Goldberg.

TELLURIDE

Consumers Packing Co.—Main St. Railroad, Rio Grande. A corporation. Capital, $2,000. Stockholders, 9. President, M. Lewis; Vice-President, C. B. Lewis. Employes, 3. Cattle, 20; hogs, 30; sheep, 20. Sausage—Fresh and smoked. Boiler capacity, 3 H. P. Retail Market—Telluride, Colo.

WALSENBURG

S-Schafer Packing Co.—608 Main St. A corporation. Capital, $25,000; issued, $21,000. Stockholders, 7. President, A. C. Schafer; Vice-President, Fred W. Schafer; Secretary, John Kirkpatrick; Treasurer, Andrew Schafer, Jr.; General Manager, A. C. Schafer. Employes, 14. Cattle, 60; hogs, 300; sheep, 450. Sausage—Fresh and smoked. Refrigeration—10-ton Baker and 10-ton Althoff; brine circulating and direct expansion. Trade Marks—Hams, "Ideal"; bacon, "Ideal"; lard, "Ideal." Two retail markets in Walsenburg.

Figures on cattle, hogs and sheep indicate AVERAGE WEEKLY KILLING CAPACITY.

CONNECTICUT

HARTFORD

East Side Hide Co.—Windsor St. Morris Later & Sons, owners.

Hartford Provision Co.—302 Pleasant St. Railroad, New York, New Haven & Hartford. A corporation. Capital, $50,000; issued, $30,000. President, Joseph Samuels; Vice-President, W. H. Haertel; Treasurer, Emil De Loreto; Secretary, Joseph Samuels; General Manager, Emil De Loreto; General Superintendent, Joseph Samuels; Sausage Department Manager, W. H. Haertel. Employes, 20. Cattle, 150; hogs, 500; sheep, 140. Sausage—Fresh and smoked. By-Products—Tankage and grease. Refrigeration—15-ton Hartford Automatic; direct expansion. Boiler capacity, 30 H. P. Trade Marks—"Hartford Brand" and "Veteran Brand."

NEW HAVEN

The Sperry & Barnes Co.—188 Long Wharf. Railroad, New York, New Haven & Hartford. A corporation. Capital, $1,300,000; issued, $1,300,000. Stockholders, 30. President, Geo. H. Swift; Vice-President, E. C. Starr; Secretary, E. H. Throm; Treasurer, H. J. Nichols; General Manager, I. D. Marshall; General Superintendent, F. H. Quinley. Employes, 357. Codes—Cross, Robinsons; code word, "Barnhug." Hogs, 2,500. Sausage—Fresh and smoked. By-Products—Tankage. Government inspection. Refrigeration—200-ton De La Vergne; direct expansion. Boiler capacity, 650 H. P. Trade Mark—"Elm City." Also exports.

DELAWARE

WILMINGTON

Beste Provision Co.—116-126 Liberty and Logan St. Railroad, Philadelphia & Reading. A corporation. Capital, $75,000; close corporation. Stockholders, 3. President, B. J. Beste; Vice-President, M. C. Beste; Secretary and Treasurer, S. E. Hinger. Employes, 20. Cattle, 200. Sausage—Fresh, smoked and summer. By-Products—Tankage. Refrigeration—Two 10-ton Brecht; direct expansion. Boiler capacity, 50 H. P.

Frederick, John—R. F. D. No. 1.

Hart & Bro. Co.—Fifth and Poplar Sts.

Johnson & Bro.—S. W. Cor. 7th and King Sts. Partnership, Wm. R. and J. E. Johnson.

Mammele & Sons, Chas. C.

Wilckens-Staats Co.—230 Tatnall St.

Wilmington Provision Co.—Front and Orange Sts. Railroads, Baltimore & Ohio, Philadelphia & Reading and Penna. A corporation. Capital, $400,000; issued, $285,000. Stockholders, 200. President, Max Matthes; Vice-President and Gen. Manager, George A. Casey; Secretary, Wm. T. Fritz; Treasurer, Mark P. Brown. Employes, 80. Cattle, 200; hogs, 3,000; sheep, 300; calves, 100. Sausage—Fresh, smoked and summer. By-Products—Dried blood, tankage. Refrigeration—200-ton Frick; direct expansion. Boiler capacity, 250 H. P. Trade Marks—Hams, bacon, lard, "Tower"; sausage, "Brandywine."

Zimmerman, Jacob—301 W. 4th St.

Figures on cattle, hogs and sheep indicate AVERAGE WEEKLY KILLING CAPACITY.

DISTRICT OF COLUMBIA

WASHINGTON, D. C.

Golden Company—928 Louisiana Ave., N. W.

Henning, Joseph—945 B St., N. W.

A. Loffler Provision Co., Inc.—Benning, D. C. Railroads, Pennsylvania and Baltimore & Ohio. A corporation. Capital, $120,000; issued, $120,000. Stockholders, 9. President, A. D. Loffler; Vice-President, A. O. Dille; Secretary and Treasurer, Chas. A. M. Loffler; General Superintendent, F. W. Vogel. Employes, 80. Cattle, 75; hogs, 1,500; sheep, 200. Slaughtered at Washington Abattoir, Benning, D. C. Sausage—Fresh and smoked. Refrigeration—Frick, brine circulating. Retail markets—Three in Washington, D. C.

Ostman, Bernard.

Swindell & Son, S.—Rr. 1229 D St., S. E.

BENNING

N. Auth Provision Company—630 Virginia Ave.

Washington Abattoir Company.

FLORIDA

BARTOW

Polk County Packing Co.—A partnership. Stockholders, 5. President, T. W. Page; Vice-President, Claude Page; Treasurer and Secretary, G. B. Skinner. Employes, 10. Cattle, 150; hogs, 100; sheep, 100. Sausage—Fresh, smoked and summer.

CHIPLEY

Chipley Packing Company—President, A. A. Myers, Jr.; Manager, C. L. Brooks.

JACKSONVILLE

Armour & Company—310-314 W. Bay St. (See Chicago.)

Farris & Company.

Jones-Chambliss Company—Forest St. Railroad, Atlantic Coast Line. A corporation. Capital, $100,000; issued, $90,000. Stockholders, 6. President, C. A. Jones; Secretary, A. H. Goedert; Treasurer, J. O. Chambliss; General Manager, C. A. Jones; General Superintendent, A. H. Goedert. Do wholesale slaughtering and packing. Refrigeration—One 8-ton York, one 8-ton Automatic and one 20-ton Frick; direct expansion. Boiler capacity, 40 H. P.

PENSACOLA

Brockman Packing Company—Davis and Gadsden Sts. President, H. Brockman.

TAMPA

Hendry Bros. & Co.—Palm River Road.

Lykes Bros., Inc.—Giddens Bldg. Branch of Lykes Bros., Inc., Havana, Cuba.

Tampa Packing Co.

Figures on cattle, hogs and sheep indicate AVERAGE WEEKLY KILLING CAPACITY.

GEORGIA

ALBANY

Municipal Abattoir.

ATLANTA

Atlanta Butchers Abattoir Co.

White Provision Co.—Howell Mill Road. Railroads, Southern and Western & Atlantic. A corporation. Capital, $600,000; issued, $600,000. President, W. H. White, Jr.; Vice-Presidents, W. L. Mewborn and E. S. Papy; Secretary, R. L. Hollbrook; Treasurer, F. C. Wilkerson; General Manager, W. H. White, Jr.; General Superintendent, E. Trimble; Provision Department Manager, J. R. Griffith; Produce Department Manager, W. E. Anderson; General Sales Manager, E. S. Papy. Employes, 200. Codes—Cross. Cattle, 600; hogs, 5,000. Sausage—Fresh and smoked. By-Products—Tankage and tallows. Refrigeration—50-ton York and 30-ton Barber; brine circulating; 25 tons ice daily. Boiler capacity, 300 H. P. Trade Marks—Hams, "Cornfield"; bacon, "Cornfield"; lard, "Grandmother's."

AUGUSTA

Augusta Packing Company—New Savannah Road. A corporation. Capital, $25,000; issued, $15,050. President, Secretary Treasurer and Manager, Charles E. Brown.

LA GRANGE

La Grange Packing Co.

MACON

Macon Packing Company—Owned by Allied Packers, Inc., Chicago, Ill.

MOULTRIE

Moultrie Packing Co.—Owned by Swift & Co. (See Chicago.)
Swift & Company—(See Chicago.)

SAVANNAH

Chatham Abattoir & Packing Co.

Savannah Abattoir & Packing Co.—Louisville Road. Railroad, S. & A. A corporation. Capital, $100,000; issued, $100,000. Stockholders, 15. President, Joe Ehrlich; Vice-President and General Manager, Isaac Meddin; Vice-President, D. Kirkland; Secretary and Treasurer, Alexander Meddin. Employes, 75. Cattle, 300; hogs, 1,000. Sausage—Fresh, smoked and summer. By-Products—Dried blood, tankage, tallow, grease and hog hair. Refrigeration—24-ton York; brine circulating and direct expansion. Boiler capacity, 80 H. P. Trade Marks—"Georgia" brand meats and lard. Retail markets—Six at Savannah, Georgia, and one at Columbus, Georgia.

STATESBORO

Statesboro Provision Co.

WAYCROSS

Waycross Packing Plant—31 Satilla Lane. Railroad, Atlantic Coast Line. Owner, Stuyvesant Fish. Investment, $125,000. Agent for owner, Ware County Light & Power Co. President, Stuyvesant Fish; Vice-President, P. R. Bomeisler; Treasurer, Chas. H. Wenman. Cattle, 20 per day; hogs, 100 per day.

Figures on cattle, hogs and sheep indicate AVERAGE WEEKLY KILLING CAPACITY.

IDAHO

BOISE

Boise Butcher Co., Ltd.—811 Idaho St. Railroad, Oregon Short Line (U. P. System). A corporation. Capital, $100,000; issued, $89,200. Stockholders, 7. President, William Lomax; Treasurer, W. M. Williams; Secretary, Walter M. Williams; General Manager, William Lomax; Superintendent, Geo. Schweitzer; Livestock Buyer, N. R. Marler. Employes, 20. Cattle, 150; hogs, 300; sheep, 200. Sausage—Fresh and smoked. Government inspection. Refrigeration—15-ton Harris, 5-ton Ideal; direct expansion. Boiler capacity, 25 H. P. Trade Marks—Hams, bacon and lard, "Arrow Rock"; compound and shortening, "Crown Brand." Retail Markets—811 Idaho St. and 813 Main St., Boise.

Davis Meat Co.—Box 186. A corporation. Capital, $500,000; issued, $414,000. Stockholders, 4. President, E. H. Davis; Secretary and Treasurer, Thos. J. Davis; General Manager, E. H. Davis; General Superintendent, Thos. J. Davis. Employes, 30. Cattle, 50; hogs, 125; sheep, 35. Sausage—Fresh, smoked and summer. By-Products—Tallow, oil and ground bone. Refrigeration—12-ton York; direct expansion. Trade Marks—"Bar O Brand Products"; hams, "Campfire"; bacon, "Campfire" and "Davis Capitol"; lard, "Bar O Brand"; compound, "Golden Glow."

Idaho Provision & Packing Co.—716 Idaho St. Railroad, Oregon Short Line. A corporation. Capital, $50,000; issued, $48,100. Stockholders, 5. President, Herbert F. Lemp; Vice-President, Mrs. Herbert F. Lemp; Secretary, Watt Piercy; Treasurer, John Maloney; General Manager, John Maloney; General Superintendent, Al. Ziegenhagen. Employes, 20. Cattle, 100; hogs, 150; sheep, 200. Sausage—Fresh, smoked and summer, lunch tongue and all other dry sausage products. Refrigeration—14-ton Ideal and 14-ton Bell & Wildman; direct expansion. Boiler capacity, 40 H. P. Trade Mark, "Lily." Retail Market—716 Idaho St., Boise.

EAGLE

Boise Valley Packing Company—A corporation. Capital, $100,000; issued, $66,583. President and Manager, E. A. Evans; Vice-President, Emil Julian; Secretary and Treasurer, Wm. Goodall.

GOODING

Gem State Packing Co.

LEWISTON

W. H. Bristol Co.—204 Main St. Railroad, Northern Pacific. Individual ownership. Capital, $50,000. Employes, 20. Cattle, 50; hogs, 100; sheep, 50. Sausage—Fresh, smoked and summer. Government inspection. Refrigeration—One 7, 3, 1, and ½-ton Armstrong, also 1-ton Cooper machine; brine and direct expansion; 1 ton ice daily. Boiler capacity, 30 H. P. Retail Markets—Two in Lewiston, Idaho, one in Clarkston, Wash., and one in Asotin, Wash.

Inland Meat Co.—832 Main St. A corporation. Capital, $20,000; issued, $12,500. Stockholders, 3. President, G. E. Beckman; Secretary, T. L. Ford; Treasurer, Alois Kalois; General Manager, T. L. Ford. Employes, 13. Cattle, 40; hogs, 80; sheep, 50. Sausage—Fresh, smoked and summer. By-Products—grease. Refrigeration—10-ton Armstrong, one 5 and one 1½-ton Racine; brine circulating. Retail Markets—Lewiston, Ida., and Clarkston, Wash.

Figures on cattle, hogs and sheep indicate AVERAGE WEEKLY KILLING CAPACITY.

MOSCOW

Hagan & Cushing Co.—West 6th St. Railroad, Northern Pacific. A corporation. Capital, $100,000; issued, $75,000. Stockholders, 12. President, C. A. Hagan; Vice-President, C. B. Holt; Secretary-Treasurer, G. H. Cushing; General Manager, C. A. Hagan; General Superintendent, G. H. Cushing. Employes, 25. Cattle, 100; hogs, 1,000; sheep, 200. Sausage—Fresh and smoked. By-Products—Tankage, hog and chicken feed. Government inspection. Refrigeration—75-ton Alaskan; brine and direct expansion. Boiler capacity, 100 H. P. Trade Marks—"Idaho Pride." Retail Market at Moscow, Idaho.

NAMPA

H. H. Keim Co.—1118 1st St. Railroad, Oregon Short Line. Individual ownership, H. H. Keim. Employes, 5. Cattle, 20; hogs, 30-60; sheep, 10. Sausage—Fresh, smoked and summer. By-Products—Tankage. Refrigeration—5-ton Armstrong; brine circulating; one ton ice daily. Boiler capacity, 10 H. P. Trade Mark—"Bar & Stirrup Brand."

King's Market—12th Ave. S. Railroad, Oregon Short Line. Individual ownership. Cattle, 15; hogs, 30; sheep, 8. Sausage—Fresh and smoked. By-Products—Tankage and tallow. Refrigeration—4-ton York; direct expansion. Boiler capacity, 16 H. P.

POCATELLO

Idaho Packing Company—A corporation. Capital, $300,000; issued, $186,000. President, S. E. Brady; Vice-President, F. H. Paradise; Secretary, F. Bentley; Treasurer, George Green.

SANDPOINT

Bonner Meat Company—First Ave. and Pine St. Phil Willi, proprietor. Wholesale butchers and packers. Government inspection.

TWIN FALLS

Independent Meat Company.

WALLACE

Wallace Meat Company—A corporation. Capital, $100,000; issued, $61,600. Stockholders, 6. President, W. W. Papesh; Vice-President, A. D. Wallace; Treasurer and Secretary, F. F. Brewer; General Manager, A. D. Wallace. Employes, 20. Cattle, 100; hogs, 150; sheep, 150. Sausage—Fresh and smoked. By-Products—Tankage. Refrigeration—Two Fricks, 10 ton each; ¾-ton Cooper; Armstrong, 1½-ton capacity; all electric driven; direct expansion. Boiler capacity, 38 H. P.

ILLINOIS

ALTON

Luer Bros. Packing & Ice Company. A corporation. Capital, $35,000. President, August Luer; Secretary and Treasurer, Wm. J. Luer; Manager, Herman F. Luer; Assistant Manager, Carl A. Luer.

AURORA

Fox River Packing Co.—Railroad, C. B. & Q. A corporation. Capital, $175,000; issued, $146,000. Stockholders, 50. President, W. H. Fitch; Vice-President, L. Mighell; Secretary, Max L. Medauer; General Manager, L. B. Huff. Employes, 45. Cattle, 150; hogs, 300; sheep, 50. Sausage—Fresh and smoked. By-Products—Dried and baled hog hair, dried blood, white and brown grease, and dry bones. Refrigeration, 30-ton York; direct expansion. Boiler capacity, 250 H. P. Trade Marks—Hams, "Acorn" and "Valley"; bacon, "Acorn" and "Valley"; lard, "Acorn"; compound, "Valley."

Figures on cattle, hogs and sheep indicate AVERAGE WEEKLY KILLING CAPACITY.

BLOOMINGTON
Holcomb Packing Company.

CAIRO
E. Bucher Packing Co.—38th to 40th on Commercial St. Railroads, Big Four and Illinois Central. A corporation. Capital stock, $250,000; issued, $150,000. Stockholders, 14. President, E. Bucher; Vice-President, Joe Bucher; Secretary, Wilbur Thistlewood; General Manager, C. A. Claus. Employes, 50. Cattle, 150 weekly; hogs, 1,000 weekly. By-Products—Tankage and hog feed. Sausage—All kinds. Government inspection. Refrigeration—125-135-ton Triumphs; brine circulating. Commercial ice plant attached. Boiler capacity, 375 H. P. Trade Mark, "Daily Service."

CHICAGO
Agar Packing & Provision Co.—4057 S. Union St. A corporation. Capital, $1,500,000. President, James S. Agar; Vice-President and Treasurer, W. S. Agar; Secretary, John G. Agar. Cattle, 2,000; hogs, 10,000. Sausage and all by-products. Trade Mark—"Orelwood."

Armour and Company—Union Stock Yards. (See list of plants.) A corporation. Capital, $400,000,000; issued, $210,000,000. Stockholders, 15,000. President, J. Ogden Armour; Vice-Presidents, Chas. W. Armour, Arthur Meeker, A. Watson Armour, F. Edson White, E. A. Valentine, Laurance H. Armour, Philip D. Armour, F. W. Croll, F. W. Waddell; Secretary, George M. Willetts; Treasurer, Frederick W. Croll; General Superintendent, John E. O'Hern. Employes, 55,000. Capacity—all plants—Cattle, 135,150; hogs, 313,500; sheep, 214,800. Sausage—Fresh, smoked and summer. By-Products—All kinds. Government inspection. Refrigeration—20,417 tons daily; 1,442 tons ice capacity. Trade Marks—Star, Shield, Oval Label, Helmet, Melrose, Veribest, Cloverbloom, Melodia, Simon Pure, Vegetole, Nutola, White Cloud, etc. Branch houses in 380 cities and towns. Also exports.

Refrigeration Data by Plants

Make	Compression or Absorption	Steam or Motor Driven	Tons Rating	Total
Chicago				
De La Vergne	Compression	Steam	1,800	
Vilter	Compression	Steam	1,200	
Linde	Compression	Steam	800	
Carbondale	Absorption	Steam	750	
York	Compression	Steam	585	
Hercules	Compression	Steam	285	
Boschardt	Compression	Steam	150	
Frick	Compression	Steam	250	
Western	Compression	Steam	200	
Consolidated	Compression	Steam	200	
Consolidated	Compression	Motor	75	6,275
Kansas City				
Frick	Compression	Steam	1,080	
Ball	Compression	Steam	1,100	
York	Compression	Steam	400	
Hutchinson	Compression	Steam	140	2,720

Figures on cattle, hogs and sheep indicate AVERAGE WEEKLY KILLING CAPACITY.

Make	Compression or Absorption	Steam or Motor Driven	Tons Rating	Total
Sioux City				
York	Compression	Steam	500	
De La Vergne	Compression	Steam	400	900
East St. Louis				
Vilter	Compression	Steam	800	
De La Vergne	Compression	Steam	750	1,550
Fort Worth				
De La Vergne	Compression	Steam	1,750	
Ball	Compression	Steam	500	2,250
Omaha				
De La Vergne	Compression	Steam	500	
Frick	Compression	Steam	400	
Linde	Compression	Steam	400	
Vilter	Compression	Steam	600	1,900
South St. Paul				
Carbondale	Absorption	Steam	300	
Vilter	Compression	Steam	1,200	1,500
Fowler, Kansas City				
Ball	Compression	Steam	500	
Linde	Compression	Steam	150	
Riverside	Compression	Steam	120	770
South St. Joseph				
Ball	Compression	Steam	750	750
Denver				
Ball	Compression	Steam	250	
Frick	Compression	Steam	75	
Linde	Compression	Steam	100	425
New York				
Vilter	Compression	Steam	400	400
Jersey City				
Vilter	Compression	Steam	150	
York	Compression	Steam	175	325
Indianapolis				
Hercules	Compression	Motor	35	
Vilter	Compression	Steam	50	
York	Compression	Steam	40	125
Jacksonville				
Ball	Compression	Steam	150	
Frick	Compression	Steam	65	
York	Compression	Steam	35	250
Spokane				
Frick	Compression	Steam	277	277

PACKERS—ILLINOIS

	Tons Ref'g	Elec. Energy K.W. Per hr.	Boiler H.P.	Air M Cu. Ft.	Air Lift	Wells—Motor Driven	Total	Tons Ice	Average Fuel Burned		Total Equiv. Tons Coal	Electric Motors	
									Tons Coal	Bbls. Oil		No.	H.P.
Chicago—													
Chicago Plant	5390	6100	21250	16593	3200	1470	4670	250	750		728	780	11863
Freedman Mfg. Co	35											25	330
Glue Works	170	3200	6820	2750	900		900		277		268	356	4369
Anglo-American Prov	680	565	2200	290					60		56	37	730
Total—Chicago	6275	9865	30570	19633	4100	1470	5570	250	1087		1052	1198	17342
Kansas City, Kan.	2720	3750	8364	1950		2500	2500	450	242	290	310	350	5000
East St. Louis, Ill.	1550	1400	3581	1200		3400	3400	160	154		135	203	3802
Fort Worth, Texas	2250	500	3166	1193	924		924	150		508	147	191	3440
South Omaha, Neb.	1900	1350	4657	2730				240	272		252	251	3036
Sioux City, Iowa	900	850	2976	10500	2180	2230	4410		127		117	99	1614
St. Joseph, Mo.	750	370	2686	2200	1000	800	1800	100	68	76	83	112	1400
Denver, Colo.	425	225	1868	1234	1270		1270	60	68		58	60	815
New York, N.Y.	400	800	1468	338					*52		53	22	1072
Fowler, Kansas City	770	500	2229	300		500	500	67	32	105	60	49	1081
Jersey City, N.J.	325	130	1300	600					*41		43	25	349
Indianapolis, Ind.	125	60	280	136		200	200		12		12	25	420
Jacksonville, Fla.	250		450			500	500			76	21	37	415
Spokane, Wash.	277		459	385		400	400	25	*10		11	37	1044
S. St. Paul, Minn.	1500	2250	4016	3062		6300	6000		182		176	245	3260
Totals	20417	22050	67752	45461	9474	18000	27474	1442	2347	1055	2530	2906	44090

*Long ton

CHICAGO

Allied Packers, Inc.—General offices, 621 Postal Telegraph Bldg. A corporation. Capital, $25,000,000; issued, $6,071,000. President, J. A. Hawkinson; Vice-Presidents, A. M. Schenk, T. F. Matthews, F. S. Snyder, Robt. Shiell; Treasurer, F. R. Warton; Secretary, A. L. Arthur; General Superintendent, A. W. Cushman. Codes—Bentleys, Western Union, Cross-Robinson, A. B. C., 5th Imprv. and private codes. Employes, 4,000. Cattle, 2,500; hogs, 50,000; sheep, 3,000. Sausage—Fresh, smoked, summer. By-Products—Dried blood, tankage, hog and chicken feed, bones, hair, tallow, grease. Trade Marks—Hams, "Good Will"; bacon, "Good Will"; lard, "Silver Medal" and "Silver Bell"; compound, "Ladina," "Dixie," "Holsom," "Mohawk." Government inspection. Plants at Chicago, Buffalo, N. Y.; Richmond, Va.; Macon, Ga.; Wheeling, W. Va.; Detroit, Mich.; Topeka, Kas.; Toronto, Montreal, Hull, Brantford, Peterboro and Winnipeg, Canada.

Arnold Bros., Inc.—660 W. Randolph St. President, H. F. Arnold; Secretary, A. W. Ewers; Treasurer, L. T. Kelly. Codes—Cross. Government inspection.

Baker Food Products Company—49th and S. Halsted Sts. President and General Manager, F. G. Baker. Canners of meats and specialties.

Beiersdorf & Bro., J. R.—932 W. 38th Place. A corporation. Capital, $150,000. Stockholders, 3. President, J. W. Beiersdorf; Treasurer and General Manager, J. W. Beiersdorf; Secretary and General Superintendent, Wm. B. Carlisle. Employes, 84. Codes—Cross and Robinson. Hogs, 1,200. Sausage—Fresh and smoked. Government inspection. Refrigeration—15-ton Healy; brine circulating. Boiler capacity, 50 H. P. Trade Marks—"Colonial," "Bell" and "Mecca."

Boyd, Lunham & Company—Union Stock Yards.

Brennan Packing Co.—3916 Normal Avenue. Railroad, Chicago Junction. A corporation. Capital, $1,100,000; issued, $1,100,000. Stockholders, 220. President, B. G. Brennan; Vice-Presidents, C. E. Herrick and H. Boore; Secretary, J. M. Haughey; Treasurer, F. J. Brennan; General Manager, B. G. Brennan; General Superintendent, P. C. Peterson. Employes, 300. Codes—Domestic: Cross and Robinson; export: A. B. C., 5th edition, Bentleys, Liebers, Western Union, Baltimore Export, Utility and Warrington. Code Word, "Brennan." Hogs, 9,000. Government inspection. Refrigeration—200-ton Vilter, 1,000-ton De La Vergne, 150-ton Wolf; direct expansion; 300 tons ice daily. Boiler capacity, 900 H. P. Trade Mark, "Brennan." Also exports.

Chicago Butchers' Packing Co.—216-222 N. Peoria St. A corporation. Capital, $300,000; issued, $297,300. Stockholders, 126. President, Frank Zahrobsky; Vice-President, W. F. Jarosh; Secretary, Joseph Liska; Treasurer, Anton Camsky; General Manager, P. J. Smidl; Sales Manager, Frank Platsky; Manager Beef Killing Department, Albert Rus; Manager Poultry Department, Vlast Velflik. Employes, 65. Cattle, 150; hogs, 350. Sausage—Fresh, smoked and summer. Cattle slaughtered at Illinois Packing Co. and hogs at Western Packing & Provision Co. Refrigeration—Two 70-ton Wolfs; direct expansion. Boiler capacity, 500 H. P. Trade Mark, "I Will."

Figures on cattle, hogs and sheep indicate AVERAGE WEEKLY KILLING CAPACITY.

Chicago Packing Company—4535 Gross Ave. Railroad, Chicago Junction. A corporation. Capital, $500,000; issued, $500,000. Stockholders, 126. President, I. Katz; Vice-President, G. D. Diebschultz; Secretary, S. T. Katz; Treasurer, Howard Pearse; General Manager, I. Katz; General Superintendent, W. J. Lake. Employes, 139. Codes—Cross. Cattle, 1,500; hogs, 1,000; sheep, 600. Government inspection. Refrigeration—75-ton Frick and 25-ton De La Vergne; direct expansion. Boiler capacity, 750 H. P.

City Abattoirs, Inc.—3840 Emerald Ave. Commission slaughterers. President, Otto Lafrentz; Vice-President, Lawrence Wall; Secretary, Emil J. Nelson; Treasurer, Otto Lafrentz. Refrigeration—50-ton Vilter; brine spray system. Boiler capacity, 200 H. P.

Cudahy Packing Company—111 W. Monroe St., executive offices. A corporation. Capital, $35,000,000; issued, $25,800,000. Stockholders, 2,500. President, E. A. Cudahy; Vice-President and General Manager, E. A. Cudahy, Jr.; Vice-President, G. C. Shepard; Secretary, A. W. Anderson; Treasurer, J. E. Wagner; General Superintendent, R. E. Yocum. Employes, 15,000. Codes—Private and all principal codes: code word, "Cudahy." Cattle, 25,000; hogs, 100,000; sheep, 25,000. Sausage—Fresh, smoked and summer. By-Products—Dried blood, tankage and hog and chicken feed. Government inspection. Refrigeration—5,000 tons, Vilter and York; 400 tons ice daily. Boiler capacity, 15,000 H. P. Trade Marks—"Puritan," "Rex," "White Ribbon," "Red Star" and "Flako." Packing plants at Omaha, Neb.; Sioux City, Iowa; Wichita, Kansas; Los Angeles, Cal.; Salt Lake City, Utah; Jersey City, New Jersey, and Memphis, Tennessee. Branch Houses—110 in principal cities of country. Also exports.

William Davies Co., Inc.—U. S. Yards. Railroad, Chicago Junction. A corporation. Capital, 166,000 shares; issued, 166,000. President and General Manager, E. C. Fox; Vice-President and Secretary, J. T. Agar. Codes—All standard codes. Cattle, 1,000; hogs, 12,000. Government inspection. Refrigeration—160-ton Vilter and 200-ton Carbondale; brine spray and brine circulating. Boiler capacity, 600 H. P. Trade Marks—"North Star" and "Dialstone." Also exports. Branch Houses—New York City, Toronto and Montreal, Canada.

Foell Supply Company—331 W. 22nd Place. Packers of canned meats.

Fuhrman & Forster Company—1841 Blue Island Ave. A corporation. Capital, $250,000; issued, $250,000. Stockholders, 6. President and General Superintendent, John Fuhrman; Vice-President, Geo. Forster; Secretary and General Manager, Andrew Fuhrman; Treasurer, Geo. Forster. Employes, 85. Hogs, 600. Sausage—Fresh and smoked. Refrigeration—35-ton Healy and 25-ton Baker; direct expansion. Boiler capacity, 90 H. P. Trade Marks—"★★★" and "Select" brands.

Gabel Packing Company—221 N. Peoria St. A corporation. Capital, $345,000. President, A. E. Falker; Vice-President, E. M. Falker; Secretary, M. E. Zuckermann; Treasurer, H. Kolier, F. Blaurock. Employes, 60. Hogs, 800. Sausage—Fresh, smoked and summer. Trade Mark—"G."

Figures on cattle, hogs and sheep indicate AVERAGE WEEKLY KILLING CAPACITY.

CHICAGO

Graver, Weller, Sachs Company—3940 Normal Ave. Railroad, Chicago Junction. A corporation. Capital, $80,000; issued, $40,000. Stockholders, 8. President, H. P. Graver; Vice-President, M. Sachs; Secretary and General Manager, L. C. Blackburn; Treasurer, M. Weller. Cattle, 200; calves, 1,000.

Guggenheim Bros.—Union Stock Yards. Railroad, Chicago Junction. A corporation. Capital, $700,000; issued, $700,000. President, Fred Guggenheim; Vice-President, Max Guggenheim; Secretary, Max Guggenheim; Treasurer, Fred Guggenheim; General Manager, B. L. Kohn. General Superintendent, Jas. Gallagher. Employes, 190. Codes—Bentley, A. B. C., 5th edition and Liebers; code word, "Gubro." Cattle, 3,000; sheep, 3,000. By-Products—Dried blood, tankage, hog and chicken feed, tallows, hides, bones and cattle hair. Government inspection. Refrigeration—50-ton Vilter and 35-ton Frick; direct expansion. Boiler capacity, 450 H. P. Branches—1-3 Fulton Mkt., 323 S. Water St., 914-916 Fulton Mkt. and 20 Fulton Mkt., Chicago. Also exports.

G. H. Hammond Co.—Union Stock Yards. Railroad, Chicago Junction. A corporation. Capital, $5,000,000; issued, $4,500,000. Stockholders, 14. President, A. B. Swift; Vice-Presidents, A. N. Benn and C. F. Stephenson; Secretary, H. Mills; Treasurer, L. A. Carton; General Manager, A. N. Benn; General Superintendent, L. B. Whitmarsh. Employes, 2,100. Codes—Peterson, A. B. C., Liebers, Western Union; code word, "Rookham-Hammond." Cattle, 12,000; hogs, 30,000; sheep, 15,000. Sausage—Fresh and smoked. By-Products—Dried blood, tankage, hog and chicken feed. Government inspection. Refrigeration—1 Linde, 1 Vilter, 2 Frick, 1 Ball and 1 Ball Booster, 1,625 tons capacity; direct expansion. Boiler capacity, 3,500 H. P. Trade Marks—Hams, "Rosebud," "Calumet," "Marca," "Hecla," "Herald"; bacon, "Rosebud," "Calumet," "Famous," "Champion," "Southern Style," "Yale"; lard, "Rosebud," "White Star"; compound, "Crescent," "Goldcrisp"; oleomargarine, "Hammond Special," "Mistletoe," "Golden Glow," "Coin Special," "Banquet," "Calumet," "White Rosebud," "Lily White," "White Star," "White Flyer," "Ko-Ko," "Olean," "Krumble Krust"; butter, "Famous"; soap, "Mohawk," "Friendship," "White Birch," "Thistle CoCo," "Elko White," "Floating." Branches—Wholesale Market, 45th and Racine Ave., Chicago; G. H. Hammond Co., 77 Hudson St., New York City; Vermont Supply Co., Boston, Mass. Also exports.

Hately Brothers Co.—37th and Iron Sts. Railroad, Chicago Junction. A corporation. Capital, $850,000; issued, $850,000. President, John A. Bunnell; Vice-Presidents, A. E. Hayes and H. G. Newhall; Secretary, A. E. Hayes; Asst. Secretary, P. Rothermel; Treasurer, H. G. Newhall; General Superintendent, E. T. Miller. Employes, 400. Codes—Cross and Robinson, A. B. C., Bentley, Lieber and private. Cattle, 75; hogs, 6,000, calves, 300. Government inspection. Refrigeration—125-ton Wolf, at Chicago, and two 25-ton Vilter at New Richmond, Wis.; brine and direct expansion. Boiler capacity, 750 H. P. Trade Marks—Hams, "Invincible"; bacon, "Invincible"; lard and others, "Diamond HP Brand." Sausage—Fresh, smoked and summer. Branch plant—New Richmond, Wisconsin. Also exports.

Illinois Meat Company—3939 Wallace St.

Figures on cattle, hogs and sheep indicate AVERAGE WEEKLY KILLING CAPACITY.

Illinois Packing Co.—911-37 W. 37th Place. A corporation. Capital, $600,000. Stockholders, 45. President, Nickolas Wolter; Vice-Presidents, H. S. Siegel and H. J. Cramer; Secretary, H. C. Levinson; Treasurer, C. Loeffler; General Manager, Nickolas Wolter; General Superintendent, W. F. Krippes; Manager Offal Department, Frank Loeffler. Employes, 125-130. Cattle, 1,800; sheep, 500. By-Products—Dried blood, blood, offal, glands, and various laboratory products, and cured hides. Government inspection. Refrigeration—80-ton Vilter; brine spray system. Boiler capacity, 325 H. P. Branch Houses—213 Green Street, Chicago, Ill.

Independent Packing Co.—41st and Halsted Sts. Railroad, Chicago Junction. A corporation. Capital, $6,000,000; issued, $2,500,000. President, Patrick Brennan; Vice-President, Thomas V. Brennan; Secretary, Thomas E. Ryan; Treasurer, Patrick Brennan; General Manager, Patrick Brennan. Employes, 800. Codes—Cross, A. B. C. and Bentley; code word—"Inpaco." Cattle, 1,000; hogs, 8,000; sheep, 500. Sausage—Fresh and smoked. By-Products—Dried blood, tankage, hair, bones, grease, tallow, etc. Government inspection. Refrigeration—Ball, Vilter and Featherstone with total capacity of 800 tons; brine circulating and direct expansion. Boiler capacity, 1,300 H. P. Trade Marks—Hams, "Finest" and "Select"; bacon, "Finest" and "Select"; lard, "Forget-me-not." Also exports.

Jourdan Packing Co.—A corporation. Capital, $250,000; issued, $250,000. Stockholders, 7. President, George Jourdan; Vice-President, William Jourdan; Secretary and Treasurer, Louis Jourdan; General Manager, Louis Jourdan; General Superintendent, William Jourdan. Employes, 85. Refrigeration—75-ton Huetteman-Cramer and 25-ton Wolf-Linde; brine circulating and direct expansion; 4 tons ice daily. Boiler capacity, 300 H. P. Trade Marks—Hams, "Rose"; bacon, "Rose"; lard, "Rose"; compound, "Roswite"; others, "Clover." Retail Market—814 W. Cullerton Street, Chicago, Ill.

David Levi & Co., Inc.—3900 Emerald Ave. Railroad, Chicago Junction. President, Henry Levi; Vice-President, Richard Levi; Secretary, Jerome Levi; Treasurer, Lawrence H. Loeb; General Manager, Richard Levi; General Superintendent, Lawrence H. Loeb. Cattle, 2,500; sheep, 5,000. Government inspection. Refrigeration—75-ton Carbondale; brine spray system.

Libby, McNeill & Libby—Union Stock Yards. Railroad, Chicago Junction. A corporation. Capital, $27,000,000; issued, $27,000,000. Stockholders, 24,000. President, E. G. McDougall; Vice-President, H. C. Carr; Secretary, Harry Williams; Treasurer, H. W. Hardy. Employes, 8,000. Codes—Private; code word, "Libby." Canners of food products. Government inspection. Branch Houses—45, in all large cities in the United States, England, Belgium and France.

Maier & Co.—2863 Lincoln Ave. A corporation. Capital, $65,000; issued, $65,000. Stockholders, 3. President, Geo. C. Maier; Secretary, Chas. F. Glaeser; Treasurer, Geo. C. Maier; General Manager, Geo. C. Maier; General Superintendent, Fremont C. Miller. Employes, 40. Hogs, 150. Sausage—Fresh, smoked and summer. Refrigeration—45-ton Almeroth and 18-ton Robsomen; direct expansion. Boiler capacity, 150 H. P. Trade Mark—"Golden Oak Brand." Retail Markets—One at 2863 Lincoln Ave.

Figures on cattle, hogs and sheep indicate AVERAGE WEEKLY KILLING CAPACITY.

CHICAGO

Manaster, Harry, & Bro.—1018 West 37th St. Railroad, Chicago Junction. A corporation. Capital, $100,000; issued, $100,000. Stockholders, 3. President and Treasurer, Harry Manaster; Vice-President, Henry Manaster; Secretary, A. G. Newman. Employes, 70. Codes—Cross. Do not slaughter, but do regular packinghouse business. Government inspection. Refrigeration—50-ton Vilter; direct expansion.

Oscar Mayer & Co.—1241 Sedgwick St. A corporation. Capital, $4,200,000; issued, $2,400,000. Stockholders, 600. President, Oscar F. Mayer; Vice-President, Max Mayer; Secretary, Oscar G. Mayer; Treasurer, Oscar F. Mayer; General Manager, Oscar G. Mayer. Employes, 650. Codes—A. B. C., 5th edition and Cross; code word, "Ofmabro." Cattle, 600; hogs, 12,000; sheep, 600. Sausage—Fresh, smoked and summer. By-Products—Digester tankage, white grease, yellow grease, prime tallow, cooked bones, casings and dried hair. Government inspection. Refrigeration—150-ton Vilter, 150-ton York and 75-ton Wolf-Linde; direct expansion and brine spray; 5 tons ice daily. Boiler capacity, 1,200 H. P. Trade Marks—Hams, "Approved" and "Moose"; bacon, "Approved," "Moose," and "Capital"; lard, "Approved," "Moose," and "White Clover"; compound, "Approved." Branches—Madison, Wis., and New York City. Also exports.

Omaha Packing Co.—2320 South Halsted St. Railroad, Chicago, Burlington & Quincy. A corporation. Capital, $500,000; issued, $500,000. President, F. R. Burrows; Vice-President, H. H. Moore; Secretary, C. A. Peacock; Treasurer, W. W. Sherman; General Manager, J. E. Glen; General Superintendent, F. W. Menge. Employes, 850. Hogs, 17,000. Sausage—Fresh, smoked and summer. Government inspection. Refrigeration—400-ton Wolf-Linde; brine circulating, direct expansion and brine spray systems. Boiler capacity, 1,400 H. P. Trade Marks—"Circle U. Underwood Brand," and "Yale." Also exports.

Morris & Company—Union Stock Yards. (See list of plants.) A corporation. Capital, $51,000,000; issued, $40,000,000. Chairman Board of Directors, Nelson Morris; President, Edward Morris; Vice-Presidents, C. M. Macfarlane, L. H. Heymann and H. A. Timmins; Secretary, T. R. Buckham; Treasurer, C. M. Macfarlane; General Superintendent, W. B. Farris. Employes, 15,000. Codes—Private, A. B. C. 5th, Lieber's Standard, Bentley's Phrase, Western Union, Scott's 10th Edition. Cable address, "Morrisco, Chicago." Cattle, 33,850; hogs, 138,000; sheep, 60,600. All packing house products, including all fresh and frozen meats, dry salt meats, smoked meats. Sausage—Fresh, smoked and summer. Canned meats, lard, compound, oils, tallows, stearines, etc. By-Products—Dried blood, tankage, hog and chicken feed, winter-strained lard oil, neatsfoot oil, fertilizer, blood albumen, eggs, poultry, butter, cheese, glue, curled hair, mince meat, etc. Branch houses in 200 cities and towns in the United States. Trade Marks—All cured products and canned meats, "Supreme"; lard, "White Leaf"; compound, "Purity." Also exports. Export houses: Morris & Company (meat importers), Ltd., London, England; Morris & Company (meat importers), Ltd., Liverpool, England, and branches in all principal English cities; Morris Packing Company, Antwerp, Belgium; Morris Packing Company, Christiania, Norway; Morris Packing Company, Rotterdam, Holland; Morris

---

Figures on cattle, hogs and sheep indicate AVERAGE WEEKLY KILLING CAPACITY.

Packing Company, Berne, Switzerland; Morris Packing Company, Hamburg, Germany; Morris Packing Company, Paris, France; Morris & Company, Havana, Cuba.

Refrigeration Data by Plants:

Make	Compression or Absorption	Steam or Motor Driven	Tons Rating	Total
Chicago				
4 Boyle	Compression	Steam	300	
2 Vilter	Compression	Steam	300	
1 Vilter	Compression	Steam	350	
1 Vilter	Compression	Steam	450	
1 Ball	Compression	Steam	350	
1 Penn.	Compression	Steam	350	2,250
Kansas City				
3 Vilter	Compression	Steam	900	
1 Penn.	Compression	Steam	150	1,150
Oklahoma City				
1 Vilter	Compression	Steam	250	
1 Vilter	Compression	Steam	250	
1 Vilter	Compression	Steam	250	750
South Omaha				
1 Vilter	Compression	Steam	450	
1 Vilter	Compression	Electric	325	
1 Wolf-Linde	Compression	Steam	150	925
St. Joseph				
2	Compression	Steam	300	
1	Compression	Steam	150	
1	Compression	Steam	50	500
St. Louis				
2 Ball	Compression	Steam	500	
1 Ball	Compression	Steam	100	
1 Featherstone	Compression	Steam	100	
2 Boyle	Compression	Steam	150	850

Louis Pfaelzer & Sons—3927 S. Halsted St. Railroad, Chicago Junction. A corporation. President, David Pfaelzer; Vice-President, Abe Pfaelzer; Secretary and Treasurer, J. L. Pfaelzer; General Superintendent, J. A. Kircher; Office Manager, Geo. Strachan; Credit Department Manager, John Falvey. Codes—Cross, A. B. C. and Bentley. Cattle, 2,500; hogs, 3,500; sheep, 3,000. Sausage—Fresh, smoked and summer. By-Products—Dried blood, tankage. Government inspection. Refrigeration—Two 150-ton Wolfs; brine circulating and direct expansion. Boiler capacity, 750 H. P. Trade Mark—"Sun." Also exports.

E. K. Pond Company—515 W. 24th Place. A corporation. Capital, $125,000; issued, $125,000. Stockholders, 4. President, W. S. Johnston; Secretary and Treasurer, F. Templeton; General Manager, G. C. Case. Employes, 80. Meat canners. Government inspection.

Figures on cattle, hogs and sheep indicate AVERAGE WEEKLY KILLING CAPACITY.

CHICAGO

Roberts & Oake—45th and Racine Ave. Railroad, Chicago Junction. A corporation. President, C. J. Roberts; Vice-President, G. T. McClean; Secretary, H. E. Rosner; Treasurer, C. J. Roberts. Employes, 450. Hogs, 9,000. Sausage—All kinds. By-Products—Tankage. Government inspection. Refrigeration—One 200-ton De La Vergne and one 85-ton Featherstone; direct expansion. Boiler capacity, 1,050 H. P. Trade Marks—"Liberty"; "Quality"; "Puritan"; "Irish"; "Glenwood"; "Virginia"; "Oakleaf" and "Butchers' Favorite." Also exports.

Rose Packing Co.—3940-58 Normal Ave. Railroad, Chicago Junction. Individual ownership. General Manager, L. A. Rose; General Superintendent, G. V. Rose. Code used—Cross. Cattle, 1,000; calves, 1,000; sheep, 500. By-Products—Dried blood and tankage. Government inspection. Refrigeration—40-ton Wolf-Linde and 30-ton Healy; direct expansion. Boiler capacity, 200 H. P.

Standard Packing Co.—3911-3919 S. Halsted St. Partnership. General Manager, Melvin F. Hirsch. Employes, 15. Cattle, 150; hogs, 100; sheep, 1,000; calves, 800. Refrigeration, 20-ton Reliance; direct expansion. Abattoir for cattle, calves, hogs and sheep.

Sterling Packing Company—3114-20 Shields Ave. A corporation. Capital, $100,000; issued, $100,000. Stockholders, 4. President and General Manager, W. A. Merryweather; Vice-President, B. R. Merryweather; Secretary, M. V. Crosby; Treasurer, C. M. Morehouse. Employes, 30. Do not kill. Meat canners. Sell only to jobbers. Government inspection. Trade Mark—"Pound Sterling." Use ice for refrigeration. Also export.

Strasser Packing Co.—1446-1452 W. 47th St. Individual ownership. Employes, 20. Hogs, 5,000. Retail Market—One at plant.

Swift & Company—U. S. Yards. (See list of plants.) A corporation. Capital, $150,000,000; issued, $150,000,000. Stockholders, 50,000. President, L. F. Swift; Vice-Presidents, Edward F. Swift, Charles H. Swift, G. F. Swift, Jr., and Harold H. Swift; Secretary, C. A. Peacock; Treasurer, L. A. Carton; Consulting Superintendent, F. J. Gardner; General Superintendent, J. Burns. Employes, 60,000. Codes—Swift & Company. Capacity all plants—Cattle, 90,000; hogs, 400,000; sheep, 225,000. Products—Dressed beef, veal, pork and mutton; cured and smoked meats, lard and all packinghouse products; dressed poultry, eggs, butter and cheese. Sausage—Fresh, smoked and summer. By-Products—Dried blood, tankage, hog and chicken feed, digester tankage, glue, gelatine, oleo oil, tallow, grease, soap, bones, horns, hoofs, casings, hides, pelts, hog hair and wool. Government inspection. Refrigeration, all plants—18,495 tons, Ball, Vilter and Wolf machines; direct expansion, 720 tons ice daily. Boiler capacity, 72,600 H. P.

Location of slaughtering plants: Andalusia, Ala.; Denver, Colo.; Moultrie, Ga.; National Stock Yards (East St. Louis), Ill.; Chicago, Ill. (3 plants); Sioux City, Iowa; Kansas City, Kan.; Boston, Mass.; South St. Paul, Minn.; St. Louis and South St. Joseph, Mo.; South Omaha, Neb.; Harrison, Jersey City and Newark, N. J.; New York City, N. Y.; Cleveland, O.; Portland, Ore.; Harrisburg, Pa.; Fort Worth, Tex., and Milwaukee, Wis.

Figures on cattle, hogs and sheep indicate AVERAGE WEEKLY KILLING CAPACITY.

Branch Houses (Wholesale Markets)—Alabama, 13; Arkansas, 5; Arizona, 3; California, 5; Colorado, 5; Connecticut, 12; Delaware, 1; District of Columbia, 1; Florida, 4; Georgia, 9; Idaho, 4; Illinois, 11; Indiana, 2; Iowa, 5; Kentucky, 1; Louisiana, 6; Maine, 12; Maryland, 2; Massachusetts, 29; Michigan, 14; Minnesota, 6; Missouri, 3; Mississippi, 11; Montana, 5; Nebraska, 2; Nevada, 2; New Hampshire, 8; New Jersey, 18; New Mexico, 3; New York, 29; North Carolina, 16; North Dakota, 2; Ohio, 6; Oklahoma, 3; Oregon, 5; Pennsylvania, 26; Rhode Island, 4; South Carolina, 9; South Dakota, 1; Tennessee, 5; Texas, 34; Utah, 1; Vermont, 8; Virginia, 11; Washington, 4; West Virginia, 6; Wisconsin, 9. Total, 381. Also exports.

U. S. Packing Company—3114 Shields Ave. A corporation. Capital, $100,000; issued, $100,000. Stockholders, 3. President, C. M. Morehouse; Vice-President, M. R. Crosby; Secretary and General Manager, F. F. Crosby; Treasurer, C. M. Morehouse. Employes, 18. Meat canners. Government inspection. Trade Mark—"Durham."

Wilson & Co., Inc.—4100 S. Ashland Ave. (See list of plants.) Railroad, Chicago Junction. A corporation. Capital, preferred, $12,000,000; common, 500,000 shares no par value; issued, $10,328,600; 202,500 com. shares no par value. President, Thomas E. Wilson; Vice-Presidents, V. D. Skipworth, Geo. H. Cowan, A. E. Petersen, A. Lowenstein, and J. Moog; Secretary, Geo. D. Hopkins; Treasurer, W. C. Buethe; General Superintendent, Seward C. Frazee. Employes, 15,060. Codes—Private. Cable Address—"Wilson-Comp." Capacity—all plants—Cattle, 29,500; hogs, 105,000; sheep, 48,500. Sausage—Fresh, smoked and summer. By-Products—Dried blood, tankage, hog and chicken feed, and all other allied to the meat packing industry. Government inspection. Trade Marks: Certified, Laurel, Advance, Economy, Wilsco, Clearbrook, Lakeview, Majestic, Oakwood.

Branch Houses—Alabama, 1; Arkansas, 2; Connecticut, 3; Delaware, 1; District of Columbia, 1; Florida, 3; Georgia, 5; Illinois, 7; Louisiana, 3; Maine, 3; Maryland, 1; Massachusetts, 8; Michigan, 3; Minnesota, 3; Missouri, 1; Nebraska, 1; New Hampshire, 1; New Jersey, 5; New York, 20; Ohio, 3; Pennsylvania, 13; Rhode Island, 1; South Carolina, 2; Tennessee, 3; Texas, 4; Vermont, 1; Virginia, 4; West Virginia, 3. Also exports.

POWER AND REFRIGERATION

Location of Plant	Cap. of Ice Mach. in Tons of Refrig.	Cap. of Electr. Generators in K. W.	Cap. of Boilers in Rated H. P.	Cap. of Wells in Cu. Ft. Water pr. 24 Hrs.	Tons of Ice. Mfd. per Week	Aver. Tons of Coal Burned per Week	Total H. P. of Electr. Motors
Chicago	2,020	2,533	6,052	2,250,000	1,400	2,500	5,690
Kansas City	1,150	450 Buy 75% of Current	3,500	1,700,000	1,000	1,260	1,955
Oklahoma City	800	1,300	3,200	925,000	700	825	1,600
New York	400	600	2,100	None	140	560	790
Los Angeles	200	Buy all Current	600	2,100,000	140	290	1,163
Albert Lea	200	Buy all Current	536	95,000	Natural Ice	300	468
Wheeling	250	200	1,016	30,000	350	265	508
Chatham, Ont., Can.	100	Buy all Current	200	Wells not used	Natural Ice	55	385
Nebraska City	300	425	1,536	None	Natural Ice	345	581

Figures on cattle, hogs and sheep indicate AVERAGE WEEKLY KILLING CAPACITY.

CHICAGO

Vette & Zuncker—220 N. Green St. Partnership. General Manager, P. M. Zuncker. Employes, 100. Hogs, 250. Kill at Western Packing & Provision Company. Sausage—Fresh, smoked and summer. Government inspection. Refrigeration—Two Wolf-Lindes, 100 tons; direct expansion. Boiler capacity, 250 H. P. Trade Marks—"Winner" and "Cabin." Also exports.

Wimp Packing Co.—1127 W. 47th Place. Railroad, Indiana Harbor Belt. A corporation. Capital, $200,000; issued, $200,000. Stockholders, 3. President, Roy Wimp; Vice-President, Elgy Wimp; Secretary, Elgy Wimp; Treasurer, Roy Wimp; General Manager, M. J. Hillon; Superintendent, M. J. Hillon. Cattle, 875; calves, 125; sheep, 500. Sausage—Fresh, smoked and summer. By-Products—Dried blood, tankage and hog and chicken feed. Government inspection. Refrigeration—25-ton Howe and 25-ton York; brine spray system. Boiler capacity, 250 H. P.

CLINTON

Rundle Bros.

DANVILLE

Campbell Bros. Company—A corporation. Capital, $75,000. President, J. B. Campbell; Secretary and Treasurer, J. Bert Phillips; General Manager, B. F. Campbell.

Ernst Packing Company—Perryville Ave. Partnership. General Manager, H. A. Ernst; General Superintendent, Andrew Hampton. Employes, 12. Cattle, 100; hogs, 250. Sausage—Fresh, smoked and summer. By-Products—Tankage, hog and chicken feed, tallow and grease. Use ice for refrigeration. Boiler capacity, 35 H. P.

DECATUR

Danzeisen Packing Co.—750 S. Main St. Railroad, Illinois Central. A corporation. Capital, $50,000; issued, $50,000. Stockholders, 3. President, G. J. Danzeisen; Vice-President, Wm. Danzeisen; Secretary and Treasurer, O. J. Danzeisen; General Manager, Wm. Danzeisen; General Superintendent, O. J. Danzeisen; Slaughtering, Fred Holpp; Sausage Department, H. Michel; Curing, Lee Durbin; Rendering, J. Chappell; Chief Engineer, J. Mildenberger. Employes, 40. Code—Robinson. Cattle, 100; hogs, 600; sheep, 25. Sausage—Fresh, smoked and summer. By-Products—Dried blood, digester tankage. Refrigeration—55-ton York, 17-ton Westinghouse; brine circulating; one ton ice daily. Boiler capacity, 450 H. P. Trade Marks—Hams, "Blue Ribbon"; bacon, "Blue Ribbon"; lard, "Monogram Brand"; "Danzeisen's Digester Tankage."

EAST ST. LOUIS

Armour & Company. (See Chicago.)

Circle Packing Co.

John Cole Packing Co.

Morris & Co.—(See Chicago.)

Sheehan Packing Co.

Swift & Company—(See Chicago.)

United Packing Co.—Second St. and Exchange Ave.

Figures on cattle, hogs and sheep indicate AVERAGE WEEKLY KILLING CAPACITY.

EAST ST. LOUIS

East Side Packing Co.—1250 N. 2nd St. Railroad, P. C. C. & St. L. A corporation. Capital, $1,020,000; issued, $850,000. Stockholders: President, Frank A. Hunter; Vice-Presidents, W. L. Hadley and Chas. I. Coy; Secretary, E. J. Delmore; Treasurer, Joseph Nebel; General Manager, Frank A. Hunter; General Superintendent, John Dreier; Plant Manager, Joseph Nebel. Employes, 350. Codes—Cross. Cattle, 500; hogs, 6,000. Sausage—Fresh, smoked and summer. By-Products—Tankage. Government inspection. Refrigeration—One 350-ton Frick and one 250-ton Ball; brine circulating, direct expansion and brine spray systems. 200 tons ice daily. Boiler capacity, 1,250 H. P. Trade Marks—"Trophy," "I. X. L." and "East Side." Also exports.

ELGIN

Kerber Packing Company—56 Grove Ave. A corporation. Capital, $430,000; issued, $380,000. President and Manager, C. A. Kerber; Vice-President, J. A. Russell; Secretary and Treasurer, Alice Kerber.

JACKSONVILLE

Powers-Begg & Co.—Johnson and Center Sts. Railroad, Wabash. A corporation. Capital, $150,000; issued, $150,000. Stockholders, 7. President, Isaac S. Powers; Secretary, Fred Begg; Treasurer, Fred Begg. Employes, 62. Codes—Cross and Robinson. Cattle, 200; hogs, 2,000. Sausage—Fresh and smoked. By-Products—Tankage. Government inspection. Refrigeration—50-ton Frick; brine circulation. Boiler capacity, 300 H. P. Also exports.

MADISON

Frangoulis, John G.—1039 Greenwood St.
Illinois Packing Co.

MILAN

Milan Produce Co.—Railroad, C. R. I. & P. A corporation, Capital, $100,000; issued, $88,000. Stockholders, 14. President, Matthew Plunkett; Secretary, Lee Plunkett; Treasurer, Matthew Plunkett; General Manager, C. A. Williamson; General Superintendent, M. F. Mullins. Employes, 22. Cattle, 100; hogs, 200; sheep, 50. Sausage—Fresh, smoked and summer. By-Products—Tankage. Refrigeration—25-ton York; brine circulating. Boiler capacity, 60 H. P. Trade Mark, "Rock River Brand." Retail Markets—4 in Rock Island, Ill.

MOUNT CARMEL

Wabash Valley Packing Co.

OTTAWA

Illinois Farmers' Packing Co.—First and Mulberry Sts. Railroad, C. R. I. & P. A corporation. Capital, $750,000; issued, $600,000. Stockholders, 1,667. President, John Eckhart; Vice-Presidents, Geo. A. Broman and Elmer Quinn; Secretary and Treasurer, Edw. T. Ryan; General Superintendent, Arthur Krause. Employes, 65. Code—Cross. Cattle, 100; hogs, 1,200. Sausage—Fresh and smoked. By-Products—Tankage. Government inspection. Refrigeration—Two 15-ton Larson and one 20-ton York. Trade Mark—"Golden Rod Brand." Retail Markets—Ottawa, Peru, Princeton, Geneseo and Streator, Illinois.

QUINCY

Quincy Packing Co.

ROCK FALLS

Pippert Bros.

Figures on cattle, hogs and sheep indicate AVERAGE WEEKLY KILLING CAPACITY.

PEORIA

Godel, E., & Sons, Inc.—A corporation. Capital, $100,000. President, E. J. Cashin; Vice-President, Phoebe B. Buehler; Secretary and Treasurer, H. D. Freeman.

McDonough Packing Company—Foot of Sanger St. Partnership, Edward J. McDonough and Chas. T. McElwee.

Wilson Provision Co.—Foot of South St. Railroad, Peoria & Pekin Union. A corporation. Capital, $200,000; issued, $200,000. Stockholders, 5. President, Everett W. Wilson; Vice-President, E. R. Wilson; Secretary, Wm. F. Schmidt; Treasurer, F. L. Wilson; General Manager, F. L. Wilson; General Superintendent, H. F. Siebert; Sales Manager, J. L. Richart. Employes, 175. Code—Cross; code word, "Wilsonrob." Cattle, 350; hogs, 3,500. Sausage—Fresh and smoked. By-Products—Dried blood and tankage. Government inspection. Refrigeration—85-ton Vilter and 60-ton Hercules; brine circulating and direct expansion. Boiler capacity, 450 H. P. Trade Marks—Hams, "Premier" and "Ideal"; bacon, "Premier, "Ideal" and "Oxford"; lard, "Premier," "Ideal" and "Oxford." Branches—1545 Winder St., Detroit, Mich. Also exports.

ROCK ISLAND

M. R. Thackaberry Co.

INDIANA

ANDERSON

Anderson Dressed Beef Company—Partnership. Charles W. Phillips and Wm. Grancey.

Hughes-Curry Packing Company.

BATESVILLE

Michael Benz.

CONNERSVILLE

John Ringloff.
Geo. E. Brown—R. R. No. 9.

CLINTON

Clinton Packing Co.

DECATUR

Fred Mutschler Packing Co.—Partnership. Employes, 25. Cattle, 25; hogs, 200. Sausage—Fresh, smoked and summer. By-Products—Tankage. Refrigeration—25-ton Wolf-Linde; brine circulating; ten tons ice daily. Boiler capacity, 300 H. P. Retail Market—219 W. Monroe St., Decatur, Indiana.

ELWOOD

J. P. Downs.
Elwood Packing Co. (Also own Federal Packing Co., Anderson, Indiana.)

EVANSVILLE

Henry Daudistel. Capital, $25,000.

Evansville Packing Co.—Morgan Ave. and Harriet St. Railroads, Chicago & Eastern Illinois and Belt. A corporation. Capital, $300,000; issued, $300,000. Stockholders, 10. President, M. Mannheimer; Vice-President, Lee Rosenbaum; Secretary, Leon Siegel; Treasurer, M. Mannheimer; Assistant Treasurer, Wm. A. Michel; General Mana-

---

Figures on cattle, hogs and sheep indicate AVERAGE WEEKLY KILLING CAPACITY.

ger, M. Mannheimer; General Superintendent, J. Henry Michel. Employes, 150. Codes—Cross and Robinson; cable code word, "Epacko." Cattle, 500; hogs, 7,500; sheep, 1,000. Sausage—Fresh, smoked and summer. By-Products—Dried blood, tankage, hog and chicken feed. Government inspection. Refrigeration—190-ton York; direct expansion; six tons ice daily. Boiler capacity, 850 H. P. Trade Marks—Hams, "Smile-Boy"; bacon, "Smile-Boy"; lard, "Smile-Boy" and "Hoosier"; sausage, "Smile-Boy." Branch Houses—Birmingham, Ala.; Montgomery, Ala., and Atlanta, Ga. Also exports.

Newman's Pork Houses, Inc.—209 3rd Ave. Railroad, Belt. A corporation. President, Samuel G. Newman, Jr.; General Superintendent, Russell Newman. Employes, 10. Hogs, 300. By-Products—Tankage and grease. Refrigeration—10-ton Creamery Package and 8-ton Barber; direct expansion. Boiler capacity, 20 H. P. Trade Marks—"Cherry Red" and "Our Own Make, Absolutely Pure Hog Lard." Retail Markets—Five in Evansville.

Schmadel Packing & Ice Company—Fulton Ave. Railroad, Illinois Central. A corporation. Capital, $250,000; issued, $70,000. Stockholders, 4. President, Henry Schmadel; Vice-President and General Manager, John Schmadel; Secretary and Treasurer, Tony Schmadel; General Superintendent, Henry Schmadel. Employes, 40. Cattle, 150; hogs, 360; sheep, 60. Sausage—Fresh and smoked. By-Products—Dried blood and tankage. Refrigeration—Three 120-ton Frick; direct expansion. 35 tons ice daily. Boiler capacity, 450 H. P.

FORT BRANCH

Peter Emge & Sons—Partnership. Stockholders, 3. President, Peter Emge; Junior Partners, Oscar Emge and Ralph Emge. Employes, 15. Cattle, 75; hogs, 300. Sausage—Fresh and smoked. By-Products—Tankage and hog feed. Refrigeration—One 10-ton York and one 15-ton Automatic. Boiler capacity, 50 H. P. Retail Markets—One at Fort Branch.

FORT WAYNE

Fred Eckart Packing Company—1825 W. Main St. A corporation. Capital, $100,000. President, Fred Eckart; Secretary, Treasurer and Manager, Henry Eckart. Retail Market at 1827 W. Main Street.

Laurents & Hartshorn Packing Co.—Dwenger Ave. Railroad, Nickel Plate. Partnership, Alexander Laurents and C. E. Hartshorn. Employes, 32. Cattle, 100; hogs, 500; sheep, 100. Sausage—Fresh and smoked. By-Products—Tankage and hog feed. Refrigeration—25-ton York, 35-ton Frick; brine circulating and direct expansion. Boiler capacity, 300 H. P.

GARY

Tittle Bros. Packing Co.

INDIANAPOLIS

Armour & Company—W. Ray and Dakota Sts. (See Chicago.)

Crescent Packing Co.—Gardner's Lane and White River.

F. Hilgemeier & Bro.—519 W. Raymond St.

Indianapolis Dressed Beef & Provision Co.—Railroad, Illinois Central. Cattle, 200; hogs, 500; calves and lambs, 100.

Moore & Co.—Union Stock Yards.

Figures on cattle, hogs and sheep indicate AVERAGE WEEKLY KILLING CAPACITY.

INDIANAPOLIS

Brown Brothers—534 West Ray St. Railroad, Pennsylvania. A corporation. Capital, $200,000; issued, $150,000. President, Sam Brown; Vice-President, A. B. Brown; Secretary, Edwin C. Weir; Treasurer, L. J. Brown. Employes, 70. Cattle, 500; hogs, 400. Sausage—Fresh and smoked. By-Products—Tankage, hog and chicken feed. Government inspection. Refrigeration—75-ton Vilter, 35-ton Wilson; brine circulating and direct expansion. Boiler capacity, 300 H. P. Trade Mark—"Wild Rose."

Indianapolis Abattoir Co.—Morris and Drover Sts. Railroad, Indianapolis Union. A corporation. Capital and surplus, $2,000,000. President, Thomas Mooney; Vice-Presidents, Henry Rauh, E. C. Merritt and W. H. Allerdice; Secretary, W. G. Axt; Assistant Secretary, T. J. Killilea; Treasurer, J. A. Thompson; Assistant Treasurer, D. W. Allerdice; General Manager, E. C. Merritt; General Superintendent, W. H. Allerdice. Employes, 1,000. Codes—Cross, Robinson, Utility and private. Cattle, 3,000; hogs, 15,000; sheep, 3,000. Sausage—Fresh, smoked and summer. By-Products—Tankage, tallow, grease, oleo oil, oleo stearine, hides, hog hair and casings. Government inspection. Refrigeration—Brine and direct expansion. Trade Marks—"Favorite." Branch Houses—New York City, Rochester, N. Y., and New Haven, Conn.

Kingan & Co., Inc.—Blackford and Maryland Sts. Railroad, White River (Private terminal). A corporation. Capital, $5,500,000; issued, $4,923,520. President, James S. Reid; Vice-Presidents, John M. Shaw, John R. Kinghan and William R. Sinclair; Secretary, A. M. McVie; Treasurer, Robert S. Sinclair; General Superintendent, Wm. H. Patrick. Employes, 3,500. Codes—Bentleys, Cross and private; code word—"Kingans Indianapolis." Cattle, 4,000; hogs, 45,000; sheep, 2,000. Sausage—Fresh, smoked and summer. By-Products—Tankage and hog and chicken feed. Government inspection. Refrigeration—One 400-ton Ball, 8 Balls totaling 350 tons, one 300-ton York, one 150-ton Frick, one 100-ton Ball and one 50-ton Frick; brine circulating, brine spray and direct expansion. 90 tons ice daily. Boiler capacity, 6,700 H. P. Trade Marks—Hams, "Reliable" and "Indiana"; bacon, "Reliable," "Indiana" and "Lily"; lard, "Reliable" and "Indiana"; compound, "Sea Foam." Branch Houses—Richmond, Va., New York City, San Francisco, Cal., Baltimore, Md., Washington, D. C., Philadelphia, Pa., Pittsburgh, Pa., Atlanta, Ga., Harrisburg, Pa., Norfolk, Va., Jacksonville, Fla., Tampa, Fla., Columbus, Ohio, Syracuse, N. Y. Retail Markets—One at Indianapolis plant. Also exports.

Meier Packing Co.—577 W. Ray St. Railroad, Pennsylvania. A corporation. Capital, $110,000; issued, $91,900. Stockholders, 12. President, Lewis Meier; Vice-President, Edmund Dietz; Secretary, Alonzo E. Schmidt; Treasurer, Bernard E. O'Connor; General Manager, Lewis Meier; General Superintendent, Jos. E. Mayer. Employes, 45. Hogs, 400. Sausage—Fresh and smoked. By-Products—Tankage. Refrigeration—38-ton Triumph and 12-ton Frick; brine circulating and direct expansion. Boiler capacity, 200 H. P. Trade Mark—"Old Fashion."

Figures on cattle, hogs and sheep indicate AVERAGE WEEKLY KILLING CAPACITY.

Riverview Packing Co.—Kentucky Ave. and White River. Partnership. General Manager, H. G. Bills. Employes, 30. Cattle, 125; hogs, 1,500; sheep, 300. Sausage—Fresh, smoked and summer. By-Products—Dried blood, tankage, hog and chicken feed. Refrigeration—20-ton United and 10-ton York; brine circulating and direct expansion. Boiler capacity, 50 H. P. Trade Marks—"Victory" and "Riverview." Retail Market—341-3 E. Washington St., Indianapolis, Ind.

Frank Schussler Packing Company.

United Butchers, Inc.—621 W. Ray St. A corporation. Capital, $100,000; issued, $95,800. Stockholders, 104. President, C. W. Sedwick; Vice-President, D. B. Darnell; Secretary, L. H. McMurray; Treasurer, John Powell; General Manager, Pat Mallon; General Superintendent, Wm. Mallon. Employes, 7. Cattle, 300; hogs, 600; sheep, 600. By-Products—Tankage. Refrigeration—20-ton York; direct expansion. Boiler capacity, 70 H. P.

Worm & Company—Ray & Dakota Sts. Railroad, Pennsylvania. A corporation. Capital, $300,000; issued, $290,000. Stockholders, 124. President, A. R. Worm; Vice-President, G. H. Evans; Secretary, L. S. Peterson; Treasurer, C. M. Jensen. Employes, 200. Codes—Cross. Cattle, 300; hogs, 200; sheep, 200. Sausage—Fresh, smoked and summer. By-Products—Tankage, hog and chicken feed. Refrigeration—55-ton Triumph; direct expansion. Trade Marks—"Crown," "Hoosier" and "Special."

LAFAYETTE

Dryfus Packing & Provision Co.—1st and Ellsworth Sts.

LOGANSPORT

W. C. Routh & Co.—Cliffe Drive. Railroad, Pennsylvania. A corporation. Capital, $350,000; issued, $287,400. President, W. L. Fernald; Vice-President, J. A. Geyer; Secretary and Treasurer, Jno. A. Geyer; Assistant Secretary, F. R. Metherd; General Manager, F. R. Neff. Employes, 60. Code—Cross. Cattle, 100; hogs, 5,000; sheep, 100. Sausage—Fresh, smoked and summer. By-Products—Dried blood, tankage, hog and chicken feed. Government inspection. Refrigeration—One 100-ton and one 40-ton Triumph; brine circulating and direct expansion. Boiler capacity, 110 H. P. Trade Marks—Hams, "Quail" and "Wabash"; bacon, "Mayflower," "Quail" and "Wabash"; lard, "Anchor," "Quail"; sausages, "Ruco." Retail Markets—Standard Market, 310 E. Broadway, Logansport, Ind. Also exports.

MADISON

Pearl Packing Co., Inc.—710-717 N. West St. Railroad, Pittsburgh, Cincinnati, Chicago & St. Louis. A corporation. Capital, $170,000; issued, $50,000. Stockholders, 5. President, Gus Yunker; Vice-President, R. Yunker; Secretary and Treasurer, Leo J. Yunker; General Manager, R. Yunker; General Superintendent, Leo J. Yunker. Employes, 35. Cattle, 15; hogs, 300; calves, 300. Sausage—Fresh and smoked. By-Products—Dried blood, tankage, hog and chicken feed and fertilizer. Government inspection. Refrigeration—One 70-ton and one 15-ton Frick; brine circulating; 25 tons ice daily. Boiler capacity, 240 H. P. Trade Mark—"Pearl." Retail Market—One at Madison, Ind.

Figures on cattle, hogs and sheep indicate AVERAGE WEEKLY KILLING CAPACITY.

MARION

Ballard Packing Co.—A corporation. Capital, $50,000; issued, $50,000. Stockholders, 8. President, J. M. Ballard; Vice-President, A. J. Ballard; Secretary and Treasurer, Otto Small; General Manager, Otto Small; General Superintendent, F. C. Lenfesty. Employes, 60. Cattle, 150; hogs, 750; sheep, 50. Sausage—Fresh and smoked. By-Products—Tankage. Refrigeration—One 30 and one 20-ton Frick; brine spray system. Boiler capacity, 250 H. P.

Faulkner-Burge Packing Co.

MISHAWAKA

Major Brothers Packing Co.—Logan St. Railroad, New York Central. A corporation. Capital, $100,000; issued, $100,000. Stockholders, 10. President, A. J. Major; Vice-President, Fred Major; Secretary, M. B. Cone; Treasurer, F. T. Major; General Manager, A. J. Major; General Superintendent, R. G. Denton. Employes, 50. Cattle, 200; hogs, 1,000; sheep, 100. Sausage—Fresh, smoked and summer. By-Products—Dried blood, tankage, and fertilizer. Refrigeration—38-ton Vilter; brine circulating. Boiler capacity, 300 H. P. Trade Mark—"Select Brand."

MUNCIE

Kuhner Packing Co.—N. Elm and 13th Sts. Railroad, Lake Erie & Western. A corporation. Capital, $30,000; issued, $30,000. Stockholders, 4. President, Henry C. Kuhner; Vice-President, Gotlip C. Kuhner; Secretary and Treasurer, Frank G. Kuhner; General Manager, Henry C. Kuhner. Employes, 50. Cattle, 120; hogs, 1,200; sheep, 120. Sausage—Fresh and summer. By-Products—Tankage. Refrigeration—One 25-ton York and one 75-ton De La Vergne; brine circulating; 45 tons ice daily. Boiler capacity, 300 H. P.

PERU

Peru Packing Co.

PORTLAND

Geo. Earhart Packing Company.

RICHMOND

Anton Stolle & Sons—40 Liberty St. Railroads, C. & O. and Pennsylvania. A corporation. Capital, $100,000; issued, $75,000. Stockholders, 4. President, Anton Stolle; Vice-Presidents, Anthony Stolle and William Stolle; Secretary and Treasurer, Joseph Stolle; General Manager, Anton Stolle; General Superintendent, Anton Stolle. Employes, 20. Cattle, 45; hogs, 225. Sausage—Fresh and smoked. By-Products—Tankage and hog feed. Government inspection. Refrigeration—15-ton Cleveland, 35-ton Frick; brine circulating. Boiler capacity, 200 H. P. Trade Mark—"Richmond Rose."

SEYMOUR

Peter Packing Co.

SHELBYVILLE

Chas. Flaitz.
C. P. Sindlinger.

THORNTON

Swine Breeders Pure Serum Co.—Slaughterers.

Figures on cattle, hogs and sheep indicate AVERAGE WEEKLY KILLING CAPACITY.

TERRE HAUTE

Ehrmann & Co.—4th and Ohio Sts. A corporation. Capital, $100,000; issued, $50,000. Stockholders, 4. President, C. H. Ehrmann; Vice-President, W. A. Elliott; Secretary, C. R. Huble; Treasurer, W. A. Elliott; General Manager, C. H. Ehrmann; General Superintendent, W. A. Elliott. Employes, 24. Cattle, 150; hogs, 500; sheep, 50. Sausage—Fresh, smoked and summer. Refrigeration—12-ton Barber; direct expansion. Boiler capacity, 50 H. P. Trade Mark—"Red, White and Blue." Retail Markets—One in Terre Haute, Indiana.

Home Packing & Ice Co.—1st and Chestnut Sts. Railroad, Big Four. A corporation. Capital, $125,000; issued, $125,000. Stockholders, 20. President, John McFall; Vice-President, Isaac Powers; Secretary and Treasurer, J. D. Royer; General Manager, Isaac Powers; General Superintendent, C. A. Congleton; Sales Manager, H. J. Larison. Employes, 250. Codes—Cross. Cattle, 100; hogs, 3,500. Sausage—Fresh and smoked. By-Products—Dried blood, tankage, hog and chicken feed. Government inspection. Refrigeration—300-ton Vogt; brine circulating; 100 tons ice daily. Boiler capacity, 1,050 H. P. Trade Marks—"Dependable" and "Wabash." Also exports.

Terre Haute Abattoir & Stock Yards Co.

Valentine & Co.—Wabash Ave. A corporation. Capital, $50,000; issued, $24,000. Stockholders, 10. President, Harvey Valentine; Secretary, I. T. McGlove; Treasurer, Edna Valentine. Employes, 15. Cattle, 75; hogs, 200; sheep, 30. Sausage—Fresh and smoked. By-Products—Tankage. Refrigeration—35-ton De La Vergne; direct expansion. Boiler capacity, 200 H. P.

VINCENNES

C. B. O'Donnell—Slaughterer. Canned meat products.

WABASH

Wabash Packing Co.—North Cass St. A corporation. Capital, $10,000. Stockholders, 4. President and Secretary, Karl P. Albes; Vice-President and Treasurer, John S. Williams. Cattle, 100; hogs, 150. Sausage—Fresh, smoked and summer. By-Products—Tankage.

IOWA

ALBIA

Albia Packing Co.—A corporation. Capital, $150,000; issued, $60,200. Stockholders, 76. President, C. W. Monroe; Vice-President, C. R. Springer; Secretary, Dorothy Monroe; Treasurer, Samuel Cooper; General Manager and Superintendent, C. W. Monroe. Employes, 21. Cattle, 90; hogs, 300. Sausage—Fresh, smoked. By-Products—Dried blood, tankage, hog and chicken feed. Refrigeration—10-ton Brecht; brine tank system. Boiler capacity, 100 H. P.

CEDAR RAPIDS

T. M. Sinclair & Co., Ltd.—South 3rd St. Railroads, Chicago & North Western, Chicago, Rock Island & Pacific, Chicago, Milwaukee & St. Paul, and Illinois Central. A corporation. Capital, $1,500,000;

Figures on cattle, hogs and sheep indicate AVERAGE WEEKLY KILLING CAPACITY.

issued, $1,300,000. President, R. S. Sinclair; Vice-President, A. C. Sinclair; Secretary, J. H. Johnson; Treasurer, R. S. Sinclair; General Manager, R. S. Sinclair; General Superintendent, B. S. Church. Codes—Robinson and Cross. Cattle, 900; hogs, 24,000. Sausage—Fresh, smoked and summer. By-Products—Tankage, hog and chicken feed, commercial and high protein tankage. Government inspection. Refrigeration—One 250-ton Vilter, three 100-ton De La Vergne, one 150-ton Carbondale and one 100-ton Carbondale; brine circulating, direct expansion and brine spray systems; 70 tons ice daily. Boiler capacity, 2,920 H. P. Trade Marks—Hams, "Fidelity" and "Cedar Rapids"; bacon, "Fidelity" and "Cedar Rapids"; lard, "Fidelity" and "White Frost"; compound, "Frosto." Branches—New York City, Boston, Mass., Portland, Oregon, and Seattle, Wash. Also exports.

DAVENPORT

Davenport Slaughtering & Rendering Co.

Kohrs Packing Co.—1337 W. 2nd St. Railroads, Chicago, Rock Island & Pacific, Chicago, Milwaukee & St. Paul and Chicago, Burlington & Quincy. A corporation. Capital, $1,000,000; issued, $730,000. Stockholders, 305. President, W. H. Gehrmann; Vice-President, John J. Kohrs; Secretary and Treasurer, Frank Kohrs; General Manager, W. H. Gehrmann. Employes, 160. Code—Cross. Hogs, 3,500. Sausage—Fresh, smoked and summer. By-Products—Dried blood, tankage, coil dried hair and grease. Government inspection. Refrigeration—130 and 90-ton Triumphs; direct expansion; five tons ice daily. Boiler capacity, 600 H. P. Trade Marks—Hams, "Crown"; bacon, "Crown" and "Arsenal"; lard, "Crown," "Arsenal" and "Pure."

DES MOINES

Iowa Packing Co.—18th and Maury Sts. Railroad, Chicago, Rock Island & Pacific. A corporation. Capital, $5,000,000; issued, $2,282,500. Stockholders, 2,017. President, F. T. Fuller; Vice-President, W. S. Hazard, Jr.; Secretary and Treasurer, Carl Muelhaupt; General Superintendent, H. D. Barrett; General Manager, F. T. Fuller; Sales Manager, W. G. Glenn. Employes, 480. Codes—Cross; cable code word, "Callfull." Cattle, 600; hogs, 12,000. Sausage—Fresh and smoked. By-Products—Dried blood, tankage and hog feed. Government inspection. Refrigeration—150-ton Vilter, 75-ton Vilter, 50-ton Hercules; brine circulating and brine spray. Boiler capacity, 600 H. P. Trade Marks—Hams, "Old Homestead"; bacon, "Old Homestead"; lard, "Lotus" and "Old Homestead"; compound, "Ipaco"; also "Fuller's Hog Digester." Meat Specialties—"Old Homestead" boiled hams and sausage. Also exports.

DUBUQUE

Belsky Packing Co.

Frith Union Slaughter House.

Geo. C. Rath & Sons—Cor. 12th St. and Central Ave. Partnership. Capital, $10,000; issued, $10,000. Employes, 8. General Manager, S. J. Rath. Sausage, fresh pork, beef, veal, lamb, poultry and produce. Refrigeration—5-ton Creamery Package; direct expansion. Trade Mark—"Alpine." Retail Markets—One at 12th and Central Ave., Dubuque.

John Strobel & Sons.

Figures on cattle, hogs and sheep indicate AVERAGE WEEKLY KILLING CAPACITY.

Corn Belt Packing Co.—3200 Jackson St. A corporation. Capital, $3,750,000; issued, $1,950,000. Stockholders, 1,857. President, F. N. Kretschmer; Vice-President, J. H. J. Stutt; Treasurer, F. N. Kretschmer; Secretary and General Manager, H. S. Rice; General Superintendent, W. C. Bower. Employes, 300. Codes—Cross; code word, "Corn Belt." Cattle, 950; hogs, 25,000; sheep, 600. Sausage—Fresh and smoked. By-Products—Dried blood, tankage, hog and chicken feed, oleo stock, edible tallow, white, yellow and brown grease. Government inspection. Refrigeration—Two 100-ton Vilters, one 135-ton Ball and one 2½-ton Clinton; direct expansion; 40 tons ice daily. Boiler capacity, 935 H. P. Trade Marks—"Corn Belt" and "Julien." Retail Markets—One at plant. Also exports.

Dubuque Packing Co.—16th and Cedar Sts. A corporation. Capital, $400,000; issued, $166,900. Stockholders, 24. President, E. J. Beach; Vice-President, C. B. Beach; Secretary and Treasurer, Chris. Schmitt; General Manager, C. Schmitt; General Foreman, Geo. A. La Preli. Employes, 60. Codes—Cross. Hogs, 1,200. Sausage—Fresh, smoked and summer. By-Products—Fertilizer. Government inspection. Refrigeration—50-ton Vilter and 25-ton De La Vergne; direct expansion. Trade Marks—Hams, "Ruby"; lard, "Our Best" and "Family." Retail Market—One at plant.

FORT DODGE

Fort Dodge Serum Company—Slaughter for serum only.

Wahkonsa Packing Company—A corporation. A subsidiary of the Fort Dodge Serum Company. Capital, $1,000,000; issued, $376,700. Stockholders, 71. President, D. E. Baughman; Vice-President, W. W. Bowen; Secretary, H. J. Shore; Treasurer, S. N. Magowan; General Manager, D. E. Baughman; General Superintendent, J. C. Schultz. Employes, 75. Hogs, 300. By-Products—Tankage. Refrigeration—8-ton Creamery Package, 20-ton Baker Ice; brine tank and expansion coils. Boiler capacity, 145 H. P.

MASON CITY

Jacob E. Decker & Sons Co., Inc.—Fifteenth St. Railroad, Minneapolis & St. Louis. A corporation. Capital, $3,000,000; issued, $2,000,000. Stockholders, 500. President, Jay E. Decker; Vice-President, Fred G. Duffield; Secretary, E. S. Selby; Treasurer, George H. Harrer; General Manager, Jay E. Decker. Employes, 700. Codes—Cross and A. B. C., 5th Edition. Cattle, 350; hogs, 15,000. Sausage—Fresh and smoked. By-Products—Dried blood, tankage, hog and chicken feed. Government inspection. Refrigeration—one 250 and one 125-ton Vilter; brine circulating, direct expansion and brine spray systems. Boiler capacity, 1,200 H. P. Trade Marks—Hams, "Iowana" and "Midland"; bacon, "Iowana," "Midland," "Iowa" and "Pigmy"; lard, "Magnet" and "Snow." Branches—Minneapolis and Duluth, Minn.; Texarkana, Dallas and Tyler, Texas. Also exports.

MUSCATINE

Farmers' Mutual Packing Co.—Corporation. Capital, $6,000,000; issued, $500,000. Stockholders, 340. President, Geo. B. Stapp; Vice-President, J. Nevin Witmer; Treasurer, R. S. McNutt; Secretary, E. A. Davis. Cattle, 300; hogs, 1,800. (Plant not completed, April, 1922.)

Figures on cattle, hogs and sheep indicate AVERAGE WEEKLY KILLING CAPACITY.

MUSCATINE

C. E. Richard & Sons—213 W. 2nd St. Railroads, Chicago, Rock Island & Pacific and Chicago, Milwaukee & St. Paul. Partnership. Employes, 25. Cattle, 200; hogs, 600. Sausage—Fresh, smoked and summer. By-Products—Tankage. Refrigeration—Three 10-ton, one 3-ton, two 2-ton Brechts; direct expansion. Boiler capacity, 130 H. P. Trade Marks—Hams, "Rich"; bacon, "Rich"; lard, "Muscatine." One wholesale and one retail market in Muscatine.

OTTUMWA

John Morrell & Co.—Hayne St. and Iowa Ave. Railroads, Chicago, Burlington & Quincy, Chicago, Rock Island & Pacific, Chicago, Milwaukee & St. Paul, and Wabash. A corporation. President, T. Henry Foster; Vice-President, W. H. T. Foster; Treasurer, M. T. McClelland; Secretary, George M. Foster; General Manager, T. Henry Foster; General Superintendent, E. Manns; General Sales Manager, M. T. McClelland. Employes, 2,500 to 2,700. Codes—Morrell, A. B. C. and Western Union, Cross and Robinson. Cattle, 2,000; hogs, 35,000. Fresh beef, pork and veal; hams, bacon and lard; canned foods, etc. Sausage—Fresh, smoked and summer. By-Products—Tankage, hog and chicken feed, hair; oils, greases, tallows, stearine; bones, glue stock, oleo, "Yorkshire" meat meal. Government inspection. Refrigeration—One 500-ton Ball, one 250-ton De La Vergne, one 200-ton Penn. Iron Works, one 150-ton Vilter, one 75-ton Arctic and one 50-ton Arctic at Ottumwa; three 500-ton Balls at Sioux Falls; one 25-ton Frick at Memphis, Tenn.; one 20-ton York at Philadelphia; one 12-ton York at St. Paul, Minn.; one 10-ton Automatic at Memphis, Tenn.; one 8-ton York at Mobile, Ala.; one 8-ton Automatic at New York City; one 6-ton York at Aberdeen, S. D.; one 6-ton Arctic at Duluth, Minn.; and one 6-ton York at Minneapolis, Minn.; 120 tons ice daily. Boiler capacity, 4,050 H. P. Trade Marks—Hams, "Morrell's Pride," "Eureka"; bacon, "Morrell's Pride," "Comet," "Wapello" and "Frontier"; lard, "Morrell's Pride," "Snow Cap," "Ye Olden Style" and "Red Letter Lard"; cheese, "Yorkshire Farm Full Cream"; mince meat, "Morrell's Pride." Branch Houses—310 3rd Ave., S. E., Aberdeen, S. D.; 75-77 Commercial St., Boston, Mass.; 616 Cherry St., Des Moines, Iowa; 108 W. Michigan Ave., Duluth, Minn.; Roberta St. & Great Northern Railway, Fargo, N. D.; Beatrice Creamery Bldg., Lincoln, Neb.; 821 Traction Ave., Los Angeles, Cal.; 30-36 Beale Ave., Memphis, Tenn.; 207 5th St., N., Minneapolis, Minn.; 120-122 N. Water St., Mobile, Ala.; 620-628 W. 36th St., New York City; 816-820 Noble St., Philadelphia, Pa.; 352 E. 6th St., St. Paul, Minn.; 107-109 N. West St., Syracuse, New York. Retail Markets—One in Ottumwa, Iowa. Motor trucks, 54; tonnage, 96 tons. Also exports. (See Sioux Falls, S. D.)

PERRY

Hausserman Packing Company.

SIOUX CITY

Armour & Company. (See Chicago.)

Cudahy Packing Co. (See Chicago.)

Midland Packing Company.

Smith Bros. Packing Company.

Swift & Company—615 S. Chambers St. (See Chicago.)

Figures on cattle, hogs and sheep indicate AVERAGE WEEKLY KILLING CAPACITY.

Sacks Bros. Packing Co.—A corporation. Capital, $50,000; issued, $37,000. Stockholders, 4. President, A. I. Kay; Vice-President, R. Sacks; Secretary and Treasurer, A. I. Sacks; General Manager, A. I. Sacks; General Superintendent, Geo. Perley. Cattle, 125. Sausage—Fresh and smoked. Refrigeration—12-ton Baker; direct expansion. Boiler capacity, 50 H. P. Trade Mark—"Sacks Royal."

WATERLOO

Rath Packing Co.—Elm and Sycamore Sts. Railroad, Illinois Central. A corporation. Capital, $2,000,000; issued, $1,300,000. President, J. W. Rath; Vice-President, F. J. Fowler; Secretary and Treasurer, E. F. Rath; General Managers, J. W. Rath and E. F. Rath; General Superintendent, John Morris; Assistant Superintendent, A. D. Donnell. Employes, 500. Codes—Cross. Cattle, 600; hogs, 12,000. Sausage—Fresh and smoked. By-Products—Dried blood, tankage, hog and chicken feed. Government inspection. Refrigeration—Three 350-ton Wolf-Lindes; brine circulating, direct expansion and brine spray systems; 30 tons ice daily. Trade Marks—Hams, "Black Hawk" and "Cedar Valley"; bacon, "Black Hawk," "Cedar Valley" and "Waterloo"; lard, "Black Hawk," "Snow Flake" and "Cedar Valley." Branches—Des Moines, Iowa, and Galesburg, Ill. Selling agencies in all large cities. Also exports.

KANSAS

ARKANSAS CITY

Henneberry & Co.—South Summit Blvd. Railroad, Atchison, Topeka & Santa Fe. A corporation. Capital, $100,000; issued, $100,000. Stockholders, 6. President, Richard T. Keefe; Vice-President, Geo. T. Bacostow; Secretary and Treasurer, A. E. Le Stourgeon; General Manager, Richard T. Keefe; General Superintendent, C. W. Brooks. Employes, 85. Codes—Cross and Robinson; code word, "Henneberry." Cattle, 75 to 100; hogs, 500 to 800. Sausage—Fresh and smoked. By-Products—Dried blood, tankage, hog feed, yellow and brown grease. Government inspection. Refrigeration—65-ton Carbondale and 50-ton Vogt; brine circulating; 35 to 40 tons ice daily. Boiler capacity, 300 H. P. Trade Marks—"Ark" and "Sweet Clover." Retail market at plant. Also exports.

FRONTENAC

Menghini Brothers.

HUTCHINSON

George Kaiser Packing Co.—81 N. First St.

Wilson & Company—Adams, Osage, Baird and Railroad. (See Chicago.)

Winchester Packing Company—Partnership. C. S. Winchester and J. H. Hadsall.

McArthur Packing Company—A corporation. Capital, $500,000; issued, $123,000. Stockholders, 30. President, V. E. McArthur; Vice-President, T. M. Madden; Secretary, F. B. Baylor; Treasurer, V. E. McArthur. Employes, 40. Cattle, 150; hogs, 250. Sausage—Fresh and smoked. By-Products—Tankage. Refrigeration—100-ton Vogt and 60-ton Frick; brine spray; 50 tons ice daily. Boiler capacity, 300 H. P.

Figures on cattle, hogs and sheep indicate AVERAGE WEEKLY KILLING CAPACITY.

IOLA

Fryer Brothers—Slaughterers.

KANSAS CITY

Armour & Company—Joy, Central and James Sts. (See Chicago.)

Jos. Baum Packing Co.—Third and Central Sts. Railroad, Kansas City Southern. Individual ownership. Employes, 35. Slaughters only beef and sheep. Government inspection. Refrigeration—16-ton York; brine circulating and direct expansion systems. Boiler capacity, 15 H. P.

Cochrane Packing Co.—Central and Water Sts. Railroad, Missouri Pacific. A corporation. Capital, $600,000; issued, $60,000. Stockholders, 5. President, James F. Cochrane; Vice-President, L. U. Faulkner; Secretary and Treasurer, Chas. Rowett; General Manager, James F. Cochrane; Superintendent, L. U. Faulkner. Employes, 75. Cattle, 600; hogs, 1,800; sheep, 600. Sausage—Fresh, smoked and summer. By-Products—Tankage and hog and chicken feed. Government inspection. Refrigeration—10-ton York; brine circulation. Boiler capacity, 140 H. P. Trade Mark—"Velvet."

Drovers Packing Co.—Fifth St. and Kaw River. Railroad, Kansas City Southern. A corporation. Capital, $3,500,000; issued, $2,243,000. Stockholders, 2,300. In bankruptcy. Taken over by Schalker Packing Co. Employes, 175. Cattle, 1,200; hogs, 2,000; sheep, 300. Government inspection. Refrigeration—One 275-ton, one 175-ton and one 55-ton York; brine circulating, brine spray and direct expansion; 100 tons ice daily. Boiler capacity, 1,050 H. P. Trade Mark—"Trophy."

Fowler Packing Co.—Branch of Armour & Co. (See Chicago.)

Means Packing Company—Second and Lyon Ave. Railroads, Missouri Pacific and Kansas City Southern. A corporation. Capital, $20,000. Stockholders, 3. President, B. J. Means; Vice-President, C. W. Means; Secretary and Treasurer, Geo. C. Means. Employes, 8. Cattle, 150; sheep, 50. Government inspection. Refrigeration—20-ton Barber; brine circulating.

Morris & Company. (See Chicago.)

Royal Packing Co.—Adams and Osage Aves. Railroad, Kansas City Southern. Subsidiary of the Royal Serum Company, serum manufacturers. Capital, $2,000; issued, $2,000. Stockholders, 3. President, Clay W. Stephenson; Vice-President, T. R. Graybill; Secretary and Treasurer, J. F. Hoaglin; General Manager, C. W. Stephenson; General Superintendent, F. H. Hueben. Employes, 15. Hogs, 250. Government inspection. Refrigeration—York, brine circulating and direct expansion systems. Note: Carcasses are sold on rail to other packing houses.

Sihler Hog Cholera Serum Company—Slaughter for serum only.

Standard Serum Company—9 S. Second St.

Swift & Company. (See Chicago.)

Wilson & Co.—(See Chicago).

Figures on cattle, hogs and sheep indicate AVERAGE WEEKLY KILLING CAPACITY.

LEAVENWORTH

Schalker Packing Co.—Third and Chocktaw Sts. A corporation. Capital, $500,000; issued $242,100. Stockholders, 140. President, John Schalker Jr.; Vice-Presidents, A. W. Schalker and C. C. Walch; Secretary, F. L. McGahen; Treasurer, A. W. Schalker; General Superintendent, C. C. Walch. Employes, 60. Code—Cross. Cattle, 200; hogs, 600; sheep, 10. Sausage—Fresh and smoked. By-Products—Tankage, hog and chicken feed. Government inspection. Refrigeration—50-ton Wolf and 25-ton Fisher; brine spray; two tons ice daily. Boiler capacity, 300 H. P. Trade Marks—Hams, "Trophy" and "Competition"; bacon, "Trophy" and "Competition"; lard, "Trophy" and "Competition"; compound, "Goldleaf" and "Leavenworth."

MANHATTAN

Manhattan Packing Co.—A corporation. Capital, $100,000; issued, $85,000. Stockholders, 15. President, A. P. Fielding; Vice-Presidents, V. V. Akin and Sam Weixelbaum; Secretary, Geo. Clammer; Treasurer, V. V. Akin; General Manager, L. H. Stenger. Employes, 12. Cattle, 25; hogs, 100. Sausage—Fresh, smoked and summer. Refrigeration—13-ton Sterling; direct expansion. Boiler capacity, 50 H. P.

OLATHE

Olathe Packing Co.—132 N. Cherry St. Railroad, Atchison, Topeka & Santa Fe. Individual ownership. Employes, 6. Cattle, 30; hogs, 180. Sausage—Fresh and smoked. By-Products—Dried blood. Refrigeration—6-ton Sterling; direct expansion. Boiler capacity, 15 H. P. Retail Market—132 N. Cherry St., Olathe.

PITTSBURG

Hull & Dillon Packing Co.—West Fourth St. Railroad, Frisco. A corporation. Capital, $150,000; issued, $100,000. Stockholders, 6. President, Louis Hull; Secretary and Treasurer, R. P. Nevin. Employes, 45. Cattle, 60; hogs, 400; sheep, 30. Sausage—Fresh and smoked. By-Products—Tankage and hog feed. Government inspection. Refrigeration—50 and 25-ton Hercules and 15-ton United Iron Works; direct expansion; 25 tons ice daily. Boiler capacity, 300 H. P. Trade Mark—"Cook Brand," "Enterprise Brand" and "Scoco."

SALINA

Butzer Packing Company—A corporation. Capital, $450,000; issued, $237,930. President, C. B. Dodge; Vice-President, J. E. Putnam; Secretary and Manager, Chas. F. Dodds; Treasurer, R. W. Reeves.

TOPEKA

Kaw Packing Company—Fred E. Barthman, General Manager. Capital, $100,000. Cattle, 250; hogs, 1,500.

Seymour Packing Co.—200 N. Kansas St.

Topeka Packing Co.

Chas. Wolff Packing Co.—Chas. Wolff, Manager. Owned by Allied Packers, Inc. (See Chicago.)

WICHITA

Jacob Dold Packing Co.—N. Lawrence Ave. and 21st St. (See Buffalo.)

Cudahy Packing Co.—N. Lawrence Ave. and 22nd St.

Figures on cattle, hogs and sheep indicate AVERAGE WEEKLY KILLING CAPACITY.

KENTUCKY

COVINGTON

C. Rice Packing Co.—Patton and Eastern Aves. A corporation. Capital, $40,000; issued, $36,700. Stockholders, 14. President, Mrs. C. Rice; Vice-President, Chas. Hegener; Secretary and Treasurer, Chas. A. Carter; General Superintendent, John Schnorbus. Employes, 15. Cattle, 200; calves, 150. By-Products—Tankage and tallow. Refrigeration—30-ton Triumph; brine circulating. Boiler capacity, 110 H. P.

HENDERSON

Eckert Packing Co.—1600 Corydon Road. A corporation. Capital, $200,000; issued, $125,000. Stockholders, 20. President, Frank F. Eckert; Vice-President, H. H. Farmer; Secretary and Treasurer, E. C. Farmer; General Manager, Frank F. Eckert; General Superintendent, E. A. Eckert. Employes, 60. Codes—Cross. Cattle, 50; hogs, 500; sheep, 100. Sausage—Fresh and smoked. By-Products—Dried blood, tankage, hog and chicken feed. Government inspection. Refrigeration—One 17-ton and one 40-ton Frick; brine circulating and direct expansion; 20 tons ice daily. Boiler capacity, 450 H. P. Trade Marks—"Echo" and "Pigs' Delight."

LEITCHFIELD

J. M. Thomas & Son.

MIDDLESBORO

New South Packing Co.—A corporation. Capital, $10,000; issued, $10,000. Stockholders, 13. President, W. P. Allen; Vice-President, ———; Secretary, E. Wieder; Treasurer, R. P. Overton. Employes, 8. Cattle, 15; hogs, 40; sheep, 15. Sausage—Fresh. Refrigeration from Middlesboro Ice & Cold Storage Co. Wholesale only.

OWENSBORO

Field Packing Co.—Dublin Lane. Railroads, Louisville, Henderson & St. Louis, Louisville & Nashville and Illinois Central. A corporation. Capital, $120,000; issued, $68,000. Stockholders, 32. President, C. E. Field; Vice-President, H. G. Smith; Secretary and Treasurer, E. G. Meisenheimer; General Manager, C. E. Field; General Superintendent, E. G. Meisenheimer. Employes, 40. Cattle, 50; hogs, 300; sheep, 100. Sausage—Fresh, smoked and summer. By-Products—Dried blood, tankage, hog and chicken feed. Refrigeration—20-ton Vogt and 10-ton York; brine circulating and direct expansion; 5 tons ice daily. Boiler capacity, 200 H. P. Trade Mark—"Chesterfield."

PADUCAH

Challenor Packing Co.—Second and Monroe Sts. Railroad, Illinois Central. Individual ownership. Employes, 16. Cattle, 100; hogs, 200; sheep, 100. Sausage—Fresh, smoked and summer. By-Products—Tankage. Refrigeration—10-ton Brecht; brine circulating. Boiler capacity, 25 H. P. Retail Markets—Three in Paducah.

Paducah Packing Co.—206 Kentucky Ave. Railroads, Chicago, Burlington & Quincy, Illinois Central and Nashville, Chattanooga & St. Louis. A corporation. Capital, $30,000; issued, $20,000. Stockholders, 10. President, J. F. Heath; Vice-President and Secretary, F. B. Heath; Treasurer and General Manager, J. B. Gray.

Figures on cattle, hogs and sheep indicate AVERAGE WEEKLY KILLING CAPACITY.

Employes, 12. Cattle, 10; hogs, 75; sheep, 10. Fresh and cured meats of all kinds. Refrigeration—Rent storage from City Consumers' Company. Wholesale and Retail Market—206 Kentucky Ave., Paducah.

LOUISVILLE

Louis P. Bornwasser Co.—921-929 Geigler St. A corporation. Capital, $100,000; issued, $100,000. Stockholders, 10. President, Chas. W. Bornwasser; Vice-President and Treasurer, Louis W. Bornwasser; Secretary, Alvin H. Bornwasser; General Manager, Chas. W. Bornwasser; General Superintendent, Alvin H. Bornwasser. Employes, 75. Cattle, 150; hogs, 1,200. Sausage—Fresh and smoked. By-Products—Tankage. Government inspection. Refrigeration—17-ton York. Boiler capacity, 250 H. P. Trade Mark—"Korn Blossom."

Emmart Packing Co.—1202-08 Story Ave. A corporation. Capital, $750,000. President, Joseph M. Emmart; Vice-President, William H. Webb; Secretary, Sydney D. Camper; Treasurer, D. Collin Ward. Employes, 125. Cattle, 3,000; hogs, 9,000. Trade Mark—Hams and bacon, "Magnolia."

Henry Fischer—1860-1862 Mellwood Ave. Individual ownership. General Manager, Henry Fischer; Assistant Manager and General Superintendent, Carl T. Fischer. Employes, 45. Cattle, 125; hogs, 600. Sausage—Fresh and smoked. Refrigeration—8-ton Brecht, 10-ton Triumph, 20-ton Vilter; direct expansion and brine circulating systems. Boiler capacity, 50 H. P.

Klarer Provision Co.—3830 W. Market St. A corporation. Capital, $20,000; issued, $20,000. Stockholders, 3. President, H. A. Broecker; Vice-President, Mrs. Dora Klarer; Secretary and Treasurer; Mrs. H. A. Broecker. Employes, 25. Cattle, 50; hogs, 200. Sausage—Fresh and smoked. By-Products—Tankage. Refrigeration—15-ton Vogt; brine circulating. Boiler capacity, 160 H. P.

Louisville Provision Co., Inc.—914-36 E. Market St. A corporation. Capital, $400,000; issued, $400,000. Stockholders, 225. President, Henry Knight; Vice-President, Chas. H. Knight; Secretary and Treasurer, J. George Miller; General Manager, Karl M. Zach. Employes, 140. Sausage—Fresh and smoked. Government inspection. Refrigeration—100-ton Vogt, 100-ton Frick compression; brine circulating and direct expansion. Boiler capacity, 310 H. P. Trade Mark—"Southern Star."

C. F. Vissman & Co.—117 Bickel Ave. Railroad, Louisville & Nashville. A corporation. Capital, $300,000; issued, $192,000. Stockholders, 6. President, R. E. Vissman; Vice-President, J. George Woerner; Secretary, George W. Vissman; General Manager, R. E. Vissman; General Superintendent, John C. Vissman. Employes, 150. Cattle, 150; hogs, 1,800. Sausage—Fresh and smoked. By-Products—Tankage. Government inspection. Refrigeration—One 100-ton and one 65-ton Vilter; brine circulating; one ton ice daily. Boiler capacity, 450 H. P. Trade Mark—"Vissman Derby Brand." Also exports.

LOUISIANA

ALEXANDRIA

Rapides Packing Company—A corporation. Capital, $65,000; issued, $65,000. President and Manager, W. D. Rush; Vice-President, Ben F. Bradford; Secretary and Treasurer, Ben F. Rush.

Figures on cattle, hogs and sheep indicate AVERAGE WEEKLY KILLING CAPACITY.

ARABI

Arabi Packing Co., Inc.—Railroad, Southern. A corporation. Capital, $242,000; issued, $125,000. President, Gregory De Reyna; Vice-President, Eugene Dours; Secretary, Georges Damiens; Treasurer, Rene Forio. Cattle, 600; hogs, 200; sheep, 100. Government inspection. Refrigeration—75 and 40-ton Frick and 20-ton Ball; direct expansion. Boiler capacity, 500 H. P. This company has taken over the Crescent City Stock Yards & Slaughter House Company, Ltd.

BATON ROUGE

City Abattoir.

NEW ORLEANS

L. A. Frey & Sons—3925-3937 Burgundy St. A corporation. President, L. A. Frey; Vice-Presidents, Andrew F. Frey, Severin L. Frey and Louis M. Frey; Secretary, Albert E. Frey; Treasurer, Chas. J. Frey; General Manager, Andrew F. Frey. Employes, 70. Code—A. B. C. Cattle, 50; hogs, 60. Sausage—Fresh and smoked. Refrigeration—15-ton Frick and 4-ton Brecht; direct expansion. Retail Market—167 French Market, New Orleans.

Hoth Bros., Ltd.—1133 Magazine St. A corporation. Stockholders, 3. President, Chas. A. Hoth; Vice-President, H. E. Hoth; Secretary and Treasurer, G. A. Hoth. Employes, 27. Wholesale butchers and packers. Refrigeration—15-ton Frick, 7-ton Remington and 2 and 4-ton Brecht; brine circulating system.

New Orleans Butchers' Co-operative Abattoir Company.

SHREVEPORT

Shreveport Packing Co., Inc.—Railroad, Texas & Pacific. A corporation. Capital, $100,000; issued, $66,000. Stockholders, 5. President, C. C. Herndon; Secretary, S. W. Dickson; General Manager, C. C. Herndon; General Superintendent, R. Noeth. Employes, 60. Cattle, 125; hogs, 250. Sausage—Fresh and smoked. By-Products—Tankage and hog and chicken feed. Refrigeration—Plant, 25-ton Frick; distributing market, two 10-ton Fricks; direct expansion. Trade Marks—Hams, "Jack Spratt"; bacon, "Jack Spratt"; lard, "Feather Flake"; compound, "Caddo Crisp." Retail Markets—Three in Shreveport.

MAINE

AUBURN

E. W. Penley—37 Knight St. Railroads, Maine Central and Grand Trunk. Individual ownership; executors of estate, C. G. Ross, E. L. Smith, R. W. Crockett. Employes, 70. General Superintendent, Fred L. Sylvester; Financial Manager, C. G. Ross; Purchasing Agent and Sales Manager, R. W. Penley. Cattle, 1,800; hogs, 20,000; sheep and lambs, 1,300; calves, 1,000 yearly. Sausage—Fresh, smoked and summer. By-Products—Dried blood, tankage, hog and chicken feed, edible and inedible tallow, hoofs, dried hog hair and bones. Government inspection. Refrigeration—Natural ice; also use public cold storage. Boiler capacity, 250 H. P. Trade Mark—"Blue Tag Brand" on pork products. Also wholesaler in Chicago dressed beef.

Merrow Packing Co.

LEWISTON

Martin Haas.

Figures on cattle, hogs and sheep indicate AVERAGE WEEKLY KILLING CAPACITY.

PORTLAND

Portland Abattoir Co.

ROCKLAND

W. M. Little Co.—Railroad, Rockland. A corporation. Capital, $40,000; issued, $22,500. Stockholders, 3. President, W. M. Little; Vice-President, M. E. Little; Secretary and Treasurer, J. E. Stevens. Employes, 12. Cattle, 35; hogs, 120; sheep, 150. Sausage—Fresh and smoked. Refrigeration—8-ton Vilter; direct expansion. Boiler capacity, 10 H. P.

MARYLAND
BALTIMORE

Baltimore Butchers' Abattoir & L. S. Co.

Consolidated Beef & Provision Co., Inc.—100-102-104 So. Exeter St. Railroad, Baltimore & Ohio. A corporation. Capital, $53,000; issued, $53,000. Stockholders, 5. President, Wolf Salganik; Vice-President, Isidor Salganik; Secretary, Louis P. Salganik; Treasurer, Harry B. Hurwitz; General Superintendent, Isidor Salganik. Employes, 35. Cattle, 50; hogs, 600; sheep, 100. Sausage—Fresh, smoked and summer. Government inspection. Refrigeration—20-ton Frick and 10-ton Remington; direct expansion; five tons ice daily. Boiler capacity, 50 H. P. Trade Marks—"Perfection Brand" and "Extra Brand." Retail Markets—925 E. Lombard St. and 559 N. Gay St., Baltimore, Md.

Corkran, Hill & Co.—Union Stock Yards. Railroads, Pennsylvania and Baltimore & Ohio. A corporation. Capital, $1,150,000; issued, $1,150,000. Stockholders, 149. Chairman of Board, Benjamin W. Corkran, Jr.; President, T. Davis Hill; Vice-President, Lloyd G. Corkran; Treasurer, J. Denny Armstrong; Secretary, Henry W. Marston; Managing Director, City Branch, H. L. Piel, Jr.; Managing Director, Beef Department, Walter B. Peppler; General Superintendent, A. T. McAllister. Employes, 230. Codes—Robinson and Cross. Cattle, 500; hogs, 10,000; sheep, 1,000. Sausage—Fresh and smoked. By-Products—Dried blood, tankage, hog and chicken feed. Government inspection. Refrigeration—One 90-ton and one 35-ton York, and one 12-ton Remington; brine circulating, direct expansion and brine spray systems. Boiler capacity, 400 H. P. Trade Marks—Hams, "Corkhill Brand" and "Orange Brand"; bacon, "Corkhill Brand" and "Orange brand"; lard, "Orange Brand" and "Busy Bee Brand"; compound, "Sunny South Brand." Branch—221-227 S. Howard St., Baltimore, Md. Also exports.

Greenwald Packing Co.—Union Stock Yards. Railroads, Pennsylvania and Baltimore & Ohio. A corporation. Capital, $250,000; issued, $250,000. Stockholders, 8. President, Solomon Greenwald; Vice-President, Isaac Greenwald; Secretary, Michael Greenwald; Treasurer, Judah Lehman; General Manager, S. R. Greenwald. Employes, 165. Codes—Cross and A. B. C., 5th edition; code word, "Greenwald." Cattle, 1,300; sheep, 1,500. By-Products—Dried blood, tankage, bones, hoofs, tallow, neatsfoot and oleo oils, oleo, stearine and beef casings. Government inspection. Refrigeration—200-ton Vilter; brine spray; 17 tons ice daily. Boiler capacity, 300 H. P. Trade Marks—"Dependable," "Balto Extra Oleo Oil,." Retail Markets—Annapolis, Md.; Hagerstown and Ellicott City, Md. Also exports.

Figures on cattle, hogs and sheep indicate AVERAGE WEEKLY KILLING CAPACITY.

BALTIMORE

John A. Gebelein—725-743 N. Castle St. Individual ownership. Hogs, 1,200. Sausage—Fresh and smoked. By-Products—Tankage. Government inspection. Refrigeration—30-ton York; brine circulating. Boiler capacity, 350 H. P. Trade Mark—"Castle Brand." Retail Market—One at Bel Air Market.

George Company, Inc.—404 S. Charles St. A corporation. Capital, $100,000; issued, $70,000. President and Manager, David Garratt; Vice-President, J. B. Stewart; Secretary, M. Raymond Roberts.

August Grebe—Slaughterer.

Haas & Fox—Packers.

C. Hohman & Sons—2138 E. Monument St. Partnership. Employes, 65. General Manager, Geo. A. Hohman. Hogs, 1,200. Sausage—Fresh and smoked. By-Products—Tankage. Government inspection. Refrigeration—One 45-ton Hendricks. Boiler capacity, 300 H. P. Trade Marks—"Blue Band Brand" and "Triangle Brand." Retail Markets—North East Market and Belair Market, Baltimore.

Martin Horn—100 Hartford St.

Jaeger Bros.—Abattoir.

Jones & Lamb Co.—6th and Lombard Sts. Railroad, Baltimore & Ohio. A corporation. Capital, $1,822,242.50; issued, $1,822,242.50. Stockholders, 400. President, W. W. Moss; Vice-President, Howard R. Smith; Secretary, R. W. Moore; Treasurer, R. W. Moore; General Manager, Howard R. Smith. Employes, 300. Hogs, 15,000. Sausage—Fresh and smoked. Dressed beef, mutton and veal. By-Products—Tankage. Government inspection. Refrigeration—One 200-ton and one 150-ton De La Vergne; direct expansion; five tons ice daily. Boiler capacity, 1,225 H. P. Trade Marks—Hams, "Extra Sugar Cured Bandaña," "Baltimore Brand" and "Eagle Brand"; bacon, "Baltimore Brand" and "Eagle Brand"; lard, "Eagle Brand"; shortening, "Peach Brand." Also exports.

Kaufman Packing Co.—Sixth St. Railroads, Pennsylvania and Baltimore & Ohio. A corporation. Capital, $100,000; issued, $8,200. Stockholders, 5. President, Harry J. Kaufman; Vice-Presidents, Walter C. Kaufman, J. Louis Kaufman and Elmer R. Kaufman; Secretary, Halver B. Kaufman; Treasurer, J. Louis Kaufman; General Manager, Walter C. Kaufman; General Superintendent, J. Louis Kaufman. Employes, 25. Cattle, 150; sheep, 75. Sausage—Fresh and smoked. By-Products—Tankage and hog and chicken feed. Government inspection. Refrigeration—15-ton Frick; direct expansion. Boiler capacity, 100 H. P. Trade Mark—"Blue Seal." Retail Markets—607-609 Lexington Market and 18-20 Hollins Market, Baltimore, Md.

Kreil, Charles G.—5221 W. Henrietta St. Individual ownership. Propr., Mrs. Hannan E. Kreil.

Kurrle Packing Co.—23-31 Taylor St. A corporation. Capital, $100,000; issued, $60,000. Stockholders, 7. President, C. F. Kurrle, Sr.; Vice-President, C. F. Kurrle, Jr.; Secretary, Wm. Vitzhum; Treasurer, Chas. Kurrle; General Manager, C. F. Kurrle, Jr. Employes, 25. Cattle, 50; hogs, 1,500; sheep, 50; calves, 200. Sausage—Fresh and smoked. Refrigeration—35-ton Remington and 18-ton York; brine circulating; 50 tons ice daily. Boiler capacity, 300 H. P. Trade Mark—"American Brand."

O. Lang Sons—Brown's Lane.

Figures on cattle, hogs and sheep indicate AVERAGE WEEKLY KILLING CAPACITY.

Raith's, Inc.—2806 Penna. Ave. A corporation. Capital, $100,000; issued, $100,000. Stockholders, 3. President, Chas. Raith; Vice-President, Wm. J. Raith; Secretary and Treasurer, Robt. M. Raith; General Manager, Chas. Raith. Employes, 30. Meat packers and sausage manufacturers. Refrigeration—12-ton Frick; brine circulating. Trade Mark—"Pine Apple." Retail Markets—2806 Penna. Ave., Belvidere Ave. and Main St., 1401 W. Lafayette Ave., 282 Community Market, Baltimore, Maryland.

L. Sellmayer & Sons—Third and Fleet Sts., Highlandtown Sta. Pork and beef packers and sausage manufacturers.

Wm. Schluderberg-T. J. Kurdle Co.—Baltimore and 6th Sts. Railroad, Baltimore & Ohio. A corporation. Capital, $2,225,000; issued, $2,225,000. Stockholders, 200. President, W. F. Schluderberg; Vice-President, Joseph Kurdle; Secretary, Theo. Schluderberg; Treasurer, Albert Kurdle; General Manager, W. F. Schluderberg; General Superintendent, A. M. Eastman. Employes, 400. Codes—A. B. C., 5th edition; code word, "Esskay." Cattle, 500; hogs, 10,000; sheep, 500. Sausage—Fresh and smoked. By-Products—Dried blood and tankage. Government inspection. Refrigeration—125-ton De La Vergne, 40-ton York and 75-ton York; direct expansion and brine spray; two tons ice daily. Boiler capacity, 900 H. P. Trade Marks—Pork products, "Esskay"; lard, "Esskay," "Oriole"; compounds, "Southern Rose," "Pearl." Branches—1727 Eastern Ave., Baltimore, Md., and Roanoke, Va. Also exports.

Jacob C. Shafer Co.—516-520 West Lexington St. A corporation. Capital, $250,000; issued, $250,000. President, Edwin G. Caver; Vice-President, C. W. Shafer; Secretary, Chas. A. Hall; Treasurer, J. Fred Shafer; General Manager, C. W. Shafer; General Superintendent, H. G. Backmiller. Employes, 125. Codes—Robinson and Cross. Hogs, 1,500. Sausage—Fresh and smoked. Government inspection. Refrigeration—30-ton Frick and 25-ton De La Vergne; brine circulating; 5 tons ice daily. Boiler capacity, 250 H. P. Trade Marks—Hams, bacon and lard, "Globe Brand"; compound, "Jacsco," "White Seal" and "Le Grand." Also exports.

Wilson-Martin Co.—Union Stock Yards. Owned By Wilson & Co. (See Chicago.)

CRISFIELD

Webb & Company—Railroads, N. Y. P. & N. Partnership, T. E. Webb and T. J. Webb. Employes, 20. Codes—Cross; code word, "Webco." Sausage—Fresh, smoked and summer. By-Products—Hog and chicken feed. Refrigeration—8-ton Sterling and 5-ton United; direct expansion. Boiler capacity, 20 H. P. Trade Mark—"Webco." Retail Markets—One at Crisfield, Md.

FREDERICK

J. A. Whitfield Company—Railroad, Baltimore & Ohio. A corporation (owned by Old Dutch Market, Inc.). President, J. A. Whitfield; Vice-President, D. B. Casley; Secretary, A. N. Mandell; Treasurer, D. B. Casley; General Manager, J. A. Whitfield; Resident Manager, H. B. Willson. General packinghouse business conducted. Retail markets—Washington, D. C.; Richmond, Va.; Frederick, Md., and Alexandria, Va. Government inspection. Trade Mark—"Blue Ridge."

Figures on cattle, hogs and sheep indicate AVERAGE WEEKLY KILLING CAPACITY.

MASSACHUSETTS
BOSTON

Batchelder & Snyder Company—47 Blackstone St. A corporation. Capital, $1,800,000. President, F. S. Snyder; Vice-Presidents, C. J. Ramsdell, J. C. Pineo, J. Buxbaum, L. R. Bolton and James Knowles; Treasurer, F. A. Burgess. Employes, 300. Wholesale provisioners, packers and poultry dressers. Government inspection. Also exports.

Boston Food Products Company—16 New St. A corporation. Capital, $750,000; issued, $105,000. President and General Manager, J. C. DeMille; Vice-President, W. A. Snecker; Secretary and Treasurer, R. B. Chace. Government inspection. Employes, 30. Do export business. Canners. Trade Marks—"Prudence" and "Amity."

Brighton Dressed Meat Co.—Railroad, Boston & Albany. A corporation. Capital, $10,000; issued, $10,000. Stockholders, 3. President, Morris Madfis; Treasurer, J. E. Maloney; General Manager, Morris Madfis. Employes, 70. Cattle, 600. Sausage—Fresh and kosher. Government inspection. Refrigeration—Rents coolers.

Cunningham Packing Company—84 South Market St. A corporation. Capital, $15,000. President, A. J. Cunningham; Treasurer, T. T. Cunningham; Clerk, Alfred H. Cunningham.

Moses Goldberg—Brighton.

John J. Kelly—Brighton.

Thomas J. Kelly—Brighton.

I. Paresky & Co.—Brighton.

Wm. Underwood Company—52 Fulton St. Voluntary association. Meat and fish canning. President, Loring Underwood; Secretary and Treasurer, F. A. Harding; Asst. Secretary, W. P. Durant. Employes, 150. Codes—A. B. C., 5th edition; code word, "Underwood." Government inspection.

BROCKTON

Joseph Curtin—Quincy St.
Philip Katz—Plain St.
Samuel Steinberg—Crescent St.

CAMBRIDGE

John P. Squire & Co.—165 Gore St. Railroads, Boston & Albany and Boston & Maine. A corporation. Capital, $5,000,000; issued, $5,000,000. President, George H. Swift; Vice-President, Edwin C. Starr; Secretary, Frank W. Crocker; Treasurer, Edward D. Whitford; General Manager, J. Frederick Hill; General Superintendent, Henry C. Fish. Employes, 1,200. Codes—A. B. C., 5th Edition; code word, "Squire-Boston." Hogs, 18,000. Sausage—Fresh and smoked. By-Products—Tankage. Government inspection. Refrigeration—One 400-ton Ball, one 300-ton Vilter and two 300-ton De La Vergnes; direct expansion; at Holyoke, Mass., two 60-ton De La Vergnes; at Rutland, Vt., one 30-ton Penn. Iron Works. Brine spray and direct expansion. Boiler capacity, total, 2,820 H. P. Trade Marks—Hams, "Arlington" and "Squire's"; bacon, "Bay State," "English" and "Cambridge"; lard, "Squire's"; compound, "Bay State Brand." Export brands, clear bellies, "Detroit Brand," "Eastern Packing Brand," "Belmont," "Harvard," "University" (barrelled pork). Branch Houses—Augusta, Bangor and Portland, Me., Fitchburg, Holyoke, Lawrence, New Bedford, Salem, and Worcester, Mass., Man-

---

Figures on cattle, hogs and sheep indicate AVERAGE WEEKLY KILLING CAPACITY.

chester, N. H., Providence, R. I., Rutland, Vermont and 39 No. Market St., Boston, Mass.

FITCHBURG
E. A. McIntire & Son—Slaughterers.

HAVERHILL
Haverhill Abattoir Co.—Plant, off Hilldale Ave. Railroad, Boston & Maine. A corporation. Capital, $15,000; issued, $15,000. Stockholders, 3. President, Edwin H. Moulton; Treasurer, Edwin A. Edgerly. Employes, 8. Cattle, 100; hogs, 200; sheep, 250. Government inspection.

NEW BEDFORD
Manuel F. Sousa—316 S. Second St. Slaughterers.

NORTH CHELMSFORD
F. W. Merrill—Westford Road. Slaughterers.

SOMERVILLE
North Packing & Provision Co.—37 Medford St. Railroads, Boston & Albany and Boston & Maine. A corporation. Capital, $5,000,000; issued, $5,000,000. President, George H. Swift; Vice-President, E. C. Starr; Secretary, F. W. Crocker; Treasurer, H. J. Nichols; General Manager, E. C. Starr; General Superintendent, W. P. Liston. Employes, 1,300. Codes—Cross and Utility; code word, "North." Capacity, 25,000 hogs. Sausage—Fresh, smoked and summer. By-Products—Hog tankage. Government inspection. Refrigeration—300-ton Vilter, 200-ton De La Vergne and two 150-ton De La Vergnes; brine circulating and direct expansion. Boiler capacity, 2,450 H. P. Trade Marks—"North Star Brand" and "White Mountain." Also exports.

SPRINGFIELD
Brightwood Dressed Beef Co.—841 North St. Slaughterers.
George Nye Co.—130 Lyman St.
Springfield Provision Co.—Plainfield St. Railroad, Boston & Maine. A corporation. Capital, $1,300,000; issued, $1,300,000. Stockholders, 30. President, George H. Swift; Vice-President, E. C. Starr; Secretary, F. W. Crocker; Treasurer and General Manager, F. A. Reed; General Superintendent, L. W. Hooker. Employes, 295. Hogs, 3,500. Sausage—Fresh and smoked. By-Products—Tankage, grease and hog casings. Government inspection. Refrigeration—150-ton Vilter and 150-ton De La Vergne; brine circulating systems. Boiler capacity, 900 H. P. Trade Marks—Hams, "Brightwood"; bacon, "Brightwood" and "Oxford"; lard, "Pearl." Also exports.

WORCESTER
White, Pevey & Dexter Co.—Putnam Lane. Railroad, Boston & Albany. A corporation. Capital, $800,000; issued, $800,000. President, George H. Swift; Vice-President, E. C. Starr; Secretary, Frank W. Crocker; Treasurer, H. J. Nichols; General Manager, A. G. Diamond; General Superintendent, P. E. Munger. Employes, 280. Hogs, 4,500. Sausage—Fresh and smoked. By-Products—Tankage. Government inspection. Refrigeration—125-ton Vilter and 75-ton De La Vergne; direct expansion and brine spray systems. Boiler capacity, 530 H. P. Trade Marks—Hams, "Worcester" and "Leicester"; bacon, "Worcester" and "Leicester"; lard, "Mother's Leaf" and "Worcester Pure"; others, "Worcester Sausage and Patties," "Leicester" and "Putnam" brands. Wholesale Markets—B. Bridge Street, Worcester, Mass. Also exports.

Figures on cattle, hogs and sheep indicate AVERAGE WEEKLY KILLING CAPACITY.

MICHIGAN

BAY CITY

Bay City Packing Company

ANN ARBOR

Ann Arbor Packing Co.

CHESANING

Farmers' Meat & Produce Co.

G. M. Peet Packing Company—Railroad, Michigan Central. A corporation. Capital, $50,000; issued, $40,000. Stockholders, 7. President, H. D. Peet; Vice-President, T. O. Jones; Secretary and Treasurer, G. M. Peet; General Manager, H. D. Peet. Employes, 20. Cattle, 25; hogs, 200. Sausage—Fresh and smoked. By-Products—Tankage. Refrigeration—One 5-ton and one 3-ton Brecht; direct expansion. Boiler capacity, 15 H. P. Retail Markets—One in Chesaning, Mich., and one at 2716 S. Washington Ave., Saginaw, Mich.

COLDWATER

Coldwater Abattoir Co.—South Clay St. Wm. Houghtaling, Prop.

DETROIT

P. A. Breitenbeck—2530 Scotten Ave. Railroad, Michigan Central. Individual ownership. General Manager, P. A. Breitenbeck; General Superintendent, H. W. Breitenbeck. Employes, 14. Cattle, 500; calves, 600. City inspection. Refrigeration—8-ton Detroit and 8-ton Brunswick; direct expansion. Boiler capacity, 40 H. P.

Bresnahan Beef Company—24th St. and Michigan Central Tracks.

Detroit Packing Co.—1120 Springwells Ave. Railroad, Michigan Central. A corporation. Capital, $3,000,000; issued, $3,000,000. Stockholders, 2,000. President, Edward F. Dold; Vice-President, Frank L. Garrison; Secretary and Treasurer, Joseph Gardulski; General Manager and General Superintendent, Edward F. Dold. Employes, 600. Code—Cross; code word, "Detroiter." Cattle, 500; hogs, 1,000; sheep, 100. Sausage—Fresh, smoked and summer. By-Products—Bone fertilizer, calf meal, 60% protein hog digester tankage, 55% to 60% protein meat scrap and 50% protein meat meal. Refrigeration—One 100-ton and two 50-ton Arctics; brine circulating and direct expansion. Boiler capacity, 400 H. P. Trade Marks—"Detroit Star Meat Food Products," "Wolverine Brand" hams and bacon.

Gunsberg Packing Co., Inc.—2384 20th St. A corporation. Capital, $750,000; issued, $375,000. Stockholders, 150. President, Louis Gunsberg; Vice-President, Sam Gunsberg; Secretary and Treasurer, Paul Gunsberg; General Manager, Louis Gunsberg; General Superintendent, Ignatz Gunsberg. Employes, 25. Cattle, 500; sheep, 500. By-Products—Tankage and grease. City inspection. Refrigeration—One 50-ton and one 10-ton Phoenix. Branch at 2460 Riopelle St., Detroit.

Hammond Standish Co.—2101-2163 20th St. Railroad, Michigan Central. A corporation. Capital, $1,050,000; issued, $859,000. Stockholders, 55. President, T. W. Taliaferro; Vice-President, S. T. Nash; Secretary, C. Van Paris; Treasurer, Walter J. Graham; General Manager, T. W. Taliaferro; General Superintendent, Chas. Bomholt. Employes, 650. Codes—Cross, Robinson and A. B. C., 5th edi-

---

Figures on cattle, hogs and sheep indicate AVERAGE WEEKLY KILLING CAPACITY.

tion; code word, "Hamstand." Cattle, 900; hogs, 12,000; sheep, 600; calves, 400. Sausage—Fresh and smoked. By-Products—Tankage. Government inspection. Refrigeration (at Detroit)—One 250-ton Wolf-Linde, one 75-ton Wolf-Linde and one 75-ton Vilter; direct expansion and brine spray systems. At Toledo plant—Two Cleveland ice machines, 75 tons each; direct expansion. One 10-ton Huettemann & Cramer at Eastern Market; direct expansion; 50 tons ice daily. Boiler capacity, 1,400 H. P. at Detroit and 600 H. P. at Toledo, Ohio. Trade Marks—Hams, "Apex" and "Excelsior"; bacon, "Apex." "Excelsior" and "Comstock"; lard, "Apex" and "Silver Star"; compound, "Best Ever"; breakfast sausage, "Greenfield." Retail Markets—One at plant, Detroit. Also exports.

Kammann Beef Company—Buchanan and Michigan Central Tracks.

Kull & Bullen Beef Company—Peterson and Michigan Central Tracks. Railroad, Michigan Central. Partnership. Employes, 10. Refrigeration—8-ton Huettemann & Cramer.

Mason Beef Company—3632 Linden St. Railroad, Michigan Central. Partnership. Employes, 6. Cattle, 150. Refrigeration—One 10-ton Huettemann & Cramer; direct expansion.

Michigan Beef Company—Waterman and Dix Aves.

Nagle Packing Company—Waterman and Dix Aves.

Newton Packing Co.—5075 14th Ave. Railroad, Michigan Central. A corporation. Capital, $1,000,000; issued, $675,000. Stockholders, 11. President, Thomas E. Newton; Vice-President, John Kelsey; Secretary, Myron S. Sempliner; Treasurer, Edwin J. Smith; General Superintendent, R. C. Blue. Employes, 120. Codes—Cross. Cattle, 1,400; hogs, 4,000; sheep, 400; calves, 300. Sausage—Fresh, smoked and summer. By-Products—Hides, tallow and tankage. Refrigeration—Brine and direct expansion. Trade Mark—"Diamond N."

Parker Webb Co.—2811 Michigan Ave. R. Schiell, Manager. Owned by Allied Packers, Inc. (See Chicago.)

Ratkofsky & Witus—4070 Deming St. Railroad, Michigan Central. Partnership. Cattle, 450. Refrigeration—25-ton Detroit.

Robinson Beef Co.—5437 12th St. Railroad, Michigan Central. Individual ownership. General Manager, W. G Thompson. Employes, 15. Cattle, 225; sheep, 150. Refrigeration—25-ton Huettemann & Cramer; direct expansion. Boiler capacity, 40 H. P.

Sullivan Packing Co.—2590 Beecher Ave. Railroad, Michigan Central. A corporation. Capital, $1,750,000; issued, $1,500,000. Stockholders, 400. President, Frank J. Sullivan; Vice-Presidents, Thos. E. Tower and Wm. D. Flanigan; Treasurer, J. A. Martin; Secretary, J. Arthur Zengerle; General Manager, Frank J. Sullivan; General Superintendent, Wm. D. Flanigan. Codes—Cross, Robinson and Utility; code word, "Sullpack." Cattle, 500; hogs, 4,000; sheep, 1,000. Sausage—Fresh, smoked and summer. By-Products—Dried blood, tankage, white grease, inedible tallow, hoofs, horns and bones. Government inspection. Refrigeration—100-ton Frick, 150-ton De La Vergne and 50-ton Great Lakes; direct expansion. Boiler capacity, 675 H. P. Trade Marks—"Shannon Brand" for export; "Cadillac," "Favorite" and "St. Clair" brands for domestic. Branches—Toledo, Ohio; 456 Riopelle St., and Riopelle and High Sts., Detroit, Michigan. Retail Markets—One in Detroit. Also exports.

Figures on cattle, hogs and sheep indicate AVERAGE WEEKLY KILLING CAPACITY.

HOWELL

H. L. Williams Provision Co.

IRON RIVER

Peninsula Packing Co.—A corporation. President, A. J. Pohland; Vice-President, Wm. J. Tully; Secretary and Treasurer, John Scalucci.

MENOMINEE

Twin City Packing Co.—210-212 Belleview St. Railroad, Chicago & North Western. A corporation. Capital, $50,000; issued, $46,000. Stockholders, 10. President, Louis Krenz; Vice-President, Jerry Madden; Secretary and Treasurer, C. J. Wuellner; General Manager, Louis Krenz. Employes, 18. Cattle, 28; hogs, 60; calves, 50. Sausage—Fresh, smoked and summer. By-Products—Dried blood and tankage. Government inspection. Refrigeration—15-ton Wolf; direct expansion. Boiler capacity, 50 H. P. Trade Marks—"Cloverland" and "Queen."

PONTIAC

Pontiac Packing Company—Railroad, Grand Trunk. A corporation. Capital, $225,000; issued, $183,200. President, F. M. Kirby; Vice-President, V. H. Hancock; Secretary, A. C. Kirby; Treasurer, H. V. Hancock; General Manager, V. H. Hancock. Employes, 25. Cattle, 100; hogs, 600; sheep, 600. Sausage—Fresh, smoked and summer. Refrigeration—15-ton York; brine circulating. Boiler capacity, 100 H. P. Trade Mark—"Bloomfield."

MINNESOTA

ALBERT LEA

Albert Lea Packing Co., Inc.—Owned by Wilson & Co. (See Chicago.)

AUSTIN

Geo. A. Hormel & Co.—Railroads, Chicago Great Western and Chicago, Milwaukee & St. Paul. A corporation. Capital, $4,200,000; issued, $3,581,000. Stockholders, 427. President, George A. Hormel; Vice-President, Jay C. Hormel; Secretary, John G. Hormel; Treasurer, J. H. Nolan; General Superintendent, M. F. Dugan; Provision Department, E. M. Doane; Sales Department, J. G. Bramham; Traffic Department, O. W. O'Berg; Credit Department, R. L. Furtney; Claim Department, Paul C. Knopf; Purchasing Department, John G. Hormel; Live Stock Department, Ben F. Hormel; Beef Department, Geo. W. Fields. Employes, 1,260. Code—Cross; code word, "Puppy." Cattle, 1,536; hogs, 30,000. Sausage—Fresh, smoked and dry. By-Products—Tankage, fertilizer, bone meal, meat scraps, blood meal and baled hair. Government inspection. Refrigeration—One 250-ton Wolf-Linde and one 150-ton Vilter; direct expansion and brine spray, at plant. Branch house refrigeration—20-ton Vilter at Minneapolis; 10-ton Wolf-Linde at Atlanta, Ga.; 15-ton Vilter at Birmingham, Ala., and 15-ton York at Dallas, Tex.; direct expansion systems. Boiler capacity, 1,300 H. P. Trade Marks—Hams, "Dairy," "Austin," "Cedar"; bacon, "Dairy," "Austin," "Cedar," "Minnesota," "Tip Top," and "Midget"; lard, "Dairy," "Austin," "Blue Tin"; compound, "Invincible." Branch Houses—225 N. 5th St., Minneapolis,

Figures on cattle, hogs and sheep indicate AVERAGE WEEKLY KILLING CAPACITY.

6th and Pine Sts., St. Paul, 14-16 W. Michigan Ave., Duluth, Minn., 52 E. Alabama St., Atlanta, Ga., 2327 1st Ave. No., Birmingham, Ala., Dallas and San Antonio, Texas. Also exports.

DULUTH

Elliott & Company—37 Ave. West. Railroad, Northern Pacific. Individual ownership. General Manager, H. R. Elliott; General Superintendent, J. A. Wentworth; Assistant Manager, H. A. Elliott. Employes, 100. Cattle, 400; hogs, 700; sheep, 100. Sausage—Fresh, smoked and summer. By-Products—Tankage. Government inspection. Refrigeration—150-ton Vilter; direct expansion. Boiler capacity, 200 H. P. Trade Mark—"Zenith."

FERGUS FALLS

Fergus Co-operative Packing Co. Railroad, Northern Pacific. A corporation. Capital, $250,000; issued, $125,100. Stockholders, 800. President, C. R. Wright; Vice-President, O. A. Moses; Secretary and Treasurer, A. R. Kitts; Manager, J. R. Kiewel. Employes, 30. Code—Cross. Cattle, 50; hogs, 500; sheep, 50. Sausage—Fresh and smoked. By-Products—Dried blood and tankage. Government inspection. Refrigeration—45-ton Vilter; direct expansion. Boiler capacity, 160 H. P. One retail market at Fergus Falls, Minn.

HIBBING

Municipal Abattoir.

NEWPORT

Farmers' Terminal Packing Co.—Railroad, Chicago, Milwaukee & St. Paul. A corporation. Capital, $10,000,000; issued, $2,000,000. Stockholders, 15,000. President, Ira M. J. Chryst; Vice-President, A. F. Polson; Vice-President, H. Hillmond; Secretary and Treasurer, H. Edmunds. Employes, 200. Code—Cross; code word, "Fapaco." Cattle, 1,250; hogs, 10,000; sheep, 500. Sausage—Fresh and smoked. By-Products—Dried blood, tankage, hog and chicken feed and fertilizers. Government inspection. Refrigeration—One 150-ton Vilter and two 15-ton Brunswick; direct expansion. Boiler capacity, 450 H. P. Trade Marks—"Merit" and "Seal of Minnesota." Branch Houses—St. Paul, Minneapolis, Duluth and Brainard, Minn. Also exports.

ST. CLOUD

The Hunstiger Co., Inc.—817 S. Germain St. Railroads, Great Northern and Northern Pacific. A corporation. Capital, $100,000; issued, $48,400. Stockholders, 21. President, Frank J. Hunstiger; Vice-President, Leo J. Hunstiger; Secretary and Treasurer, John J. Spaniol; General Manager, Frank J. Hunstiger. Employes, 23. Cattle, 50; hogs, 300; sheep, 100. Sausage—Fresh, smoked and summer. By-Products—Tankage, hog and chicken feed and tallows. Refrigeration—15-ton and 5-ton York; brine circulating and direct expansion. Boiler capacity, 80 H. P. Trade Mark—"Premier Brand." Retail Markets—Three in St. Cloud and one in Waite Park, Minn.

ST. PAUL

Armour & Company. (See Chicago.)
Hertz & Rifkin.
R. J. King—Slaughterer.
Midway Abattoir Company.
Swift & Company. (See Chicago)

Figures on cattle, hogs and sheep indicate AVERAGE WEEKLY KILLING CAPACITY.

ST. PAUL

Katz & Horn Packing Co.—Individual ownership. General Manager, R. N. Katz. Employes, 25. Cattle, 350; hogs, 100; sheep, 100. By-Products—Tankage. Refrigeration—50-ton Vilter; direct expansion. Branch Houses—St. Paul and Minneapolis.

Luley Abattoir Co.—567-569 No. Cleveland Ave. Railroad, Minnesota Transfer Ry. A corporation. Capital, $150,000; issued, $40,000. Stockholders, 3. President, Frederick E. Luley; Vice-President, Martha A. Luley; Secretary, Martha A. Luley; Treasurer, M. L. Corneveaux; General Manager, Frederick E. Luley. Employes, 15. Cattle, 300. By-Products—Tallow, bones, neatsfoot oil and tankage. Refrigeration—None; use ice.

J. T. McMillan Co.—St. Clair and Spring Sts. Railroad, Chicago, St. Paul, Minneapolis & Omaha. A corporation. Capital, $600,000; issued, $485,000. Stockholders, 112. President, J. T. McMillan; Vice-President, Mrs. Annie McMillan; Secretary and Treasurer, Myron McMillan; General Superintendent, R. Seastrand. Employes, 225. Codes—Cross and Robinson. Hogs, 4,500. Sausage—Fresh, smoked and summer. By-Products—Dried blood and tankage. Government inspection. Refrigeration—One 120 and one 65-ton Vilter; direct expansion. Five tons ice daily. Boiler capacity, 300 H. P. Trade Marks—"Family Seal," "Paragon" and "Wheatland." Also exports.

WINONA

Interstate Packing Co.—Railroads, Chicago, Burlington & Quincy, Chicago, Milwaukee & St. Paul and Chicago & North Western. A corporation. Capital, $700,000. President, P. A. Jacobson; Vice-President, W. L. Gregson; Secretary and Treasurer, G. J. Rohweder; Assistant Secretary and Treasurer, C. P. Fehring; General Manager, P. A. Jacobson. Employes, 175. Codes—Cross and Robinson. Cattle, 200; hogs, 6,000; sheep, 500. Sausage—Fresh and smoked. By-Products—Dried blood, tankage, hog and chicken feed. Government inspection. Refrigeration—225-ton Wolf-Linde; 15-ton Remington at Minneapolis branch. Boiler capacity, 400 H. P. Trade Marks—"Bell," "Interstate," "Winona," "North Star" and "Acme." Branch House—Minneapolis, Minn. Also exports.

Winona Packing & Soap Co. Partnership—Benj. J. Kaiser and Walter C. Kaiser.

Lang Packing Co.

MISSISSIPPI

COLLINS
Wood-Canfield Company.

GREENWOOD
Farmers' Meat Packing Company.

GREENVILLE
Greenville Abattoir Company.

JACKSON
Sam Raines.
Geo. Gramp.

Figures on cattle, hogs and sheep indicate AVERAGE WEEKLY KILLING CAPACITY.

MISSOURI

CHILLICOTHE

Boehner's Slaughter House.

COLUMBIA

Hetzler Packing Co.—708 Broadway. Railroad, Missouri, Kansas & Texas. A corporation. Capital, $150,000; issued, $150,000. Stockholders, 23. President, W. J. Hetzler; Vice-President, J. P. Hetzler; Secretary, S. F. Conley; Treasurer, J. P. Hetzler. Employes, 45. Hogs, 50. Sausage—Fresh, smoked and summer. Refrigeration—100-ton York and 60-ton De La Vergne; direct expansion; 50 to 70 tons ice daily. Boiler capacity, 600 H. P. Trade Marks—Hams, "Old Log Cabin"; bacon, "Honey Suckle"; lard, "White Clover." Retail Markets—Two at Columbia, Mo.

EDINA

Krueger Packing Co.

HANNIBAL

Hannibal Packing Co.

JOPLIN

Boyd, Pipkin & Neal.

Trecker Bros.

KANSAS CITY

(See Kansas City, Kansas.)

KIRKSVILLE

A. J. Burk Meat Company.

MARYSVILLE

Forsyth Packing Co.

ST. JOSEPH

Armour & Company. (See Chicago.)

Morris & Company. (See Chicago.)

Seitz Packing & Mfg. Co.—Garfield Ave., 15th to 16th Sts. Railroad, Atchison, Topeka & Santa Fe. A corporation. Capital, $50,000. Stockholders, 4. President, A. J. Seitz; Secretary, Henry J. Glaser; Treasurer, Max A. Mang; General Manager, A. J. Seitz; General Superintendent, Max A. Mang; Sales Manager, Henry J. Glaser. Employes, 45. Cattle, 200; hogs, 500. Sausage—Fresh and smoked. By-Products—Tankage, hog and chicken feed and fertilizer. Refrigeration—10-ton Brecht and 50-ton Wolf-Linde; direct expansion and brine circulating. Boiler capacity, 180 H. P.; electric motor, 114 H. P. Trade Marks—"Best Products," "Lovers' Lane Brand" and "Seico Brand."

Swift & Company. (See Chicago.)

ST. LOUIS

American Packing Co.—3842 Garfield Ave. A corporation. Capital, $350,000; issued, $300,000. Stockholders, 90. President, Wm. G. Mueller; Vice-President, Edw. Olszewski; Secretary and Treasurer, Eugene F. Olszewski; General Manager, Karl Heim; Sales Manager, Chas. W. Honegger. Employes, 135. Cattle, 150; hogs, 1,500; sheep, 50. Sausage—Fresh, smoked and summer. By-Products—Tankage, hog and chicken feed. Refrigeration—50-ton Ruemmeli-Dawley; brine circulating. Boiler capacity, 300 H. P.

Figures on cattle, hogs and sheep indicate AVERAGE WEEKLY KILLING CAPACITY.

ST. LOUIS

Banner Packing Company—135 Russell Ave. Railroad, Missouri Pacific. A corporation. Capital, $300,000; issued, $225,000. Stockholders, 22. President, John Feldwisch; Vice-President, Henry Schreff; Secretary, Rich. Pechmann; Treasurer, Geo. F. Brueggemann. Employes, 38. Cattle, 200; hogs, 600; sheep, 100. Sausage—Fresh and smoked. By-Products—Tankage. Refrigeration—20-ton American and 20-ton Consolidated; direct expansion. Boiler capacity, 500 H. P.

J. H. Belz Provision Co.—3601 S. Broadway. Railroad, Manufacturers Railway Co. A corporation. Capital, $200,000; issued, $200,000. Stockholders, 3. President, John H. Belz; Vice-President, H. A. Belz; Secretary, A. von Brunn; General Manager, J. H. Belz. Employes, 150. Cattle, 200; hogs, 2,000; sheep, 50. Sausage—Fresh, smoked and summer. By-Products—Tankage, hog and chicken feed. Government inspection. Refrigeration—75-ton Maynard and 25-ton Wolf-Linde; direct expansion. Boiler capacity, 800 H. P. Trade Mark—"Belz Brand." Retail Market—One at plant.

John Bettendorf—1730 S. 9th St.

Carondelet Packing Company—8000 Ivory St.

Cox & Gordon Packing Co.—1019 S. 3rd St. Railroads, Missouri Pacific and Iron Mountain. Corporation. Capital, $250,000. Stockholders, 9. President, Chas. A. Cox; Vice-President and Treasurer, Sam Gordon; Secretary, Whitman B. Daniels. Employes, 65. Codes—Cross and Robinson. Government inspection. Boiler capacity, 270 H. P. Trade Marks—"Missouri" brand hams and bacon, "C & G" brand lard.

Gerst Brothers.

F. W. Haas Provision Company.

Heil Packing Co.—2216 La Salle St. Railroad, Missouri Pacific. A corporation. Capital, $500,000; issued, $500,000. President, Geo. L. Heil; Vice-President, Geo. L. Lauth; Secretary, Geo. L. Lauth; Treasurer, Geo. L. Heil; Sales Manager, Eugene Urban. Code—Cross. Cattle, 150; hogs, 3,000. Sausage—Fresh, smoked and summer. By-Products—Tankage. Government inspection. Refrigeration—One 100-ton Frick and one 50-ton Ball; brine circulating and direct expansion. Boiler capacity, 1,000 H. P. Also exports. Employes, 200.

Kehr Packing Company.

Hy. Klockmann—4222 Natural Bridge Road.

Knapp Packing Company.

Krey Packing Co.—2100 Bremen Ave. A corporation. Capital, $990,000; issued, $990,000. President, Fred Krey; Secretary, H. W. Wahlert; Treasurer, C. G. Breier. Employes, 500. Stockholders, 21. Codes—Cross, Bentley's, A. B. C., 5th, Robinson, Utility; Utility livestock and private; code word, "Krey-St. Louis." Cattle, 400; hogs, 12,000. Sausage—Fresh and smoked. By-Products—Dried blood, tankage and hog feed. Government inspection. Refrigeration—One 250 and two 60-ton Ball; direct expansion with fan system. Boiler capacity, 900 H. P. Trade Marks—"Xray" and "Pride" brands. Also exports.

Kruckemeyer's Sons.

Figures on cattle, hogs and sheep indicate AVERAGE WEEKLY KILLING CAPACITY.

Laclede Packing Company—2025 Shenandoah Ave.

Morris & Company. (See Chicago.)

Remley Packing Company—8201 Olive Street, Clayton, St. Louis County.

St. Louis Independent Packing Co.—3817 Chouteau Ave. Railroad, Missouri Pacific. A corporation. Capital, $3,500,000; issued, $3,500,000. President, Gustave Bischoff, Sr.; Vice-President, Gustave Bischoff, Jr.; Secretary and Treasurer, L. E. Dennig; General Manager, Gustave Bischoff; Beef Department Manager, W. W. Krenning. Codes—Cross, Griffin, Utility, and A. B. C., 5th edition. Cattle, 3,000; hogs, 20,000; sheep, 500. Sausage—Fresh, smoked and summer. By-Products—Dried blood, tankage and hog feed. Government inspection. Refrigeration—One 600-ton Ruemmeli, one 450-ton American and one 350-ton York; direct expansion; 60 tons ice daily. Boiler capacity, 3,400 H. P. Trade Marks—Hams, "Independent"; bacon, "Independent," and lard, "Rock Springs." Branch Houses—6349 East Station Street, Pittsburgh, Pa. Also exports.

Sartorius Provision Company—2734 Arsenal St.

Wm. Schmidt—7511 Michigan Ave.

Seiloff Packing Co.—4319-39 Natural Bridge Ave. A corporation. Capital, $400,000; issued, $350,000. Stockholders, 60. President, Emil Seiloff; Vice-President, Geo. Hohmann; Secretary, Simon Zeitler; Treasurer, Simon Zeitler; General Manager, Emil Seiloff; General Superintendent, Simon Zeitler. Employes, 125. Cattle, 175; hogs, 2,400; sheep, 50; calves, 100. Sausage—Fresh, smoked and summer. Refrigeration—Two 12½ and one 32-ton York; brine circulating and direct expansion. Boiler capacity, 200 H. P. Trade Mark—"Honey-Dew." Retail Market—One at 4339 Natural Bridge Avenue, St. Louis.

Swift & Company—3919 Papin St. (See Chicago.)

Waldeck Packing Co.—Montrose and La Salle Sts. Branch of St. Louis Independent Packing Company. General Manager, J. C. C. Waldeck. Government inspection. Refrigeration—One 150-ton Ball, one 65-ton York, one 40-ton Menard and one 25-ton Featherstone; brine circulating; 50 tons ice daily. Boiler capacity, 600 H. P. Trade Mark—"Waldeck."

SIKESTON

Walpole Packing Co.

SPRINGFIELD

Welsh Packing Co.—Box 56, So. Side Station. Railroad, Frisco. A corporation. Capital, $200,000; issued, $200,000. Stockholders, 22. President, Thos. N. Welsh; Vice-President, Ed. V. Williams; Secretary, Thos. J. Glynn; Treasurer, L. C. Kennedy; General Manager, Thos. J. Glynn; General Superintendent, J. J. Glynn. Employes, 45. Cattle, 75; hogs, 600; sheep, 25. Sausage—Fresh, smoked and summer. By-Products—Dried blood, tankage and hog feed. Refrigeration—50-ton United Iron Works and 25-ton York; direct expansion. Boiler capacity, 80 H. P. Trade Marks—Hams, "Ozark"; bacon, "Ozark"; lard, "Ozark" and "Special."

Tegarden Packing Co.

Banfield Packing Co.

Figures on cattle, hogs and sheep indicate AVERAGE WEEKLY KILLING CAPACITY.

TIPTON

White's Market. Railroad, Missouri Pacific. Individual ownership. Employes, 5. Cattle, 25; hogs, 20; sheep, 6. Sausage—Fresh and smoked. Retail Markets—One in Tipton.

WEST PLAINS

Southern Packing Company—Pork packers and sausage manufacturers.

West Plains Serum Co.—Springfield Road.

MONTANA

BILLINGS

Crosser's Meat Co.

Yellowstone Packing Company—Meat packers and provisioners. Capital, $1,250,000; issued, $772,000. President and General Manager, John B. Henderson; Vice-President, Freeman Philbrick; Secretary, O. F. Goddard; Treasurer, Richey Young. Trade Marks—"Old Faithful" and "Yellowstone" hams, bacon, lard. Government inspection.

BUTTE

Hansen Packing Company—A corporation. President and General Manager, Walter Hansen; Secretary, Minnie Hansen.

GREAT FALLS

Great Falls Meat Co.—310 Central Ave. Railroad, Great Northern and Chicago, Milwaukee & St. Paul. A corporation. Capital, $150,000. Stockholders, 4. President, H. P. Brown; Vice-President, F. B. Brown; Secretary, C. N. Dickinson.

HAVRE

J. G. Pedersen & Co.—216 1st St. A corporation. Capital, $50,000; issued, $16,000. Stockholders, 4. President, J. G. Pedersen; Vice-President, Paul F. Krezelak; Secretary, E. F. Montgomery; Treasurer, E. F. Montgomery; General Manager, J. G. Pedersen. Employes, 20. Cattle, 50; hogs, 100; sheep, 100. Sausage—Fresh, smoked and summer. Refrigeration—8-ton Niebling; direct expansion. Trade Mark—"Woodside." Retail Markets—Pioneer Market and Gerry Market, East Havre, Mont.; Pioneer Groc. Co., Havre, Mont.

HELENA

Northwestern Packing Co.—Railroad, East Helena. A corporation. Capital, $400,000; issued, $100,000. President, O. A. Anderson; Vice-President, Andrew Boyd; Secretary and Treasurer, Claude C. Hay.

LIVINGSTON

Retallick & Baumgart—112 S. Main St. Partnership. Employes, 4. Cattle, 25; hogs, 50; sheep, 15. Sausage—Fresh, smoked. Refrigeration—5-ton Brecht; direct expansion. Boiler capacity, 8 H. P. Retail Market at Livingston, Mont.

NEBRASKA

GRAND ISLAND

K. W. H. Company. (Successors to Loup Valley Packing Co.)

Figures on cattle, hogs and sheep indicate AVERAGE WEEKLY KILLING CAPACITY.

HASTINGS

Kauf & Rinderspacher Co.—613 W. 2nd St. Railroad, Chicago, Burlington & Quincy. A corporation. Capital, $150,000; issued, $100,000. Stockholders, 5. President, W. J. Rinderspacher; Vice-President, August Rinderspacher; Secretary and Treasurer, George Rinderspacher. Employes, 40. Cattle, 300; hogs, 600. Sausage—Fresh, smoked and summer. Refrigeration—Two 8-ton Brecht and 14-ton York; brine circulating. Boiler capacity, 80 H. P. Trade Mark—"Countrymaid."

LINCOLN

Lincoln Packing Co.—320 N St. Railroad, Chicago, Burlington & Quincy. A corporation. Capital, $300,000; issued, $139,500. Stockholders, 14. President, V. E. McArthur; Vice-President, T. M. Madden; Secretary, J. P. Murphy; General Manager, T. M. Madden; General Superintendent, W. A. Mechling. Employes, 45. Cattle, 350; hogs, 550. Sausage—Fresh and smoked. Refrigeration—75 and 50-ton Vilter; brine circulating and brine spray systems. Boiler capacity, 175 H. P. Trade Mark—"Lancaster."

Weiler Packing Company—214 N. 10th St. Individual ownership. Employes, 7. Cattle, 30; hogs, 50; sheep, 5. Sausage—Fresh, smoked and summer. Refrigeration—2-ton Baker; direct expansion. Boiler capacity, 100 H. P.

NEBRASKA CITY

Morton-Gregson Company—Railroad, Chicago, Burlington & Quincy. A corporation. Capital, $500,000; issued, $350,000. Stockholders, 2. President, C. F. Burrell; Vice-President, C. M. Aldrich; Secretary, H. B. Goff; Treasurer, F. J. Penn; General Manager, C. M. Aldrich; General Superintendent, E. F. Williams. Employes, 450. Codes—Private; code word, "Gregmore." Hogs, 7,500. Sausage—Fresh and smoked. By-Products—Dried blood, tankage, hog feed, hog hair and casings. Government inspection. Refrigeration—75-ton De La Vergne, 50-ton Vilter, 75-ton Wolf-Linde and 100-ton Wolf-Linde; direct expansion and brine circulating. Boiler capacity, 1,800 H. P. Trade Marks—Hams, "Certified," "Otoe," "Red Top," "White Breast," and "Coupon"; bacon, "Certified," "Otoe," "Red Top," "White Breast," and "Coupon"; lard, "Certified," "Laurel," and "Coupon." Branch House—Philadelphia, Penn. Also exports.

OMAHA

Dold Packing Co.—27th and Y Sts. Railroad, Union Pacific. A corporation. Capital, $1,000,000; issued, $1,000,000. President, J. C. Dold; Vice-Presidents, R. S. and J. P. Dold; Secretary, P. O. Rial; Treasurer, J. L. Carson; General Manager, R. S. Dold; General Superintendent, J. J. Cuff. Employes, 1,000. Codes—Cross-Yopps, Western Union and A. B. C., 5th; code word, "Doldqual." Cattle, 500; hogs, 2,500; sheep, 1,000; calves, 500. Sausage—Fresh and smoked. By-Products—Dried blood, tankage, hog and chicken feed and greases. Government inspection. Refrigeration—One York and one De La Vergne, total tonnage 650; brine circulating, direct expansion and brine spray system; 40 tons ice daily. Boiler capacity, 1,000 H. P. Trade Marks—Hams, "Niagara" and "Sterling"; bacon, "Niagara" and "Sterling"; lard, "Niagara" and "White Rose"; compound, "Sterling." Branch Houses—215 N. Green St., Chicago, Ill., and 107-9 S. Front St., Memphis, Tenn. Also exports.

Figures on cattle, hogs and sheep indicate AVERAGE WEEKLY KILLING CAPACITY.

OMAHA

Armour & Company—South Side Station. (See Chicago.)

Higgins Packing Co.—36th and L Sts. Railroad, Chicago, Burlington & Quincy. A corporation. Capital, $5,000,000; issued, $910,803. Stockholders, 1,100. President, Walter V. Hoagland; Vice-President, Florian Fuchs; Secretary and Treasurer, E. E. Howell; General Manager, John W. Pepperdine; General Superintendent, G. C. Pironnet. Employes, 75. Codes—Cross. Cattle, 300; hogs, 2,000; sheep, 200. Sausage—Fresh, smoked and summer. By-Products—Dried blood, tankage, hog and chicken feed. Government inspection. Refrigeration—25, 15 and 10-ton Baker; direct expansion; 25 tons ice daily. Boiler capacity, 40 H. P. Trade Marks—"Higgins Special Dry Cured Bacon," "Midvale Sweet Pickle," "Okay," "Security," and "Mascot Brand." Branch house—Missouri Valley, Iowa.

Hoffman Brothers.

Midwest Packing Co.—26th and P Sts. Railroads, Union Pacific and Missouri Pacific. Partnership, J. P. Mailender and J. S. Hoffman. Employes, 8. Cattle, 100; hogs, 50; sheep, 10. Sausage—Fresh and smoked. By-Products—Tankage. City inspection.

Morris & Company—29th and O Sts. (See Chicago.)

Omaha Packing Company—Branch plant. (See Chicago.)

Skinner Packing Co.—12th and Douglas Sts. A corporation. Capital, $8,000,000; issued, $7,250,000. Stockholders, 4,000. President, Keith Neville; Vice-President, Robert Gilmore; Secretary and Treasurer, D. C. Robertson; General Manager, Keith Neville. Government inspection. Trade Marks—"Skinners." NOTE: This company's plant is leased to Dold Packing Company, Omaha, Neb.

Swift & Company—27th and Q Sts., South Side Station. (See Chicago.)

NEW JERSEY

CAMDEN

A. Schlorer & Sons—800 Chestnut St. A corporation. Capital, $250,000; issued, $99,300. Stockholders, 4. President, Adam Schlorer; Vice-President, John A. Schlorer; Secretary, Wm. E. Schlorer; Treasurer, John A. Schlorer; General Manager, John A. Schlorer. Employes, 20. Cattle, 90; sheep, 100. By-Products—Tankage. Refrigeration—12-ton York; direct expansion. Boiler capacity, 50 H. P. Retail Market—800 Chestnut St., Camden.

HARRISON

Swift & Company. (See Chicago.)

JERSEY CITY

Armour & Company—324 Seventeenth St. (See Chicago.)

Brainard Bros.—15 Exchange Place.

H. Heilbrunn Co.—Foot of Sixth St.

Thos. A. Hughes Co.—Exchange Place.

Jersey City Stock Yards Co.—Foot of 6th St. Railroads, Pennsylvania, Erie, New York Central and Lehigh Valley. A corporation. Capital, $500,000; issued, $500,000. Stockholders, 7. President, R. C. Bonham; Secretary, F. A. Cassidy; Treasurer, H. L. Pope; General Manager, R. C. Bonham; Superintendent, F. N. Greenlaw; Traffic

---

Figures on cattle, hogs and sheep indicate AVERAGE WEEKLY KILLING CAPACITY.

Manager, F. A. Cassidy. Cattle, 400; hogs, 20,000; sheep, 10,000. By-Products—Tankage, hides, calf skins, lard, tallow and grease. Government inspection. Refrigeration—150-ton York; brine circulating and brine spray systems. Boiler capacity, 815 H. P.

Swift & Company—154 Ninth St. (See Chicago.)

Nagle Packing Co.—681-691 Henderson St. Railroad, Delaware Lackawanna & Western, Hoboken. A corporation. Capital, $500,000; issued, $500,000. Stockholders, 4. President, M. H. Nagle; Vice-President, E. A. Cudahy, Jr.; Secretary, A. W. Anderson; Treasurer, J. E. Wagner; Plant Manager, L. F. Gerber. Employes, 300. Codes—Utility and Griffiths. Cattle, 1,500; sheep, 5,000; calves, 1,000. By-Products—Oils, fertilizers, tallows, wools and fats. Government inspection. Refrigeration—130-ton Vilter, brine spray system. Boiler capacity, 300 H. P. Trade Marks—"Henderson Brand Extra Oleo Oil," and "Palisade Brand Oleo Oil." Branch houses—Hoboken, N. J., and Detroit, Mich. Also exports.

NEWARK

Bimbler Company—30 Plane St. Railroads, Erie, Delaware Lackawanna & Western and Pennsylvania. A corporation. Capital, $25,000; issued, $25,000. Stockholders, 10. President, James A. Brady; Vice-Presidents, George J. Edwards and R. B. Neff; Secretary and Treasurer, E. W. Meyer; General Manager, James A. Brady; General Superintendent, James T. Boyle. Employes, 150. Code—Utility. Hogs, 20,000. By-Products—Grease, casings and hair. Government inspection. Also exports. Slaughterers only.

A. Fink & Sons—810 Frelinghuysen Ave. Railroad, Pennsylvania. A corporation. Capital, $1,000,000; issued, $331,000. Stockholders, 68. President, August C. Fink; Vice-President, Adolph Fink; Secretary, Louis F. Keller; Treasurer, August C. Fink; General Manager, Louis F. Keller. Employes, 350. Codes—Cross and Utility; code word, "Finkco." Cattle, 200; hogs, 5,000. Sausage—Fresh and smoked. By-Products—Greases, dried blood, tankage and hair. Government inspection. Refrigeration—100-ton De La Vergne, 35-ton Vilter and 35-ton Ruemmeli-Dawley; brine circulating, direct expansion and brine spray systems; manufactures ice. Boiler capacity, 650 H. P. Trade Marks—Hams, bacon and lard, "Finkco Brand"; "Superior Quality." Retail Market—127 Belmont Ave., Newark, N. J. Also exports.

Schloss, Held & Schloss—Ft. of Astor St. Railroad, Pennsylvania. Individual ownership. General Manager and Superintendent, Leo Schloss; H. Singer, Accountant. Employes, 40. Cattle, 100; sheep, 2,000. Government inspection. By-Products—Tankage. Boiler capacity, 30 H. P.

Swift & Company—Harrison Ave., Kearney Station. (See Chicago.)

NEW BRUNSWICK

H. Keller.

PASSAIC

Henry Muhs Company—41 Central Ave., cor. Monroe St. A corporation. Capital, $500,000. President, Henry C. Muhs; Vice-President, Geo. Muhs, Sr.; Secretary and Treasurer, Herbert Rumsey.

Figures on cattle, hogs and sheep indicate AVERAGE WEEKLY KILLING CAPACITY.

PATERSON

D. Fullerton & Co.—306 River St. Railroad, Erie. A corporation. Capital, $200,000; issued, $200,000. Stockholders, 7. President, Emmons T. Fullerton; Vice-President, David Fullerton; Secretary, Emmons P. Fullerton; Treasurer, Ethelbert G. Fullerton; General Manager, Ethelbert G. Fullerton; General Superintendent, Ethelbert G. Fullerton; Manager Packing Department, L. J. Cook. Employes, 100. Codes—The Utility Live Stock Cipher. Cattle, 100; hogs, 1,000; sheep, 100. Sausage—Fresh, smoked and summer. By-Products—Dried blood, tankage and inedible tallow. Government inspection. Refrigeration—30-ton Vesterdahl, 25-ton Ball and 15-ton Vesterdahl; direct expansion. Boiler capacity, 450 H. P. Trade Marks—Hams, bacon and lard, "Fullerton's Special."

WEST HOBOKEN

Charles R. Miller Company—Secaucus Road.

NEVADA

RENO

Nevada Packing Co.—4th and Alameda Sts. Railroads, Western Pacific and Southern Pacific. A corporation. Capital, $500,000; issued, $353,400. Stockholders, 5. President, J. W. Blum; Vice-President, F. E. Humphrey; Secretary, Wm. Henderson; General Manager, J. W. Blum; General Superintendent, A. Devine. Employes, 120. Code—Cross. Cattle, 350; hogs, 1,200; sheep, 2,000. Sausage—Fresh and smoked. By-Products—Dried blood, tankage, hog and chicken feed. Government inspection. Refrigeration—35 and 50-ton Vulcan and 80-ton Ball; direct expansion; 10 tons ice daily. Boiler capacity, 275 H. P.

Reno Supply Co.—645 Sierra St.

ELY

Ely Packing Company—A corporation. Capital, $100,000. President, W. N. McGill; Secretary and Treasurer, V. J. Carruthers; General Manager, J. H. Eager.

FALLON

Fallon Slaughtering Company.

GOLDFIELD

T. & G. Meat Company.
United Cattle & Packing Company.

NEW HAMPSHIRE

GARFIELD

Joseph Feld Co.
Alfred Fusco.

PORTSMOUTH

Herman A. Brackett—368 South St. Slaughterers.

NEW YORK

ALBANY

Lewis Newhof & Son—410 S. Pearl St. Railroad, Delaware & Hudson. Partnership. Cattle, 200. By-Products—Tankage and hog and chicken feed. Government inspection. Refrigeration—15-ton York. Boiler capacity, 70 H. P.

Figures on cattle, hogs and sheep indicate AVERAGE WEEKLY KILLING CAPACITY.

BROOKLYN

Gotham Packing Co., Inc.—352 Johnson Ave. A corporation. Capital, $100,000; issued, $100,000. President, Leo S. Joseph; Vice-President, Samuel Plaut; Secretary, Arthur A. Plosscowe; General Manager, Samuel Plant; General Superintendent, Arthur A. Plosscowe. Employes, 30. Cattle, 500. By-Products—Tallow and grease. Government inspection. Refrigeration—30-ton York and 20-ton Mayer; direct expansion. Boiler capacity, 75 H. P.

International Provision Company—33-43 DeGraw St. A corporation. Capital, $10,000; issued, $10,000. President, P. J. Sweeney; Vice-President, T. J. Sweeney; Treasurer, E. Patten; General Manager, P. J. Sweeney. Employes, 75. Codes—Robinson and A. B. C., 5th edition; code word, "Hazelyork." Hogs, 3,000. Government inspection. Refrigeration—150-ton Allen; brine circulation; 10 tons ice daily. Boiler capacity, 300 H. P.

C. Lehmann Packing Co., Inc.—321 Johnson Ave. Railroad, Long Island. A corporation. Capital, $250,000; issued, $250,000. Stockholders, 5. President, Camille Lehmann; Vice-President, Leon Lehmann; Secretary, Florence Lehmann; Treasurer, Maurice Lehmann; General Manager, Maurice Lehmann; General Superintendent, Jacob Romer. Employes, 15. Cattle, 500. Government inspection. Refrigeration—16-ton Buffalo; direct expansion.

Hugo Strauss Packing Co., Inc.—284 Johnson Ave. Railroad, Bushwick Station, Long Island R. R. A corporation. President, Philip Valk; Secretary, M. C. Abusa; Treasurer, Hugo Strauss; General Manager, Philip Valk; General Manager, Max Valk. Employes, 25. Codes—A. B. C., 5th, Bentley and Lieber; code word, "Philvalk-New York." Horses, 200. Sausage—Fresh, smoked and summer. By-Products—Hog and chicken feed, dog food, ground bones and polished bones. Government inspection. Refrigeration—12-ton Automatic; brine circulating. Packs horse meat products. New York office, 305 Broadway. Also exports.

BUFFALO

Jacob Dold Packing Co.—745 William St. Railroad, New York Central. A corporation. Capital, $6,000,000; issued, $6,000,000. President, J. C. Dold; Vice-Presidents, F. W. Dold, J. P. Dold and R. S. Dold; Secretary, J. J. Dolphin; Treasurer, J. L. Carson; General Manager, J. P. Dold; General Superintendent, J. J. Cuff. Employes, 1,800. Codes—Cross, Yopps, Western Union, A. B. C., 5th; code word, "Doldqual." Cattle, 6,000; hogs, 30,000; sheep, 6,000, at Buffalo plant. Cattle, 1,500; hogs, 15,000; sheep, 3,000, at Wichita, Kans., plant. Sausage—Fresh and smoked. By-Products—Dried blood, tankage, hog and chicken feed, glue and greases. Government inspection. Refrigeration—Two Keystone, three Arctic, one Hercules, two Carbondale, totalling 1,070 tons; 45 tons ice daily. Boiler capacity, 3,175 H. P. Trade Marks—Hams, "Niagara" and "Sterling"; bacon, "Niagara" and "Sterling"; lard, "Niagara" and "White Rose"; compound, "Sterling." Branch Houses—Wichita, Kansas; Perry and East Market St., Buffalo; 245-261 Walton St., Syracuse, N. Y.; 1204-5 Metz St., Brooklyn, N. Y., 431 Main St., Utica, N. Y.; 77-79 S. Market, Boston, Mass., and Matthew St., Liverpool, England. Also exports.

Figures on cattle, hogs and sheep indicate AVERAGE WEEKLY KILLING CAPACITY.

BUFFALO

Danahy Packing Co.—25 Metcalf St. Railroad, New York Central. A corporation. Capital, $150,000; issued, $150,000. Stockholders, 5. President, John M. Danahy; Vice-Presidents, Arthur T. Danahy and Raymond G. Danahy; Secretary, S. Edgar Danahy; General Manager, John M. Danahy. Employes, 210. Codes—Cross. Cattle, 250; hogs, 4,000; sheep and lambs, 1,200. Sausage—Fresh and smoked. By-Products—Tankage. Government inspection. Refrigeration—90 and 60-ton York and 30-ton Case; direct expansion. Trade Marks—"Easter Brand" and "Tastefine." Also exports.

Klinck & Schaller, Inc.—620-630 Babcock St. Railroad, Erie. A corporation. Capital, $185,000; issued, $185,000. Stockholders, 4. President, Jacob C. Schaller; Secretary, A. C. Klinck; Treasurer, R. L. Klinck; General Manager, R. L. Klinck. Employes, 25. Cattle, 350; lambs, 1,000. By-Products—Dried blood, tankage, chicken feed and tallow. Refrigeration—Two 35-ton Fricks; brine spray system. Boiler capacity, 60 H. P. Trade Mark—"Quality First."

Klinck Packing Co.—G. W. Klinck, Manager. Owned by Allied Packers, Inc. (See Chicago.)

Klinck Bros., Inc.—588 Howard St. and 107 E. Market St. Partnership; Christian and Charles Klinck.

Laux & Edbauer—40 Spencer St.

Jacob Moschel's Sons, Inc.—153 Peckham St. A corporation. Capital, $75,000; issued, $75,000. President and Treasurer, Charles Moschel; Vice-President, Conrad Moschel; Secretary, Louis Moschel.

Neber & McGill—16 Hanna St.

New England Dressed Meat Company—Howard and Babcock Sts. (Swift & Company plant.)

Frank Rausch—618 Howard St.

Edward J. Smith Packing Company—247 Lewis St.

Sahlen Packing Co.—318 Howard St. Co-partnership composed of following members of firm: Edw. C. Sahlen, Jos. W. Sahlen, Wm. Sahlen, Frank J. Sahlen, Alexander Sahlen, Elizabeth S. Sahlen and Jos. Rast. Employes, 50. Codes—Cross. Hogs, 1,500. Sausage—Fresh and smoked. By-Products—Dried blood and tankage. Government inspection. Refrigeration—50-ton Frick and 25-ton Case; direct expansion. Boiler capacity, 300 H. P. Trade Mark—"Sahlen." Also exports.

CANAJOHARIE

Beech-Nut Packing Co.—Church St. Railroad, West Shore. A corporation. Capital, $3,000,000; issued, $2,124,500. Stockholders, 700. President, Barlett Arkell; Vice-Presidents, F. E. Barbour and Edward S. Moore; Secretary, W. C. Arkell; Treasurer, John S. Ellithorp; Superintendent, Col. Davis. Employes, 1,250. Government inspection. Trade Mark—"Beech-Nut." Do not slaughter. Pack meat specialties, bacon, etc.

CHESTER

Frank J. Murray Co., Inc.—Railroad, Erie. A corporation. Capital, $100,000; issued, $75,000. Stockholders, 5. President, Frank J. Murray; Vice-President, C. A. Koelsch; Secretary, F. J. Murray; Treasurer, C. A. Koelsch. Employes, 15. Cattle, 100; hogs, 250; sheep, 250. By-Products—Dried blood and tankage. Government inspection. Refrigeration—15-ton York; brine circulating. Boiler capacity, 40 H. P.

Figures on cattle, hogs and sheep indicate AVERAGE WEEKLY KILLING CAPACITY.

HUDSON

C. A. Van Deusen Co.—13-23 North 7th St. A corporation. Capital, $120,000; issued, $81,300. Stockholders, 8. President, Charles A. Van Deusen; Vice-President, Leslie M. Van Deusen; Secretary, C. Werter Van Deusen; Treasurer, D. H. Van Deusen. Employes, 40. Sausage—Fresh and smoked. Do not slaughter. Wholesale curers and sausage makers. Government inspection. Refrigeration—10-ton Carbondale.

KINGSTON

Jacob Forst Packing Co.—Partnership. J. & M. B. Forst. Capital $250,000.

NEW YORK CITY

Allied Packers, Inc.—Owned by Allied Packers, Inc. (See Chicago.)

Armour & Co.—39th St. and 11th Ave. (See Chicago.)

F. A. Ferris & Co., Inc.—262-272 Mott St. A corporation. President, F. A. Ferris, Jr.; Vice-Presidents, E. S. Hand and W. P. Uhler; Secretary, Merritt L. Stewart; Treasurer, Harris H. Uhler; Plant Superintendent, W. P. Uhler. Government inspection. Trade Mark—"Ferris." Cure hams and bacon.

Figge & Hutwelker Co., Inc.—621-7 W. 40th St. A corporation. Capital, $100,000. President and Treasurer, Frederick Figge; Vice-President, Charles Hutwelker; Secretary, Alexander H. Figge. Pork packers.

Thomas Halligan—549 W. 40th St. Individual. Sheep, 1,000. Government inspection. Refrigeration—Rents cooler space for 300 lambs. Only slaughters lambs.

A. Lester Heyer—318-320 East 39th St. Individual ownership. Government inspection. Refrigeration—20-ton Voss; brine circulating. Pork packer.

New York Butchers' Dressed Meat Co.—495 11th Ave. Railroad, New York Central. A corporation. Owned by Armour & Co. Capital, $2,000,000; issued, $2,000,000. President, Frederick Joseph; Vice-President, Leo S. Joseph; Secretary, Edward Bacon; Treasurer, Frederick Croll; General Manager, Leo S. Joseph; General Superintendent, Alfred Fryatt. Employes, 550. Codes used—Armours. Cattle, 3,360; sheep, 7,200; calves, 3,000. By-Products—Tankage, oleo oil, oleo stearine and tallow. Government inspection. Refrigeration—100 and two 150-ton Vilter; brine spray system. Boiler capacity, 1,468 H. P. Also exports.

Rohe & Brother—527 West 36th St. A corporation. President, Chas. Rohe; Vice-President, A. T. Rohe; Secretary, Wm. Rohe; Treasurer, O. F. Rohe. Pork packers. Also export.

J. W. & P. Scanlan, Inc.—613-619 West 40th St. A corporation. Capital, $75,000; issued, $75,000. Stockholders, 16. President, Harry Scanlan; Vice-President, J. H. Scanlan; Secretary, O. F. England; Treasurer, M. E. Scanlan; General Manager, Harry Scanlan; General Superintendent, F. O. Sullivan. Employes, 109. Sheep, 5,000. Government inspection. Refrigeration—75-ton De La Vergne and 20-ton York; direct expansion. Boiler capacity, 300 H. P. Branch Houses—169 Fort Greene Place, Brooklyn, and 14-16 Thompson St., New York City.

Figures on cattle, hogs and sheep indicate AVERAGE WEEKLY KILLING CAPACITY.

NEW YORK CITY

Joseph Stern & Sons, Inc.—616 W. 40th St. A corporation. Capital, $3,000,000; issued, $3,000,000. President, F. L. Bisbee; Vice-President, J. H. Burns; Secretary and Treasurer, M. E. Smith; Assistant Secretary, E. C. Hartman; General Manager, F. L. Bisbee; General Superintendent, M. S. Mandle. Employes, 800. Codes—Scotts, Bentley, Liebers, W. U., A. B. C., 5th; code word, "Jostern." Cattle, 2,000; hogs, 12,000; calves, 1,000. Sausage—Fresh, smoked and summer. By-Products—Dried blood, tankage, oleo oil, prime oleo stearine, tallows, greases, hides and cattle tails. Government inspection. Refrigeration—500-ton De La Vergne; direct expansion. Boiler capacity, 1,050 H. P. Trade Marks—Sausage, "Foresthill" and "Anchor." Also exports.

Strauss & Adler—607-11 W. 40th St. Beef butchers.

Swift & Company—778 First Ave. (See Chicago.)

United Dressed Beef Co. of New York—778 First Ave. A corporation. Capital, $3,000,000. President, Walter Blumenthal; Vice-President and Treasurer, Irving Blumenthal; Secretary, Martin Rothschild. Slaughterers, packers, manufacturers of oleo oil, tallow, tankage and blood and cured meats. Government inspection. Also exports.

Wilson & Co.—45th and First Ave. (See Chicago.)

PINE PLAINS

Roy W. Pulver.

ROCHESTER

Rochester Packing Company, Inc.—Maple St. Railroad, New York Central and Hudson River. A corporation. Capital, $500,000; issued, $379,800. Stockholders, 40. President, Frederick M. Tobin; Vice-President and General Superintendent, O. E. Espey; Secretary, P. J. Vaeth, Jr.; Treasurer, J. J. Burke; General Manager, Frederick M. Tobin. Employes, 112. Hogs, 1,500. Sausage—Fresh and smoked. Refrigeration—60 and 35-ton York; direct expansion; 10 tons ice daily. Boiler capacity, 275 H. P.

Rochester Abattoir Co.—Kerr St. Owner, Joseph Amdoursky.

SCHENECTADY

Miller Bros.—Slaughterers.

SYRACUSE

A. C. Hofmann & Sons—301-327 Free St. Railroads, New York Central, West Shore. A corporation. Capital, $500,000. Stockholders, 3. President and General Manager, A. C. Hofmann, Jr.; Treasurer and Assistant Manager, N. L. Hofmann; Secretary and Sales Manager, T. P. Considine. Employes, 60. Cattle, 100; hogs, 1,000. Sausage—Fresh and smoked. By-Products—Dried blood and tankage. City inspection. Refrigeration—30-ton Automatic, 75-ton Carbondale. Boiler capacity, 200 H. P. Trade Marks—"Peerless" and "Diamond H."

TONAWANDA

Schwinger Packing Co.

H. S. Golde Packing Co.

TROY

Greylock Packing Co.

Figures on cattle, hogs and sheep indicate AVERAGE WEEKLY KILLING CAPACITY.

UTICA

C. A. Durr Packing Co., Inc.—A corporation. Capital, $1,000,000. President, Herman A. Amberg; Vice-President, Albert C. Durr; Secretary, Jacob F. Ammann; Treasurer, John M. Snyder.

Spath Bros.—Canal St.

Jutz & Pfluke Packing Co., Inc.—Foot of Schuyler St. Railroad, New York Central. A corporation. Capital, $100,000; issued, $55,100. Stockholders, 5. President, L. M. Pfluke; Vice-President, Anton Jutz; Secretary, N. J. Dilker; Treasurer, G. A. Hoetzer; General Manager, L. M. Pfluke. Employes, 25. Hogs, 300; lambs, 75; cattle, 30; calves, 25. Sausage—Fresh and smoked. Refrigeration—12-ton York; direct expansion. Boiler capacity, 35 H. P.

YONKERS

Siebert Bros.

NEW MEXICO

KOEHLER

Blossburg Merc. Company—Railroad, Koehler Junction. A corporation. General Manager, A. W. Wildenstein. Cattle, 100; hogs, 200. Sausage—Fresh and smoked. Refrigeration—40-ton Huettemann & Cramer; brine circulating system. Retail Markets—Koehler, Brilliant, Gardiner, Van Houter and Sugarite, New Mexico.

NORTH CAROLINA

ASHEVILLE

Asheville Packing Company—428 Depot St. Partnership. General Manager, F. Zimmerman. Railroad, Southern. Cattle, 100; hogs, 20; sheep, 12. By-Products—Dried blood and tankage. Refrigeration—8-ton Brecht; direct expansion.

RALEIGH

Raleigh Abattoir.

WILMINGTON

Cape Fear Packing Co.—Railroad, Atlantic Coast Line. A corporation. Capital, $500,000; issued, $400,000. Stockholders, 5. President, G. Herbert Smith; Vice-President, J. Corbett; Assistant Secretary, D. C. Whitted; Secretary and Treasurer, W. L. Griffith; General Manager, C. T. Ruhl. Cattle, 180; hogs, 5,000. Sausage—Fresh, smoked and summer. By-Products—Dried blood and tankage. Government inspection. Refrigeration—Two 20 and one 10-ton York; brine circulating and direct expansion; 200,000 cu. ft. cold storage capacity. Boiler capacity, 300 H. P. Also exports.

Carolina Packing Co.—Railroad, Atlantic Coast Line. A corporation. Capital, $200,000; issued, $72,000. Stockholders, 150. President, W. W. Love; Vice-President, J. W. Brooks; Secretary and Treasurer, Milton Calden. Employes, 20. Cattle, 50; hogs, 150; sheep, 100. Sausage—Fresh, smoked. By-Products—Fertilizer tankage. Government inspection. Refrigeration—20-ton York; brine circulating. Boiler capacity, 65 H. P. Trade Mark—"Old Hickory."

WINSTON-SALEM

Winston-Salem Abattoir Company.

Figures on cattle, hogs and sheep indicate AVERAGE WEEKLY KILLING CAPACITY.

CHARLOTTE
Charlotte Packing Company.

ELIZABETH CITY
Pasquotank Packing Co.

NORTH DAKOTA

WEST FARGO
Equity Co-Operative Packing Co.—Railroad, Northern Pacific. A corporation. Capital, $10,000,000; issued, $2,500,000. Stockholders, 17,000. President, C. W. Reichert, Carrington; First Vice-President, U. L. Burdick, Williston; Second Vice-President, John L. Mikklethun, Wimbledon; Third Vice-President, C. P. Peterson, Bisbee; Treasurer, George Brastrup, Courtenay. Employes, 255. Codes—Cross. Cattle, 1,500; hogs, 5,000; sheep, 500. Sausage—Fresh, smoked, summer. By-Products—Dried blood and tankage. Government inspection. Refrigeration—75-ton Vogt; absorption system; brine spray system; 25 tons ice daily. Boiler capacity, 1,200 H. P. Trade Marks—Hams, "Fargo" and "Valley" brand; bacon, "Valley" and "Cheyenne" brand; lard, "Fargo." Branch Houses—110 W. Michigan St., Duluth, and 339 E. Fifth St., St. Paul, Minn.

GRAND FORKS
City Abattoir Co.
Northern Packing Company.

OHIO

ATHENS
F. C. Stedman Company—A corporation. Capital, $1,000,000; issued, $600,000. President, J. H. Winder; Vice-President, A. C. Russi; Secretary and Treasurer, R. E. Slaughter.

AKRON
C. A. Schell Provision Co.—504 Locust St. Partnership. President, C. A. Schell; Vice-President, Secretary and Treasurer, Ray C. Piero. Employes, 20. Sausage—Fresh and smoked. Wholesale meat packers and sausage makers. Refrigeration—15-ton Arctic; direct expansion. Boiler capacity, 50 H. P. Trade Mark—"Shell Brand."

BELLEVUE
A. Ruedy—Union Stock Yards.
Zehner Brothers Packing Co.—A corporation. Capital, $1,000,000. President, Charles Zehner; Vice-President, W. C. Carr; Secretary and Treasurer, Carl Zehner.

BEREA
J. N. Curtis & Son.

BUCYRUS
Dobbins & Geiger.

CANTON
Canton Provision Co.—Carnahan Ave. and Penn St. A corporation. Capital, $500,000. President and General Manager, Curtis N. Wade; Vice-President, Frank Wade; Secretary and Treasurer, M. J. Rank.
E. W. Renner & Sons.

Figures on cattle, hogs and sheep indicate AVERAGE WEEKLY KILLING CAPACITY.

CHILLICOTHE

Tobias Edinger—11 E. Main St. Railroads, Cincinnati, Hamilton & Dayton and Baltimore & Ohio. Owned by Tobias Edinger and five sons. General Manager, Tobias Edinger, Jr.; General Superintendent, George Edinger. Cattle, 12; hogs, 90; sheep, 10. Sausage—Fresh and smoked. By-Products—Dried blood, tankage and fertilizer. Refrigeration—20-ton Triumph; brine circulating; two tons ice daily. Boiler capacity, 135 H. P. Retail Market—One at plant.

Frank & Charles Hun.

CINCINNATI

Jacob Bauer Sons—2932 Massachusetts Ave. Partnership, Emil Bauer, Prop. Employes, 10. Cattle, 100; calves, 125. Government inspection. Refrigeration—12-ton York.

Butchers' Packing Company—Owned and operated by the E. Kahn's Sons Company, Cincinnati, Ohio.

Cincinnati Abattoir Co.—Spring Grove and Alabama Ave. Railroad, Baltimore & Ohio. A corporation. Capital, $1,750,000; issued, $1,350,000. Stockholders, 425. President, Joseph Ryan; Vice-Presidents, Fred E. Edmonds and Daniel H. Loewenstein; Chairman Board of Directors, Michael Ryan; Secretary, Fred A. Dietrich; Asst. Secretary, Henry H. Brockhoff; Treasurer, Chas. R. Hubbard; General Sales Manager, A. C. Huneke; General Superintendent, D. H. Loewenstein; General Manager, Michael Ryan. Employes, 900. Codes—Cross, Robinson, Griffin, Utility, A. B. C., 5th, and W. U.; code word, "Cinabbat." Cattle, 4,000; hogs, 15,000; sheep, 2,000. Sausage—Fresh, smoked and summer. By-Products handled by Joslin-Schmidt Co. Government inspection. Refrigeration—350-ton Ball, two 100-ton Triumph, 80-ton York and 80-ton Wolf; brine circulating, direct expansion and brine spray system. Boiler capacity, 1,600 H. P. Trade Marks—Hams, "Pheasant"; bacon, "Imperial Club," "Pheasant" and "Queen City"; lard, "Pheasant" and "Imperial Club"; compound, "Caco"; canned meats, "Imperial Club," "Pheasant" and "Buckeye." Branch Houses—New York City, Ashland and Lexington, Kentucky, and Chattanooga, Tenn. Also exports.

Cincinnati Packing Co.—2011 Branch St. A corporation. Capital, $25,000; issued, $16,700. Stockholders, 6. President, E. J. Schwarz; Secretary, D. T. Hackett; Treasurer, Otto Hirschfeld; General Manager, Otto Hirschfeld Employes, 4. Codes—Liebers and A. B. C.; code word, "Nurto." Horses, 30. Government inspection. Refrigeration—12-ton Triumph; brine circulating. Also exports. Note—Nature of business is killing of horses and pickling of horsemeat.

G. Ehrhart & Sons.

Charles A. Freund—1215 W. Liberty St. Individual ownership. Employes, 20. Cattle, 125; hogs, 900. Sausage—Fresh and smoked. Government inspection. Refrigeration—12-ton Frick; brine circulating. Boiler capacity, 150 H. P.

John Hilberg & Sons.

John Hoffman's Sons Co.—2162 Colerain Ave. Capital, $150,000; issued, $150,000. President, J. A. Wiederstein; Vice-President, Mrs. M. Hoffman; Secretary, A. Lammers; Treasurer, Jacob Hoffman.

Figures on cattle, hogs and sheep indicate AVERAGE WEEKLY KILLING CAPACITY.

CINCINNATI

Ideal Packing Co.—2141 Baymiller St. A corporation. Capital, $150,000; issued, $150,000. Stockholders, 17. President, Charles Hauck; Vice-President and Treasurer, Chas. A. Buehler; Secretary, John B. Mueller; General Manager, Charles Hauck. Employes, 70. Codes—Cross and Robinson. Cattle, 90; hogs, 1,200. Sausage—Fresh and smoked. By-Products—Tankage. Government inspection. Refrigeration—50 and 37½-ton Linde and 60-ton Vilter; direct expansion and brine circulating; 25 tons ice daily. Boiler capacity, 500 H. P. Trade Marks—"Ideal Brand" and "Liberty Brand."

Gus Juengling—2871 Massachusetts Ave. Individual ownership, Gus Juengling. Employes, 10. Cattle, 150; sheep, 75. Government inspection. Refrigeration—12-ton York.

E. Kahn's Sons Co.—517-523 Livingston St. A corporation. Capital, $750,000; issued, $438,200. Stockholders, 115. President, Louis W. Kahn; Vice-Presidents, Eugene Kahn and Nathan Kahn; Secretary, Henry Hellwitz; Treasurer, Albert H. Kahn; Sales Manager, J. L. Grauman; General Superintendent, E. L. Bertram. Employes, 300. Code—Cross. Cattle, 750; hogs, 3,500; sheep, 1,000. Sausage—Fresh, smoked and summer. By-Products—Hides, grease, tallow, tankage and hog feed. Government inspection. Refrigeration—50-ton Featherstone, 40-ton Niebling and 150-ton Frick; brine circulating and brine spray. Boiler capacity, 450 H. P. Trade Mark—"American Beauty."

Kroger Grocery & Baking Co.—817 Main St. A corporation. President, B. H. Kroger; Vice-Presidents, W. H. Albers and A. L. Nagel; Secretary, Geo. Meiners; Treasurer, B. H. Kroger, Jr.; General Manager, A. L. Nagel; General Superintendent, Wm. Wetta. Employes, 150. Codes—Cross. Cattle, 300; hogs, 1,200. Sausage—Fresh, smoked and summer. By-Products—Unground tankage. Government inspection. Refrigeration—50-ton Triumph and two 25-ton Fricks; direct expansion and brine circulating. Boiler capacity, 200 H. P. Trade Marks—"Country Club," "Old Colony" and "Compass." Retail Markets—St. Louis, Mo., Detroit, Mich., and Cincinnati, Columbus, Dayton and Hamilton, Ohio. Also exports.

A. Loewenstein's Sons Co.—1713-15 John St. A corporation. Capital, $40,000; issued, $40,000. Stockholders, 5. President, David Loewenstein; Vice-President, Melvyn G. Loewenstein; Secretary, Albert V. Ehlen; Treasurer, Emanuel Loewenstein; General Manager, David Loewenstein. Employes, 15. Cattle, 200; sheep, 100. Government inspection. Refrigeration—One 35-ton Triumph.

Lohrey Packing Co.—2827 Massachusetts Ave. Partnership. Geo. Lohrey and Henry Moellering. Employes, 25. Cattle, 10; hogs, 250. Sausage—Fresh and smoked. By-Products—Tankage. Government inspection. Refrigeration—20-ton York and 20-ton Triumph; brine circulating and direct expansion. Boiler capacity, 100 H. P. Trade Mark—"Brighton Belle."

Maescher & Co.—1754-56 Central Ave. Partnership, Harry W. Maescher and Arthur V. Maescher. Curers and packers of meats. Government inspection. Refrigeration—22-ton Huettemann & Cramer; direct expansion. Trade Mark—"Crescent Brand."

Figures on cattle, hogs and sheep indicate AVERAGE WEEKLY KILLING CAPACITY.

H. H. Meyer Packing Co.—2118 Linn St. A corporation. Capital, $207,350; issued, $207,350. President, N. R. Meyer; Secretary, R. A. Meyer; Treasurer, B. O. Gebred; General Manager, R. A. Meyer. Employes, 150. Codes—Robinson and Cross. Hogs, 3,000. Sausage—Fresh and smoked. By-Products—Tankage and grease. Government inspection. Refrigeration—60-ton Triumph and 60-ton Hercules; brine circulating. Boiler capacity, 405 H. P. Trade Marks—Hams, bacon, lard and compound, "Partridge," "Golden Corn," "Economy Bacon." Also exports.

I. Oscherwitz & Sons—569 W. Sixth St. Individual ownership. General Manager, Max B. Oscherwitz. Cattle, 40. Sausage—Kosher and smoked. Government inspection. Refrigeration—One 8-ton and one 5-ton York; direct expansion.

Wm. G. Rehn's Sons—452-454 Bank St. Partnership. Employes, 7. Cattle, 200. Government inspection. Refrigeration, 14-ton Triumph; brine circulating and direct expansion. Retail Market—23-25 Pearl St., Cincinnati.

A. Sander Packing Co.—1024 Gest St. A corporation. Capital, $200,000; issued, $200,000. Stockholders, 9. President and Treasurer, Armin Sander; Secretary, George Kaufmann; General Manager, Armin Sander; General Superintendent, George Kaufmann. Employes, 100. Codes—Cross and Robinson. Hogs, 3,000. Sausage—Fresh, smoked and summer. By-Products—Dried blood, tankage, liquid stick. Government inspection. Refrigeration—100-ton Triumph and 50-ton De La Vergne; direct expansion. Boiler capacity, 450 H. P. Trade Mark—"Morning Glory."

Jacob Schlachter's Sons—2831-2841 Colerain Ave. Partnership. General Manager, Henry Schlachter. Employes, 15. Cattle, 100; sheep, 300. Government inspection. Refrigeration—20-ton Triumph; brine circulating.

J. & F. Schroth Packing Co.—Cormany Ave. and Township St. A corporation. Capital, $500,000; issued, $500,000. Stockholders, 16. President, Fred Schroth; Vice-Presidents, Elmore M. and Michael Schroth; Secretary, Elmore M. Schroth; Treasurer, Fred Schroth; General Manager and General Superintendent, Fred Schroth; Live Stock Buyer, Michael Schroth. Employes, 130. Cattle, 150; hogs, 4,800. Sausage—Fresh and smoked. By-Products—Dried blood and tankage. Government inspection. Refrigeration—One 80-ton Niebling and one 50-ton Linde. Trade Mark—"Fountain Brand." Also exports.

Valley Packing Co.—3673 Colerain Ave. Owned by Jos. Geringer Estate. Jos. C. Geringer, General Manager. Cattle, 60; hogs, 200; sheep, 25. Sausage—Fresh, smoked and summer. By-Products—Hog and chicken feed. Government inspection. Refrigeration—35-ton York; brine circulating; 10 tons ice daily. Boiler capacity, 160 H. P.

Jacob Vogel & Son—2600-2614 Colerain Ave. Individual ownership. Employes, 35. Cattle, 30; hogs, 1,500. Sausage—Fresh, smoked and summer. Government inspection. Refrigeration—100 and 50-ton Frick; direct expansion. Boiler capacity, 550 H. P. Trade Marks—"Star Brand" and "Maple Leaf Brand."

Figures on cattle, hogs and sheep indicate AVERAGE WEEKLY KILLING CAPACITY.

CIRCLEVILLE

John Groce & Son.

Hoser Packing Co.

CLEVELAND

Blumenstock & Reid Co.—3261 West 65th St. Railroad, Big Four. A corporation. Capital, $200,000; issued, $121,000. Stockholders, 46. President, George Blumenstock; Vice-President, Jno. Amersbach; Secretary and Treasurer, Thos. Reid. Employes, 80. Code—Cross. Hogs, 3,000. Sausage—Fresh and smoked. By-Products—Dried blood and tankage. Government inspection. Refrigeration—50-ton Frick; brine spray system. Boiler capacity, 60 H. P. Trade Mark—"Golden Rule."

Cleveland Provision Company, The—2527 Canal Road. Capital, $4,500,000; issued, $1,000,000. President, S. T. Nash; Vice-President, W. F. Nash; Secretary and Treasurer, Thos. H. Nash.

Federal Packing Co.—Packinghouse, 3207 W. 65th St. Railroad, Big Four. A corporation. Capital, $850,000; issued, $276,000. Stockholders, 150. President, F. C. Thornton; Vice-President, G. H. Hall; Secretary, A. V. Cannon; Treasurer, A. E. Nelson; General Manager, A. E. Nelson. Employes, 60. Codes—Cross. Cattle, 500; sheep and calves, 1,000. Sausage—Fresh and smoked. By-Products—Tankage and tallows. Government inspection. Refrigeration—60-ton Vilter; direct expansion. Boiler capacity, 150 H. P. Trade Mark—"Fedco." Retail Market—3207 W. 65th St., Cleveland. Branch House—2303-7 E. Fourth St., Cleveland.

Lake Erie Provision Co.—3112 West 63rd St. Railroad, Big Four. A corporation. Capital, $60,000; issued, $60,000. Stockholders, 12. President, N. O. Newcomb; Vice-President, John Beck; Secretary, C. G. Newcomb; Treasurer, N. O. Newcomb; General Manager, M. T. Morgan. Employes, 150. Code—Cross. Cattle, 750; hogs, 3,000; sheep, 750. Sausage—Fresh, smoked and summer. By-Products—Dried blood and tankage. Government inspection. Refrigeration—125-ton York, 75 and 60-ton Arctic; direct expansion and brine spray system. Boiler capacity, 536 H. P. Trade Mark—"Meadowbrook Brand." Also exports.

Long Dressed Beef Co.—2351 E. Fourth St. Railroad Big Four. Pack beef and pork.

Ohio Provision Company—Clark Ave. & Big Four Crossing. Railroad, Big Four. A corporation. Capital, $400,000; issued, $326,800. Stockholders, 14. President and General Manager, J. B. McCrea; Secretary and Treasurer, E. L. Schneider; Plant Superintendent, Philip Scheuermann; Live Stock Buyer, J. J. Gallagher. Employes, 125. Codes—Cross; code word, "Purple." Cattle, 300; hogs, 2,000; sheep, 300. Sausage—Fresh and smoked. By-Product—Tankage. Government inspection. Refrigeration—100-ton York; 25-ton Case and 25-ton Arctic; direct expansion. Boiler capacity, 225 H. P. Trade Mark—"Ohio Brand." Retail Market—One at plant. Branch Houses—One in Cleveland. Also exports.

Swift & Company—3241 W. 65th St. (See Chicago.)

Figures on cattle, hogs and sheep indicate AVERAGE WEEKLY KILLING CAPACITY.

Theurer-Norton Provision Co.—W. 63rd St., near Big Four. A corporation. Capital, $80,000; issued, $80,000. President and Treasurer, M. C. Teufel; Vice-President, T. J. Holmden; Secretary, A. F. Lucht.

COLUMBUS

Fred Schmidt Packing Co.—253 E. Kossuth St. Railroad, Pennsylvania. Individual ownership. General Manager, Geo. L. Schmidt. Employes, 30. Hogs, 400. Sausage—Fresh, smoked and summer. Refrigeration—20-ton York; direct expansion. Trade Mark—"Montrose." Retail Market—Central Market, Columbus.

Blumer-Sartain Packing Co.—Sandusky & River Sts. Railroad, Pennsylvania. A corporation. Capital, $125,000; issued, $75,800. Stockholders, 12. President, R. A. Blumer; Vice-President, A. E. Sartain; Secretary, W. E. Langdon; Treasurer, W. H. Sartain; General Manager, W. H. Sartain. Employes, 40. Cattle, 100; hogs, 400; sheep, 50. Sausage—Fresh, smoked and summer. By-Products—Tankage, hog and chicken feed. Refrigeration—25-ton York; direct expansion. Boiler capacity, 125 H. P. Trade Mark—"Ohio Pride Brand."

Columbus Packing Co.—South High St. Railroad, Hocking Valley. A corporation. Capital, $250,000; issued, $250,000. Stockholders, 180. President, Frank Schmidt; Vice-President, William C. Frech; Secretary, O. P. Lamb; Treasurer, J. F. Diebel; General Manager, J. F. Diebel; General Superintendent, O. R. Canter; Superintendent hog killing, Sam Miller. Employes, 150. Codes—Cross. Cattle, 250; hogs, 3,000; sheep, 200. Sausage—Fresh and smoked. By-Products—Dried blood, tankage, hog and chicken feed. Government inspection. Refrigeration—Three 80-ton Fricks; direct expansion. Boiler capacity, 800 H. P. Trade Marks—Hams and bacon, "Capital"; lard, "Royal" and "Capital." Also exports.

David Davies—616 W. Mound St. Individual ownership. General Manager, Bruce Culp; General Superintendent, Wm. Bauer. Employes, 50. Cattle, 125; hogs, 350; sheep, 50. Sausage—Fresh, smoked and summer. By-Products—Tankage. City inspection. Refrigeration—One 35-ton Creamery Package.

Denton Bros.

COSHOCTON

Coshocton Provision Company.

DAYTON

Henry Burkhardt Packing Co.—235 S. Irwin St. Railroad, Baltimore & Ohio. A corporation. Capital, $35,000; issued, $35,000. Stockholders, 5. President, G. Burkhardt; Vice-President, A. William Freund; Secretary and Treasurer, L. J. Burkhardt; General Manager, L. J. Burkhardt; General Superintendent, A. Wm. Freund. Employes, 100. Code—Cross. Cattle, 100; hogs, 1,200. Sausage—Fresh and smoked. By-Products—Tankage. Government inspection. Refrigeration—50-ton Vogt; absorption system; brine circulating. Boiler capacity, 375 H. P. Trade Marks—Hams and bacon, "Pig Brand" and "Honey Bee Brand." Also exports.

Dayton Abattoir Co., Inc.—1022 Valley St. Capital, $50,000. Owners: Jesse Jacobs, Harry J. Jacobs, Alvin Jacobs and H. D. Mayer

---

Figures on cattle, hogs and sheep indicate AVERAGE WEEKLY KILLING CAPACITY.

DAYTON

Gem City Packing Co.—Union Stock Yards. A corporation. Capital, $100,000; issued, $85,000. Stockholders, 12. President, Wm. M. Adelberger; Secretary and Treasurer, Oscar C. Ott; General Superintendent, Fritz Laurinet. Employes, 15. Cattle, 30; hogs, 200; sheep, 20. Sausage—Fresh and smoked. By-Products—Tankage. City inspection. Refrigeration—20-ton York; brine circulating: Boiler capacity, 100 H. P.

Chas. Sucher Packing Co.—N. Western Ave. Railroad, Baltimore & Ohio. A corporation. President, Chas. Sucher; Vice-President, C. F. Sucher; Secretary, J. F. Sucher; Treasurer, Chas. Sucher; General Manager, Chas. F. Sucher, Jr. Codes—Cross and Robinson. Cattle, 150; hogs, 1,500. Sausage—Fresh, smoked and summer. By-Products—Dried blood, tankage, hog and chicken feed. Government inspection. Refrigeration—30-ton Frick and 60-ton York; direct expansion. Trade Marks—"Victory Brand."

FOSTORIA

Fostoria Provision Company—Columbus Ave. Railroad, New York, Chicago & St. Louis. A corporation. Capital, $200,000. Stockholders, 60. President, J. M. Myers; Vice-President, W. J. Hakes; Secretary and Treasurer, Chas. Fultz. Employes, 45. Codes—A. E. Cross. Cattle, 200; hogs, 100; sheep, 200. Sausage—Fresh and smoked. By-Products—Tankage. Government inspection. Refrigeration—35-ton Arctic; brine circulating and direct expansion. Makes own ice. Boiler capacity, 100 H. P.

GREENVILLE

Chas. G. Buchy—North Broadway. Individual ownership. Employes, 5. Cattle, 15; hogs, 100. Sausage—Fresh, smoked and summer. By-Products—Tankage. Refrigeration—8-ton York; direct expansion. Boiler capacity, 25 H. P. Trade Mark—"Buchy's Prize Brand."

HAMILTON

Slifer Packing Co.—111 Main St. A corporation. Capital, $20,000; issued, $20,000. Stockholders, 5. President, Ross R. Slifer; Vice-President, Emma Slifer; Secretary and Treasurer, F. G. Slifer; General Manager, F. G. Slifer. Employes, 25. Cattle, 40; hogs, 150; sheep, 10. Sausage—Fresh and smoked. By-Products—Tankage. Refrigeration—25-ton Triumph; brine circulating. Makes own ice. Boiler capacity, 100 H. P. Retail Markets—Four in Hamilton.

Chas. Manche.

W. P. Eaton Packing Co.—Railroad, Chicago, Indiana & Western. A corporation. Capital, $500,000; issued, $300,000. Stockholders, 100. President, T. E. Slade; Vice-President, Fred Beisswinger; Secretary, H. J. Iske; Treasurer, W. P. Eaton; General Manager, W. P. Eaton; General Superintendent, Gus Mueller. Employes, 50. Cattle, 100; hogs, 1,000; sheep, 25. Sausage—Fresh and smoked. By-Products—Dried blood, tankage, hog and chicken feed. Refrigeration—50-ton Wolf-Linde, 20-ton Triumph; direct expansion. Boiler capacity, 160 H. P. Trade Marks—"Checkers," "Rainbow" and "Eato." Retail Markets—565 South Front St., Hamilton.

Rupp Packing Co.

KENMORE

Ohio Packing & Provision Co.—Partnership, Frank S. Canfield and Jack Canter.

Figures on cattle, hogs and sheep indicate AVERAGE WEEKLY KILLING CAPACITY.

KENMORE

Zimmerly Bros. Co.—25 S. Manchester Rd. Railroad, Erie. Individual ownership. Capital, $70,000. Stockholders, 5. President, Jacob Zimmerly; Vice-President, G. Zimmerly; Secretary and Treasurer, John Zimmerly; General Manager and Superintendent, Herman Zimmerly. Employes, 20. Cattle, 60; hogs, 200. Sausage—Fresh, smoked and summer. By-Products—Tankage, hog and chicken feed. Refrigeration—25-ton Phoenix; direct expansion. Boiler capacity, 180 H. P. Retail Market—One at Kenmore, Ohio.

LANCASTER

Bauman Bros.

Lancaster Packing Co.

LIMA

Lima Packing Co.—215 South Central Ave. Railroad, Baltimore & Ohio. A corporation. Capital, $150,000; issued, $127,500. Stockholders, 27. President, B. F. Thomas; Vice-President, W. C. Bradley; Secretary, D. W. Leichty; Treasurer, W. C. Bradley; General Manager, B. F. Thomas; General Superintendent, F. D. Bradley. Employes, 60. Code—Cross. Cattle, 60; hogs, 550. Sausage—Fresh and smoked. By-Products—Dried blood and hard cake. Refrigeration—60-ton Carbondale and 30 and 20-ton Linde; direct expansion; 12 tons ice daily. Boiler capacity, 300 H. P.

MANSFIELD

W. A. Kearns Packing Co.

MARIETTA

Frank Webber.

MARION

Marion Packing Co.—West Center St. Railroad, Erie. A corporation. Capital, $500,000; issued, $325,000. Stockholders, 140. President, L. H. Guthery; Vice-President, Josiah Bindley; Secretary, French Crow; Treasurer, D. R. Crissinger; General Manager, L. H. Guthery; General Superintendent, Ferd Frey. Employes, 60. Cattle, 300; hogs, 1,200; sheep, 100. Sausage—Fresh and smoked. By-Products—Dried blood, tankage, hog and chicken feed. Government inspection. Refrigeration—75-ton Triumph and 30-ton York; direct expansion. Boiler capacity, 200 H. P.

MIDDLETON

Reiner Brothers.

NEWARK

C. W. Miller Co.—Daniel Ave. Railroad, Baltimore & Ohio. Partnership. Cattle, 150; hogs, 900; sheep, 50. Sausage—Fresh and smoked. By-Products—Tankage and hog and chicken feed. Refrigeration—30, 25 and 12-ton Triumph; direct expansion. Boiler capacity, 50 H. P. Trade Mark—"Newark Brand."

Independent Packing Co.—141 Wilson St.

PIQUA

Val Decker Packing Co.—East Ash St. and River, Railroads, Baltimore & Ohio and Pittsburgh, Cleveland, Cincinnati & St. Louis. A corporation. Capital, $50,000; issued, $50,000. Stockholders, 7.

Figures on cattle, hogs and sheep indicate AVERAGE WEEKLY KILLING CAPACITY.

President, Val Decker; Vice-President, L. F. Decker; Secretary, W. J. Decker; Treasurer, Geo. H. Decker; General Manager, Val Decker; General Superintendent, Val Decker; Sales Manager, Geo. H. Decker; Purchasing Agent, Walter J. Decker; Stock Buyers, Wm. J. Decker, Wm. C. Catterlin. Employes, 75. Cattle, 125; hogs, 600; sheep, 25. Sausage—Fresh, smoked and summer. By-Products—Hog feed. Refrigeration—20 and 35-ton Frick; direct expansion. Boiler capacity, 250 H. P. Trade Mark—"Our Pride."

Kugelman Packing Co.—E. Ash St. Railroads, Pittsburgh, Cleveland, Cincinnati & St. Louis and Baltimore & Ohio. A corporation. Capital, $50,000; issued, $40,000. Stockholders, 5. President, Fred Goeke; Vice-President, Edw. J. Kugelman; Treasurer, Leon M. Goeke; General Manager, Edw. J. Kugelman. Employes, 20. Cattle, 50; hogs, 250; sheep, 25. Sausage—Fresh and smoked. By-Products—Tankage. Refrigeration—20-ton Arctic; direct expansion. Boiler capacity, 100 H. P.

SANDUSKY

Sandusky Packing Co.—Perkins and Campbell Rd. Railroad, Baltimore & Ohio. A corporation. Capital, $100,000; issued, $78,000. Stockholders, 5. President, William C. Routh; Vice-President, A. C. Routh; Secretary and Treasurer, Guy Manaugh; General Manager, W. C. Routh; General Superintendent, A. C. Routh. Employes, 30. Codes—Cross. Cattle, 200; hogs, 600; sheep, 200. Sausage—Fresh and smoked. By-Products—Tankage, hog and chicken feed. Refrigeration—30-ton Arctic; brine circulating. Boiler capacity, 130 H. P. Trade Mark—"Sandusky Brand."

SIDNEY

Sidney Packing Co.—East Sidney St. Railroad, Cleveland, Cincinnati, Chicago & St. Louis. A corporation. Capital, $100,000; issued, $81,000. Stockholders, 32. President, H. W. Robinson; Vice-President, B. T. Bull; Secretary and Treasurer, S. V. Willcutts; General Manager, E. Collins. Employes, 15. Cattle, 100; hogs, 500; sheep, 20. Sausage—Fresh and smoked. By-Products—Tankage. Refrigeration—15-ton Phoenix and 7-ton York; direct expansion. Boiler capacity, 50 H. P.

SPRINGFIELD

Finck & Heine Packing Co.

TOLEDO

Carr Bros.—Lagrange St. and M. C. R. R.

Jacob Folger—Phillips Ave. and New York Central R. R.

Phillip Provo—Home Packing Co.

N. Rassel Sons Co.—Lagrange St. and M. C. R. R.

Ruedy Bros.—4108 Lagrange St.

TROY

Braun Bros. Packing Co.

Grunlich Bros.

Troy Packing Co.

URBANA

Urbana Packing Co.—Corner Railroad and Church Sts. Railroad, Pittsburgh, Cleveland, Cincinnati & St. Louis. A corporation. Capital, $500,000; issued, $222,300. President, W. R. Wilson; Vice-President, J. P Neer; Secretary and Treasurer, F. C. Wilson; General

---

Figures on cattle, hogs and sheep indicate AVERAGE WEEKLY KILLING CAPACITY.

Manager and Superintendent, H. A. Colvin. Stockholders, 118. Employes, 55. Cattle, 200; hogs, 600; sheep, 150. Sausage—Fresh and smoked. By-Products—Tankage and hog and chicken feed. Refrigeration—75-ton Frick; direct expansion; 25 tons ice daily. Boiler capacity, 500 H. P. Trade Marks—"Nox-All" and "Urban." Retail Markets—Two at Urbana, O., two at Springfield, O., and one at Marysville, O.

WAPAKONETA

Jacob Werner & Son—Partnership. Employes, 10. Cattle, 20; hogs, 150; few sheep. Sausage—Fresh, smoked, summer. Refrigeration—One 10-ton York; direct expansion. Boiler capacity, 15 H. P.

XENIA

City Market Co.—E. Main St.

YOUNGSTOWN

Youngstown Packing & Provision Co.—Railroad, Pennsylvania. A corporation. Capital, $200,000; issued, $150,000. Stockholders, 25. President, J. Calvin Ewing; Vice-President, Randall Anderson; Secretary and Treasurer, Wm. Sampson; General Manager, C. M. Bell. Employes, 70. Codes—Robinson. Cattle, 150; hogs, 1,500; sheep, 500. Sausage—Fresh, smoked and summer. By-Products—Tankage. Refrigeration—60-ton Wolf-Linde; brine circulating; 3 tons ice daily. Boiler capacity, 100 H. P. Trade Marks—Hams, "Wickliffe"; bacon, "Wickliffe"; lard, "Wickliffe."

ZANESVILLE

Henry Kessler.

Max Taylor.

Rittberger Bros.

New Zanesville Provision Company—Railroad, Baltimore & Ohio. A corporation. Capital, $300,000; issued, $300,000. Stockholders, 7. President, F. G. Groce; Vice-President, A. P. Rogge; Secretary and Treasurer, F. Boyd. Employes, 110. Cattle, 100; hogs, 1,000. Sausage—Fresh, smoked and summer. By-Products—Dried blood, tankage, hog and chicken feed. Refrigeration—100-ton Frick, 50-ton De La Vergne and 15-ton Huettemann & Cramer; direct expansion and brine spray systems. Boiler capacity, 400 H. P. Trade Marks—"Groce," "Scioto" and "Sunset."

OKLAHOMA

OKLAHOMA CITY

W. H. Butcher Packing Co.—100 E. Choctaw St. A corporation. Capital, $50,000; issued, $38,200. Stockholders, 3. President, W. H. Butcher; Vice-President, Anna Butcher; Secretary and Treasurer, T. Butcher. Employes, 15. Cattle, 60; hogs, 200. Sausage—Fresh and smoked. Refrigeration—10-ton York; brine circulating and direct expansion. Boiler capacity, 12 H. P.

Morris & Company. (See Chicago.)

Riverside Packing Co.

Wilson & Company. (See Chicago.)

SHAWNEE

Graf Packing Co. A corporation. Capital, $50,000; issued, $30,100. Stockholders, 4. President, Geo. B. Graf; Vice-President, Julius

---

Figures on cattle, hogs and sheep indicate AVERAGE WEEKLY KILLING CAPACITY.

Greenlee; Secretary and Treasurer, Geo. Graf, Jr.; General Manager, Geo. B. Graf; General Superintendent, E. L. Graf. Employes, 7. Cattle, 50; hogs, 150. Sausage—Fresh and smoked. Refrigeration—12-ton United-Kenosha; brine circulating. Boiler capacity, 16 H. P.

TULSA

Independent Packing Co.—Olympia and Katy Sts. Railroad, Missouri, Kansas & Texas. A corporation. Capital, $100,000; issued, $50,000. Stockholders, 5. President, U. Holderman; Secretary and Treasurer, I. F. Crow; General Manager, U. Holderman; General Superintendent, Harry Wollbrinck. Employes, 23. Cattle, 100; hogs, 400. Sausage—Fresh and smoked. Refrigeration—50-ton United Iron Works; brine circulating and direct expansion. Boiler capacity, 60 H. P. Trade Mark—"Crown."

OREGON

ALBANY

D. E. Nebergall Meat Co.—Railroad, Oregon Electric. A corporation. Capital, $100,000; issued, $70,000. Stockholders, 14. President, D. E. Nebergall; Vice-President, H. L. Nebergall; Secretary and Treasurer, A. R. Tartar; General Manager, D. E. Nebergall. Employes, 45. Cattle, 200; hogs, 1,200; sheep, 400. Sausage—Fresh and smoked. Government inspection. Refrigeration—12½, 7 and 1 ton Armstrong, 1½-ton Frick, 2-ton Automatic and 2½-ton Remington; direct expansion and brine circulating. Boiler capacity, 50 H. P. Trade Mark—"Quality." Retail Markets—Two in Albany, one at Corvallis and one at Eugene, Ore.

BAKER

Baker Packing Co.—1921 Main St. Railroad, Union Pacific. A corporation. Capital, $20,000; issued, $20,000. Stockholders, 5. President, Jas. A. Russell; Vice-President, H. L. Mohr; Secretary and Treasurer, E. W. Cox; General Manager, H. L. Mohr. Employes, 8. Cattle, 50; hogs, 300; sheep, 75. Sausage—Fresh, smoked and summer. By-Products—Tankage and hog and chicken feed. Refrigeration—5-ton Armstrong and 3-ton Harris; direct expansion. Boiler capacity, 35 H. P. Trade Mark—"Nugget Brand." Retail Markets—2424 Center St. and 1921 Main St., Baker, Ore.

Smith Packing Co.—1920 Main St. Railroads, O. W. R. & N. Co. Partnership. Stockholders, 3. Manager, W. P. Smith; Live Stock Purchaser and Slaughter House Manager, Chris Smith; Office, A. J. Durr. Cattle, 100; hogs, 150; sheep, 125. Sausage—Fresh and smoked. By-Products—Tankage and tallow. Refrigeration—5-ton Armstrong; direct expansion. Boiler capacity, 15 H. P. Trade Mark—"Bee Hive." Retail Market—1740 Main St., Baker, Ore.

CORVALLIS

Corvallis Meat Co.—138 2nd St. Railroad, S. P. & O. E. Individual ownership. Employes, 14. Cattle, 60; hogs, 90; sheep, 90. Sausage—Fresh and smoked. By-Products—Chicken feed. Refrigeration—5-ton Armstrong and 2-ton Isko; direct expansion; ½ ton ice daily. Boiler capacity, 17 H. P. Retail Markets—Two in Corvallis.

THE DALLES

The Dalles Meat Company—300 East 3rd St. Railroad, O. W. R. & N. Co. A corporation. Capital, $20,000; issued, $14,000. Stockholders, 3. President, A. S. Roberts; Vice-President, V. H. French; Secre-

---

Figures on cattle, hogs and sheep indicate AVERAGE WEEKLY KILLING CAPACITY.

tary, E. P. Roberts; General Manager, A. Salter. Employes, 6. Cattle, 10; hogs, 12; sheep, 8. Sausage—Fresh and smoked. Government inspection. Refrigeration—8-ton Murray; direct expansion.

EUGENE

Eugene Packing Co.

KLAMATH FALLS

Klamath Packing Co.—526 Main St. A corporation. Capital, $125,000; issued, $104,000. Stockholders, 5. President, L. E. Walker; Vice-President, C. E. Drew; Secretary and Treasurer, J. S. Kent; General Manager, C. A. Pauley; General Superintendent, C. A. Pauley. Employes, 15. Cattle, 200; hogs, 600; sheep, 400. Sausage—Fresh, smoked and summer. By-Products—Tankage, hog and chicken feed. Refrigeration—2-ton Stephens, 2-ton Brecht and 3 and 5-ton Cyclops; brine circulating and direct expansion. Boiler capacity, 25 H. P. Trade Mark—"Diamond Brand." Retail Market—Klamath Falls.

LA GRANDE

Grande Ronde Meat Co.—1116 Adams St. Railroad, O. W. R. & N. Co. A corporation. Capital, $100,000; issued, $100,000. Stockholders, 10. President, J. A. Russell; Vice-President; W. P. Mohr; Secretary and Treasurer, F. A. Epling; General Manager, J. A. Russell; General Superintendent, W. P. Mohr. Employes, 42. Cattle, 25; hogs, 80; sheep, 25. Sausage—Fresh, smoked and summer. By-Products—Tankage, hog and chicken feed. Government inspection. Refrigeration—20, 10, 5 and 3-ton Harris; brine circulating; 50 tons ice daily. Boiler capacity, 80 H. P. Trade Marks—"Mount Emily" and "Royal Brand." Retail Markets—La Grande, Elgin and Baker, Ore.

LEBANON

Carl Seibel & Otto Eichentopf Packing Company.

PENDLETON

Pendleton Packing & Provision Co.—Corporation. Capital, $30,000. Directors: H. P. Whitman, Chas. J. Greulich, Geo. Singer.

PORTLAND

Adams Brothers—Panama Bldg.

Geo. W. Donaldson—Kenton Station.

Gelinsky Packing Co., Inc.—271 Yamhill St. A corporation. Capital, $25,000; issued, $25,000. Stockholders, 3. President and General Manager, W. E. Gelinsky; Vice-President, Alex. B. Cameron; Secretary and Treasurer, Axel Anderson. Employes, 21. Cattle, 30; hogs, 150; sheep, 40. Sausage—Fresh and smoked. Refrigeration—8-ton and 10-ton Fricks; direct expansion. Retail Market—271 Yamhill St., Portland. Wholesale Department—168 4th St. Trade Marks—"Honeysuckle" and "Lily."

Swift & Company—North Portland. (See Chicago.)

Schlesser Brothers—1644 Peninsula Ave.

RAINIER

Columbia River Meat Co.—Water St. Individual ownership. Capital, $12,000. Prop., C. R. Hallberg. Wholesale slaughterer and packer. Sausage—Fresh, smoked and summer. Refrigeration—1-ton A. & S.; brine circulating. Trade Mark—"Unique." Retail Markets—One at St. Helena and one at Rainier.

Figures on cattle, hogs and sheep indicate AVERAGE WEEKLY KILLING CAPACITY.

SPRINGFIELD

Swarts & Washburn.

SALEM

Valley Packing Co.—Portland Road. Railroad, Southern Pacific. A corporation. Capital, $200,000; issued, $200,000. Stockholders, 5. President, F. W. Steusloff; Vice-President, W. H. Steusloff; Secretary and Treasurer, Curtis B. Cross; General Manager, Curtis B. Cross; General Superintendent, F. W. Steusloff. Employes, 50. Cattle, 150; hogs, 600; sheep, 300. Sausage—Fresh, smoked and summer. By-Products—Dried blood, tankage and hog and chicken feed. Government inspection. Refrigeration—12-ton Automatic; direct expansion. Boiler capacity, 100 H. P. Trade Marks—"Cascade Brand" and "Opal Brand."

PENNSYLVANIA

ALLENTOWN

Arbogast & Bastian Co.—21 Hamilton St. Railroad, Lehigh Valley. A corporation. Capital, $1,250,000. Stockholders, 10. President, Wilson Arbogast; Vice-President, Walter E. Bastian; Secretary, William J. Moessner; Treasurer, Morris C. Bastian; General Manager, William J. Moessner. Employes, 240. Codes—Cross. Cattle, 250; hogs, 2,800; sheep, 40. Sausage—Fresh, smoked and summer. By-Products—Dried blood and tankage. Government inspection. Refrigeration—100 and 150-ton Vilter; direct expansion; 18 tons ice daily. Boiler capacity, 2,000 H. P.

Zach Boyer.

ALTOONA

Confederated Home Abattoir Corporation—Capital, $500,000. President, John G. Sellers; Vice-President and Secretary, C. L. Brumbaugh; Treasurer, Jacob G. Snyder.

United Home Dressed Meat Co.—9th Ave. and 31st St. Railroad, Penna. A corporation. Capital, $250,000; issued, $250,000. Stockholders, 5. President, Frank Endress; Vice-Presidents, W. A. Mattern and Peter Gutwald; Secretary and Treasurer, Chas. G. Mattas; Asst. Secretary and Treasurer, C. L. Salyards; General Sales Manager, C. L. Salyards. Employes, 85. Cattle, 180; hogs, 1,000; sheep, 100. Sausage—Fresh, smoked and summer. By-Products—Tankage, hog and chicken feed, tallow, grease and cracklings. Refrigeration—60-ton Vilter; direct expansion. Boiler capacity, 350 H. P.

BALLIETTSVILLE

Wm. D. George—Railroad, Ironton. Individual ownership. General Manager, Wm. D. George. Employes, 20. Cattle, 25; hogs, 50; calves, 10. Sausage—Fresh and smoked. By-Products—Tankage. Refrigeration—Two 10-ton Brecht and one 1½-ton York; direct and brine circulating. Boiler capacity, 60 H. P. Retail Market—One at Catasauqua, Pa., and one at Bethlehem, Pa.

BLAIRSVILLE

Brown Packing Co.—Liberty and Brown Sts. Railroad, Penna. Individual ownership—Thos. C. Brown. Employes, 15. Code—Cross. Cattle, 25; hogs, 150. Sausage—Fresh and smoked. By-Products—Tankage. Refrigeration—35-ton Boyle; brine circulating; 12 tons ice daily. Boiler capacity, 250 H. P.

Figures on cattle, hogs and sheep indicate AVERAGE WEEKLY KILLING CAPACITY.

BRIDGEPORT

A. H. March Packing Co.—Front and DeKalb Sts. Railroad, P. & R. A corporation. Capital, $100,000; issued, $100,000. Stockholders, 3. President, Paul March; Vice-President, A. H. March; Secretary and Treasurer, Warren Geiger; General Manager, Paul March; General Superintendent, Chas. Vielhauer. Employes, 75. Hogs, 2,500. Sausage—Fresh and smoked. By-Products—Tankage, cracklings, hog hair and grease. Government inspection. Refrigeration—30-ton Pennsylvania and 80-ton Huettemann & Cramer; direct expansion. Boiler capacity, 600 H. P.

BUTLER

Butler Packing Company—A corporation. Capital, $75,000; issued, $31,250. President, A. F. Myers; Vice-President, George McGary; Secretary, W. A. Gibson; Treasurer, H. J. Daum.

CARNEGIE

Abbott Packing Co.—Foot of Walnut St. Railroads, Pennsylvania and P. C. & Y. A corporation. Capital, $50,000; issued, $50,000. Stockholders, 6. President, Edward Abbott, Sr.; Vice-President, Albert Abbott and Edward Abbott, Jr.; Secretary, Christian Abbott; Treasurer, Florence H. Abbott. Employes, 20. Cattle, 100; hogs, 250; sheep, 500. Sausage—Fresh and smoked. By-Products—Tankage. Refrigeration—60-ton Frick; direct expansion; 30 tons ice daily. Boiler capacity, 675 H. P.

CHESTER

John J. Buckley Company—18-40 W. 2nd St. A corporation. Capital, $250,000; issued, $100,000. President, J. J. Buckley; Secretary, J. E. McDonough; Treasurer, J. J. Buckley, Jr.

COATESVILLE

Beiswanger Packing Company—139-143 Chestnut St. Individual ownership. Proprietor, Jacob Beiswanger.

EASTON

Easton Abattoir—131 Delaware Drive. S. S. Ealer.

ERIE

Schaffner Bros. Co.—15th and Reed Sts. Railroad, N. Y. C. A corporation. Capital, $500,000; issued, $285,000. Stockholders, 43. President, Morris Schaffner; Vice-Presidents, Jacob and Alfred S. Schaffner; Secretary, H. G. Schaffner; Asst. Secretary, M. D. Levy; Treasurer, Milton Schaffner; General Manager, Morris Schaffner; General Superintendent, Martin D. Levy. Employes, 125. Code—Cross. Cattle, 500; hogs, 3,000; sheep 1,000. Sausage—Fresh, smoked and summer. By-Products—Tankage. Government inspection. Refrigeration—80-ton York and 35-ton Wagner; direct expansion and brine spray; 5 tons ice daily. Boiler capacity, 250 H. P. Trade Marks—Hams, "Sovereign" and "Keystone"; bacon, "Sovereign" and "Keystone"; lard, "Full White Brand."

South Erie Slaughter House—31st and Peach Sts.

HARRISBURG

Harrisburg Abattoir Company—Sayford and Currant Sts.

Swift & Company—Seventh and North Sts. (See Chicago.)

Figures on cattle, hogs and sheep indicate AVERAGE WEEKLY KILLING CAPACITY.

INDIANA

Indiana Beef & Provision Co.—13th St. extension. Railroads, Penna. and B. R. & P. A corporation. Capital, $125,000; issued, $80,000. Stockholders, 58. President, Joseph Shearer; Vice-President, C. M. Wortman; Secretary, C. W. McNaughton; Treasurer, Elmer Ellis; General Manager, C. W. McNaughton. Employes, 25. Cattle, 100; hogs, 500; calves, 50; sheep, 25. Sausage—Fresh and smoked. Hams, bacon and lard. By-Products—Dried blood, tankage, hog and chicken feed. Refrigeration—13-ton Frick; direct expansion. Boiler capacity, 50 H. P.

JOHNSTOWN

Ferguson Packing Co.—Sixth Ave. Railroad, Pennsylvania. A corporation. Capital, $1,000,000; issued, $277,000. Stockholders, 13. President, C. L. Ferguson; Vice-Presidents, Ralph L. Swank and Louis Zang; Secretary, Frank A. Clark; Treasurer, Frank D. Phillip; General Manager, C. L. Ferguson; General Superintendent, R. J. Dower. Employes, 85. Cattle, 300; hogs, 1,500; sheep, 300. Sausage—Fresh and smoked. By-Products—Dried blood and tankage. Refrigeration—70-ton Frick; direct expansion. Boiler capacity, 400 H. P. Trade Marks—"O-so-good" and "Ferguson's Best."

Edward Hahn—Hickory St. and B. & O. R. R. Railroad, B. & O. Individual ownership. General Manager, Edward Hahn; General Superintendent, William Bird. Employes, 38. Cattle, 150; hogs, 500; sheep, 100. Sausage—Fresh and smoked. By-Products—Dried blood, tankage, hog and chicken feed. Refrigeration—40-ton Frick and 40-ton York; brine circulating. Boiler capacity, 150 H. P. Trade Marks—"Blue Ribbon Brand."

C. A. Young Co.—Railroads, Pennsylvania and B. & O. A corporation. Capital, $100,000; issued, $100,000. President, C. A. Young; Vice-President, B. J. Picking; Secretary and Treasurer, R. M. Putman; General Manager, C. A. Young. Employes, 35. Trade Marks—Ham and bacon, "Bob White"; sausage, "Dixie."

LANCASTER

Ch. Kunzler Co.—652 Manor St. A corporation. Capital, $400,000; issued $190,700. Stockholders, 3. President, Christ Kunzler; Secretary, G. W. Birrell; Treasurer, G. W. Birrell; General Manager, Christ Kunzler. Employes, 26. Cattle, 100; hogs, 100; sheep, 200. Refrigeration—35-ton Vogt and 10-ton Vilter; brine circulating; 4½ tons ice daily. Boiler capacity, 300 H. P. Retail Markets—One in Lancaster, Pa.

Chas. Falk & Bros.—509 St. Joseph St.

Lancaster Abattoir Co.—820 N. Christian St.

D. W. Shaeffer—416 N. Pine St.

LEHIGHTON

Joseph Obert Co., Inc.—Railroads, L. V. and C. of N. J. A corporation. Capital, $150,000; issued, $150,000. Stockholders, 5. President, Wm. H. Olbert; Vice-President, ———; Secretary and Treasurer, H. B. Kennell. Employes, 30. Cattle, 100; hogs, 600. Sausage—Fresh, smoked and summer. Hog and beef products. Refrigeration—50-ton Vilter; direct expansion. Boiler capacity, 200 H. P. Retail Markets—One in Lehighton.

Figures on cattle, hogs and sheep indicate AVERAGE WEEKLY KILLING CAPACITY.

LEBANON

David O. Bomberger.
E. L. Brooks—155 N. 10th St.
Keystone Abattoir Co.
Lebanon Abattoir Co.
J. B. Sheaf.
Walter W. Rittle.
J. K. Troutman.

MEDIA

Clement C. Allen Company.

Habbersett Bros.—Railroad, Pennsylvania. Partnership. Employes, 8. Hogs, 100. Sausage—Fresh. Pork products. Boiler capacity, 30 H. P.

McKEESPORT

Peters Packing Company—922 Rose St. A corporation. Capital, $500,000. President, E. E. Peters; Vice-President and General Manager, C. F. Peters; Secretary, F. J. Shuch.

MONT CLARE

Benj. F. Wagner.

MT. CARMEL

L. W. Weissinger—Railroads, Pennsylvania and Lehigh Valley. Individual ownership. Employes, 36. Hogs, 500. Refrigeration—75-ton Vilter and 35-ton Penn. Iron Works; direct expansion; 30 tons ice daily. Boiler capacity, 300 H. P.

PHILADELPHIA

Louis Burk—N. W. Cor. 3rd and Girard Ave. Individual ownership, Louis Burk.

Consolidated Dressed Beef Company—30th and Race Sts. A corporation. Capital, $500,000; issued, $315,000. President, Charles Harlan; Vice-President and Treasurer, J. Noble, Jr.; Secretary, Wm. A. Haines.

B. Ernst Bros.—2920 N. 6th St.

John J. Felin & Co., Inc.—4142-66 Germantown Ave. Railroad, P. & R. A corporation. Capital, $1,500,000; issued, $850,000. President, John J. Felin; Vice-President, Wm. C. Felin; Secretary, Wm. D. Reilly; Treasurer, Wm. C. Felin; General Manager, Wm. E. Felin; General Superintendent, Pat Ford. Hogs, 8,000. Sausage—Fresh and smoked. By-Products—Dried blood and tankage. Government inspection. Refrigeration—150, 50 and 25-ton De La Vergne and 12-ton Vilter; direct expansion; 10 tons ice daily. Boiler capacity, 1,500 H. P. Trade Marks—"IXL" and "Gold Medal." Branch House—New York City.

Geo. Hausmann & Sons, Inc.—5111 Westminster Ave. A corporation. Capital, $50,000; issued, $50,000. Stockholders, 7. President, Geo. Hausmann; Vice-President, J. F. Hausmann; Secretary and Treasurer, A. G. Hausmann; General Manager, Adam G. Hausmann. Employes, 40. Hogs, 600. Sausage—Fresh and smoked. By-Products—Tankage. Government inspection. Refrigeration—30-ton Frick; direct expansion. Boiler capacity, 80 H. P. Trade Mark—"Penn City Brand."

G. F. Pfund & Son—3945 Germantown Ave. Pork packers. Individual ownership, Carl F. Pfund.

Figures on cattle, hogs and sheep indicate AVERAGE WEEKLY KILLING CAPACITY.

PHILADELPHIA

Philadelphia Abattoir Company—30th and Race Sts. A corporation. Capital, $100,000. President, Jos. M. Harlan; Vice-President and Treasurer, E. B. Shriver; Secretary, H. E. Leutner.

Pusey-Maynes Breish Company, Inc.—3034-36 Market St. Railroad, Pennsylvania. A corporation. Capital, $100,000; issued, $53,800. President, R. J. Maynes; Vice-President, H. J. Pusey; Secretary and Treasurer, P. J. Breish, Sr. Calves, 600; sheep, 2,500. Government inspection. Refrigeration—24-ton York; direct expansion.

F. G. Vogt & Sons, Inc.—30th and Race Sts. Railroad, Penna. A corporation. Capital, $850,000; issued, $700,000. Stockholders, 250. President, Frederick A. Vogt; Secretary, Gustave L. Vogt; Treasurer, Charles H. Vogt; General Manager, Frederick A. Vogt; General Superintendent, Chas. H. Vogt. Employes, 350. Codes—Cross. Cattle, 50; hogs, 6,000. Sausage—Fresh and smoked. By-Products—Tankage, hog and chicken feed. Government inspection. Refrigeration—125-ton Vilter; direct expansion and brine spray system; makes own ice. Boiler capacity, 600 H. P. Trade Mark—"Liberty Bell Brand." Also exports.

Wilson-Martin Company—3000 Market St. A corporation. Capital, $5,000,000; issued, $500,000. Stockholders, 2. President, T. E. Wilson; Vice-President, D. Moog; Secretary, S. D. Shannon; General Manager, D. Moog. Codes—All standard; code word, "Wilsmarco." Cattle, 4,000; hogs, 20,000; sheep, 6,000. Sausage—Fresh and smoked. By-Products—Dried blood and tankage. Government inspection. Refrigeration—1,630 tons, Vilter, De La Vergne and Featherstone machines; brine circulating, brine spray and direct expansion. Boiler capacity, 5,521 H. P. Trade Marks—"Blue Ribbon," "National" and "Martinette Shortening." Branch Houses—Wilmington, Del.; Baltimore, Md.; Grays Ferry, Penn., and Camden, N. J. Also exports.

PHILIPSBURG

Philipsburg Beef Co.—Railroad, Penn. A corporation. Capital, $50,000; issued, $50,000. Stockholders, 2. President, A. B. Curtis; Vice-President, A. O. Curtis; Secretary and Treasurer, A. B. Lutz; General Manager, A. B. Curtis; General Superintendent, John Condon. Employes, 25. Cattle, 50; hogs, 300. Sausage—Fresh and smoked. By-Products—Tankage. Refrigeration—24-ton York; brine circulating. Trade Marks—"A. B. C."

PHOENIXVILLE

Christ F. Bader.

Weiland Packing Company, Inc.—Rear 493-557 Bridge St. Railroad, Pennsylvania. A corporation. Family ownership. Capital, $100,000. Stockholders, 4. President, Carl H. Weiland; Vice-President, J. Alvin Weiland; Secretary-Treasurer and General Manager, Frank B. Weiland; General Superintendent, Carl H. Weiland. Employes, 70. Cattle, 50; hogs, 2,500. Sausage—Fresh and smoked. By-Products—Tankage, grease, cracklings, hog and chicken feed, pigskins, tallow, hides. Refrigeration—40-ton York; 18-ton Frick; direct expansion system. Boiler capacity, 250 H. P. 10 tons ice daily.

PINEVILLE

Van Pelt Co.

Figures on cattle, hogs and sheep indicate AVERAGE WEEKLY KILLING CAPACITY.

PITTSBURGH

J. M. Denholm Bros. Co.—5th and Frankstowns Aves. Railroad, Penna., East Liberty Sta. A corporation. Capital, $25,000; issued, $25,000. President, J. M. Denholm; Vice-President, B. M. Denholm; Secretary, H. K. McJunkin; Treasurer, J. C. Williams; General Manager, B. M. Denholm. Employes, 75. Stockholders, 8. Cattle, 150; hogs, 500; sheep, 200. Sausage—Fresh. By-Products—Dried blood and tankage. Refrigeration—40 and 15-ton York; brine circulating. Boiler capacity, 600 H. P. Trade Mark—"Keystone Brand."

Dunlevy-Franklin Co.—6500 Hamilton Ave. A corporation. Capital, $500,000; issued, $500,000. Stockholders, 163. President and General Manager, Geo. L. Franklin; Vice-President, M. J. Hennessey; Secretary and Treasurer, W. G. Horne; General Superintendent, P. C. Walker. Employes, 400. Codes—Cross; code word, "Dunlevy." Cattle, 400; hogs, 6,000; sheep, 200. Sausage—Fresh and smoked. By-Products—Tankage and dried blood. Government inspection. Refrigeration—One 50-ton Frick and two 30-ton York; direct expansion; 30 tons ice daily. Boiler capacity, 800 H. P.

Fried & Reineman Packing Co.—East Ohio St. Blvd., N. S., opp. Union Stock Yards. A corporation. Capital, $750,000. Stockholders, 185. President, E. A. Reineman; Secretary and Treasurer, G. N. Meyer. Employes, 215. Hogs, 8,400; cattle, 900. Sausage—Fresh and smoked. Dairy products. By-Products—Hides, tallow, tankage, grease and hog hair. Refrigeration—100 and 50-ton York; direct expansion. Boiler capacity, 500 H. P. Trade Mark—"Fort Pitt."

F. J. Kuhn Company—2108 East St. A corporation. Capital, $150,000; issued, $100,700. Stockholders, 5. President, F. J. Kuhn, Sr.; Vice-President, F. J. Kuhn, Jr.; Secretary and Treasurer, M. Nedwedeck. Employes, 20. Cattle, 100; hogs, 500; sheep, 200. Sausage—Fresh and smoked. Refrigeration—One 20-ton York; brine circulating; two tons ice daily. Retail Markets—2108 East St., Pittsburgh.

Henry Lohrey Co.—2234-44 East St., N. S. A corporation. Capital, $125,000; issued, $125,000. President, B. O. Lohrey; Vice-President and Treasurer, Walter G. Lohrey; Secretary, Joseph E. Lohrey.

North Side Packing Company—2200-14 Spring Garden Ave., N. S. A corporation. Capital, $40,000; issued, $40,000. President and Treasurer, J. G. Hofman; Vice-President, Wm. Mall; Secretary, Chas. Stigh.

Oswald & Hess Co., Inc.—855 Spring Garden Ave. A corporation. Capital, $50,000; issued, $30,000. Stockholders, 3. President, George A. Hess; Vice-President and Treasurer, Wilbert W. Oswald; Secretary, G. J. Hess. Employes, 48. Hogs, 1,000; sheep, 150. Sausage—Fresh and smoked. Refrigeration—15-ton York; direct expansion. Boiler capacity, 180 H. P. Trade Mark—"Delicious."

Pittsburgh Provision & Packing Co.—Union Stock Yards. Railroads, Pennsylvania and B. & O. A corporation. Capital, $600,000; issued, $600,000. President, R. Allerton; Vice-President, Jas. S. McFadyen; Secretary and Treasurer, John Anderson; General Manager, Chas. H. Ogden. Employes, 550. Codes—Cross and Robinson. Cattle, 1,200; hogs, 4,000; sheep, 1,000; calves, 1,200. Sausage—Fresh and

---

Figures on cattle, hogs and sheep indicate AVERAGE WEEKLY KILLING CAPACITY.

smoked. By-Products—Dried blood, tankage, hog and chicken feed, tallow, greases, oleo stearine and oleo oils. Government inspection. Refrigeration—Two 100, one 80 and one 40-ton Frick; 25 tons ice daily. Branch Houses—Johnstown, Pa.; Cumberland, Md.

Sun Packing Company—1912 Spring Garden Ave. Capital, $1,000,000. Manager, Albert L. Brahm, Jr.

William Zoller Co.—Spring Garden Ave. A corporation. Capital, $300,000; issued, $300,000. President, Wm. Zoller; Secretary, H. O. Fisher; Treasurer, Edw. Wettach; General Manager, E. Wettach; General Superintendent, A. J. Riester. Cattle, 250; hogs, 3,800; sheep, 300. Sausage—Fresh and smoked. By-Products—Dried blood and tankage. Government inspection. Refrigeration—150 and 75-ton Frick; direct expansion. Boiler capacity, 750 H. P. Trade Marks—Hams, "Rosevale"; bacon, "Rosevale"; lard, "Clover" and "Pearl."

POTTSTOWN

Pottstown Abattoir—17 E. 3rd St. Railroads, Philadelphia & Reading and Pennsylvania. Individual ownership. Proprietor, H. F. Himmelberger. Employes, 15. Codes—Cross. Cattle, 100; hogs, 500. Sausage—Fresh and smoked. Refrigeration—20-ton Frick; brine circulating. Boiler capacity, 160 H. P.

POTTSVILLE

Seltzer Packing Co.—Water and Temple Sts. Railroad, Phila. & Reading. A corporation. Sausage—Fresh and smoked. By-Products—Dried blood and tankage. Government inspection. Refrigeration—Two 35-ton Pennsylvania Iron Works.

Geo. Weissinger & Bro.—250 Peacock St. Partnership, Geo. Weissinger and Harry Weissinger.

Jacob Ulmer Packing Co.—Front and East Railroad. Railroads, P. & R. and Pennsylvania. A corporation. Capital, $100,000; issued, $100,000. President, Jacob S. Ulmer; Secretary, A. B. McCool; Treasurer, Louis F. Ulmer; General Manager, Jacob S. Ulmer. Codes—Cross. Cattle, 100; hogs, 1,000. By-Products—Tankage. Sausage—Fresh, smoked and summer. Government inspection. Refrigeration—130-ton Frick; brine circulating and direct expansion; 25 tons ice daily. Retail Market—Pottsville, Pa.

PRESTO

J. W. Stewart Co.—Railroad, P. C. & Y. A corporation. Capital, $40,000. President, J. W. Stewart; Vice-President, Chas. E. Eaton; Secretary-Treasurer, W. S. Walsh; Manager and General Superintendent, J. W. Stewart; Sales Manager, Chas. E. Eaton. Employes, 14. Cattle, 50; hogs, 150; calves, 50; lambs, 50. Sausage—Fresh and smoked. Refrigeration—25-ton Frick; brine circulating and direct expansion; 5 tons ice daily. Boiler capacity, 95 H. P.

PUNXSUTAWNEY

Punxsutawney Beef & Provision Co.—Railroad, Pennsylvania. A corporation. Capital, $40,000; issued, $30,000. Stockholders, 5. President, H. H. Beezer; Vice-President, John A. Philliber; Secretary and Treasurer, H. A. Philliber. Employes, 35. Cattle, 125; hogs, 1,000. Sausage—Fresh and smoked. By-Products—Dried blood, tankage, hog feed, tallow and grease. Refrigeration—20-ton Wilson & Snyder and 20-ton Vilter; direct expansion; 4 tons ice daily. Boiler capacity, 160 H. P.

Figures on cattle, hogs and sheep indicate AVERAGE WEEKLY KILLING CAPACITY.

READING

Central Abattoir Company—2nd and Chestnut Sts. Railroad, Phila. & Reading. Employes, 25. Cattle, 100; hogs, 150. Sausage—Fresh and smoked. By-Products—Dried blood and tankage. Refrigeration—25-ton Frick; brine circulating. Boiler capacity, 60 H. P.

John F. Forner—828 Bingaman St. Abattoir.

D. Isecovitz & Bro.—135 Grape St.

Reading Abattoir Co.—216 Pine St. Railroad, Penna. A corporation. President, Howard DeLong; Vice-President, Geo. H. Rader; Secretary, O. G. Mull; Treasurer, Howard DeLong; General Manager, Geo. H. Rader; General Superintendent, Aug. Garis. Cattle, 300; hogs, 2,500. Sausage—Fresh, smoked and summer. By-Products—Dried blood and tankage. Government inspection. Refrigeration—75 and 35-ton York; brine circulating and direct expansion. Boiler capacity, 400 H. P. Trade Mark—"Raco."

RICHLAND

A. Forry & Son.

ROYALTON

Vogt Farm Meat Product Co.—Railroad, Pennsylvania. A corporation. Capital, $500,000; issued, $150,000. Stockholders, 700. President, Guy S. Vogt; Vice-President, Harry Bonholtzer; Secretary and Treasurer, R. I. Mahler.

SALTSBURG

W. B. Serene & Sons.

SCALP LEVEL

Baumgardner Meat Co.

SCRANTON

Brizer & Bernstein.

Frank L. Carr—1200-1212 Remington Ave. Railroad, Del. & Hudson. Individual ownership. Employes, 25. Cattle, 100; hogs, 200. Sausage—Fresh and smoked. By-Products—Dried blood, tankage, tallow, grease, hog hair and dry bones. Refrigeration—Two 20-ton Remington; direct expansion. Boiler capacity, 130 H. P. Retail Market—108 Penn Ave., Scranton.

Stowers Pork Packing & Prov. Co.—2-56 Greens Lane. Railroad, Delaware & Lackawanna and Delaware & Hudson. A corporation. Capital, $100,000; issued, $100,000. President, J. R. Schlager; Secretary, H. A. Benson; Treasurer, J. R. Schlager; General Manager, J. R. Schlager; General Superintendent, Charles Clarke. Cattle, 25; hogs, 500. Sausage—Fresh and smoked. Refrigeration—25-ton Frick; brine circulating. Boiler capacity, 200 H. P.

SHAMOKIN

Croninger Packing Company, Inc.—A corporation. Capital, $200,000; issued, $100,000. President, Jas. H. Straub; Vice-President, J. A. Wert; Secretary and Treasurer, E. M. Moyer.

SHENANDOAH

Shenandoah Abattoir Co.—Market and Poplar Sts. Railroad, Lehigh Valley. A corporation. Capital, $200,000; issued, $139,600. Stockholders, 75. President, Andrew Meluskey; Secretary, Stanley F. Bauser; Treasurer, Albin A. Meluskey; General Manager, Albin A.

Figures on cattle, hogs and sheep indicate AVERAGE WEEKLY KILLING CAPACITY.

Meluskey, Sr.; General Superintendent, Albin Meluskey, Jr. Employes, 100. Codes—Cross. Hogs, 2,500. Sausage—Fresh and smoked. By-Products—Dried blood, tankage, hog feed, grease, cracklings, hog hair and dried bones. Refrigeration—50-ton Frick and 35-ton Nagle; direct expansion. Boiler capacity, 300 H. P. Trade Mark—"Nonpareil Brand."

SOUTH BETHLEHEM
Gottlieb & Hulbier—503 E. Third St.

SOUTHAMPTON
George Stockburger.

UNIONTOWN
Union Provision, Ice & Cold Storage Company—Beef and Pork Packers. Coffee St. Railroad, Baltimore & Ohio. A corporation. Capital, $200,000; issued, $164,300. Stockholders, 67. President, T. S. Lackey; Vice-President, W. E. Vansickle; Secretary, E. E. Dupray; Treasurer and General Manager, E. J. Creamer. Employes, 15. Do not slaughter at present. Refrigeration—20-ton Frick; direct expansion. Boiler capacity, 90 H. P.

WASHINGTON
Washington Packing Company.

WAYNESBORO
Union Abattoir Co.

WEST GROVE
Pusey & Jones.

WEST READING
Thomas Rahn & Son.

WILLIAMSPORT
John Peters—1320 East Third St. Railroad, Phila. & Reading. Individual ownership. Employes, 40. Codes—Cross. Cattle, 100; hogs, 600; sheep, 200. Sausage—Fresh, smoked and summer. By-products—Dried blood and tankage, hair, bones and cracklings. Refrigeration—Two 75-ton Vilter and one 150-ton De La Vergne; direct expansion and brine circulating; 75 tons ice daily. Boiler capacity, 400 H. P.

YORK
York Storage & Ice Co.

RHODE ISLAND

MIDDLETON
Samuel Berman—Paradise Road.

NEWPORT
William H. Easton & Co.—67 W. Broadway.
Wm. A. Stoddard—The Boulevard.

PAWTUCKET
Anderton Bros.
Bateman & Tarpy—71 Dexter St.
Harry Bramham—56 Upton Ave.
Comstock & Company—Concord St.
Chas. S. Johnson—197 Garden St.
H. S. Johnson—266 Weedon St.
Lecht Bros.—Abattoir.

Figures on cattle, hogs and sheep indicate AVERAGE WEEKLY KILLING CAPACITY.

PORTSMOUTH
Wm. A. S. Cummings.
Nahum Green.
L. Levine.

PROVIDENCE
Kimball & Colwell Co.—459 Washington St. A corporation. Capital, $25,000; issued, $25,000. Stockholders, 3. President, M. K. Hadley; Secretary and Treasurer, C. D. Kimball; General Manager, H. C. Gagner. Employes, 85. Hogs, 900. Sausage—Fresh and smoked. Government inspection. Refrigeration—25-ton Linde; brine circulating. Boiler capacity, 200 H. P. Trade Marks—"K. C. Brand" and "What Cheer Brand." Also exports.

Confederated Home Abattoirs Co.—A corporation. Capital, $750,000. President, Emil Schierhodtz; Vice-President and General Manager, Louis Berman; Treasurer, C. I. Bigney; Secretary, George F. McCanna.

SOUTH CAROLINA

ANDERSON
Anderson Abattoir Company.

BEAUFORT
Seacoast Packing Company.

CHARLESTON
Charleston Abattoir Company.

GREENVILLE
W. H. Balentine—208 E. Court St. Railroad, Charleston & Western Carolina. Individual ownership, W. H. Balentine. Capital, $75,000. General Superintendent, J. W. Gilreath. Employes, 30. Codes—Cross and Telegraphic. Cattle, 30; hogs, 100; sheep, 25. Sausage—Fresh and smoked. Refrigeration—Two 15-ton Yorks; direct expansion. Boiler capacity, 30 H. P.

Greenville Abattoir Co.—109 S. Main St. Railroad, Southern. Individual ownership, M. H. Goodlett. General Manager, M. H. Goodlett; General Superintendent, John Pettiette. Employes, 25. Cattle, 300; hogs, 1,200; sheep, 800. By-Products—Dried blood, tankage, hog and chicken feed and tallow and grease. City inspection. Refrigeration—20-ton Frick; direct expansion. Boiler capacity, 200 H. P.

GREENWOOD
Greenwood Abattoir—457 Newmarket St. A corporation. Capital, $10,000. Stockholders, 2. President, W. P. Carley; Vice-President, Dr. W. A. Barnett; Secretary, Dr. W. A. Barnett; Treasurer, W. P. Carley. Employes, 5. Cattle, 220; hogs, 300; sheep, 200. By-Products—Dried blood and tankage.

ORANGEBURG
Carolina Packing Company of South Carolina—Railroads, Atlantic Coast Line and Southern. A corporation. Capital, $500,000; issued, $223,800. Stockholders, 225. President, W. W. Love; Vice-President, Niels Christensen; Secretary and Treasurer, Geo. C. Dixon; General Manager, Geo. C. Dixon; General Superintendent, Ladi Tobias. Employes, 50. Cattle, 50; hogs, 500. Sausage—Fresh and smoked. By-Products—Tankage. Government inspection. Refrigeration—80-ton Carbondale; absorption system; brine and direct expansion; 6 tons ice daily. Boiler capacity, 300 H. P. Trade Mark—"Carolina Brand."

Figures on cattle, hogs and sheep indicate AVERAGE WEEKLY KILLING CAPACITY.

SOUTH DAKOTA
HURON
Farmers' Co-operative Packing Company—A corporation. Capital, $2,000,000; issued, $1,230,000. President, B. F. Meyers; Vice-President, Henry Scott; Secretary, G. W. Wright; Treasurer, H. G. Spratt.

RAPID CITY
Rapid City Packing Co.

SIOUX FALLS
John Morrell & Co.—Foot of Nesmith Ave. General Superintendent, Geo. M. Foster. (See Ottumwa, Iowa.)

TENNESSEE
CENTERVILLE
R. E. Shouse & Company.

CHATTANOOGA
J. H. Allison & Co.—Middle St. Railroads, Belt Line and Nashville, Chattanooga & St. Louis. A corporation. Stockholders, 3. President, J. H. Allison; Vice-President and Treasurer, H. W. McCall; Secretary, B. M. Allison. Employes, 75. Cattle, 900; hogs, 1,800; sheep, 500. Sausage—Fresh, smoked and summer. By-Products—Dried blood, tankage and steamed bones. Government inspection. Refrigeration—150 and 50-ton Linde; brine circulating and spray; 100 tons ice daily. Boiler capacity, 1,000 H. P. Trade Marks—Hams, "Allison's" and "St. Elmo"; Bacon, "Allison's Best" and "St. Elmo"; lard, "Hollyleaf" and "St. Elmo"; sausage, "East Tennessee All Pork."

CLARKSVILLE
Kleeman & Co.—Franklin St. Partnership. Cattle, 25; hogs, 25; sheep, 25. Sausage—Fresh and smoked. By-Products—Tallow. Refrigeration—10-ton Triumph; brine circulating and direct expansion. Trade Marks—"Trilby," "Our Own Make." Retail Market—212 Franklin St., Clarksville, Tenn.

GOODSPRINGS
O. J. Tomerlin.

GREENEVILLE
Greeneville Packing Co.—Railroad, Southern. Partnership. Employes, 24. Hogs, 300. Sausage—Fresh and smoked. By-Products—Tankage. Government inspection. Refrigeration—6½-ton Automatic; direct expansion. Operates only three or four months in the year. Dry salt all meats, country style. Specialize in pure pork country style sausage, put up in paraffine bags.

JOHNSON CITY
A. N. Feathers & Son.

KNOXVILLE
East Tennessee Packing Co.—A corporation. Capital, $200,000; issued, $100,000. Stockholders, 8. President, J. B. Madden; Vice-President, D. G. Madden; Treasurer, H. H. Slattery. Employes, 50. Codes—All. Cattle, 700; hogs, 2,100; sheep, 700. Sausage—Fresh and smoked. By-Products—Tankage. Refrigeration—100-ton Frick, 150 and 50-ton De La Vergne; brine circulating, direct expansion and

---

Figures on cattle, hogs and sheep indicate AVERAGE WEEKLY KILLING CAPACITY.

brine spray; 25 tons ice daily. Boiler capacity, 450 H. P. Trade Marks—Hams, "Selecto"; bacon, "Selecto"; lard, "Lily"; compound, "Invincible." Retail Markets—Knoxville and Maryville, Tenn.

T. L. Lay Packing Co.—400-402 E. Jackson Ave. Railroad, Southern. A corporation. Capital, $150,000; issued, $100,000. Stockholders, 5. President, T. L. Lay; Vice-Presidents, W. T. Lay and L. M. Lay; Secretary, Frank B. Hall; Treasurer, Ira V. Lay; General Manager, W. T. Lay; General Superintendent, Frank B. Hall. Employes, 50. Cattle, 300; hogs, 2,000. Sausage—Fresh and smoked. By-Products—Dried blood and tankage. Refrigeration—Two 20-ton and one 6-ton York; direct expansion. Retail Market—141 S. Central Ave., Knoxville.

Knoxville Abattoir Co.

MEMPHIS

Memphis Packing Corporation.—Riverside and Trigg Sts. Railroad, Frisco. A corporation. Capital, $2,000,000; issued, $1,990,745. President, Joseph Newburger; Vice-Presidents, S. M. Williamson, L. K. Salisbury and J. L. McCabe; Secretary and Treasurer A. S. Nordlinger; General Manager, J. L. McCabe; General Superintendent, F. F. Rueping. Employes, 200. Code—Cross. Cattle, 750; hogs, 3,600. Sausage—Fresh and smoked. By-Products—Dried blood, tankage and hog feed. Government inspection. Refrigeration—Two 50-ton Vogt; absorption system; brine circulating and brine spray; 12 tons ice daily. Boiler capacity, 450 H. P.

Scott's Abattoir Co.

MORRISTOWN

Donaldson & Holtsinger.
Morristown Packing Company.

NASHVILLE

Nashville Abattoir H. & M. Ass'n—1416 Adams St. Railroads, Louisville & Nashville and Nashville, Chattanooga & St. Louis. A corporation. Capital, $32,500; issued, $32,500. Stockholders, 65. President, N. Lahart; Vice-President, W. R. Bruce; Secretary, G. S. Jacobs; Treasurer, E. C. Fox; General Manager, O. W. Climer. Employes, 55. Cattle, 500; hogs, 750; sheep, 375; calves, 175. By-Products—Dried blood, tankage, grease, tallow and casings. City inspection. Refrigeration—100-ton Vilter and 90-ton York; brine circulating and direct expansion systems; 16 tons ice daily. Boiler capacity, 450 H. P.

Neuhoff Packing Company—1310 Adams St. Lorenz Neuhoff, Treasurer and Manager.

Power Packing Plant—1101 1st Ave., N. Railroad, Louisville & Nashville. A corporation. Capital, $500,000; issued, $500,000. Stockholders, 45. President, R. E. Power; Vice-President, H. E. Warner; Secretary and Treasurer, H. A. Tenbrunsel; General Manager, Chris. J. Power. Employes, 125. Code—Cross. Cattle, 200; hogs, 1,500; sheep, 500. Sausage—Fresh and smoked. By-Products—Dried blood and tankage. Government inspection. Refrigeration—60 and 40-ton Frick; direct expansion. Boiler capacity, 300 H. P. Trade Mark—"Power."

PULASKI

R. W. George.

Figures on cattle, hogs and sheep indicate AVERAGE WEEKLY KILLING CAPACITY.

UNION CITY

Reynolds Packing Co.—S. 5th St. Railroad, Mobile & Ohio. A corporation. Capital, $100,000; issued, $77,700. Stockholders, 209. President, W. G. Reynolds; Vice-President, G. W. Stovall; Secretary and Treasurer, G. B. White; General Manager, W. G. Reynolds; General Superintendent, Arnold Glott. Employes, 20. Cattle, 35; hogs, 140; calves, 10; sheep and lambs, 50. Sausage—Fresh and smoked. Government inspection. Refrigeration—16-ton Brecht; direct expansion and brine circulating. Trade Mark—"Reynolds' Quality Products."

TEXAS

AUSTIN

Walker Properties Association—3rd and San Antonio Sts. Railroad, International & Great Northern. Mfgrs. of chili con carne and Mexene Chili Powder. President, F. W. Catterall; General Manager, W. F. Gohlke. Government inspection.

BEAUMONT

Zummo Packing Co.

DALLAS

Armstrong Packing Co.—Cockrell Ave. and Alma St. Railroads, Missouri, Kansas & Texas and Cotton Belt. A corporation. Capital, $1,000,000; issued, $1,000,000. Stockholders, 3. President, E. L. Flippen; Vice-President, H. E. Prather; Active Vice-President, E. H. Terrell; Secretary and Treasurer, W. Worrill. Employes, 500. Codes—All. Cattle, 1,000; hogs, 5,000; sheep, 500. Sausage—Fresh, smoked and summer. By-Products—Full line. Government inspection. Refrigeration—Two 190-ton De La Vergne and one York; 25 tons ice daily. Boiler capacity, 2,000 H. P. Also exports.

Max Hahn Packing Co.—North End Alamo St. Railroad, Missouri, Kansas & Texas. A corporation. Capital, $50,000; issued, $50,000. Stockholders, 6. President, Max Hahn; Vice-President, E. D. Slaughter; Secretary and Treasurer, W. A. Currens; General Manager, Max Hahn; General Superintendent, Carl Hahn. Employes, 25. Cattle, 500; hogs, 1,000; sheep, 350. Sausage—Fresh, smoked and summer. By-Products—Tankage. Refrigeration—25-ton Frick; brine circulating. Boiler capacity, 120 H. P. Trade Mark—"Rooster Brand."

EL PASO

Border Packing Co., Inc.—104 South Florence St. A corporation. Capital, $20,000; issued, $20,000. President, Mack Camp; Vice-President, A. W. Graham; Secretary and Treasurer, H. W. Lackland; General Manager, A. W. Graham. Employes, 8. Slaughterers only. Buy refrigeration.

Nations Packing Co.—301 S. Kansas St. Railroad, Texas & Pacific. A corporation. Capital, $250,000; issued, $150,000. Stockholders, 5. President, J. H. Nations; Vice-President, C. B. Ardoin; Secretary and Treasurer, O. A. Danielson; General Manager, O. A. Danielson; General Superintendent, Ben Meyer. Employes, 27. Code—Private; code word, "Nations." Cattle, 100; hogs, 75; sheep, 50. Sausage—Fresh and smoked. Refrigeration—Buys brine from ice factory.

National Meat Company.

Figures on cattle, hogs and sheep indicate AVERAGE WEEKLY KILLING CAPACITY.

Peyton Packing Co.—Railroads, El Paso & South Western and Texas & Pacific. A corporation. Capital, $1,000,000; issued, $400,000. Stockholders, 25. President, J. C. Peyton; Vice-President, Robt. Rhea; Secretary, P. S. C. Davis; Treasurer, Robt. Rhea; General Manager, J. C. Peyton; General Superintendent, W. A. Kessler. Employes, 150. Code—Cross. Cattle, 800; hogs, 1,000; sheep, 500. Sausage—Fresh and smoked. By-Products—Dried blood and tankage. Government inspection. Refrigeration—125-ton De La Vergne and 40-ton Vilter; brine spray. Boiler capacity, 320 H. P. Trade Marks—"Circle Star" and "Chesterfield." Also exports.

Henry G. Schneider—Railroad, El Paso & South Western. Individual ownership. General Superintendent, H. G. Schneider. Employes, 3. Cattle, 250; hogs, 500; sheep, 500. Only an abattoir—do not pack. Sausage—Fresh and smoked. By-Products—Hog and chicken feed. Refrigeration—None. Boiler capacity, 20 H. P. Retail Market—506 Mesa Ave., El Paso.

Sanitary Slaughter House—H. Peper, Prop.

FORT WORTH

Armour & Company—(See Chicago.)

Fort Worth Packing Co.—301 E. 22nd St. Railroad, Fort Worth Belt. A corporation. Capital, $125,000; issued, $125,000. Stockholders, 4. President, E. W. Gruendler; Vice-President, T. F. Maurin; Secretary, R. N. Dumble; Treasurer, Alvin T. Lange; General Manager, R. N. Dumble. Employes, 85. Codes—Cross and Robinson. Cattle, 100; hogs, 300. Sausage—Fresh and smoked. By-Products—Tankage. Refrigeration—50-ton Frick, 30-ton Frick and 14-ton Wolf-Linde; direct expansion; 33 tons ice daily. Boiler capacity, 450 H. P. Trade Mark—"Fairy Brand."

Swift & Company—(See Chicago.)

GALVESTON

Rosenthal Packing Co., Inc.—61st and Avenue J. Railroad, Southern Pacific. Individual ownership. President, Louis Rosenthal; Vice-President, J. J. Niedermann; Secretary and Treasurer, Chas. F. Hildenbrand. Cattle, 100; calves, 150; hogs, 300. Sausage—Fresh, smoked and summer. Refrigeration—25-ton York; direct expansion. Boiler capacity, 150 H. P. Trade Mark—"Oleander Brand."

HOUSTON

Houston Packing Co.—Engelke and Roberts. Railroads, International & Great Northern and San Antonio & Aransas Pass. A corporation. Capital, $800,000. President, R. E. Paine; Vice-President, P. B. Timpson; Treasurer, W. J. Hyde; Secretary, E. W. Guendlen; General Manager and General Superintendent, T. F. Maurin. Employes, 350. Codes—Robinson, Utility, Cross, Yopp, Armsby and Western Union. Cattle, 1,800; hogs, 10,000. Sausage—Fresh, smoked and summer. Edible oils and shortenings; ice. By-Products—Dried blood, tankage, chicken feed, bone meal. Government inspection. Refrigeration—250-ton Frick; 250-ton York. Direct expansion. Boiler capacity, 1,500 H. P. Branch houses at Beaumont, Brownsville and Galveston, Tex.; and Lake Charles, La.

Texas Packing Company—108 Milam St. Individual ownership. Employes, 20. Cattle, 250; hogs, 100. Sausage—Fresh, smoked and summer. Refrigeration—Furnished by contract from ice plant.

Figures on cattle, hogs and sheep indicate AVERAGE WEEKLY KILLING CAPACITY.

PARIS

Paris Municipal Abattoir.

SAN ANTONIO

Alamo Dressed Beef Co.—Roosevelt Ave. and Mitchell St. Railroad, San Antonio & Aransas Pass. A corporation. Capital, $50,000; issued, $50,000. Stockholders, 11. President, J. D. Oppenheimer; Vice-President, A. H. Halff; Secretary, Frederick Terrell; Treasurer, J. D. Oppenheimer; General Manager, Chas. R. Bergstrom; General Superintendent, Chas. R. Bergstrom. Employes, 55. Cattle, 250; hogs, 250; sheep, 50. Sausage—Fresh, smoked and summer. By-Product—Tankage. Refrigeration—60-ton York; brine circulating; 20 tons ice daily. Boiler capacity, 300 H. P.

Union Meat Company—Vice-President, J. A. Gallagher.

SHERMAN

Sherman Slaughtering & Rendering Co.—Railroad, Missouri, Kansas & Texas. A corporation. Capital, $10,000; issued, $10,000. Stockholders, 4. President, R. E. Paine; Vice-President and Manager, Chas. Knapp; Treasurer, E. W. Guendlen. Employes, 4. Wholesale slaughterers, custom killers and renderers.

TAFT

Taft Packing House—Railroad, San Antonio & Aransas Pass. A corporation. (Coleman Fulton Pasture Co.) Capital, $849,700; issued, $849,700. President, Charles P. Taft; Vice-President, Joseph T. Green; Secretary, H. V. Fetick; General Manager, H. V. Fetick; General Superintendent, J. G. Vann. Employes, 40. Cattle, 250; hogs, 250; sheep, 150. Sausage—Fresh, smoked and summer. By-Products—Tankage, hog and chicken feed. Refrigeration—25-ton Wolf-Linde; 25-ton Frick; brine circulating; 25 tons ice daily. Boiler capacity, 1,100 H. P. Trade Mark—"Taft's Crystal Shortening." Retail Market—Taft, Texas. Also exports.

UTAH

EDEN

George McDonald.

MURRAY

Jensen Bros.
McMillan & Son.

OGDEN

Ogden Packing & Provision Co.—West 24th St. Railroad, Union Pacific. A corporation. Capital, $3,425,000; issued, $2,000,000. Vice-President, W. H. Sherman; Secretary, Eva C. Erb; Treasurer, Charles H. Barton; General Manager, James Brennan; General Superintendent, Geo. F. Madsen. Employes, 250. Codes—A. B. C., 5th, Cross and Bentley; code word, "Opaco." Cattle, 750; hogs, 5,000; sheep, 2,000. Sausage—Fresh and smoked. By-Products—Dried blood, tankage, hog and chicken feed, meat meal. Government inspection. Refrigeration—175-ton Vulcan; direct expansion and brine spray. Boiler capacity, 650 H. P. Trade Marks—Hams, "Mountain" and "Nectar" brands; bacon, "Mountain," "Nectar" and "Sego"; lard, "Mountain" and "Sego"; compound, "Chefo" and "Canyon." Branch Houses—1150 East First, South Los Angeles, and 25 Crook St., San Francisco, Cal.; 700 South Utah Ave., Butte, Mont.; 370 West First, South Salt Lake, Utah; 306 North Main, Pocatello, Idaho. Also exports.

Figures on cattle, hogs and sheep indicate AVERAGE WEEKLY KILLING CAPACITY.

SALT LAKE CITY

Block & Guss—1672 Berk St. Railroad, Oregon Short Line. Partnership—Louis Block and Sam. Guss. Employes, 15. Cattle, 180; hogs, 50; sheep, 250. Sausage—Fresh and smoked. By-Product—Tankage. City inspection. Refrigeration—6-ton Niebling; direct expansion. Boiler capacity, 100 H. P.

Archie McFarland & Son—2922 S. State St. Railroads, Denver & Rio Grande and Oregon Short Line. Partnership. General Manager, Archie McFarland; Buyer, Rae McFarland. Employes, 18. Cattle, 200; hogs, 100; sheep, 1,000. City inspection. Refrigeration—6-ton Vilter; direct expansion. Boiler capacity, 80 H. P.

VERMONT

HYDE PARK

John Miner—Railroad, Boston & Maine. Individual ownership. Cattle, 200; hogs, 300; sheep, 500. Government inspection. Refrigeration—4-ton Brunswick; direct expansion. Boiler capacity, 10 H. P. Retail Market—Hyde Park, Vt.

BELLOWS FALLS

J. J. Cray Packing Company.

ESSEX JUNCTION

Fletcher & Co.

VIRGINIA

BRISTOL

Virginia Packing Co., Inc.—Railroad, Southern. A corporation. Capital, $50,000; issued, $50,000. Stockholders, 3. President, W. S. Lindsay, Vice-President, S. E. Harkrader; Secretary and Treasurer, C. E. Sarver. Employes, 28. Cattle, 100; hogs, 500. Sausage—Fresh and smoked. By-Products—Tankage. Government inspection. Refrigeration—15-ton Frick; direct expansion. Boiler capacity, 100 H. P.

CHARLOTTESVILLE

Charlottesville Abattoir.
Charlottesville Ice Co.

DANVILLE

McGuire & Co.

IVOR

L. H. Brantley.
Ivor Grocery Co.
L. C. Pulley & Co.

LYNCHBURG

M. R. Scott Co.

NEWPORT NEWS

Levinson Packing Co., Inc.—2610-2612 Jefferson Ave. Railroad, Chesapeake & Ohio. A corporation. Capital, $100,000; issued, $35,000. Stockholders, 3. President, Max Levinson; Secretary, C. H. Carpenter; Treasurer, Ben Levinson; General Manager, Max Levinson. Employes, 12. Codes—Cross; code word, "Levson." Cattle, 25; hogs, 200. Sausage—Fresh. By-Products—Tankage and dried blood. Refrigeration—25-ton York; brine circulating. Trade Marks—"Levson" and "Smithfield Hams."

Figures on cattle, hogs and sheep indicate AVERAGE WEEKLY KILLING CAPACITY.

NORFOLK

Banks Bros. Packing Co.—Chapel St. Railroad, Norfolk & Western. Partnership. Employes, 50. Cattle, 200; hogs, 500; sheep, 200. Sausage—Fresh and smoked. By-Product—Tankage. Refrigeration—65-ton York; brine circulating; 40 tons ice daily. Boiler capacity, 350 H. P. Trade Marks—"Snow Bank" and "Sunshine Brand."

Interstate Packing & Ice Co.—Railroad, Norfolk & Western. A corporation. Capital, $150,000; issued, $50,000. President, J. T. Lynch; Vice-President, G. C. Culpepper; Treasurer, E. L. Cote; Secretary, Chas. R. Barnes. Cattle, 300; hogs, 1,200; sheep, 120. Sausage—Fresh, smoked and summer. By-Product—Tankage. Refrigeration—90-ton Frick and 30-ton Frick; brine circulating system. Not yet operating—April, 1922.

NORTH TAZEWELL

Tazewell Packing Company.

NORTON

Norton Packing Co., Inc.—Railroads, Louisville & Nashville, Norfolk & Western and Interstate. A corporation. Capital, $100,000; issued, $60,000. Stockholders, 5. President, E. A. Harner; Secretary and Treasurer, S. A. McCluen; General Manager and Superintendent, E. A. Harner. Employes, 15. Cattle, 200; hogs, 3,000; sheep, 200. Sausage—Fresh, smoked and summer. By-Products—Dried blood, tankage and fertilizer. Refrigeration—50-ton York; brine circulating. Boiler capacity, 150 H. P.

PETERSBURG

Bowman Bros.—31 Sycamore St.

PORTSMOUTH

Codd & Co.
Freedman Packing Co.
Norman Packing Co.
Portsmouth Provision & Packing Co.
T. O. Williams.

RICHMOND

Brown Abattoir Co.
W. S. Forbes & Co., Inc.—10th and Byrd Sts. G. E. Mengel, Manager. Owned by Allied Packers, Inc., Chicago, Ill.
R. Kastelberg's Sons, Inc.
Patrick-Young Co.—Hermitage Rd. near Leigh.
Richmond Abattoir—(Valentine's Meat Juice Co., Proprietors.) Chamberlayne Parkway. Railroad, Seaboard Air Line. A corporation. Capital, $100,000; issued, $100,000. President, G. G. Valentine; Vice-President, F. S. Valentine; Secretary and Treasurer, H. L. Valentine. Cattle, 200; calves, 200. By-Products—Dried blood, tankage and hog and chicken feed. Government inspection. Employes, 40. Boiler capacity, 90 H. P. Refrigeration.
Union Abattoir Co.
Valentine's Meat Juice Co.—Chamberlayne Parkway. Railroad, Seaboard Air Line. A corporation. Capital, $100,000; issued, $100,000. President, G. G. Valentine; Vice-President, F. S. Valentine; Secretary and Treasurer, H. L. Valentine. Employes, 60. Cattle, 125. By-Products—Dried blood, tankage and hog and chicken feed. Government inspection. Refrigeration.

Figures on cattle, hogs and sheep indicate AVERAGE WEEKLY KILLING CAPACITY.

ROANOKE

Brown Abattoir Co., Inc.—A corporation. Capital, $25,000; issued, $16,000. Stockholders, 10. President, Frank E. Brown; Vice-President, George E. Markly; Treasurer, Secretary and General Manager, Frank E. Brown; General Superintendent, R. P. Vandegrift. Employes, 4. Cattle, 40; hogs, 30; sheep, 10. By-Product—Tankage. Refrigeration—15-ton Automatic; direct expansion. Boiler capacity, 50 H. P.

Frank Brown Company.

Griggs Packing Co., Inc.—Franklin Road. Railroad, Norfolk & Western. A corporation. Capital, $150,000; issued, $100,000. Stockholders, 7. President, R. B. Griggs; Vice-Presidents, B. P. Huff and A. J. Huff; Secretary and Treasurer, C. M. Griggs; General Manager, R. B. Griggs; General Superintendent, C. M. Griggs. Employes, 25. Hogs, 1,000. Sausage—Fresh, smoked and summer. By-Products—Dried blood and tankage. Refrigeration—230-ton Frick and 70-ton York; brine circulating and direct expansion; 100 tons ice daily. Boiler capacity, 1,350 H. P. Trade Mark—"Lily Leaf."

ROSSLYN

Rosslyn Packing Co.—Railroads, Pennsylvania, delivery at Rosslyn, and Baltimore & Ohio at Georgetown, D. C. A corporation. Capital, $25,000; issued, $25,000. Stockholders, 6. President, Chas. G. Pfluger; Vice-President, Wm. G. Carter; Secretary and Treasurer, Robt. G. Carter. Employes, 50. Cattle, 300; hogs, 1,000; sheep, 500. Sausage—Fresh and smoked. By-Products—Tankage and hog and chicken feed. Government inspection. Refrigeration—310-ton Vilter; brine circulating and direct expansion; 75 tons ice daily. Boiler capacity, 400 H. P. Retail Markets—Riggs Market, Center Market, Eastern Market, O Street Market and Western Market, Washington, D. C.

SMITHFIELD

J. B. Grimes & Son.
P. D. Gwaltney, Jr., & Co.
V. W. Joyner & Co.
Tazewell T. Spratley Co.
J. W. Stott.
J. Waverly Thomas.

SUFFOLK

W. R. Frasier.
Smithfield Meat Co.
Smithfield Packing Co., Inc.

WASHINGTON

ABERDEEN

Erickson Meat Co.

BELLINGHAM

Sanitary Meat Co.—1017 Elk St. Railroads, Northern Pacific and Great Northern. A corporation. Capital, $20,000; issued, $18,000. Stockholders, 3. President, Hans Oberleitner; Vice-President, A. W. Pierce; Secretary and Treasurer, P. M. Johnson; General Manager, Hans Oberleitner. Employes, 7. Cattle, 15; hogs, 70; sheep, 30. Sausage—Fresh, smoked and summer. By-Products—Tankage, hog

Figures on cattle, hogs and sheep indicate AVERAGE WEEKLY KILLING CAPACITY.

and chicken feed. Refrigeration—5-ton Armstrong; brine circulating and direct expansion; one-half ton ice daily. Boiler capacity, 30 H. P. Retail Market—Bellingham, Wash.

SEATTLE

Barton & Co.—Spokane St. and East Waterway. A corporation. Capital, $1,000,000; issued, $650,000. President, P. Burns; Vice-President, G. I. C. Barton; Secretary and Treasurer, R. J. Ferguson; General Manager, G. I. C. Barton; General Superintendent, A. B. Musser. Employes, 150. Cattle, 250; hogs, 1,000; sheep, 1,500. Sausage—Fresh, smoked and summer. By-Products—Dried blood, tankage, hog and chicken feed and bone meal. Government inspection. Refrigeration—40-ton Vulcan and two 50-ton Vilter; direct expansion and brine circulating; 5 tons ice daily. Boiler capacity, 100 H. P. Trade Marks—"Circle W" and "Carnation." Retail Markets—Tacoma, Bellingham and Port Angeles, Wash. Also exports.

Frye & Company—S. 9th and Walker Sts. Railroad, Northern Pacific. A corporation. Capital, $1,500,000; issued, $1,500,000. Stockholders, 3. President, Charles H. Frye; Vice-President, Frank F. Frye; Secretary, John C. Higgins; Assistant Secretary, F. A. Danielson; Treasurer, Charles H. Frye; General Manager, Charles H. Frye; General Superintendent, John Klinkham. Employes, 750. Codes—A. B. C., 5th, 6th, Bentley's and Cross. Cattle, 250; hogs, 3,000; sheep, 500. Sausage—Fresh and smoked. By-Products—Dried blood, tankage, chicken feed, bone, ground bone, bone meal, etc. Government inspection. Refrigeration—Two 150-ton De La Vergne and one 80-ton De La Vergne, also 30 at branch houses with capacity of 150 tons; direct expansion; 50 tons ice daily. Boiler capacity, 600 H. P. Trade Marks—"Frye's Delicious," "Wild Rose" and "Frye's Shortening." Branch Houses—San Francisco, Cal.; Portland, Ore.; Tacoma, Wash.; Nampa, Idaho; Sedro Valley, Bremerton, Aberdeen, Bellingham, Everett, Centralia and Buckley, Wash. Retail Markets—Seattle, 8; Tacoma, 2; Portland, 2, and Bremerton, Centralia, Raymond, Aberdeen, Hoquiam, Sedro Valley, Buckley, Bellingham, Anacostes, New Castle, Seward, Port Townsend, Wash.; Juneau, Ketchikan and Skagway, Alaska. Also exports.

James Henry Packing Co.—2025-2029 9th Ave. South. Individual ownership. General Manager, O. B. Joseph; General Superintendent, Wm. Moran. Employes, 65. Cattle, 250; hogs, 600; sheep, 1,000. Sausage—Fresh and smoked. By-Products—Dried blood and tankage. Refrigeration—40-ton Baker; direct expansion. Boiler capacity, 100 H. P. Trade Mark—"Diamond 'H' Brand." Retail Market—One at Seattle.

SHELTON

Shelton Meat & Ice Co.—J. F. Bichsel.

SNOHOMISH

Bruhn & Henry, Inc.
Columbia Packing Co.
A. Evoy.

SPOKANE

Adam Brown Packing Co.—116 Havana St. Railroad, Chicago, Milwaukee & St. Paul. A corporation. Capital, $100,000; issued, $65,000. Stockholders, 12. President, Adam Brown; Vice-Presidents, Chas. Dezell and T. Horton; Secretary and Treasurer, D. M. Brown; General Manager, Adam Brown. Employes, 34. Cattle, 250; hogs,

---

Figures on cattle, hogs and sheep indicate AVERAGE WEEKLY KILLING CAPACITY.

300; sheep, 150. Sausage—Fresh, smoked and summer. By-Products—Hog and chicken feed. Refrigeration—One 9-ton Armstrong, three 2-ton Frick and one 1½-ton Cooper; direct expansion; one ton ice daily. Boiler capacity, 80 H. P. Trade Marks—"Eagle Brand" and "Union Brand." Retail Markets—5 in Spokane.

Armour & Company—Virginia and E Sts. (See Chicago.)

John Lewis & Company—Pork and Beef Packers. 4103 S. Mission St. Railroad, Spokane Eastern. A corporation. Capital, $200,000. President, John Lewis; Vice-President, J. F. Williams; Treasurer, C. P. Ray. Cattle, 75; hogs, 300; sheep, 250. Sausage—Fresh and smoked. Refrigeration—15-ton Armstrong; direct expansion. Retail Markets—6 in Spokane. Boiler capacity, 200 H. P.

Spokane Packing Co.

Trefry & Demeester.

TACOMA

Carstens Packing Co.—1623 East J St. Railroads, Oregon-Washington Railway & Navigation and Northern Pacific. A corporation. Capital, $1,500,000; issued, $1,150,000. Stockholders, 9. President, Thomas Carstens; Vice-President, Henry Wolff; Secretary and Treasurer, O. F. Kuhl; Superintendent, J. H. Beidler. Employes, 300. Codes—W. U. and A. B. C., 5th edition; code word, "Carstens." Cattle, 1,500; hogs, 15,000; sheep, 3,500. Sausage—Fresh, smoked and summer. By-Products—Dried blood, tankage, hog and chicken feed, bone meal, granulated bone, fish meal and sheep guano. Government inspection. Refrigeration—50, 35 and 10-ton Vilter, and 120-ton De La Vergne; brine spray system and direct expansion; 10 tons ice daily. Boiler capacity, 1,400 H. P. Trade Marks—Hams, "Diamond T. C.," "Washington"; bacon, same; lard, same; compound, "Snow Cap Brand," "Autumn Leaf Lard Compound," "Silver Star Lard." Branch Houses—Seattle, Aberdeen, Everett, Bellingham, Camp Lewis and Spokane, Wash., and Portland Oregon. Retail and Wholesale Markets—Total of forty in Washington, Idaho and Alaska. Also exports.

WALLA WALLA

Walla Walla Meat & Cold Storage Co.—Railroad, Oregon Railway & Navigation. A corporation. Capital, $150,000; issued, $150,000. Stockholders, 20. President, R. H. Johnson; Vice-President, J. J. Kauffman; Secretary and Treasurer, F. M. Lowden, Jr.; General Manager, F. M. Lowden, Jr.; General Superintendent, Robt. Linder. Employes, 70. Cattle, 100; hogs, 200; sheep, 300. Sausage—Fresh and smoked. By-Products—Dried blood, tankage, hog and chicken feed. Government inspection. Refrigeration—50-ton Harris, 35-ton Frick and 7-ton Harris; direct expansion and brine circulating; 30 tons ice daily. Trade Mark—"Holly Brand." Retail Markets—Three in Walla Walla and one in Pasco, Wash.

WENATCHEE

Schrock-Nelson Co.

YAKIMA

H. & S. Meat Co., Inc.—24 N. 2nd St. A corporation. Capital, $20,000; issued, $20,000. Stockholders, 2. President, Peter Hansen; Secretary and Treasurer, F. J. Herberger. Employes, 16. Cattle, 100; hogs, 300; sheep, 300. Sausage—Fresh, smoked. By-Products—Tankage. City inspection. Refrigeration—12-ton Armstrong; direct expansion. Boiler capacity, 25 H. P. Trade Mark—"H. & S." Retail Markets—Two in Yakima, Wash.

Figures on cattle, hogs and sheep indicate AVERAGE WEEKLY KILLING CAPACITY.

YAKIMA

Independent Meat Co.
Yakima Meat Co.

WEST VIRGINIA

WHEELING

C. Kalbitzer Packing Co.—1128 Water St. Railroad, B. & O. A corporation. Capital, $400,000; issued, $188,700. Stockholders, 45. President, Geo. W. Kalbitzer; Vice-President, Wm. Fette; Secretary and Treasurer, Geo. J. Weiskircher; General Manager, Geo. W. Kalbitzer; General Superintendent, Wm. J. Sullivan. Employes, 85. Cattle, 100; hogs, 1,000. Sausage—Fresh. By-Products—Unground tankage. Government inspection. Refrigeration—100-ton De La Vergne; direct expansion; 30 tons ice daily. Boiler capacity, 500 H. P. Trade Marks—Hams, "Royal"; bacon, "Royal"; lard, "Kalbitzer's Special" and "Wheeling Brand." Retail Markets—1327 Market St. and 2201 Market St., Wheeling, W. Va., and 316 33d St., Bellaire, Ohio.

Paul O. Reymann Co.—Wetzell and Warren Sts. Owned by Wilson & Co. (See Chicago.)

F. Schenk & Sons Co.—J. O. Schenk, Manager. Owned by Allied Packers, Inc., Chicago, Ill.

Weimer Packing Co.—1033 Main St. A corporation. Capital, $100,000; issued, $59,000. Stockholders, 7. President, Wm. G. Weimer; Vice-President, A. F. Schairer; Secretary and Treasurer, H. G. Weimer; General Manager, Wm. G. Weimer. Employes, 27. Cattle, 25; hogs, 200; sheep, 40; calves, 25. Sausage—Fresh and smoked. By-Products—Tankage, hog and chicken feed. Refrigeration—25-ton Wilson-Snyder, two 10-ton Yorks; brine circulating and direct expansion; one-half ton ice daily. Boiler capacity, 160 H. P. Trade Mark—"Fort Henry." Retail Market—1033 Main St., Wheeling, W. Va.

John Wenzel Company—4320 Jacob St. Charles Norteman, owner.

Wheeling Butchers' Association—Fourth and Center Sts. Railroads, Pennsylvania and B. & O. A corporation. Capital, $50,000; issued, $20,000. Stockholders, 21. President, Geo. W. Weimer; Vice-President, Fred W. Neininger; Secretary and Treasurer, Wm. M. Schenck; General Manager and Superintendent, Wm. M. Schenck. Employes, 8. Cattle, 75; hogs, 400; sheep, 75; calves, 150. By-Products—Tankage, inedible yellow grease, inedible tallow, and also dealers in hides, calf skins, pelts and horse hides. Refrigeration—15-ton York; brine circulating. Boiler capacity, 250 H. P. Note—Abattoir only.

HUNTINGTON

Fesenmeier Packing Co.—14th and Madison. Railroads, C. & O. and B. & O. A corporation. Capital, $200,000; issued, $200,000. Stockholders, 35. President, J. L. Fesenmeier; Vice-Presidents, Paul R. Riddlemoser and A. J. Fesenmeier; Secretary, A. V. Ward; Treasurer, C. M. Gohen; General Manager, Paul R. Riddlemoser; General Superintendent, M. J. Kearney. Employes, 40. Codes—Cross. Cattle, 100; hogs, 2,000. Sausage—Fresh and smoked. By-Products—Tankage and hog and chicken feed, and grease. Government inspection. Refrigeration—200-ton York; brine circulating. Boiler capacity, 950 H. P. Trade Marks—"Apple Blossom Brand" and "Wesva Brand."

Figures on cattle, hogs and sheep indicate AVERAGE WEEKLY KILLING CAPACITY.

WISCONSIN

CUDAHY

Cudahy Brothers Company—Railroad, Chicago & North Western. A corporation. Capital, $3,200,000; issued, $3,200,000. Stockholders, 30. President, Michael F. Cudahy; Vice-President, John Cudahy; Secretary, James W. Bryden; Treasurer, M. J. Connell. Employes, 1,200. Codes—Cross, Utility and Summit; code word, "Cudahy-Milwaukee." Cattle, 1,000; hogs, 40,000; sheep, 600. Sausage—Fresh, smoked and summer. By-Products—Dried blood, tankage and hog and chicken feed. Government inspection. Refrigeration—Four Vilter, 825 tons capacity; brine circulating, direct expansion and brine spray system; 75 tons ice daily. Boiler capacity, 3,000 H. P. Branch Houses—Milwaukee, Wis.; Chicago, Ill.; Detroit, Mich., and Akron, O. Trade Marks—Hams, "Peacock" and "Cream City"; bacon, "Peacock," "Cream City" and "Jack Spratt"; lard, "Peacock," "White Champion" and "Cream City"; compound, "Snowball."

EAU CLAIRE

Drummond Packing Co.—Railroads, C. & N. W. and C. M. & St. Paul. A corporation. Capital, $125,000; issued, $125,000. Stockholders, 19. President, John Drummond; Vice-President, D. G. Calkins; Secretary, H. H. Smith; Treasurer, F. W. Thomas; General Superintendent, F. B. Drummond. Employes, 125. Codes—Cross. Cattle, 200; hogs, 5,000; sheep, 1,000. Sausage—Fresh and smoked. By-Products—Dried blood and tankage. Government inspection. Refrigeration—Wolf-Linde; 65 tons; direct expansion. Boiler capacity, 125 H. P. Trade Mark—"Arbutus."

GREEN BAY

Acme Packing Co.—A corporation. Meat canners. Capital, $12,000,000. President and Treasurer, Thos. O. Gibbs; General Manager, J. C. Nielson; Secretary, Ely M. Aaron; General Superintendent, J. B. Rogers. Codes—A. B. C., Bentley's and W. U.; code word, "Apaco-Ipco." Have facilities, but do not kill. Government inspection. Refrigeration—One 80-ton Triumph and one 35-ton Frick; direct expansion. Trade Marks—"Red Crown" and "Council." Also exports.

LA CROSSE

John Niedercorn & Son.
Schams Bros.

MADISON

Oscar Mayer & Company—Branch plant. (See Chicago.)
Madison Packing Co.

MILWAUKEE

Layton Co.—Railroad, C. M. & St. P. A corporation. Capital, $100,000; issued, $100,000. Stockholders, 11. President, Wm. A. Dawson; Secretary, F. O. Streckewald; Treasurer, Chas. F. Dickens; General Manager, Wm. A. Dawson; General Superintendent, Richard Steilow. Employes, 100. Codes—Cross, A. B. C. and Private; code word, "Laytonco." Hogs, 2,000. Sausage—Fresh, smoked and summer. By-Products—Tankage. Government inspection. Refrigeration—50-ton Vilter; brine circulating. Boiler capacity, 300 H. P. Also exports.

Figures on cattle, hogs and sheep indicate AVERAGE WEEKLY KILLING CAPACITY.

MILWAUKEE

Plankinton Packing Co.—Muskegon Ave. and Canal St. Railroad, C. M. & St. P. A corporation. Capital, $5,000,000; issued, $5,000,000. President, H. C. Carr; Vice-Presidents, W. C. Nicholson and H. McLerie; Secretary, C. A. Peacock; Assistant Secretary, C. P. Hobson; Treasurer, W. W. Sherman; Assistant Treasurer, J. J. McGuire; General Manager, W. C. Nicholson; General Superintendent, H. J. Kurtz. Employes, 1,200. Cattle, 2,750; hogs, 20,000; sheep, 5,000; calves, 20,000. Sausage—Fresh, smoked and summer. By-Products—Dried blood, tankage and hog and chicken feed. Government inspection. Refrigeration—500-ton Ball, two 250-ton Vilter and two 160-ton Hercules; brine circulating, brine spray and direct expansion. Boiler capacity, 3,000 H. P. Branch Houses—3018 Meinecke Ave., Milwaukee, Wis., and Calumet, Mich. Also exports.

Swift & Company—Muskegon Ave. and Grand Canal St. (See Chicago.)

NEW RICHMOND

Hately Brothers Company—Hog slaughtering plant. Capacity, 6,000 hogs. (See Chicago.)

RACINE

Rowley Packing Co.—Stannard St. Railroad, C. M. & St. Paul. A corporation. Capital, $150,000; issued, $75,000. Stockholders, 7. President, A. J. Rowley; Vice-President, J. H. Rowley; Secretary, Jessie Rowley; Treasurer, W. B. Rowley; General Manager, A. J. Rowley; General Superintendent, J. H. Rowley. Employes, 60. Kill pork and veal. Hogs, 600. Pork and lard. Sausage—Fresh, smoked and summer. Refrigeration—30-ton York; brine circulating and direct expansion; 2 tons ice daily. Boiler capacity, 100 H. P.

WATERTOWN

Dobbratz & Mussfeldt—10 Third St. Corporation. Capital, $20,000; issued, $10,000. Stockholders, 4. President, G. E. Dobbratz; Vice-President, L. Dobbratz; Secretary and Treasurer, B. J. Mussfeldt; Assistant Secretary, M. Mussfeldt; General Manager, G. E. Dobbratz; General Superintendent, B. J. Mussfeldt. Hogs, 50. Hams, bacon, lard. Sausage—Fresh, smoked, summer.

WAUSAU

Wisconsin Packing Co.—1009 Town Line Road. Railroads, C. & N. W. and C. M. & St. P. A corporation. Capital, $450,000; issued, $289,400. Stockholders, 1,989. President, J. D. Christie; Vice-President, W. R. Happe; Secretary, F. N. Blecha; Treasurer, Ben Lang; General Manager, Aug. G. Anderson; General Superintendent, Aug. G. Anderson. Employes, 50. Codes—Cross. Cattle, 300; hogs, 1,500; sheep, 300. Sausage—Fresh, smoked and summer. By-Products—Tankage, inedible tallows and brown grease. Government inspection. Refrigeration—Two 25-ton Vilter; brine circulating and direct expansion. Boiler capacity, 150 H. P. Trade Mark—"Wisconsin Pride." Retail Markets—Wausau Cash & Carry Market, 201 Wash St., and Sixth Street Cash & Carry Grocery & Market, 1910 Sixth St., Wausau, Wis.

Figures on cattle, hogs and sheep indicate AVERAGE WEEKLY KILLING CAPACITY.

WYOMING

CASPER

Casper Packing Co.
Sheridan Meat Co.

CHEYENNE

Hammond Packing Co.—Railroad, Union Pacific. A corporation. Stockholders, 3. General Manager, C. E. Hammond. Employes, 12. Sausage—Fresh and smoked. Government inspection. Refrigeration—One 65-ton York. Boiler capacity, 225 H. P. Trade Mark—"Diamond H."

LARAMIE

Wyoming Packing & Provision Co.—Laramie Stock Yards. Railroad, Union Pacific. Trust syndicate: S. E. Smith, Louis J. Neuner, Henry Reimers, Frank H. McGinnis and Victor E. Hall. Capital, $25,000. General Manager, S. E. Smith. Employes, 20. Cattle, 75; hogs, 500; sheep, 150. Sausage—Fresh, smoked and summer. By-Products—Tankage and chicken feed. Boiler capacity, 75 H. P. Government inspection applied for. Export.

ROCK SPRINGS

Rock Springs Butchering Co.—P. O. Box 486. Railroad, Union Pacific. A corporation. Stockholders, 4. General Manager, Otto Schnauber. Employes, 8. Cattle, 15; hogs, 20; sheep, 40. Sausage—Fresh, smoked and summer. By-Products—Hog and chicken feed. Refrigeration—8-ton Althoff; brine circulating. Boiler capacity, 30 H. P.

THERMOPOLIS

Central Market—528 Broadway. Railroad, C. B. & Q. A corporation. Capital, $50,000; issued, $30,000. Stockholders, 3. President, Harry C. Vail; Vice-President, Clifford Wilson; Secretary, J. A. Thompson; Treasurer and General Manager, Harry C. Vail. Employes, 10. Cattle, 15; hogs, 40; sheep, 20. Sausage—Fresh and smoked. By-Products—Hog feed. City inspection. Refrigeration—5-ton Vilter. Retail Market—Thermopolis.

PANAMA CANAL ZONE

CRISTOBAL

Panama R. R. Co., Commissary Division, Supply Department.—Railroad, Panama. United States Government ownership, through Panama Railroad Co. Operated by Supply Department, Panama Canal. R. K. Morris, Chief Quartermaster, Panama Canal; General Manager, C. A. Gilmartin; General Foreman, F. L. Miller. Employes, 237. Codes—A. B. C., 5th edition, Western Union and Bentley's; code word, "Commissary Colon." Cattle, 600; hogs, 2,400. Sausage—All kinds. By-Products—Dried blood, tankage, hog and chicken feed, grease, dried tankage, ground bone, bone fertilizer. Government inspection. Refrigeration—Three 200-ton York and one 15-ton York; brine circulating, brine spray, direct expansion; 150 tons ice daily. Boiler capacity, 1,050 H. P. Retail Markets—Cristobal, Gatun, Gamboa, Culebra, Red Tank, Paraiso, Pedro Miguel, Balboa, Ancon, Ancon Market and La Boca.

Figures on cattle, hogs and sheep indicate AVERAGE WEEKLY KILLING CAPACITY.

CANADA

ALBERTA

CALGARY

P. Burns & Company, Ltd.—East Calgary. Railroads, Canadian Pacific, Grand Trunk Pacific and Canadian National Railway. A corporation. Capital, $10,000,000; issued, $5,000,000. Stockholders, 10. President, Patrick Burns; Vice-President, W. J. Wilson; Secretary, W. E. Corlet; Treasurer, N. Hindsley; General Manager, John Burns; General Superintendent, R. B. Musser. Employes, 1,500. Codes—W. U. (Universal and five letter), A. B. C., 5th edition, Bentley's, Cross and Utility; code word, "Burns" (in Canada), "Burcana" (in London and Liverpool, England). Cattle, 5,000; hogs, 20,000; sheep, 10,000. Sausage—Fresh, smoked and summer. By-Products—Dried blood, tankage, hog and chicken feed, commercial fertilizers, tallows, oleo stock and neatsfoot oil. Government inspection. Refrigeration—Two Linde-Canadian; 75-ton capacity; one Ball, 250 tons; one 110 and one 80-ton Niebling, one 60-ton Vulcan, one 20-ton Triumph and one 20-ton Linde; direct expansion and brine spray systems. Boiler capacity, 2,550 H. P. Trade Marks—Hams, "Shamrock" and "Dominion"; bacon, "Shamrock" and "Dominion"; lard, "Shamrock"; shortening, "White Carnation"; sausage, "Shamrock" and "Dominion." Export Brands—Hams and bacon, "Dominion" and "Colonial." Branch Houses—Calgary, Vancouver, Montreal, Victoria, Vernon, Prince Rupert and Prince George. Retail Markets—Calgary, Edmonton, Lethbridge, Blairmore, MacLeod, Vancouver, Nelson, Cranbrook, Revelstoke, Fernie and others. Also exports. Foreign Offices—London and Liverpool, England.

Union Packing Company, Ltd.

Calgary Abattoir & Packers, Ltd.

EDMONTON

Swift Canadian Company.

Gainer's, Limited.

P. Burns & Company, Ltd.—(See Calgary.)

STETTLER

Everhardy Packing Co., Ltd.—Railroad, Canadian National. A corporation. Capital, $75,000. Stockholders, 3. President, E. F. Everhardy; Treasurer and Secretary, F. J. Everhardy. Employes, 16. Cattle, 90; hogs, 250; sheep, 250. Sausage—Fresh and smoked. Refrigeration, Linde British and York; direct expansion system. Boiler capacity, 125 H. P. Trade Marks—"National" and "Golden Harvest." Retail Markets—Stettler and Nordegg.

STRATHCONA

Gainer's, Ltd.

BRITISH COLUMBIA

NEW WESTMINSTER

Vancouver-Prince Rupert Meat Company.

Swift Canadian Company.

Figures on cattle, hogs and sheep indicate AVERAGE WEEKLY KILLING CAPACITY.

PRINCE RUPERT

Prince Rupert Meat Company.

VANCOUVER

Reid & Miller.

P. Burns & Company, Ltd. (See Calgary.)

VICTORIA

Kirkham, H. O., & Company, Ltd.

MANITOBA

ST. BONIFACE

Dawson Road Abattoir.
Farmers' Packing Company, Ltd.
Manitoba Abattoir & Packers, Ltd.
Union Abattoir Company—Union Stock Yards. Partnership. General Manager, John Innes. Employes, 50. General packing. Cattle, 500; hogs, 1,000; sheep, 500. Sausage—Fresh, smoked and summer. By-Products—Dried blood, tankage and hog and chicken feed. Government inspection. Refrigeration—Two 35-ton York; direct expansion. Boiler capacity, 100 H. P. Trade Mark—"Union Brand."

WINNIPEG

Gallagher Holman, Limited—Logan Ave. West. Railroad, Canadian Pacific. A corporation. Capital, $1,000,000; issued, $490,000. Stockholders, 38. President, J. Q. Gallagher; Vice-President, B. H. Holman; Managing Director, J. D. Cameron; Treasurer and Secretary, Arch. McLean; General Manager, J. D. Cameron; General Superintendent, G. W. Fink. Employes, 125. Codes—A. B. C., Cross and Private; code word, "Holman" (Winnipeg). Cattle, 500; hogs, 2,500; sheep, 500. Sausage—Fresh, smoked and summer. By-Products—Dried blood, tankage, hog and chicken feed and fertilizers. Government inspection. Refrigeration—40-ton York and 40-ton Linde; brine spray and direct expansion. Boiler capacity, 200 H. P. Trade Marks—"Prairie" and "Security." Branch Houses—Fort William and Kenora, Ontario, and Montreal, P. Q. Retail Markets—Gibson Gage, Limited, Winnipeg. Also exports.

Gordon-Ironside & Fares Packers, Ltd.—Logan Ave. Railroad, Canadian Pacific. A corporation. Capital, $2,000,000; issued, $2,000,000. President, W. H. Fares; Vice-President, James Harris; Secretary and Treasurer, J. S. McLean; Assistant Secretary, Mr. Ranson; General Manager, A. G. Hall. Employes, 500. Codes—Cross; code word, "Gifco." Cattle, 3,000; hogs, 6,000. Sausage—Fresh and smoked. By-Products—Tankage and all other classes. Government inspection. Refrigeration—200-ton Wolf-Linde; direct expansion. Trade Marks—(Domestic) "Sweet Clover" and "Sterling"; (Export) "Gifco" and "Amada." Branch Houses—Regina, Sask.; Saskatoon, Sask., and Fort William, Ont. Also exports.

Manitoba Abattoir & Packers, Ltd.
Societe S. P. A.
Swift Canadian Company.
Western Packing Co. of Canada, Ltd.

Figures on cattle, hogs and sheep indicate AVERAGE WEEKLY KILLING CAPACITY.

NEW BRUNSWICK

RENFREW

Jamieson Meat Company, Ltd.

ST. JOHN

John Hopkins—186 Union St.
Slip & Flewelling—340 Main St.
F. E. Williams Company, Ltd.—80 Charlotte St.

ST. STEPHENS

Wry Pork Packing Company.

NOVA SCOTIA

HALIFAX

Davis & Fraser—Railroad, Charlottetown P. E. I. Partnership. Employes, 50. Codes—Cross; code word, "Dofraser." Pork packers. Government inspection. Refrigeration—40-ton Linde and 40-ton York; direct expansion and air circulation. Boiler capacity, 200 H. P.

ONTARIO

BRANTFORD

Canadian Packing Co., Ltd.—Owned by Allied Packers, Inc., Chicago, Ill.

CHATHAM

Wilson-Canadian Company, Ltd.—(See Chicago, Ill.)

HAMILTON

W. A. Freaman Company, Ltd.
John Duff & Son.

INGERSOLL

Ingersoll Packing Company, Ltd.—Cor. Wonham and Victoria Sts. Railroads, Grand Trunk and Canadian Pacific. A corporation. Capital, $1,000,000; issued, $400,000. Stockholders, 5. President, T. L. Boyd; Vice-President and Treasurer, H. C. Wilson; Secretary, C. H. Sumner; General Manager, H. C. Wilson. Employes, 100. Codes—A. B. C., 5th edition, Cross and W. U. Hogs, 2,000. Sausage—Fresh and smoked. By-Products—Tankage. Government inspection. Refrigeration—Two 75-ton Vilter; direct expansion. Boiler capacity, 450 H. P. Trade Marks—"Beaver," "Pearl" and "Lilac." Also exports.

KITCHENER

Dumart Packing Company, Limited—Guelph St. Railroads, Grand Trunk and Canadian Pacific. Individual corporation. Capital, $1,000,000; issued, $225,000. Stockholders, 173. President, W. H. Dumart; Vice-Presidents, A. Jansen and M. Wunder; Secretary, E. G. Schierholtz; Treasurer, W. J. Smith. Employes, 75. Pork packers and slaughterers. Hogs, 700. Sausage—Fresh, smoked and summer. By-Products—Tankage. Refrigeration—35-ton Canadian Linde; direct expansion. Boiler capacity, 125 H. P.

J. M. Schneider & Sons, Ltd.—63 Courtland Ave. East. Railroad, Grand Trunk. A corporation. Capital, $200,000; issued, $100,000. Stockholders, 5. President and Manager, J. M. Schneider; Vice-President, C. A. Schneider; Secretary and Treasurer, F. H. Schneider. Em-

---

Figures on cattle, hogs and sheep indicate AVERAGE WEEKLY KILLING CAPACITY.

ployes, 105. Cattle, 50; hogs, 800. Sausage—Fresh, smoked and summer. Refrigeration—20-ton York and 10-ton Sawyer; direct expansion. Boiler capacity, 200 H. P. Retail Market—One at plant.

OTTAWA

Oscar Leclair.

PETERBOROUGH

Canadian Packing Co., Ltd.—Owned by Allied Packers, Inc., Chicago, Ill.

ST. THOMAS

St. Thomas Packing Co.

SIMCOE

Dominion Canners, Limited.

SMITH'S FALLS

Jones Packing & Provision Company, Ltd.—(See list of Refiners.)

STRATFORD

Whyte Packing Company, Limited.

TORONTO

Canadian Packing Company, Limited—Bathurst St. Owned by Allied Packers, Inc., Chicago, Ill.

William Davies Company, Limited—521 Front St. East. Railroads, Canadian Pacific and Canadian Government. A corporation. Capital, $5,000,000; issued, $3,000,000. President and General Manager, E. C. Fox; Vice-President and Assistant General Manager, E. O. Mitchell; Vice-President, R. N. Watt; Vice-President, A. F. Sheed; Secretary and Treasurer, A. F. Park; Assistant Treasurer, C. S. Leckie. Codes—Bentley's; code word, "Daveben." Cattle, 5,750; hogs, 22,500; sheep, lambs and calves, 9,500. Government inspection. Does export business.

Gunn's, Limited—Gunn's Road. Railroads, Grand Trunk and Canadian Pacific. Joint stock company. Capital, $5,000,000; issued, $3,260,000. Stockholders, 850. President, J. A. Gunn; Secretary and Treasurer, O. L. Waite; General Manager, F. M. Moffat; General Superintendent, H. T. Horton. Employes, 1,000. Codes—Yopp, Cross and A. B. C. Cattle, 2,500; hogs, 6,000; sheep, 2,000. Sausage—Fresh and smoked. By-Products—Dried blood, tankage, hog and chicken feed, cattle feed, casings and fertilizers. Government inspection. Refrigeration—Two 100 and two 75-ton Linde-Canadian and one 125-ton York; direct expansion, brine circulating and brine spray; 42 tons ice daily. Boiler capacity, 1,150 H. P. Trade Mark—"Mapleleaf." Branch Houses—Montreal, Que.; St. John, N. B., and Wingham, Ont., and St. Lawrence Market, Ont. Also exports.

Harris Abattoir Company, Ltd.—Union Stock Yards. Railroads, Grand Trunk and Canadian Pacific. A corporation. Capital, $5,000,000; issued, $2,000,000. Stockholders, 11. President and Sales Manager, W. T. Harris; Vice-President and Managing Director, Jas. Harris; Secretary and Treasurer, J. S. McLean; General Superintendent, C. H. Pringle; Assistant Secretary, S. G. Brock. Employes, 1,500.

Figures on cattle, hogs and sheep indicate AVERAGE WEEKLY KILLING CAPACITY.

Codes—Bentley's; code word, "Harrab." Cattle, 6,000; hogs, 6,000; sheep, 9,000. Sausage—Fresh, smoked and summer. By-Products—Dried blood, tankage and hog and chicken feed. Government inspection. Refrigeration—Two 130-ton British Linde and two 130-ton Ball; brine circulating, brine spray and direct expansion; 30 tons ice daily. Boiler capacity, 1,200 H. P. Branch Houses—Montreal and Quebec, Que.; Toronto, London, Ottawa, Windsor, Sudbury and Sault Ste. Marie, Ont.; Sidney and Halifax, Nova Scotia; Charlottetown, Prince Edward Island and St. John, N. B. Also exports.

Municipal Abattoir & Cold Storage—Foot of Tecumseh St. Railroad, Grand Trunk. Owned by city, municipally operated. General Manager, W. R. Corneil; General Superintendent, John H. Smith. Cattle, 3,000; hogs, 2,000; sheep, 3,000. By-Products—Tankage, dried blood, steam cooked bones, casings, tallow and grease and neatsfoot oil. Refrigeration—110-ton York and 110-ton Linde; direct expansion. Boiler capacity, 450 H. P.

W. Harris & Company, Ltd.
Ruddy's, Ltd.
Stanbury-Fuller Co.
Swift Canadian Co.
W. Wight & Company.

PRINCE EDWARD ISLAND

CHARLOTTETOWN

Sims Packing Company, Ltd.
Canadian Packing Company, Ltd.—Owned by Allied Packers, Inc., Chicago, Ill.
Island Cold Storage Company.

QUEBEC

HULL

Canadian Packing Company, Ltd.—Owned by Allied Packers, Inc., Chicago, Ill.

MONTREAL

Standard Beef Co., Ltd.
Canadian Packing Company, Limited—Owned by Allied Packers, Inc., Chicago, Ill.
W. Clark, Limited—83 Amherst. Railroad, Canadian Pacific. A corporation. Capital, $500,000. President, W. Clark, Esq.; Vice-President, H. Clark, Esq.; Secretary, H. Clark, Esq. Codes—A. B. C., 5th edition; code word, "Clarkfoods." Canned meats. Government inspection. Refrigeration—30 and 20-ton Linde-Canadian; brine circulating and direct expansion. Boiler capacity, 300 H. P. Branch Houses—St. Remi, Napiereville, Quebec; and Harrow, Ontario. Also exports.
Irwin Davies, Ltd.
Montreal Abattoirs, Ltd.—139 Mill St. Railroad, Grand Trunk. A corporation. Capital, $1,500,000; issued, $1,080,000. Stockholders, 100. President, A. B. Colville; Vice-President, W. B. Strachan; Secretary and Treasurer, D. Brogan; General Manager, G. C. Silcock; General Superintendent, E. C. Rettig. Employes, 250. Codes—Bentleys and Cross; code word, "Monabaco." Cattle, 3,000; hogs, 5,000; sheep,

---

Figures on cattle, hogs and sheep indicate AVERAGE WEEKLY KILLING CAPACITY.

4,000. Sausage—Fresh, smoked and summer. By-Products—Dried blood and tankage. Government inspection. Refrigeration—150-ton Triumph and 50-ton Linde. Boiler capacity, 450 H. P. Trade Mark—"Mabco." Also exports.

Oxo Limited—232 Lemoine St. Manufacturers of concentrated foods. Canadian head office at Montreal—Branches at Toronto and Winnipeg. This business is the Canadian Head Office of Oxo Limited, of London, England, and the packing houses of this firm are in Uruguay and Argentina. Manager for Canada, A. Mossman.

Wm. Davies Company, Ltd.—(See Toronto.)

QUEBEC

Emond & Cote—20 St. Peter St.
L. Baller & Son—304 St. John St.

SHERBROOKE

Alex. Ames & Sons, Ltd.
Hovey Bros. Packing Company.

SASKATCHEWAN

MOOSE JAW

Mid-West Packing Company.
Gordon-Ironside & Fares Packers Limited—(See Winnipeg.)

PRINCE ALBERT

P. Burns & Company, Ltd.—(See Calgary.)

REGINA

Hugh-Armour & Company, Ltd.

CUBA

HAVANA

Lykes Brothers, Inc.—A corporation. Capital, $3,000,000; issued, $3,000,000. Stockholders, 7. Presidents, F. E. Lykes; Vice-Presidents, H. T. Lykes, J. M. Lykes, F. A. Morris, L. G. Lykes, T. M. Lykes, J. T. Lykes; Secretary, H. T. Lykes; Treasurer, F. E. Lykes. Codes—Scotts, Bentley's, A. B. C., 5th Edition; code word, "Lykes." Cattle, 600; hogs, 750. Sausage—Fresh, smoked and summer. By-Products—Tankage. Government inspection. Refrigeration—Nine 20-ton Yorks; direct expansion. Boiler capacity, 300 H. P. Trade Mark—"F-6." Branches at Tampa, Fla.; Galveston, Tex.; New York City; New Orleans.

MEXICO

SONORA

CANANEA

Cananea Cattle Co.—A corporation. Capital, $25,000; issued, $25,000. Stockholders, 11. President, M. G. Wiswall; Vice-President, C. E. Wiswall; Secretary and Treasurer, A. L. Houck; Manager, C. E. Wiswall. Cattle, 250. Retail Market—Cananea.

Figures on cattle, hogs and sheep indicate AVERAGE WEEKLY KILLING CAPACITY.

SOUTH AMERICA

ARGENTINA

Anglo South American Meat Company—Calle Sarmiento 443, Buenos Aires.
Cia. Sansinena de Carnes Congeladas—Calle San Martin 132 Buenos Aires.
Cia. Frigorifica Argentina de Tierra del Fuego—Rio Grande, Tierra del Fuego.
Dickinson & Co.—Calle San Martin 186, Buenos Aires.
Establecimientos Argentinos de Bovril—Calle 25 de Mayo 182, Buenos Aires.
Establecimiento Bovril—Santa Elena, Entre Rios.
Frigorifico Armour de La Plata—Calle Sarmiento 443, Buenos Aires.
Frigorifico Le Blanca—Calle Cangallo 499, Buenos Aires.
Frigorifico Swift de La Plata—Calle 25 de Mayo 195, Buenos Aires.
Frigorifico Wilson—Reconquista 314, Buenos Aires.
Frigorifico Wilson—Pente Alsina, Buenos Aires.
Frigorifico Campana—Campana, F. C. C. A.
Frigorifico Swift—Rio Gallegos, Costa Sud.
Frigorifico Anglo Zarate—Zarate, F. C. C. A.
Liebig's Extract of Meat Company, Ltd.—Calle Lavalle 754, Buenos Aires.
Liebig's Extract of Meat Company, Ltd.—Colon, Entre Rios.
"La Forestal" Limitada—Calle Alsina 261, Buenos Aires.
Las Palmas Produce Co.—Calle Reconquista 314, Buenos Aires.
Soc. Anon. "Menendez Behety"—Calle San Martin 296, Buenos Aires.
Smithfield & Argentine Meat Company—Calle Reconquista 46, Buenos Aires.
Smithfield & Argentine Meat Company—Zarate, F. C. C. A.

BRAZIL

Cia. Frigorifica de Santos—Santos.
Continental Products Company—Osassco, Sao Paulo.
Dickinson & Co.—"Saladero Miranda," Matto Grosso.
Dickinson & Co.—Calle San Antonio 25, Santos.
Frigorifico Armour—Santa Anna de Livramento.
Frigorifico Swift—Rio Grande do Sul.
Frigorifico Mendez—Mendez, Rio de Janeiro.
Frigorifico Barretos—Estado Sao Paulo.
Saladero Bagge—Bagge, Rio Grande do Sul.

CHILE

Braun & Blanchard—Punta Arenas.
Cia. Frigorifica Puerto Natales—Puerto Natales, Punta Arenas.
Frigorifico Bories—Ultima Esperanza, Magallanes.
Frigorifico Punta Arenas—Hocneisen, Punta Arenas.
Frigorifico Armour—Santa Cruz, Costa Sud.
Frigorifico Rio Seco—Punta Arenas.
Frigorifico Puerto Montt—Puerto Montt, Costa Sud.
Lailhacar Hnos—Santiago.

Mataderos Modelos—Valparaiso.
Margezzini Hnos—Santiago.

COLOMBIA

Colombia Products Co.—Cartagena. President, Don Carlos Valez.

FALKLAND ISLANDS

Falkland Islands Company—Port Stanley.

MAGELLAN ISLANDS

Cia. Frigorifica de la Patagonia—Puerto Sara, Estrecho de Magallanes.

PARAGUAY

Central Products Company—Asuncion.

URUGUAY

Cia. Argentina de Frigorificos—Calle Ituzaingo 1467, Montevideo.
Cia. Liebig's—Fray Bentos, Montevideo.
Dickinson & Company—Calle Piedras 385, Montevideo.
Dickinson & Co.—"La Caballada" Salto.
Frigorifico Artigas—Calle Ituzaingo 1467, Montevideo.

AFRICA

UNION OF SOUTH AFRICA

CAPETOWN

Imperial Cold Storage Company, Ltd.—Dock Road.
Southern Meat Supply Company—Chiappini St.
Smithfield Cold Storage & Export Co., Ltd.—Woodstock.

AUSTRALIA

[Cold Storage, Meat Packing and Freezing Plants]

NEW SOUTH WALES

BROKEN HILL

Chilled Meat Supply Company—Argent St.

CANTERBURY

J. C. Hutton Pty., Ltd.

TASMANIA

LAUNCESTON

J. C. Hutton & Co., Ltd.

NEW SOUTH WALES

NEW CASTLE

Clarencetown Curing Company.

RAMORNIE

Australian Meat Company.

SYDNEY

Baynes Bros. Meat Export Company.
Bergl Australian Company, Ltd.—63 Pitt St.

Thos. Borthwick & Sons—79 Pitt St.
Birt & Company, Ltd.—4 Bridge St.
John Cooke & Co., Pty., Ltd.—253 George St.
Field, T. A.—678 Harris St.
Flemington Meat Preserving Company.
Gollin & Company, Ltd.
J. C. Hutton Pty. Co., Ltd.
Arthur Kidman—Chamber of Commerce.
National Meat Preserving Company.
N. S. W. Canning Factory.
Nevanas, V., & Company.
Pastoral Finance Ass'n, Ltd.
Paterson, John.
Phoenix Meat Company.
Queensland Meat Export & Agency Company—Liverpool St.
Richards, B., & Son, Ltd.
Sydney Meat Preserving Company.
Walker, F. J., & Company.
Yuill, G. S., & Co., Ltd.

Angliss & Co., N. S. W., Prop., Ltd., W.—Daroobalgie Works, near Forbes. 400 sheep and 250 cattle per day, and 4,000 crates rabbits weekly; storage capacity, 100,000 carcasses lamb.

Metropolitan Stores—Harris St., Sidney. 1,200 sheep per day; storage capacity, 40,000 carcasses mutton.

Australian Chilling & Freezing Company, Ltd.—Aberdeen Works, near New Castle. 100 cattle and 3,000 sheep per day; storage capacity, 60,000 carcasses mutton.

Cooke and Co., Prop., Ltd., John—Sandown Works, near Parramatta. 7,000 sheep, or 400 cattle and 3,000 sheep per day; storage capacity, 255,000 carcasses mutton.

Crystal Ice & Cold Storage Company, Ltd.—Works, Rozelle, Sidney. 1,500 sheep per day; storage capacity, 40,000 carcasses mutton.

Little & Co., R.—Byron Bay Freezing Works, Byron Bay. 120 cattle per day.

Metropolitan Meat Industry Board—State Abattoir, Homebush Bay. 8,000 sheep per day; storage capacity, 250,000 carcasses mutton. Meat distributing depot, Pyrmont. Storage capacity, 3,000 carcasses mutton.

Municipal Cold Storage Works—Sidney. Receiving 2,000 sheep per day; storage capacity, 100,000 carcasses mutton.

New South Wales Fresh Food & Ice Company, Ltd.—Darling Harbour Works, Sidney. 6,000 sheep per day; storage capacity, 170,000 carcasses mutton.

Pastoral Finance Association, Ltd.—Kirribilli Point Works, Sidney. 2,500 sheep per day; storage capacity, 75,000 carcasses mutton.

Riverstone Meat Company, Ltd.—Riverstone Works, Riverstone. Killing capacity, 5,000 sheep per day.

Sidney Ice Skating Rink & Cold Storage Company, Ltd.—Cold stores, Harris St., Ultimo, Sidney. 2,500 sheep per day; storage capacity, 60,000 carcasses mutton.

Walker & Co., F. J.—Tibberena Meat Works, Tibberena. 1,000 sheep and 80 cattle per day; storage capacity, 5,000 carcasses mutton.

Waterside Cold Stores, Ltd.—Darling Harbour, Sidney. Storage capacity, 100,000 carcasses mutton. Storage capacity for 200,000 carcasses mutton in course of erection.

Rabbit Packing Plants

Angliss & Co., N. S. W., Prop., Ltd., W.—Have the following rabbit freezing and packing works: Balderogery, storage capacity, 1,000 crates; Condobolin, storage capacity, 1,000 crates.

Thomas Borthwick & Sons (Australasia), Ltd.—Have the following rabbit freezing and packing works: Gulgong, storage capacity, 3,000 crates; Mudgee, freezing 600 crates per day, storage capacity, 5,000 crates.

Country Freezing Company, Ltd.—Works in the following towns: Blayney, storage capacity about 35,000 carcasses mutton; Boggabri, storage capacity, about 1,000 carcasses mutton; Boorowa, storage capacity, about 1,500 carcasses mutton; Crookwell, storage capacity, about 2,500 carcasses mutton; Dubbo, killing 800 sheep per day, storage capacity, about 5,000 carcases mutton; Gilgandra, storage capacity, about 500 carcasses mutton; Gunnedah, storage capacity, about 3,000 carcasses mutton; Harden, storage capacity, about 2,000 carcasses mutton; Millthorpe, storage capacity, about 1,200 carcasses mutton; Molong, storage capacity, about 1,000 carcasses mutton; Muswellbrook, storage capacity, about 2,000 carcasses mutton; Orange, storage capacity, about 3,000 carcasses mutton; Wellington, storage capacity, about 1,000 carcasses mutton; Yass, storage capacity, about 2,000 carcasses mutton; Yeoval, storage capacity, about 1,500 carcasses mutton; Young, storage capacity, about 3,000 carcasses mutton. (Used for rabbit freezing.)

Curtis & Curtis, Ltd.—Have the following freezing works: Binnaway, storage capacity, 3,000 carcasses mutton; Camden, storage capacity, 5,000 carcasses mutton; Crookwell, storage capacity, 4,000 carcasses mutton; Dunedoo, storage capacity, 5,000 carcasses mutton; Mundooran, storage capacity, 3,000 carcasses mutton; Rylstone, storage capacity, 3,000 carcasses mutton; Wagga, storage capacity, 2,000 carcasses mutton. (Used for rabbit freezing.)

Earle, John C.—Freezing and packing plant at Albury.

Little & Co., R.—Rabbit freezing works at Millthorpe, storage capacity, 2,000 carcasses mutton.

Nevanas & Co., Prop., Ltd.—Rabbit packing works at Bathurst, with storage capacity of 3,000 crates.

O'Brien Bros., Ltd.—Have the following rabbit freezing works: Braidwood, storage capacity, 6,000 crates; Nimmitabel, storage capacity, 3,000 crates; Moruya, storage capacity, 3,000 crates.

Paterson, Ltd., John—Rabbit freezing works at Cowra, with a storage capacity of about 4,000 crates; Coonabarabran, with storage for 4,000 crates; Albury with a storage capacity of 4,000 crates; and at Baradine, for rabbit chilling only.

Scandrett, L. A.—Controls freezing works in the following country centers: Bombala, storage capacity, 5,500 carcasses mutton; Gravesend, storage capacity, 5,000 carcasses mutton; Woolbrook, storage capacity, 5,000 carcasses mutton; Darby's Falls, storage capacity, 5,000 carcasses mutton; Yamble, storage capacity, 4,000 carcasses mutton; Boggabri, storage capacity, 3,500 carcasses mutton;

Binalong, storage capacity, 2,000 carcasses mutton; Newbridge, storage capacity, 2,000 carcasses mutton; Quirindi, storage capacity, 2,000 carcasses mutton; Warialda, storage capacity, 2,000 carcasses mutton. The combined output of these works would be about 300,000 carcasses per annum, or about the same quantity in crates of rabbits.

White, Ltd.—Have the following rabbit freezing works: Cootamundra, storage capacity, about 4,000 carcasses mutton; Tumut, storage capacity, about 2,000 carcasses mutton; Holbrook, storage capacity, about 1,500 carcasses mutton.

Wilson & Flood—Have the following works used for rabbit freezing: Bathurst, storage capacity, about 2,000 carcasses mutton; Galong, storage capacity, 9,000 carcasses mutton; Goulburn, storage capacity, 15,000 carcasses mutton; Rockley, storage capacity, 2,000 carcasses mutton.

QUEENSLAND

Australian Meat Export Company, Ltd.—Brisbane River Works, Brisbane. 600 cattle and 3,000 sheep per day; storage capacity, 40,000 quarters beef and 135,000 carcasses mutton; total 400 tons. Alligator Creek Works, Townsville. 750 cattle and 1,500 sheep per day; storage capacity, 40,000 quarters beef and 135,000 carcasses mutton; total, 4,000 tons.

Bergl Australia, Ltd.—Bowen Meat Freezing Works, Bowsen, 150 cattle per day; storage capacity, 1,500 tons.

Birt & Co., Ltd.—Musgrave Wharf Cold Stores, South Brisbane. 100 cattle or 1,100 sheep per day; storage capacity, 750 tons. Murarrie Works, near Brisbane. 150 cattle, or 1,500 sheep, or 110 cattle and 1,100 sheep per day; storage capacity, 1,100 tons.

Biboohra Freezing Works—Biboohra, near Cairns. 60 cattle per day; storage capacity, 500 tons.

Borthwick & Sons (Australasia), Ltd., Thos.—Moreton Freezing Works, Brisbane. 300 cattle or 4,000 sheep per day; storage capacity, 1,000 tons.

Burdekin River Meat Preserving Company, Ltd.—Burdekin Works, Sellheim. 260 cattle or 2,000 sheep per day; storage capacity, 1,200 tons.

Central Queensland Meat Export Company, Ltd.—Lakes Creek Works, Rockhampton. 500 cattle or 3,000 sheep per day; storage capacity, 3,000 tons.

Cooke & Co., Prop., Ltd., John—Redbank Works, via Brisbane. 450 cattle, or 280 cattle and 1,600 sheep per day; storage capacity, 2,800 tons.

Gladstone Meat Works of Queensland, Ltd.—Gladstone Works, Gladstone. 400 cattle or 3,500 sheep per day, or 200 cattle and 2,000 sheep per day; storage capacity, 2,500 tons.

Queensland Meat Export Company, Ltd.—Eagle Farm Works, Brisbane. 330 cattle and 300 sheep, or 150 cattle and 2,000 sheep per day; storage capacity, 32,000 quarters beef, or 105,000 carcasses mutton; total, 2,000 tons. Ross River Works, Townsville. 750 cattle, or 700 cattle and 1,000 sheep, or 600 cattle and 2,000 sheep, or 550 cattle and 2,500 sheep per day; storage capacity, 88,000 quarters beef, or 290,000 carcasses mutton; total, 5,500 tons.

BRISBANE

Baynes Bros. Meat Export Company.
Foggit, Jones & Co., Ltd.—Turbot St.

J. C. Hutton & Co., Ltd.
Queensland Co-operative Bacon Co., Ltd.
Redbank Freezing Works—62 Creek St.
Rosewarne Queensland, Ltd.
Traills, Ltd.

SOUTH AUSTRALIA

South Australian Government, State Refrigerating Works, Port Adelaide—120 cattle and 8,000 sheep or lambs per day; storage capacity, 300,000 carcasses lamb (600,000 cubic feet).

Light Square Works, Adelaide—3,000 carcasses mutton per day; storage capacity, 65,000 carcasses mutton (150,000 cubic feet).

Borthwick & Sons (Australasia), Ltd.—Have rabbit freezing works at Blue Hill, Mt. Gambier, with freezing capacity of 800 crates per week, and storage of 1,500 crates.

TASMANIA

North-West Freezing & Canning Company, Ltd.—Burnie Freezing Works, Burnie. 600 sheep per day; storage capacity, 60,000 carcasses mutton.

Murdoch Bros.—Hobart.

Earle, John C.—Has freezing and packing plants for rabbits located at Hobart and Launceton.

NORTHERN TERRITORY

North Australian Meat Company, Ltd.—Darwin. 500 cattle per day; storage capacity, 6,000 tons beef.

VICTORIA

MELBOURNE

Clark, A., & Sons., Pty., Ltd.
Flemington Meat Preserving Company.
Geelong Freezing Company—414 Collers St.
Hutton, J. C., & Co., Ltd.
Kensington Preserving Company.
Riveriva Frozen Meat Company—475 Collin St.
Victorian Meat Preserving Co.
Western Murray Districts Co-operative Bacon Curing Company—522 Flanders Lane.

Angliss & Co., Prop., Ltd., Wm.—Works, Bourke St. Melbourne. 2,000 lambs per day; storage capacity, 25,000 carcasses lamb.

City Market Cool Stores & Ice Works—King St. Melbourne. 200 carcasses lamb per day; storage capacity, 120,000 carcasses lamb.

Imperial Freezing Works—Footscray, near Melbourne. 11,000 sheep or lambs and 250 cattle per day; storage capacity, 250,000 carcasses lamb or equivalent in beef.

VICTORIA

Borthwick & Sons (Australasia), Ltd.—Portland Freezing Works, Portland. 1,100 lambs per day; storage capacity, 24,000 carcasses lamb.

Brooklyn Freezing Works—Footscray. Killing 6,000 lambs and 50 cattle per day; freezing 4,000 lambs per day; storage capacity, 100,000 carcasses lamb.

Cooke & Co., Prop., Ltd., John—Newport Freezing Works, near

Melbourne. Kill 6,000 and freeze 4,000 lambs per day; storage capacity, 120,000 carcasses lamb.

Fletcher, Ltd., W. and R.—North Shore Freezing Works, Geelong. 1,600 lambs per day; storage capacity, 28,000 carcasses lamb.

Goulburn Valley Industries Company, Ltd.—Shepparton. 2,500 sheep or 3,000 lambs per day; storage capacity, 90,000 carcasses lamb.

Melbourne Cool Stores & Ice Works (W. Dunkerley, Propr.)—Flinders Lane, Melbourne. Storage only; capacity, 60,000 carcasses lamb.

Melbourne Ice Skating & Refrigerating Company—The Glaciarium, Melbourne. 2,500 carcasses lamb per day; storage capacity, 45,000 carcasses lamb.

Metropolitan Ice & Cold Store Works (W. Dunkerley, Propr.)—Blackwood St., North Melbourne. Storage only, capacity, 50,000 carcasses lamb.

Sennitt & Son, Prop., Ltd.—Works, Miller St., Melbourne. 1,000 carcasses lamb per day; storage capacity, 50,000 carcasses lamb.

Sims, Cooper & Co. (Aust.), Prop., Ltd.—Geelong Harbour Trust's Corio Freezing Works, Geelong. 9,000 lambs per day; storage capacity, 200,000 carcasses lamb.

Victorian Butter Factories Co-Operative Co.—Works, Flinders St. Extension, Melbourne. 3,000 carcasses lamb per day; storage capacity, 80,000 carcasses lamb.

Victorian Government—Victoria Docks Stores, Melbourne. 15,000 carcasses mutton or 20,000 carcasses lamb per day; storage capacity, 320,000 carcasses lamb (total refrigerated space 1,350,000 cubic ft., nearing completion).

Western & Murray Districts Co-Operative Bacon Curing Company, Ltd.—Works, Geelong road, West Footscray. 2,500 lambs per day; storage capacity, 80,000 carcasses lamb.

Wimmera Inland Freezing Company, Ltd.—Murtoa Works, Murtoa. 2,600 sheep or lambs per day; storage capacity, 60,000 carcasses lamb. Plant being enlarged to enable treating 4,000 sheep or lambs per day and storing 80,000 carcasses lamb.

Ballarat & District Co-Operative Freezing Company, Ltd.—Ballarat Freezing Works, Ballarat. 2,000 lambs per day; storage capacity, 60,000 carcasses lamb.

Donald Inland Freezing Company, Ltd.—Donald Freezing Works, Donald. 2,500 lambs per day; storage capacity, 90,000 carcasses lamb.

Victorian Co-Operative Freezing Company, Ltd.—Bendigo Freezing Works, Bendigo. 2,500 lambs per day; storage capacity, 45,000 carcasses lamb.

Benalla & District Co-Operative Freezing Company, Ltd.—Benalla.

Echuca District Co-Operative Freezing Company, Ltd.—Echuca.

Rabbit Packing Plants

Angliss & Co., Prop., Ltd., W.—Rabbit freezing and packing works: Ouyen, storage capacity, 1,000 crates; Horsham, storage capacity, 6,000 crates.

Earle, John C.—Rabbit freezing and packing plants at Alexandria, Bairnsdale, Bendigo, Drouin, Leongatha, Maryborough, Mirboo North, Mildura, Poowong, Stawell, Sale, Tallangatta, Warragul, Wangaratta, Wodonga and Yea.

Nevanas & Co., Prop., Ltd., S. V.—Rabbit freezing and packing works; Hamilton, storage capacity, 4,000 crates; Euroa, storage capacity, 3,000 crates.

WESTERN AUSTRALIA

Baker Bros. Works—Fremantle. 100-150 cattle per week; storage capacity, 26,140 cubic feet.

Nor'-West Meat Works, Ltd.—Babbage Island Works, Carnarvon. Killing 1,600, freezing 500 sheep per day; storage capacity, 30,000 carcasses mutton.

Western Australian Government—Wyndam Works, Wyndam. 300 cattle per day; storage capacity, 1,200 tons.

Western Ice Company, Ltd.—Works, Fremantle. 1,000 carcasses lamb per day; storage capacity, 53,000 cubic feet. Storage capacity about to be increased to 4,000 tons.

West Australian Meat Exports Company, Ltd.—Fremantle Freezing Works, Fremantle. 200 sheep and 100 cattle per day; storage capacity, 60,000 carcasses mutton.

Westralian Meat Works, Ltd.—Geraldton.

NEW ZEALAND

Auckland Farmers' Freezing Company, Ltd.—Horotiu Works, Horotiu, Waikato. 3,000 sheep or 220 cattle per day; storage capacity, 260,000 carcasses mutton.

Southdown Works—Penrose, Auckland. 3,000 sheep or 220 cattle per day; storage capacity, 250,000 carcasses mutton. Auckland Works, Auckland. Storage only, 220,000 carcasses mutton.

Borthwick & Sons (Australasia), Ltd., Thos.—Paki Paki Works, Hastings, Hawkes Bay. 2,000 sheep and 30 cattle per day; storage capacity, 60,000 carcasses mutton. Canterbury Works, Belfast, near Christchurch. 4,000 sheep per day; storage capacity, 105,000 carcasses mutton. Waitara Works, Waitara, Taranaki. 2,000 sheep and 200 cattle per day; storage capacity, 80,000 carcasses mutton.

Canterbury Frozen Meat & D. P. Export Company, Ltd.—Belfast Works, Belfast, Canterbury. 6,000 sheep and 100 cattle per day; storage capacity, 380,000 carcasses mutton. Fairfield Works, Fairton, Ashburton. 4,000 sheep and 25 cattle per day; storage capacity, 150,000 carcasses mutton. Pareora Works, Pareora, near Timaru. 5,000 sheep and 50 cattle per day; storage capacity, 350,000 carcasses mutton.

East Coast Co-Operative Freezing Company, Ltd.—Whakatane Works, Whakatane, Bay of Plenty. 1,000 sheep and 200 cattle per day; storage capacity, 140,000 carcasses mutton.

Feilding Farmers' Freezing Company, Ltd.—Feilding Works, Aorangi, Feilding. 2,000 sheep per day; storage capacity, 250,000 carcasses mutton.

Gear Meat Preserving & Freezing Company of N. Z., Ltd.—Petone Works, Petone, near Wellington. 10,000 sheep and 100 cattle per day; storage capacity, 400,000 carcasses mutton.

Gisborne Sheepfarmers' Frozen Meat & Mercantile Company, Ltd.—Kaiti Works, Kaiti, Gisborne. 4,500 sheep and 120 cattle per day; storage capacity, 422,000 carcasses mutton.

Hawkes Bay Farmers' Meat Company, Ltd.—Whakatu Works, Whakatu, near Hastings. 3,000 sheep and 80 cattle per day; storage capacity, 155,000 carcasses mutton.

National Mortgage & Agency Company of N. Z., Ltd.—Longburn Works, Longburn, near Palmerston North. 1,500 sheep and 70 cattle per day; storage capacity, 100,000 carcasses mutton.

Nelson Bros., Ltd.—Tomoana Works, Tomoana, Hawkes Bay. 3,000 sheep and 80 cattle per day; storage capacity, 185,000 carcasses mutton. Taruheru Works, Gisborne. 2,500 sheep and 85 cattle per day; storage capacity, 100,000 carcasses mutton.

Nelson Freezing Company, Ltd.—Stoke Works, Stoke, near Nelson. 500 sheep and 30 cattle per day; storage capacity, 50,000 carcasses mutton.

New Zealand Meat Packing & Bacon Company (Co-Op.), Ltd.—Eltham Works, Taranaki. 150 pigs and 60 cattle per day; storage capacity, 25,000 carcasses mutton. Frankton Junction Works, Waikato. 300 pigs per day; storage capacity, 25,000 carcasses mutton. Ngahauranga Works, near Wellington. 1,200 pigs, 120 cattle and 3,000 sheep per day; storage capacity, 120,000 carcasses mutton. Te Aroha Works, Thames Valley. 150 pigs per day.

New Zealand Refrigerating Company, Ltd.—Islington Works, Islington, near Christchurch. 7,000 sheep and 50 cattle per day; storage capacity, 362,000 carcasses mutton. Smithfield Works, Smithfield, near Timaru. 6,000 sheep and 50 cattle per day; storage capacity, 304,000 carcasses mutton. Pukeuri Works, Pukeuri, near Oamaru, 3,000 sheep per day; storage capacity, 229,000 carcasses mutton. Burnside Works, Burnside, near Dunedin. 3,500 sheep and 50 cattle per day; storage capacity, 216,000 carcasses mutton. Picton Works, Marlborough. 1,000 sheep and 30 cattle per day; storage capacity, 23,000 carcasses mutton. Imlay Works, near Wanganui. 6,000 sheep and 200 cattle per day; storage capacity, 217,000 carcasses mutton.

North British & Hawkes Bay Freezing Company, Ltd.—Spit Works, Port Ahuriri, Napier. 2,500 sheep per day; storage capacity, 40,000 carcasses mutton.

North Canterbury Sheepfarmers' Co-Operative Freezing E. & A. Company, Ltd.—Kaiopoi Works, Kaiopoi, near Christchurch. 4,000 sheep and 100 cattle per day; storage capacity, 200,000 carcasses mutton.

Otaihape Farmers' Meat & Produce Company, Ltd.—Winiata Works, Winiata, near Taihape. 1,200 sheep and 50 cattle per day; storage capacity, 120,000 carcasses mutton.

Patea Farmers' Co-Op. Freezing Company, Ltd.—Patea Works, Canville, near Patea. 1,500 sheep and 150 cattle per day; storage capacity, 180,000 carcasses mutton.

Poverty Bay Farmers' Meat Company—Waipaoa Works, near Gisborne, Poverty Bay. 4,000 sheep and 200 cattle per day; storage capacity, 300,000 carcasses mutton.

Southland Frozen Meat & Produce Export Company, Ltd.—Bluff Works, Bluff, Southland. Storage only, 115,000 carcasses mutton. Mataura Works, Mataura, Southland. 2,000 sheep and 100 cattle per day; storage capacity, 105,000 carcasses mutton. Wallacetown Works, Makarewa, Southland. 2,000 sheep and 120 cattle per day; storage capacity, 74,000 carcasses mutton.

South Otago Freezing Company, Ltd.—Finegand Works, Balclutha, South Otago. 2,500 sheep and 60 cattle per day; storage capacity, 200,000 carcasses mutton.

Taranaki Farmers' Meat Company, Ltd.—Smart Road Works, New Plymouth, Taranaki. 2,500 sheep and 150 cattle per day; storage capacity, 150,000 carcasses mutton.

Tokomaru Sheepfarmers' Freezing Company, Ltd.—Tokomaru Works, Tokomaru Bay. 2,000 sheep and 50 cattle per day; storage capacity, 130,000 carcasses mutton.

Wairoa Farmers' Co-Operative Meat Company, Ltd.—Wairoa Works, Wairoa, Hawkes Bay. 2,000 sheep and 50 cattle per day; storage capacity, 165,000 carcasses mutton.

Wanganui Meat Freezing Company, Ltd.—Castlecliff Works, Castlecliff, Wanganui River. 2,200 sheep and 80 cattle or 3,000 sheep per day; storage capacity, 110,000 carcasses mutton.

Ward & Co., Ltd., J. G.—Ocean Beach Freezing Works, Bluff. 2,000 sheep and 120 cattle per day; storage capacity, 115,000 carcasses mutton.

Wellington Farmers' Meat Company, Ltd.—Waingawa Works, Waingawa, near Masterton. 7,000 sheep and 250 cattle per day; storage capacity, 350,000 carcasses mutton.

Wellington Meat Export Company, Ltd.—Ngahauranga Works, Ngahauranga, near Wellington. 7,000 sheep and 130 cattle per day; storage capacity, 240,000 carcasses mutton. Kakariki Works, Kakariki near Marton. 2,000 sheep and 50 cattle per day; storage capacity, 100,000 carcasses mutton.

Westfield Freezing Company, Ltd.—Westfield Works, near Auckland. 2,000 sheep and 200 cattle per day; storage capacity, 225,000 carcasses mutton.

Whangarei Freezing Company, Ltd.—Whangarei Works, Teotahi, Whangarei Heads. 1,200 sheep or 150 cattle per day; storage capacity, 90,000 carcasses mutton.

Auckland Farmers' Freezing Company, Ltd.—Waipuna Works, Waipuna, North Auckland.

Hicks Bay Farmers' Meat Company—Hicks Bay. 1,200 sheep per day; storage capacity, 40,000 carcasses mutton.

Wright, Stephenson & Co., Ltd.—34 Customhouse Quay, Wellington. Commission house. Handle wool, livestock, hides, pelts, tallow, oil, etc. Export. Also operate fertilizer plants.

AUCKLAND

Auckland Meat Company.
Hellaby, R. & W., Ltd.—53 Shortland Ave.
New Zealand Farmers Co-Operative Bacon & M. P. Company.
Whangerei Freezing Co., Ltd.

BELFAST

Belfast Freezing Works.

MATURA

Frozen Meat & Provision Export Co., Ltd.
Matura Freezing Works.

NGAHAURANGA

Banks Co-Op. Meat Distribution Co.

PALMERSTON NORTH

Kiwi Bacon Factory.
Manawatua Meat & C. S. Company.

WELLINGTON

Dimock & Co., W.
Freezing & Fertilizer Industries, Chemical Service Factory.
Great Meat Preserving & Freezing Co.—Petone.
N. Z. Farmers Co-Op. Bacon & Meat Company, Ltd.
Wellington Farmers Meat Company.

WHOLESALE SAUSAGE MAKERS, WHOLESALE MEAT DEALERS AND WHOLESALE PROVISIONERS

(Key—WMD=Wholesale Meat Dealer. WSM=Wholesale Sausage Maker. WP=Wholesale Provisioner.)

ALABAMA

BIRMINGHAM—Legg Sausage & Prov. Co., 108 N. 22nd St. (WSM) (WP).

ARIZONA

GLOBE—The Gila Meat Co. (WMD).
MIAMI—The Gila Meat Co. (WMD).
PHOENIX—Phoenix Meat Company. (WMD).
TUCSON—Tucson Meat & Provision Company. (WMD).
YUMA—Yuma Wholesale Meat Company. (WMD).
JEROME—Jerome Meat Co. (WMD).

ARKANSAS

FORT SMITH—W. A. Stanton, 815 Garrison Ave. (WMD) (WSM).
LITTLE ROCK—Becker Provision Co. (WSM).
RUSSELLVILLE—Lawson & Son, B. F. (WSM).

CALIFORNIA

BAKERSFIELD—California Market Co., Inc., 1618 19th St. Railroad, So. Pac. and A. T. & S. F. A corporation. Capital, $30,000; issued, $30,000. Stockholders, 4. President, P. Smith; Vice-President, Treasurer and Secretary, L. P. Keester. Employes, 20. Refrigeration—York. Buy ready-dressed meats. (WSM).
    Biggs, Dan. (WMD).
EMERYVILLE—Sunset Packing Co., 65th and Bay Sts. Railroad, S. P. Corporation. Capital, $74,000. President, M. J. Chadwick; Vice-President, A. Kenney; Treasurer and Secretary, H. W. Lotsprice; General Manager, M. J. Chadwick. Employes, 9. (WSM) (WMD).
HOLLISTER—Ernest Weller, 410 San Benito St.
LOS ANGELES—New Market Company (WSM) (WMD).
    Kancy & Co., 307 Aliso St. (WSM).
OAKLAND—Bright & Miller, 224-230 Second St. Partnership. General Manager, Ben Miller. Employes, 15. Refrigeration—20-ton York; brine circulating; one ton ice daily. Boiler capacity, 80 H. P. (WSM).
    Chin Fook & Company (WMD).
RIVERSIDE—Garner, John T. (WSM) (WMD).
SAN FRANCISCO—Jacob Alterberger, 1451 Haight St. (WSM).
    A. O. Angerman, 1550 Guerrero St. (WSM).
    Bayle-LaCoste & Co., 444 Pine St. Railroad, S. P. A corporation. Capital, $150,000; issued, $150,000. Stockholders, 6. President, John LaCoste; Vice-President, Alfred LaCoste; Secretary and Treasurer, P. Bareilles. Employes, 35. Dealers in packinghouse by-products;

tripe packers; do not slaughter. Refrigeration—4-ton Cyclops; direct expansion. Boiler capacity, 200 H. P. (WMD).

Cariani Sausage Factory, 226 Jackson St. (WSM).

Fromme & Fisher, 1789 Mission St. (WSM).

Giordani & Resinili, 434 Castor St. (WSM).

Heineman & Stern, 1040 McAllister St. Individual ownership. Employes, 15. General Manager, M. Stern. Refrigeration, 5-ton Cyclops; direct expansion and brine tanks. Boiler capacity, 50 H. P. (WSM).

San Francisco Sausage Factory, 804 Montgomery St. (WSM).

Anchor Packing Co., 149 Clay St. (WMD).

California Meat Co., 576 Clay St. (WMD).

Ecklon Bros., 444 Jessie St. (WMD).

J. Hoffman & Sons Co., 264 Sixth St. (WMD).

Horstmann Holm & Co., 264 Sixth St. (WMD).

J. Meyers & Co., California Market. (WMD).

Union Sheep Co., 1 Montgomery St. (WMD).

F. Ury & Co., 521 Clay St. (WMD).

Home Sausage Co., 1179 Sutter St. (WSM).

Italian Sausage Manufacturing Co., 804 Montgomery St. (WSM)

James Co., J. G. (WSM).

Ligure Sausage Factory, 1325 Grant St. (WSM).

Melano & Maggiora, 1402 Grant St. (WSM).

Milano Sausage Factory, 1337 Grant St. (WSM).

Molinari & Sons, P. G., 373 Columbus St. (WSM).

Muller, John, 517 Castro St. (WSM).

Nieri, A., 530 Washington St. (WSM).

Pucinelli, E., 1465 Powell St. (WSM).

Rathjens & Kupfer, 1331 Pacific St. (WSM).

Verga & Company, 1301 Grant St. (WSM).

Wilfert's Sausage Co., 118 Turk St.

Hawaii Meat Company (WSM) (WMD).

Hilo Meat Company (WSM) (WMD).

SANTA MARIA—**Santa Maria Meat Company** (WSM).

SANTA ROSA—**Noonan Meat Company** (WMD).

SANTA CRUZ—**Walter Schilling & Co.** (WMD).

SELMA—**Selma Meat Company** (WSM) (WMD).

STOCKTON—**Wagner Meat Company** (WMD).

TAFT—**Pioneer Market.** (WMD).

VENTURA—**Hobson Bros.** (WSM) (WMD).

WATSONVILLE—**The Tuttle Company** (WSM).

COLORADO

DENVER—**Kansas City Pkg. & Prov. Co.,** 1536 Wazee St. (WMD) (WSM).

Keogh-Doyle Meat Company, 1410 Market St. (Wholesale and retail meats and packinghouse products.) A corporation. Capital, $20,000; issued, $20,000. President, J. N. Doyle; Vice-President, P. N. Doyle; Secretary and Treasurer, Fred M. Runstetler. Employes, 12. Stockholders, 3. Refrigeration—5-ton York; direct expansion. (WMD).

K. & B. Packing Co., 1525 Blake St. (WSM) (WMD).

Lindner Packing & Provision Co., 1523 Market St. A corporation. Capital, $150,000; issued, $100,000. Stockholders, 3. President, H. Lindner; Vice-President, H. G. Timmins; Secretary and Treasurer, Albert Klingstein. Distributors for Morris & Co. (WP).

Raymond Packing & Prov. Co., 1331 Fifteenth St. (WMD).

Standard Meat & L. S. Co., Ulster, between 39th and 40th (WMD).

MONTROSE—**Gatterer & Miles** (WMD).

PUEBLO—**Herman Mercantile Company** (WMD) (WP).

CONNECTICUT

BRIDGEPORT—**Peter Huron, Inc.**, 1211 Stratford St. (WSM).

McNamara & Sons, T. J., 368 Coleman St. (WSM).

BRISTOL—**Williams, R. W.** (WSM).

HARTFORD—**Grote & Weigel**, 38 Spruce St. (WSM).

Hartford Centre Bologna Factory, 1197 Main St. (WSM).

Independent Packing Co., 660 Windsor St. (WMD) (WSM).

MIDDLETOWN—**Rogers & Hubbard Company** (WMD) (WSM).

NEW HAVEN—**The Alois Schwab Company**, 201-205 State St. (WSM). Railroad, N. Y., N. H. & H. A corporation. President, Alois Schwab; Vice-President, Chas. A. Anderson; Secretary, A. K. Anderson; Treasurer, A. Schwab. Employes, 30. Refrigeration—12-ton York; direct expansion.

Herster, Mrs. Charles, 118 Lawrence St. (WSM).

Hugo & Sons, S., Crown St. (WSM).

Rossler, Charles, 96 Pearl St. (WSM).

NEW LONDON—**Brigham, J. B.** (WSM) (WMD).

STAMFORD—**Klein & Gmahle**, 556 Pacific St. (WSM).

WATERBURY—**Herman, J.**, 173 S. Main St. (WSM).

Hullstnick, J., 27 W. Clay St. (WSM).

Sachsenhauser, Inc., 506 W. Main St. (WSM).

DISTRICT OF COLUMBIA

WASHINGTON, D. C.—**Harry B. Denham & Co., Inc.**, 929 D St., N. W. A corporation. Capital, $150,000; issued, $150,000. Stockholders, 3. President, H. B. Denham; Secretary, A. M. Hoover; Treasurer and General Manager, H. B. Denham. Employes, 23. Government inspection. Refrigeration—7-ton York; direct expansion (WSM) (WMD).

Keane, T. T., Co., Inc., 619 B St. (WP).

Jos. Phillips Co., 10 Wholesale Row (WSM) (WMD).

FLORIDA

PERRY—**Perry Packing Company** (WMD).

GEORGIA

ATHENS—**Cutler Sausage Company** (WSM).

ATLANTA—**Enterprise Sausage Factory** (WSM).

Willie & Sam Reisman (WMD).

MACON—**Nash Sausage Company**, 222 Cotton Ave. Railroads, C. of Ga., So. Ry., G. S. & F. Individual ownership, T. M. Nash. Refrigeration—4-ton Frick; direct expansion. Operates in connection with L. W. Rogers' chain of grocery stores. Twelve retail meat markets—9 in Macon, 1 in Ft. Valley, 1 in Americus and 1 in Milledgeville, Ga. (WSM).

SAVANNAH—**Hester Co., J. W.**, City Market. Railroad, C. of Ga. A corporation. Capital, $20,000; issued, $20,000. Stockholders, 4. President, J. W. F. Hester; Vice-President, D. C. Hester; Secretary, R. A. Laird; Treasurer, J. W. F. Hester. Employes, 57. Refrigeration—20-ton Cincinnati Refg. Co. and 8-ton Frick; direct expansion. Boiler capacity, 15 H. P. (WMD) (WSM).

Hohnerlein, Joe, 1101 Wheaton St. Individual ownership. Employes, 4. Refrigeration—3-ton Brecht; direct expansion. Boiler capacity, 8 H. P. (WSM).

IDAHO

MONTPELIER—**Hoff Meat Co., H. H.** (WMD).

NAMPA—**Cold Storage Market** (WSM) (WMD).

POCATELLO—**Zweigart Brothers** (WSM).

ILLINOIS

BENTON—**Werner, Wm.** (WMD).

CARMI—**Brown & Nauert** (WMD).

Union Market (WMD).

CHICAGO—**Agar Provision Company**, Fulton and Green Sts. (WMD) (WP).

Alexander Pkg. & Prov. Co., East 92nd and Calumet River (WMD).

Anderson Bros. Stores, Inc., 10004 Avenue L. A corporation. Capital, $75,000; issued, $75,000. Stockholders, 17. President, D. F. Anderson; Secretary, B. D. Anderson; Treasurer, H. N. Anderson. Employes, 30. Refrigeration—Xcell; brine circulating. (WSM).

Anglo American Provision Company, 208 S. La Salle St. (WP).

Atlas Sausage & Prov. Co., 2305-9 W. Taylor St. A corporation. Capital, $20,000; issued, $12,500. President, R. Souta; Vice-President, Frank Sedlar. Employes, 5. (WSM) (WP).

Bert Packing Company, 176 N. Green St. Individual ownership. Employes, 5. General Manager, Henry Bertolotti. Refrigeration—2-ton Toledo; brine circulating. (WSM).

Blumenhagen, F., 1533 Augusta St. (WSM).

Blum, I., 942 Fulton Ave. Railroads, C. M. & St. P., Pennsylvania and C. & N. W. Individual ownership. General Manager, N. Blum; General Superintendent, L. Blum. Government inspection. Employes, 45. Codes, Cross. Refrigeration—15-ton Baker; direct expansion. Boiler capacity. (WMD).

David Berg & Company, 449 West 37th St. A corporation. Capital, $250,000; issued, $250,000. Stockholders, 4. President, Max Weinberg; Secretary and Treasurer, C. Lowenstein. Employes, 75. Government inspection. (WMD) (WP).

Brunner Provision Company, 3737 S. Halsted St. (WP).

C. A. Burnette Co., 827 W. 22nd St. A corporation. Capital, $100,000; issued, $60,000. Stockholders, 6. Employes, 25. Codes—Cross and Robinson. President, H. F. Wilkins; Vice-President, C. A. Burnette; Secretary, Sue Wilkins; Treasurer, C. A. Burnette. Government inspection. Refrigeration—20-ton York; direct expansion. (WMD) (WP).

Central Provision Co., 854 Fulton St. Corporation. Capital, $30,000. (WMD) (WP).

Chicago Butchers' Packing Co., 216 N. Peoria St. (WSM).

Chicago Sausage Company, 2910 Armitage Ave. (WSM).

Cicero Sausage Company, 4912 W. 25th St. (WSM).

Citti Bros., 309 W. Illinois St. (WSM).

Corn Belt Packing Company, 3120 E. 92nd St. (WSM) (WMD).

Dahmke Pkg. Co., J. A., 2334 W. Lake St. (WMD).

Drexel Packing Co., 852 W. Grand Ave. (WMD).

Duntz, Herman, 3820 S. Ashland Ave. A corporation. Capital, $25,000; issued, $24,300. Stockholders, 7. President, Herman Duntz; Vice-President, A. Pigaesch; Secretary and Treasurer, Frank Duntz, Jr. Employes, 15. Refrigeration—10-ton Baker; brine circulating. (WSM).

Englewood Packing Co., 6845 S. Halsted St. (WMD).

Fulton Packing Co., 820 Fulton St. (WMD) (WSM).

Hetzel & Co., 1743 Larrabee St. A corporation. Capital, $600,000. Stockholders, 3. President and Treasurer, John Hetzel; Vice-President, H. B. Hetzel; Secretary, J. P. Hetzel, Jr. Employes, 250. Refrigeration—One Filter & Stowel and one Wolf. (WMD) (WSM).

Hoffman & Co., J. S., 219 N. Franklin (WSM).

Hollenbach & Sons, G., 1100 Marquette Rd. (WSM).

Hollenbach, C., 1317 S. Oakley Ave. (WSM).

Home Made Sausage Company, 4500 W. 22nd St. (WSM).

Italian Sausage Works, 442 W. Chicago Ave. (WSM).

Jefferson Pkg. Co., 4915 Milwaukee Ave. (WMD).

Kleppel Packing Co., 3059 Arthington St. Railroad, Great Western. A corporation. Stockholders, 5. Employes, 20. Capital, $25,000; issued, $25,000. Refrigeration—10-ton Creamery Package Mfg. Co.; direct expansion. (WMD) (WSM).

Kosher Star Sausage Co., 1010 Maxwell St. (WSM).

Lawndale Sausage Co., 3617 W. 27th St. Partnership. General Manager, Otto Pelikan. Employes, 12. (WSM).

Lincoln Packing Company, 3804 S. Halsted St. (WMD) (WSM).

Mickelberry's Food Products Company, 801 W. 49th Place. A corporation. Capital, $50,000; issued, $50,000. President, C. M. Mickelberry; Vice-President, C. W. Mickelberry; Secretary, M. W. Mickelberry; Treasurer, O. C. Mickelberry. Refrigeration—10-ton Baker; brine circulating. (WSM).

Mutual Sausage Company, 2713 Quinn St. Railroad, C. & A. A corporation. Capital, $100,000; issued, $100,000. Stockholders, 13. President, John L. Hotka; Vice-President, A. Schiack; Secretary, W. H. Gausselin; Treasurer, W. H. Gausselin. Refrigeration—15-ton Rubsamen & Almroth; direct expansion. (WSM).

National Home Made Sausage Co., 4714 S. Paulina St. (WSM).

Nitzschke, G., 2639 S. Harding Ave. (WSM).

North Western Packing Co., 1012 Milwaukee Ave. (WMD) (WSM).

Pfaelzer Brothers, 936 West 38th Place. Partnership. (WMD).

Purity Packing Co., 1320 W. 21st St. (WSM) (WMD).

Real Sausage Company, 2710 Poplar Ave. A corporation. Capital, $100,000; issued, $77,500. Stockholders, 14. President, N. C. Poczulp; Vice-President, J. A. Mitrius; Secretary and Treasurer, H. E. Ward. Employes, 45. Refrigeration—25-ton Larson-Bader and 8-ton Healy; direct expansion. Boiler capacity, 60 H. P. (WSM).

Rohn, Theodore, 314 W. 32nd St. (WSM).

Ruprecht & Company, 560 W. Randolph. Partnership. Refrigeration—30-ton Howe; direct expansion. Boiler capacity, 400 H. P. (WSM) (WMD).

Schley Packing Co., 2332 W. Lake St. (WMD) (WSM).

Sinai Kosher Sausage Company, 3351 S. Halsted St. A corporation. Capital, $100,000; issued, $100,000. President, Jacob Levin. Employes, 35. Government inspection. Boiler capacity, 35 H. P. Also does export business. (Wholesale Kosher Sausage Manufacturer.)

Siegel, Sidney, 3804 S. Halsted St. (WMD).

Singer & Co., M. D., 3451 Forrest Ave. (WSM).

Sotir & Sorich Company, 3831 S. Halsted St. A corporation. Capital, $50,000; issued, $29,000. Stockholders, 3. President, Sotor Arangelorich; Secretary and Treasurer, J. Sorich. Government inspection. (WSM).

South Side Packing Company, 4817 S. Ashland Ave. (WSM) (WMD).

Standard Sausage Company, 963 W. 37th St. A corporation. Capital, $110,000; issued, $110,000. Stockholders, 11. President, H. Arndt; Vice-President, H. Baerenz; Secretary and Treasurer, A. W. Klingbeil. Employes, 60. Refrigeration—One 50 and one 35-ton Linde; direct expansion. Boiler capacity, 150 H. P. (WSM).

Union Stock Yards Pkg. & Prov. Co., 1750 W. 63rd St. (WP).

United Butchers' Packing Co., 913 Fulton St. (WMD).

Vienna Sausage Mfg. Co., 1215 S. Halsted St. A corporation. Capital, $250,000; issued, $250,000. Stockholders, 7. President, Emil Reichl; Secretary, Abraham Loebe; Treasurer, Samuel Ladanyi. Employes, 75. Codes—A. B. C., 5th edition; code word, "Veribus." Government inspection. Refrigeration—One 15 and one 13-ton Rubsamen & Almroth; direct expansion. Boiler capacity, 100 H. P. Also does export business. (WSM).

Vittori, L., 1057 W. Harrison St. (WSM).

Zeiger Company, G. W., 515 W. Chicago Ave. A corporation. Capital, $50,000; issued, $50,000. Stockholders, 10. President, G. W.

Zeiger; Vice-President, G. F. Reis; Secretary, E. Zeiger; Treasurer, G. W. Zeiger. Refrigeration—18-ton De La Vergne; direct expansion. Boiler capacity, 160 H. P. (WSM) (WMD).

Zuegel-Rieger Co., 2413 W. Roosevelt Rd. (WSM).

Also the following (WSM):

Acme Sausage Co., 3738 S. Ashland Ave.

Alfred Altschul, 3451 Giles St.

Frank Binkowski, 1731 W. Chicago Ave.

John Brzuskiewicz, 4850 S. Ashland Ave.

Arnold Busch, 6151 Ravenswood Ave.

Oscar Casperson, 2138 N. Cicero Ave.

Crowe & Company, 1803 W. Chicago Ave.

Jos. Demuth, 3509 Janssen Ave.

Dicks & Fischer, 1704 Belmont Ave.

Englewood Sausage Co., 6161 Wentworth Ave.

Julius Fennig, 1702 Milwaukee Ave.

John Fiorenti, 940 N. Wells St.

Frank & Co., 157 W. Kinzie St.

Italo-America Sausage Mfg. Co., 1514 N. Wells St.

W. J. Kaltwasser, 716 W. 43rd St.

Jos. Kaminski, 11754 S. Michigan Ave.

Frank Kozyra, 1375 W. Austin Ave.

John K. Kwiatkowski, 1749 N. Hermitage Ave.

Leviton Co., 1064 Argyle Ave.

Marquette Packing Co., 6780-82 South Chicago Ave.

McCune & Sneed, 3033 Vernon Ave.

Meadowbrook Farm Sausage & Prov. Co., 716 W. 43rd St.

Mike Mikolajczyk, 1737 W. Diversey Ave.

Milan Sausage Mfg. Co., 339 Kensington St.

Chas. Murawske, 1800 Eddy St.

J. F. Nadzieja, 1941 W. Division St.

National Sausage Co., 2336 Blue Island Ave.

Northwest Side Kosher Sausage Mfg. Co., 1741 W. Division St.

Panovich Bros., 339 W. 25th Place.

Pure Sausage Co., 6821 Keefe Ave.

Emil Reichl, 1215 S. Halsted St.

Relation Packing Co., 1838 N. Leavitt St.

Warsaw Home Made Sausage Co., 2153 W. 21st St.

JOLIET—Adler, J. C. (WSM) (WMD).
  Fritz, Fred H., 209 South Joliet St. Individual ownership. Employes, 9. Refrigeration—5-ton Wolf-Sayer & Heller; brine circulating. Boiler capacity, 15 H. P. (WSM).

MATTOON—Steidle Brothers (WMD).

MENDOTA—Geo. Erbes (WMD).

PEORIA—Central City Provision Company, 3111 South Adams St. (WSM) (WP).
PARIS—Steidl Bros. (WSM) (WMD).
QUINCY—Auck Brothers. (WMD).
RED BUD—Burgdorf, H. F. (WSM).
ROCKFORD—Carty-Dever Co., 321 W. State St. Corporation. Capital, $20,000. Stockholders, 3. President, D. J. Dever; Secretary-Treasurer, F. J. Phinney. Three retail markets. (WMD).
SPRINGFIELD—Mohay, John. (WSM).
  Richter & Son, Ed. H. (WSM).
SULLIVAN—Lovelless & Wagoner. (WSM).
STERLING—Horn & Morath. (WSM).
SAVANNA—Law, M. A. (WSM).
ZION CITY—Zion Institution and Industrial Meat Market, 2712 Elijah Ave. Railroad, C. & N. W. Individual ownership (W. G. Voliva). Manager, S. G. Biddle. Employes, 8. Refrigeration—4-ton Howe; brine circulating.

INDIANA

HAMMOND—Calumet Packing & Provision Co. (WMD) (WP).
WINCHESTER—Earl W. Wise. (WMD).
BLOOMINGTON—Bloomington Packing Co. Owned by Whisenand & Burns. (WSM).
FRANKFORT—Milner Provision Co. (WMD).

IOWA

AMES—Briley & Sons, H. E. (WSM).
BOONE—Steve Anderson. (WSM).
  Boone Market Company. (WMD).
CALMER—Wesserman & Brocar. (WMD).
CARROLL—Beiter Brothers. (WSM).
CHARITON—Yengle Brothers. (WSM).
COLFAX—Terpestra & Bridge. (WSM).
CUMBERLAND—Band & Store. (WSM).
DUBUQUE—H. Trenkle & Co., 1227 Central Ave. A corporation. Capital, $75,000; issued, $50,000. President, H. Trenkle; Vice-President, A. W. Neuwolhner; Secretary, Wm. H. Trenkle; Treasurer, A. W. Neuwolhner. Employes, 17. Refrigeration—8-ton Baker; direct expansion. Boiler capacity, 25 H. P. (WSM).
  Wimmer Sausage Shop, 1533 Central Ave. (WSM).
EMMETTSBURG—Wardus & Allen. (WSM).
GRINNELL—Grinnell Provision Co. (WMD) (WSM).
HARLAN—Kemp & Davis. (WSM).
MANNING—Central Meat Market. H. Timmerman, owner. (WMD).
MASON CITY—Stott Meat Market. (WSM).
NEW HAMPTON—Cross Market. (WSM).
OELWEIN—Fred Voelker. (WSM).

OSAGE—Palmer Bros. (WSM).
  Huesselman Brothers. (WSM) (WMD).
WASHINGTON—Pankwordt & Kurtz (WSM) (WMD).
WATERLOO—A. J. Wittich. (WSM).

KANSAS

LEAVENWORTH—Sannish Bros. (WMD).
ROSEDALE—Far Famed Meat & Sausage Co. (WSM).
WICHITA—Wichita Dressed Beef Co. (WMD).
KANSAS CITY—Loschke & Zercher, Southwest Blvd. and 45th St. (WSM).

KENTUCKY

ASHLAND—Yungkau, L. & A., 112 18th St. Partnership. Government inspection. Codes—Cross, Cipher. Wholesale meats and provisions. (Processers.)
COVINGTON—Sprunk, Peter, 16th and Water St. (WSM) (WMD).
  Haehnle Provision Co. (WMD).
NEWPORT—Becker Bros. Co., 942 Monmouth St. A corporation. Capital, $50,000; issued, $30,000. Stockholders, 3. President, F. B. Bassmaun; Secretary and Treasurer, Henry Becker. Employes, 30. Government inspection. Refrigeration—20-ton Niebling; brine circulating and direct expansion. Boiler capacity, 50 H. P. Retail Markets—942 Monmouth St., Newport, Ky., and 203 W. Sixth St., Cincinnati, O. (WMD) (WSM).
PADUCAH—Jones, W. R. (WSM).
  Jones, T. A. (WSM).

LOUISIANA

BATON ROUGE—J. J. Sanchez. (WSM).
  Webre Bros. (WSM).
NEW ORLEANS—L. Artigne. (WSM).
  Geo. & Wm. Schott, 524 Howard Ave. (WSM) (WMD).
  G. A. Weigand & Co., 523 Poydras St. (WSM) (WP).
  Western Meat Co., Alabo and N. Peters Sts. J. A. Hillery. (WMD).
ST. BERNARD, ARABI P. O.—Claverie & Company. Aug. J. (Co-partnership, Aug. J. Claverie and E. Victor Fassman.) Railroad, Louisiana Southern. Codes—Western Union. Government inspection. Employes, 45. (WSM) (WMD).

MAINE

AUBURN—Littlefield & Sons Co. (WSM).
BANGOR—Bean & Sons, W. A. (WSM).
  Joy Co., A. W. (WSM).
  Rice Company, C. H., 193 Broad St. Railroad, Maine Central. A corporation. Capital, $200,000. Stockholders, 3. President, A. F. Rice; Vice-President and Treasurer, E. H. Rice; Secretary, Grace M. Rice. Employes, 22. Refrigeration—8-ton Automatic; direct expansion. Sausage manufacturers and jobbers of beef and pork products.

HOULTON—Houlton Meat Supply Co., Pleasant St. (WSM).
PORTLAND—**Kern & Son, John.** (WSM).
    Schonland Bros., 10 Union St. Railroads, Boston & Maine and Maine Central. A corporation. Capital, $30,000; issued, $30,000. Stockholders, 4. President, R. R. Schonland; Treasurer, Chas. Schonland. Employes, 12.
    Cummings Bros. (WP).

MARYLAND

BALTIMORE—**Fred Haas,** Brown's Lane. (WSM).
    Ottenheimer Bros., Inc., 2308 Frederick Ave. A corporation. Capital, $1,500,000; issued, $600,000. Stockholders, 20. President, R. E. Ottenheimer; Vice-President, B. M. Ottenheimer; Secretary and Treasurer, S. M. Ottenheimer. Employes, 15. Government inspection. Refrigeration—15-ton Remington; brine circulating. (WSM).
    Rettberg, L. H. (WSM).
    Charles W. Leydecker. (WMD).
    John W. & Wm. F. Lower. (WMD).
    Albert H. Plitt. (WMD).
    Charles Plitt. (WMD).
    Chas. E. Plitt. (WMD).
    E. Wilbur E. Plitt. (WMD).
    J. W. Plitt. (WMD).
    Charles Rettberg's Sons. (WSM).
    Geo. G. Rappersberger & Son. (WSM).
    Gustav H. Ruppersburger Sons. (WMD).
    Chas. Schmidt. (WMD).
    John Shuppner. (WMD).
    W. W. Snoot. (WSM).
    John Truss. (WMD).
    Wetzelberger Bros. (WSM).
    Bloecher & Schaaf. (WMD).
    H. L. Caplan & Co., 916 Lombard St. (WP).
    John H. Eichner. (WMD).
    J. Frederick. (WMD).
    John G. Frederick. (WMD).
    Gengnagle Meat Co. (WMD).
    Howard F. Greasley. (WMD).
    A. Hansen. (WMD).
    Henry Heil. (WMD).
    Charles & L. G. Messersmith. (WMD).

MASSACHUSETTS

BOSTON—**Baldau, F. W.,** 5 Fulton Place. (WSM).
    Boston Sausage & Provision Company, 161 Blackstone St. (WSM).
    Chamberlain & Company, 24 South Market St. Railroad, Union Freight. A corporation. Capital, $500,000; issued, $500,000. Presi-

dent, G. N. Chamberlain; Vice-Presidents, G. A. Chamberlain, L. B. Crandon and C. W. Chamberlain; Secretary and Treasurer, A. A. Huse. Employes, 325. Government inspection. Fourteen retail markets in Boston. (WMD) (WSM).

Munro-Sexton Co., 43-44 S. Market St. President, Everett W. Munro; Treasurer, Alfred M. Sexton; Secretary, George E. Sexton. (WMD) (WP).

Parks Sausage & Provision Company, 200 State St. A corporation. Capital, $5,000. President, H. W. Taylor; Vice-President, H. S. Taylor; Secretary, E. E. Baldwin; Treasurer, I. R. Taylor. Employes, 25. (WSM) (WMD).

Rounsevell, P. W., 103 Blackstone. Individual ownership. Employes, 30. (WSM) (WP).

Weitz, C. A., 105 Elliott St. (WSM).

Wattendorf & Co., Frank M., 36 North St. (WSM) (WMD).

The Hanover Co., 52 Blackstone St. A corporation. Capital, $25,000; issued, $25,000. President, F. Batchelder; Vice-President, A. R. Shepardson.

CAMBRIDGE—**Schultz, J. A.**, 345 Main St. Railroad, Boston & Albany. Individual ownership. Employes, 3. Government inspection. (WSM).

HAVERHILL—**E. H. Moulton Co.**, 132 Essex St. Railroad, Boston & Maine. A corporation. Capital, $100,000; issued, $100,000. Individual stockholders, 3. President, E. A. Edgerly; Treasurer, E. H. Moulton. Employes, 35. Government inspection. Refrigeration—12 and 22-ton Remington and 8-ton Westinghouse; direct expansion. (WSM).

HOLYOKE—**Moskel Prov. Co., John.** (WSM).
Zasadzinski & Rzeszstarski. (WSM).

LAWRENCE—**John Holt.** (WSM).
H. J. Furneaux, 649 Essex St. Railroad, Boston & Maine. Individual ownership. Employes, 13. General Manager, S. E. Furneaux. Boiler capacity, 20 H. P. (WSM) (WP).

NEW BEDFORD—**Davidson & Son, J.**, 280 Austin St. (WSM).
Kuechler Bros., 337 South Second St. (WSM).
Rezendes, J. F., 433 South Second St. (WSM).
Schmidt, J. H., 424 South Second St. (WSM).

ROXBURY—**Claus, A.**, 312 Roxbury St. (WSM).
David Cohen, 59 Prentiss St. (WSM).

SOMERVILLE—**Sturtevant & Haley Beef & Supply Co.** (WSM).

SOUTHBORO—**Deerfoot Farm Company.** Individual ownership. (WSM).

SPRINGFIELD—**H. L. Handy Co.**, 41 Hampden St. (WSM).
A. C. Hunt & Co., 16 to 32 Sanford St. Partnership. Government inspection. Refrigeration—30-ton De La Vergne; brine circulating. (WSM) (WMD).

WORCESTER—**Bertels, B. J.** (WSM).
Boepple, Geo., Co., Inc., 600 Millbury St. A corporation. Capital,

$10,000; issued, $10,000. Stockholders, 3. President, Jacob Baum; Vice-President, Fritz Baum; Secretary, Fritz Baum; Treasurer, John Renhert. Employes, 28. Refrigeration—20-ton York. (WSM).

Chicago Beef & Produce Company. (WSM) (WP).

L. B. Darling. (WSM).

Geo. Geiger. (WSM).

MICHIGAN

DETROIT—**Eastern Market Sausage Mfg. Co.**, 2472 Riopelle St. (WSM).

Kelley & Company, 3449 Michigan Ave. A corporation. Capital, $10,000; issued, $9,100. Stockholders, 9. President, H. Kramer; Vice-President, Robert Hay; Secretary and Treasurer, D. P. Kelley. Employes, 20. Refrigeration—8-ton Frick; direct expansion. Boiler capacity, 12 H. P. (WSM) (WP).

Miotke & Co., Jos., 1834 E. Forest Ave. Partnership. Stockholders, 2. President, J. A. Miotke; Treasurer, J. L. Miotke. Employes, 16. Refrigeration—8-ton Brunswick; brine circulating. Boiler capacity, 14 H. P. (WSM).

Orling Brothers, 3142 Elmwood Ave. A corporation. Capital, $75,000; issued, $75,000. Stockholders, 5. President, Fritz Orling; Vice-President, Louise Orling; Secretary and Treasurer, E. R. Orling. Refrigeration—6-ton Brunswick; direct expansion. Boiler capacity, 10 H. P. (WSM).

Peschke & Killian, 6031 Rivard St. (WSM).

Peters, John A., 5454 Dix Ave. Individual ownership. Employes, 50. City and state inspection. Refrigeration—25-ton Detroit; brine circulating. Boiler capacity, 75 H. P. (WSM) (WMD).

Standard Sausage Company, 502 E. Forest Ave. (WSM).

Swope, C. A., 7806 Mack Ave. (WSM).

FLINT—**Flint Sausage Works**, 1210 Avenue A. Emil Salay, Proprietor. Capital, $30,000. Employes, 8. Refrigeration—5-ton York; brine circulating. Boiler capacitl, 8 H. P. (WSM).

HANCOCK—**Richard Vollwerth**, 207 Franklin St. (WSM).

IRONWOOD—**Swanson & Deronci**, 223 S. Curray St. (WSM).

JACKSON—**Jackson Sausage Company**, 502 N. Mechanic. Individual ownership. President, G. A. Stoldt. (WSM).

PONTIAC—**Harger Beef Co.** (WMD).

PORT HURON—**Port Huron Sausage & Provision Company**, Fourth and Wall Sts. President, H. H. oodward; Vice-President, Gus Hill. (WSM) (WP).

SAGINAW—**Henning & Son**, 407 N. Water St. Partnership. Employes, 35. General Manager, L. A. Henning. (WSM).

THREE RIVERS—**Three Rivers Packing Co.** (WMD) (WSM).

MINNESOTA

NEW ULM—**Andrew Saffert.** (WSM).

ST. PAUL—**Twin City Sausage Co.**, 169 W. 63d St. (WSM).

MISSISSIPPI

JACKSON—**Pearl Market Co.** (WSM).

MERIDIAN—Lutz & Bebeze. (WSM).
 Michel, Joe. (WSM).
McCOMBS—Stokes Brothers. (WSM) (WMD).

MISSOURI

KANSAS CITY—Frankfort Sausage Co., 1208 Forest Ave. (WSM).
 Kansas City Sausage Co., 14 E. Missouri Ave. (WSM).
 Neuer Bros. Meat Co., 1326 Main St. A corporation. Capital, $50,000. President, Ernst Neuer. Employes, 75. Government inspection. Refrigeration—25-ton Linde and 15-ton Triumph. (WSM).
ST. LOUIS—Sweet Provision Co., Chas. A., 813 Spruce St. A corporation. Capital, $50,000; issued, $50,000. Stockholders, 5. President, S. H. Kleinschmidt; Vice-President, P. C. Ziemer; Secretary and Treasurer, Wm. H. Schnecko. Codes—Cross and Robinson. Government inspection. (WSM) (WP).
 Missouri Packing Company, 2734 Franklin Ave. (WSM) (WMD).
WARRENSBURG—Roseland Farm & Mfg. Co. (WSM).

MONTANA

DILLON—Montana Market. (WMD).
HELENA—Helena Meat Market. (WMD).
LEWISTOWN—Abel Brothers Co. (WSM) (WMD).
 Lewistown Packing Co. (WSM) (WMD).
LIVINGSTON—Iten's Cold Storage Meat Market. (WSM) (WMD).
MISSOULA—Daily Co., J. R. (WMD).

NEBRASKA

OMAHA—Omaha Sausage Company, 4726 South 27th St. A corporation. Capital, $10,000; issued, $10,000. Stockholders, 3. Employes, 12. President, H. Goldenberg. Refrigeration—8-ton York.
 Purity Provision Company, 2424 O St. Individual ownership. Employes, 15. Refrigeration—4-ton York; direct expansion.
 H. Glassburg, 2401 Leavenworth St. (WSM).
 Kroger Wholesale Provision Co., 2405 Woolworth Ave. (WSM).

NEW HAMPSHIRE

NASHUA—Nashua Packing Co. (WMD) (WSM).
KEENE—Cheshire Beef & Produce Co. (WSM) (WP).

NEW JERSEY

ASBURY PARK—Marx, A., Asbury Ave. (WSM).
ELIZABETH—Krinzman & Jaffee, Morris Ave. (WSM).
HOBOKEN—Empire Bologna & Prov. Co., 700 1st St. (WSM).
JERSEY CITY—Egner, Geo., 67 Central Ave. (WSM).
 Wm. Everett's Sons' Co., foot of Sixth St. (WMD) (WSM).
 Edw. Reckenstein & Sons, 112 Griffith St. (WSM).
 Lafayette Provision Co., 384 Pacific Ave. (WMD).
NEWARK—Pfeiffer, Hy., 57 Napoleon St. (WSM).
 Reinfield & Sons, M., 98 Prince St. (WSM).
 Frank J. Cloran, 51 Ward St.. (WMD) (WSM).

WHOLESALERS AND SAUSAGE MAKERS

Van Wagenen & Schickhaus Co., 30 Plane St. A corporation. Capital, $200,000; issued, $200,000. Stockholders, 11. Employes, 308. President, J. A. Brady; Vice-President, G. J. Edwards; Secretary and Treasurer, E. W. Meyer. Government inspection. Refrigeration—90-ton Vilter and 60-ton Carbondale; brine and direct expansion. Boiler capacity, 425 H. P. (WSM) (WP).

J. W. Beardsley's Sons, 690 Frelinghuysen Ave. (WMD).

Herbst-Moch Co., 222 Frelinghuysen Ave. (WMD).

Holcombe Prov. Co., Inc., 74 N. Canal St. (WMD).

Fred Horns, 114 Mulberry St. (WMD).

Maybaum Packing Co., 14 Ward St. (WMD).

Newark Packing Co., 217 Astor St. (WMD).

Schreihofer Bros., Inc., 32 Center Market. (WMD).

NEW BRUNSWICK—Hy. Frank, 144 Paterson St. (WSM).

PENN'S GROVE—Matthew Mitchell & Son, 3 W. Main St. (WMD) (WSM).

PERTH AMBOY—Kellner Bros., Hall and Elizabeth Ave. (WSM).

PHILLIPSBURG—J. R. Shimer Co., 16 First St. A corporation. Capital, $200,000; issued, $67,600. Stockholders, 5. Employes, 28. Code—Cross. President, J. E. Carpenter; Vice-President, W. W. Bryan; Secretary, J. C. Duffin; Treasurer, J. E. Carpenter. Government inspection. Refrigeration—25-ton Huettemann; direct expansion. Boiler capacity, 100 H. P. (WSM) (WMD).

TRENTON—Ketterer's Son, Fred., 670 S. Broad St. (WSM).

Margerman Provision Co., 2 S. Broad St. (WSM).

Wagner & Sons, Chas., Chestnut and Toebling St. (WSM).

NUTLEY—Raritan Prov. Company. (WSM) (WP).

RIVERTON—A. M. Ellsworth, Inc., 100 Main St. (WMD) (WSM).

NEW MEXICO

GALLUP—Gallup Meat & Provision Co. (WSM).

ALBUQUERQUE—Farr Company, Wm. (WMD) (WSM).

CARLSBAD—Lowenbruck Bros., U. S. Market. (WSM) (WMD).

NEW YORK

ALBANY—Moch, F. E., 835 Broadway. (WSM).

AUBURN—Cayuga County Sausage Co. (WSM).

Myers Sausage Company. (WSM).

BINGHAMTON—Gruschwitz, Bruno, 130 Washington St. (WSM). J. P. Maxwell. (WSM).

BROOKLYN—Jacob Cohen, 137 Franklin St. (WSM).

Comer & Pollack, 179 Fort Green Place. (WSM).

Jacob Dangler & Son, 722 Myrtle Ave. Partnership. General Manager, Herman Sticht. Employes, 40. Government inspection. Refrigeration—20-ton Isabell-Porter; brine circulation. Boiler capacity, 125 H. P. (WSM).

Fenchs, Adolf, 231 Meserole St. (WSM).

Eatmore Provision Co., Inc., 488-490 Broadway. (WSM) (WP).

Adolf Gobel, Inc., Morgan Ave. and Rock St. (WSM).
Grozinger, Inc., Chris., Stockholm St. and Wilson Ave. (WSM).
Samuel Heymann, 5703 Fifth Ave. (WMD).
Chas. Hutwelker, 14-22 Hall St. (WMD).
Hygrade Provision Co., 131 S. 8th St. (WSM).
A. H. Lotz, 11 Chauncey St. (WSM).
Merkel Bros., Jamaica, L. I. (WSM).
L. Meyer Co., 374 Flushing Ave. (WSM) (WMD).
Musser & Co., 183 Fort Green Place. (WSM).
Ors & Co., Wyckoff Ave. and Greene St. (WSM).
Schaeffer & Deueke, 576 Woodward Ave. (WSM).
Sunshine Provision Co., 1988 Bergen St. (WSM).
Trunz, Max., 25 Lombard St. (WSM).
A. Aron, Inc., 335 Johnson Ave. (WMD).
D. Blumberg & Son, Dumont and Christopher Ave. (WMD).
Bushwick Pork Packing Co., 31 Bushwick Ave. (WMD).
Chieffetz & Greenberg, 264 Hudson Ave. (WMD).
Fred Figge, 285 Atlantic St. (WMD).
Franklin Provision Co., Inc. (WMD).
Aaron Levy & Co., 262 Hudson Ave. (WMD).
M. & D. Levy, 224 N. Ninth St. (WMD).
Philip B. Newmark, 1202 Metz St. (WMD).
Abraham Plaut, 307 Johnson Ave. (WMD).
George Schaefer, 575 Johnson Ave. (WMD).
Strauss, Schick & Strauss, 300 Johnson Ave. (WMD).
Wallabout Market Packing Co., Inc. (WMD).
Weill & Isaacs, 24 Hudson Ave. (WMD).

BUFFALO—Gerber, S. R. (WSM).
Kammann Company, John B., 445 Elliott St. Capital, $35,000; issued, $35,000. Stockholders, 3. Employes, 150. Codes—Cross and Utility. President, J. H. Kammann; Vice-President, L. E. Wilson; Secretary, L. M. Haas. Refrigeration—Two 50-ton Frick; direct expansion. Boiler capacity, 150 H. P. Wholesale meat and provisioners, and wholesale sausage makers. (WSM).
Klein, Andrew. (WSM).
Koehler Company, George, 109 Lovejoy St. Capital, $30,000; issued, $28,000. Stockholders, 4. President, Geo. Koehler; Secretary, Conrad Koehler; Treasurer, Mac Lunkenheimer. Employes, 20. Refrigeration—15-ton York; brine circulating. (WSM).
Lang, Gerhardt. (WSM).
Lang, Frank. (WSM).
Scherer, Frank. (WSM).
C. J. D. Packing Co., Inc., 88 Holt St. A corporation. Capital, $35,000. President and Treasurer, Christian J. Dressel; Secretary, William F. Dressel. (WMD).
Empire Beef & Provision Co. (WSM) (WP).
Fuhrmann, Louis P.—1014 Clinton St. Individual ownership. (WSM) (WMD).

Hoffman & Klinck, 526 Howard St. (WMD).
Everett C. Horlein, 16 Hannah St. (WMD).
Henry A. Kammann, 447 Bailey Ave. (WMD).
John H. Kammann Co., 445 Elliott St. (WMD).
John Klein, 940 Smith St. (WMD).
Alfred Milson, 1109 William St. (WMD).
Paul J. Schober, 618 Howard St. (WMD).
Seeger & Co., Inc., 231 Lewis St. (WMD).
M. L. Wallens, 176 Guildford St. (WMD).
B. Wertheimer, Lewis St. (WMD).
Edward Zier, William St. (WMD).

DUNKIRK—E. H. Group & Co. (WSM).

JAMAICA—Chas. Trautmann Co., 28 Division St. Capital, $55,000. President, Chas. Trautmann; Secretary and Treasurer, Peter Trautmann. (WMD).
Merkel Bros. (WSM).

KINGSTON—Roach Brothers, 38 Ann St. (WSM).

NEW YORK CITY—Brenzinger, Inc., G., 811 East 180th St., New York City. A corporation. Capital, $12,000; issued, $12,000. Stockholders, 4. Employes, 14. President, Gottlob Brenzinger; Vice-President, R. Recknagel; Secretary and Treasurer, Chas. Klotz and John Lang. (WSM).
Bronx Pkg. Co., 3257 3rd Ave. (WSM).
Bronx Provision Corporation, 143rd St. and 3rd Ave. (WSM).
Aaron Buchsbaum Co., 729 Ninth Ave. (WMD).
Burkle Bros., 45th St. and 11th Ave. (WSM).
Conron Bros. Co., 40 Tenth Ave. A corporation. Capital, $500,000; issued, $500,000. Stockholders, 10. President, Jos. Conron; Vice-President, Thos. Nash; Secretary and Treasurer, J. E. Conron and John J. Fitzgerald. Government inspection. Employes, 200. Codes—Cross and A. B. C. (WMD) (WSM).
Derby Co., H. C., 626 West 39th St. (WSM).
Cullman, Philipp, 515 East 19th St. Individual ownership. Employes, 6. Government inspection. (WMD).
Ederle Bros., 110 Amsterdam Ave. (WSM).
Ershowsky & Sons, Inc., 173 E. Houston St. Capital, $60,000; issued, $42,000. Stockholders, 8. President, Samuel Ershowsky; Vice-President, D. J. Ershowsky; Secretary, Chas. E. Ershowsky; Treasurer, S. Ershowsky. Employes, 50. Refrigeration—15-ton Voss; direct expansion. (WSM).
Bacharach, Milton, Inc., 370 Greenwich Ave. (WSM).
Chicago Sausage & Provision Co., 37 Ninth Ave. (WSM).
Fleck, Henry, 1679 Avenue A. (WSM).
E. Greenebaum Co., 328 E. 103rd St. (WMD) (WSM).
Hebrew National Sausage Factory, 155 E. Broadway. (WSM).
Marcus, Isidor, 44 Rivington St. (WSM).
Metzger, Felix, 1044 Second Ave. (WSM).
Sohn, L., 157 Broome St. (WSM).

Abramovitz, Isaac L., 144 West Ave. (WMD).
Abrams, Jos., 11 Thompson Ave. (WMD).
Abrams, M., 560 Brook Ave. (WMD).
Abt-Bernet, Inc., 626 Hegney Place. (WMD).
Adler, Gustave, 3 Grace Ave. (WMD).
Adler, Isidor, 176 2d Ave. (WMD).
Arlington Beef Co., 450 2d Ave. (WMD).
Astruck, Felix, 3 Thompson Ave. (WMD).
Back, Ernest, & Son, 452 Westchester Ave. (WMD).
Beinecke & Co., 184 Duane and 33 Great Jones Sts. (WMD).
Berliner, Nathan, 3 Hewitt Ave. (WMD).
Breidenbach, Gus., 511 E. 152d St. (WMD).
Christian, L. M., West Wash. Market. (WMD).
Crystal Market, 2776 Eighth Ave. (WMD).
Daitch & Marcus, 422 11th Ave. (WMD).
Danzig, A. & L., 502 E. 153d St. (WMD).
Davis, E. W., & Co., Foot of W. 39th N. R. (WMD).
Donovan & Meyer, 3 Loew Ave. (WMD).
Duncan, Clarence W., 20 Loew Ave. (WMD).
Empire City Beef Co., 48 Tenth Ave. (WMD).
Frank, A., & Son, 789 First Ave. (WMD).
Frank Emanuel, 18 Grace Ave. (WMD).
Frank & Co., 10 Loew Ave. (WMD).
Gansevoort Beef & Provision Co., 1 Lawton Ave. (WMD).
Geier, Harry, 6 Clinton St. (WMD).
Geier, Wm., 166 Suffolk St. (WMD).
Gillen, John, West Wash. Mkt. (WMD).
Gold, S., & Co., 3 Thompson Ave. (WMD).
Goldberg, H., 15 Lawton Ave. (WMD).
Golden Packing Co., Inc., 53 Little W. 12th St. (WMD).
Goodman, A., 924 Sixth Ave. (WMD).
Greenbaum & McKelvey, Washington Market. (WMD).
Greenberg, Morris, West Washington Markt. (WMD).
Greenwald, Al., 665 Brook Ave. (WMD).
Greenwald & Marcuse, Inc., 528 Westchester Ave. (WMD).
Gurry, Patk., Inc., 84 Barclay St. (WMD).
Halem, J., 211 E. Houston St. (WMD).
Heim, I. Julius, 619 Ninth Ave. (WMD).
Interborough Beef Supply Co., 367 Hudson St. (WMD).
Kahl, A., Washington Market. (WMD).
Kahn, Adolf, 18 Thompson Ave. (WMD).
Kansas Beef & Provision Co., 475 Ninth Ave. (WMD).
Kestenbaum & Newmark, 1696 Washington St. (WMD).
Kestler, John, 68 First Ave. (WMD).

Klein & Cedrone, 495 E. 152d St. (WMD).
Kohn & Sklansky, 77 E. 106th St. (WMD).
Korn, G., 185 Avenue C. (WMD).
Kornblum, Meyer, Grace and West Sts. (WMD).
Kotler, David, & Sons, 350 Madison St. (WMD).
Lazarowitz, I., 61 E. 95th St. (WMD).
Levy Co., 969 First Ave. (WMD).
Levy, Milton, 17 Thompson St. (WMD).
Levy, Sol., 102 Gansevoort St. (WMD).
Lewis, G. H., & Sons, West Washington Market. (WMD).
Lowenstein, J., & Son, Inc., 252 Ninth Ave. (WMD).
Manhattan Beef & Provision Co., 472 Ninth Ave. (WMD).
Maybruck & Heller, 120 E. 108th St. (WMD).
Metropolitan Beef & Supply Co., 2705 Third Ave. (WMD).
John Minder & Son, Inc., 97 Barclay St. (WMD) (WSM).
Moritz, Ferd., 7 Thompson Ave. (WMD).
Moritz, Simon, 7 Thompson Ave. (WMD).
Mosner, L., 561 Brook Ave. (WMD).
Mosner, Max., 532 Webster Ave. (WMD).
Murray, Frank J., Co., 78 Barclay St. (WMD).
North River Beef Co., 687 Ninth Ave. (WMD).
O'Mara, M. T., Co., Inc., Washington Market. (WMD).
L. Oppenheimer, Inc., 609 W. 130th St. (WMD).
Port Morris Packing House, 686 E. 134th St. (WMD).
Preiser, Samuel M., 795 E. 158th St. (WMD).
Reitman, Aron, 143 Orchard St. (WMD).
Reitman, Jacob, 86 Ridge St. (WMD).
Sayles-Zahn Co., Sixth Ave. and Tenth St. (WMD).
Schrag & Muth, 647 W. 39th St. (WMD).
Silberman, L. I., Grace Ave. (WMD).
Silverman & Silverman, 155 E. 110th St. (WMD).
Sperling & Worshoufsky, 107 E. 110th St. (WMD).
Strassburger, H., 497 E. 152d St. (WMD).
Strauss, A., New West Washington Market. (WMD).
Strauss, J., 28 Loew Ave. (WMD).
Tanklefsky, A. and D., 3898 3d Ave. (WMD).
Tauky, Henry, 1537 Avenue A. (WMD).
Third Avenue Beef Co., 63 Third Ave. (WMD).
Tremont Packing House, 826 E. Tremont Ave. (WMD).
Wald, J., 561 Brook Ave. (WMD).
Wallace, J. B., Co., 94 Barclay St. (WMD).
Warburg, P., West Washington Market. (WMD).
West Side Beef Co., 432 Amsterdam Ave. (WMD).
Winant, D., Inc., 178 Front St. (WMD).
Zwiren, Jos., 451 Westchester Ave. (WMD).

Ernst & Son, L., 670 Morris Park Ave. (WSM).

Gordon, Sam, 62 East 110th St. Individual ownership. Government inspection. Refrigeration—10-ton Wolf-Linde; brine circulation.

Greenbaum, Inc., Edw., 328 E. 103d St. (WSM).

Guckenheimer & Hess, Inc., 81 Third Ave. A corporation. Capital, $250,000; issued, $30,000. Stockholders, 2. President, Edw. Guckenheimer; Vice-President, Adolph Hess; Secretary, Sidney Hess; Treasurer, Adolph Hess. Employes, 85. Government inspection. Refrigeration—9-ton Vesterdahl; brine circulating.

Habich & Son, Adolf, 640 Tenth Ave. (WSM).

Kast, Hy., 277 Greenwich St. (WSM).

Kern, Inc., Geo., 344 West 38th St. A corporation. Capital, $350,000; issued, $250,000. Stockholders, 7. President, Geo. Kern, Sr.; Vice-President, Geo. Kern, Jr.; Secretary, J. B. Hallinan; Treasurer, Frank M. Firor. Employes, 120. Refrigeration—50-ton Wolf and 25-ton Voss; brine circulating. Boiler capacity, 750 H. P.

Keiser & Sons, J., 1507 Tenth Ave. (WSM).

Manthe Bros., 42d St. and Tenth Ave. (WSM).

Meier & Son, B., 516 Westchester Ave. Partnership. General Manager, E. F. Meier. Government inspection. (WSM) (WMD).

Melchner, John, 16 Tenth Ave. (WSM).

Oceanic Cheese & Sausage Co., Inc., 46 Jay St. A corporation. Capital, $50,000; issued, $50,000. Stockholders, 3. Employes, 14. Codes—A. B. C., 5th edition, and Bentley's; code word, "Cinaeco." President, E. R. Milhisen; Vice-President, H. I. May; Secretary, A. Abrahams; Treasurer, H. I. May. (WSM).

Ottmann & Co., Wm., 207 Water St. (WMD) (WP).

Frank Moe, Eleventh Ave. and 39th St. (WMD) (WSM).

Otto Stahl, Inc., 155 E. 126th St., 167 E. 127th St., and 2333 Third Ave. A corporation. Capital, $250,000. President, Otto Stahl; Vice-President, August Stahl; Secretary, Treasurer and General Manager, Geo. A. Schmidt. (WSM) (WMD).

Anderson & Tarbon, 482 Austin Place, Bronx. Government inspection. (WSM).

ROCHESTER—Zwiegle Bros., 210 Joseph Ave. Employes, 6. (WSM).

PORT CHESTER—Vahsen, Chas., 359 Willett Ave. (WSM).

POUGHKEEPSIE—Knauss Bros. (WMD).

ROME—Darlington, Geo. E., 320 W. Thomas St. (WMD).

SCHENECTADY—Behan, E. M., 766 State St. (WSM).

Geiser, A., 1118 Albany St. (WSM).

SYRACUSE—Hansen & Co., P. (WSM).

UTICA—Henry Hoffman, 707 South St. Individual ownership. Employes, 14. Refrigeration—2-ton Brunswick; direct expansion. (WSM) (WMD).

Lee & Sons, A. (WSM).

A. Scala & Son, 713 Bleecker St. (WP).

WATERTOWN—**Empire Provision Co.**, 430 Light & Power Bldg. Individual ownership. Codes—Cross, Robinson and Millers. (WP).

WEST ALBANY—**Bennett Brothers.** (WSM).

LITTLE FALLS—**Zoller, Jacob, Co.**, East Mill St. A corporation. Capital, $100,000. President, J. I. Zoller; Vice-President, Abram Zoller; Secretary and Treasurer, Tom J. Zoller. (WSM) (WMD).

TROY—**Fritz Helmbold.** (WSM).

NORTH CAROLINA

ASHEVILLE—**J. A. Baker Packing Co.**, 334 W. Haywood St. (WSM) (WMD).

CHARLOTTE—**Weber, J. G.** (WSM).

KINGSTON—**Hooker & Company.** (WMD).

KINGSTREE—**H. A. Miller.** (WSM) (WMD).

WILSON—**Ed. Lamm.** (WMD) (WP).

OHIO

AKRON—**Akron Sausage Co.**, 713 Bowry St. A corporation. Capital, $25,000. President, A. Steidle; Vice-President, Frank Abrahams; Treasurer, Mike Kapels. (WSM).

ALLIANCE—**Alliance Cold Storage & Packing Co.** (WMD).
Fairmount Provision Co. (WMD).

CANTON—**Stark Provision Company**, 1018 McKinley, S. W. Railroad, Pennsylvania. Ownership, individual. General Manager, Harry Lavin. Dept. Head, Arthur Lavin. Trade Marks, "Sugardale" and "Stark." (WSM).

CINCINNATI—**John Blackburn**, 2124 Baymiller St. (WMD).
Chas. A. Fruend, Rachel and Henshaw. (WMD).
Sam Gall, 2121 Freeman Ave. (WMD).
Herman Kemper, 3068 Sidney Ave. (WMD).
Robert Meyer & Son, 3095 Colerain Ave. (WMD).
Henry Meyer's Sons, 2855 Sidney Ave. (WMD).
People's Packing House Co., 123 W. Elder St. (WMD).
Albert F. Settelmayer, 690 Riddle Road, Clifton Heights. (WMD).
H. F. Busch Co., Vine and Fourteenth Sts. (WSM).
Tom Jones Products Co., E. Third St. Trade Mark, "Tecjay." (WSM).

CLEVELAND—**Brookside Sausage Co.** (WSM).
Buechler, Henry. (WSM).
Cleveland Delicatessen Co. (WSM).
Gutscher Co., Theodore, 2129 W. Nineteenth St. A corporation. Capital, $25,000; issued, $17,900. Stockholders, 12. Employes, 30. President, Theo. Gutscher; Secretary and Treasurer, O. L. Fricke. Refrigeration—7-ton York. (WSM).
A. Habermann Provision Company, 2302 Broadway. A corporation. Capital, $25,000; issued, $25,000. Stockholders, 5. Employes, 45. President, Mrs. F. Habermann; Secretary, C. L. Habermann; Treasurer, J. J. Naegele. Refrigeration—One 42 and one 10-ton Frick; direct expansion. (WSM) (WP).
Hildebrandt Provision Company, 3620 Clark Ave. A corporation.

Capital, $65,000; issued, $64,000. Stockholders, 7. President, C. R. Hildebrandt; Secretary and Treasurer, Hugo A. Hildebrandt. Employes, 120. Code—Cross. Refrigeration—40-ton Huettemann & Cramer and 75-ton Triumph; direct expansion. (WSM) (WP) (WMD).

Pavelka Bros. (WSM).
Anglo American Pork Products Co., 63 Wade Bldg. (WMD).
John H. Bennett, 3261 W. 65th St. (WMD).
Benson & Co., 1211 St. Clair Ave. (WMD).
Buechler-Jaeger Sausage Co., 3675 Fulton Road. (WSM).
E. O. W. Castle, 68th and Big Four. (WMD).
Citizens Provision Co., 2291 E. Fourth St. (WP) (WMD).
Cleveland Hotel Supply Co., 421 Woodland Ave. (WMD).
East Cleveland Provision Co., 2602 Payne Ave. (WP) (WMD).
J. J. Flick Dressed Beef Co., 3378 W. 65th St. (WMD).
Fromson & Davis, 3261 W. 65th St. (WMD).
A. Hammond, 624 Bolivar Road. (WMD).
J. H. & R. Hartman, 7 Bolivar Road. (WMD).
Hartman Provision Co. (WMD).
Home Packing Co., 3979 W. 25th St. (WMD).
Oliver C. Hughes Co., W. 68th St. (WMD).
Hughes Provision Co., 2291 E. Fourth St. (WMD).
Hughes & Castle, 3207 W. 65th St. (WMD).
Koblenzer Bros., 2315 E. Fourth St. (WMD).
Phil Null & Co., 424 Bolivar Road, S. E. (WMD).
Pinkett Commission Co., 417 Bolivar Road. (WMD).
Retail Butchers' Protective Ass'n Co., 3199 W. 65th St. (WMD).
Schlichting Meat Products Co., 3596 E. 48th St. (WMD).
A. E. Schultz Co., 509 Bolivar Road. (WMD).
Superior Provision Co., W. 79th and Nickel Plate R. R. (WMD).
Webb Bros., 3261 W. 65th St. (WMD).

COLUMBUS—Kosher Packing Co., 750 Elmore Ave. (WMD).
Herman Falter, 378 Greenlawn Ave. (WMD).

DAYTON—G. & O. Baumkechel. (WMD).
Henry Blust. (WMD).
Bueker Packing Co., Rappee Ave. (WMD).
Champion Meat Co., W. Riverview. (WMD).
Charles Hambrecht, 415 N. Brandt St. (WMD).
A. Hasenstab Sons, 433 N. Valley St. (WMD).
Joseph Schmeider, 376 E. Pruden. (WMD).
Alvin Jacobs, 316 S. Williams St. (WMD).

HUBBARD—Wm. Weitz. (WMD).
H. B. Phillips. (WMD).

LORAIN—Lorain Provision Co. (WP).
MASSILLON—Graber, Lee. (WSM).
SIDNEY—Sidney Packing Co. (WMD) (WSM).

TOLEDO—Jacob Folger, Phillips Ave. (WMD).
YOUNGSTOWN—Moog, Karl. (WSM).
NEW PHILADELPHIA—The Only Sausage Company. (WSM).
STEUBENVILLE—Buckeye Sausage Company. (WSM).
XENIA—Anderson Wholesale Meat Co., Bellbrook Ave. (WMD).

OKLAHOMA

OKLAHOMA CITY—Harris Meat Co., Grand and Western Aves. (WSM).
    Huerding Bros., 35 Harrison Ave. (WSM).
    Schwab & Co., 1101 Linwood Ave. (WSM).

OREGON

PORTLAND—Albina Cash Market, Russell and Vancouver Sts. (WSM) (WMD).
    Howitt Co., T. R., Front and Adler Sts. (WMD).
    Montvilla Market, 80th and Stark Sts. (WMD).
PORTLAND—United Meat Co., Box 117, Kenton Station. (WSM).
SALEM—Wait, E. D. (WMD).

PENNSYLVANIA

EASTON—M. E. Sampson Estate, 149 Nesquehoning St. (WSM).
    Schafer, H. S., Coal and Belmont Sts. Government inspection. Refrigeration—6-ton Brecht; direct expansion. (WSM).
ERIE—Schlaudecker Bros. (WSM).
FARRELL—Sam Schermerl. (WMD) (WSM).
HAZELTON—Edwin W. Reese & Sons. (WMD).
JOHNSTOWN—Berkebile, R. L. (WSM).
LEBANON—Lebanon Bologna & Provision Company, Eleventh Ave. and Lebanon St. A corporation. Capital, $30,000; issued, $30,000. Stockholders, 3. President, Ira A. Newman; Vice-President, E. H. Rieser; Secretary, D. B. Buck; Treasurer, J. H. Hilbert. Employes, 20. Government inspection. (WSM).
    Bucks & Co., Inc. (WSM).
    Weaver, John S. (WSM).
PALMYRA—Palmyra Bologna Company. Railroad, Philadelphia & Reading. Employes, 12. General Manager, H. L. Seltzer. Government inspection. (WSM).
PHILADELPHIA—Pincus, B. S., 222 N. Delaware Ave. Individual ownership. Employes, 40. Government inspection.
    Kabisch & Company, Inc., 54th and Wyalusing Ave. (WMD) (WSM) (WP).
    Moland's Sons, Wm., 120 Market St. Partnership. (WP).
    Penn Beef Company, 48 N. Delaware Ave. A corporation. Capital, $50,000; issued, $50,000. President, F. D. Ellis. (WSM) (WMD).
    Standard Provision Company, 212 N. Front St. (WSM).
    Karl Seiler & Sons, 4051 N. Fifth St. Employes, 25. (WMD).
    Wilson & Rogers, Inc., 134 W. Delaware Ave. A corporation. Capital, $100,000; issued, $100,000. Employes, 15. President, M. M. Jones; Vice-President, B. Drake; Secretary, H. Williams; Treasurer, H. W. Hardy. Government inspection. (WSM).

PITTSBURGH—Italian Sausage & Provision Co. (WSM) (WP).
LOCK HAVEN—F. L. Winner. (WSM).
MASONTOWN—Lofstead, Frank. (WSM).
WILKES-BARRE—A. Percy Brown, 26 E. Northampton St. (WMD).
  Diamond City Beef Co., 54 S. Pennsylvania Ave. (WMD).
  Lehigh Beef Co., 59 N. Pennsylvania Ave. (WMD).
  Wilkes-Barre Beef Co., 128 E. Market St. (WMD).
  Wyoming Valley Beef Co., 48 S. Pennsylvania Ave. (WMD).

RHODE ISLAND

PROVIDENCE—Meinel, C. F., 35 Dike St.
  Lippman Bros., 230 Union Ave. Railroad, New York, New Haven & Hartford. Partnership. Employes, 16. (WSM).
  Saugy, Inc., A., Canal St. (WSM).
  Schott Suter Co., 52 Randall St. (WSM).

SOUTH CAROLINA

CHARLESTON—Avenue Market, 210 Rutledge Ave. A corporation. Capital, $16,000; issued, $16,000. Stockholders, 3. Employes, 25. President, I. Weinberg; Vice-President, L. Weinberg; Secretary, C. L. Pearlstine; Treasurer, I. Weinberg.
GREENVILLE—C. B. Osborne & Co. (WP).

SOUTH DAKOTA

ABERDEEN—Welsh Market, A. H. Hardt, Prop. (WMD) (WSM)
HURON—The Lampe Company, Inc., 262 Dakota Ave. President, Albert Lampe, Sr.; Vice-President, Fred Lampe; Secretary-Treasurer, Henry Lampe. (WMD).

TENNESSEE

CHATTANOOGA—Manz, E. H. (WSM).
  Stolz, Eugene. (WSM).
FAYETTEVILLE—Gray Bros. (WSM).
JACKSON—Barnes, G. W. (WSM).
KNOXVILLE—Acker, J. (WSM).
MEMPHIS—Memphis Sausage Works. (WSM).
NASHVILLE—Jacob's Bros. (WSM).
  Thompson Bros. (WSM).

TEXAS

DALLAS—Empire Beef & Provision Co., 515 S. Envoy. (WMD).
HOUSTON—Dixon Packing Co., Inc., 106 Milam St. A corporation. Capital, $20,000; issued, $20,000. Stockholders, 4. Employes, 25. President, S. F. Dixon; Vice-President, A. Charney; Secretary and Treasurer, T. K. Dixon. (WSM) (WMD).
  Texas Union Packing Company. (WMD).
SAN ANTONIO—Hammer, J. T., 109 W. Pecan St. (WSM).
  Apache Packing Co., 1200 Tampico St. (WMD).
  Cohen-Bible Meat Co., S. San Jacinto St. (WMD).
  Ducoz & Martinez, 834 S. Laredo St. (WMD).
  Mission Provision Co., Yoakum Bend. (WMD).
  S. A. Packing Co., Tampico St. (WMD).

Wiegand Sausage Works, 1100 S. Laredo St. (WSM).
Joe Monsalvo, Union Stock Yards. (WMD).
VICTORIA—Simmons Sanitary Meat Market, 504. S. Main St (WSM).
FORT WORTH—Texas Dressed Beef Co. (WMD).
WACO—Brazos Packing Co. (WMD).
Robinson Packing Co. (WMD).
F. A. Waldrop. (WMD).

UTAH

OGDEN—The Keller Dressed Meat Company. (WMD).
MIDVALE—Joseph S. Wells. (WMD) (WSM).
PROVO—Provo Meat & Packing Co. (WMD).
Utah County Wholesale Meat Company. (WMD).
SALT LAKE CITY—Modern Sausage Factory, 229 S. Second St., West. (WSM).
Salt Lake Sausage Factory, 224 E. Fifth St., South. (WSM).
Success Market, Inc., 26-28 W. First South St. (WMD) (WSM).

VIRGINIA

BEDFORD—Bedford Market Co. (WMD).
RICHMOND—V. Heckler Packing & Commission Co., Seventh and Canal Sts. Individual ownership. Employes, 12. (WMD).
DANVILLE—Haraway, J. W. (WSM) (WMD).
NORFOLK—Virginia Lard & Provision Co., Inc., 210 Water St. (WMD).

WASHINGTON

ANACORTES—Anacortes Meat Company. (WSM) (WMD).
BURLINGTON—Wollen & Son, C. E. (WSM).
CENTRALIA—Centralia Meat Company. (WSM).
CHEWELAH—Chewelah Meat Co. (WSM).
EAST STANWOOD—Rygg Bros., Inc. Capital, $30,000. (WMD).
LIND—Pool, E. C. (WSM).
SEATTLE—Augustine & Kyer. (WSM) (WMD).
Pacific Meat Company. (WSM).
Yukon Meat Company. (WSM) (WMD).
SPOKANE—Welch's Market. (WSM).
TACOMA—Fern Hill Market, Fern Hill Station. (WMD).
WENATCHEE—Inland Meat Company. (WMD).
Wenatchee Meat Company. (WMD).

WISCONSIN

EAU CLAIRE—A. F. Schwahn & Sons Co., 320 Barstow St. (WSM).
LA CROSSE—La Crosse Sausage Factory. (WSM).
GREEN BAY—Platten Bros., 413 Dousman St. Partnership. Railroad, Chicago & North Western. Employes, 30. (WMD).
MILWAUKEE—Frank & Company, 742 Market St. A corporation. Capital, $150,000; issued, $150,000. Stockholders, 40. Employes, 50.

Codes—Cross, A. B. C. and W. U. President, W. Frank; Vice-President, Edgar Herzberg; Secretary and Treasurer, C. S. Perego. Government inspection. Refrigeration—50-ton Vilter; direct expansion. (WSM) (WMD).

Hess, Chas., 802 Third St. (WSM).

Milwaukee Sausage Company, 926 Center St. (WSM).

Schaaf, Frank, 71 Second St. (WSM).

Usinger, Fred, 302 Third St. (WSM).

Weisel & Company, foot of Humboldt St. A corporation. Capital, $200,000; issued, $200,000. Stockholders, 20. Employes, 70. President, Carl Weisel; Secretary, C. Friedrich; Treasurer, A. Weisel. Government inspection. Refrigeration—20-ton Vilter. (WSM).

Born & Son, August, 794 Teutonia St. (WMD).

Donnar, Oscar, 1055 Third Ave. (WMD).

Cross & Bros. Co., F. C., Muskego Ave. (WMD).

Gumz & Co., R., 125 Muskego Ave. (WMD).

Elschner, Louis, 408 Walker St. (WSM).

Grubecki, B., 647 Third Ave. (WMD).

Heusler Sausage Factory, 2030 Chambers St. (WSM).

Haefner, Wm. E., 921 Scott St. (WMD).

Hermann, G. E., 485 American Ave. (WMD).

Hiller, W. C., 2011 Vliet Ave. (WMD).

Jordan, M. G., 31 Martin Ave. (WMD).

Lins, Fred E., 1202 Burleigh. (WSM).

Luck & Kamesar, 715 Fourteenth St. (WMD).

Milwaukee Kosher Sausage Co., 1002 Galena Ave. (WSM).

Milwaukee Boiled Ham Co., 901 Holton Ave. (WMD).

Nieske & Son, Fred, 1522 Center St. (WSM).

Noebre & Co., Emil F., 1044 National Ave. (WSM).

Peck & Son, B. (WMD).

Quality Products Co., 638 Arthur Ave. (WSM).

Wagner, Alvin, 1010 Concordia. (WSM).

Wisconsin Sausage Mfg. Co., 418 National Ave. (WSM).

Zastrow, Henry, 1318 Teutonia Ave. (WSM).

Zitron Bros., Muskego and Canal. (WMD).

OSHKOSH—The Waas-Cain Co. (WSM).

WEST VIRGINIA

CHARLESTON—Fisher & Fruth Meat Co. (WMD).

HUNTINGTON—Tri-State Sausage & Provision Co., 1501 Jefferson Ave. (WSM).

MARTINSBURG—The Weller Bros., N. Queen St. (WMD).

MARTIN'S FERRY—Heil Packing Co. (WMD).

MORGANTOWN—Lough Brothers Co. (WMD).

WELLSBURG—West Packing Co. (WMD).

WHEELING—F. Weimer Sons. (WMD).

RENDERERS

[The majority of renderers are also dealers in country hides and tallows]

ALABAMA

BIRMINGHAM—Birmingham Rendering Co.
Birmingham Hide & Tallow Co.
DECATUR—L. Simrell.
HUNTSVILLE—Huntsville City Abattoir.
NEW DECATUR—R. W. Holland.
ANNISTON—Anniston Hide & Tallow Company.

CALIFORNIA

ALBANY—Chas. Willinger & Co.
AZUZA—George W. Fuhr.
BENECIA—Kullman-Salz Co.
CHICO—H. R. Meade.
COLFAX—D. A. Russell.
COLMA—San Mateo Reduction Works.
EAST BAKERSFIELD—Geo. W. Foo.e.
EMERYVILLE—Chin Took & Cook.
EMERYVILLE STOCKYARDS—Bayle-LaCoste Co.
Peterson Tallow Co.
FRESNO—Fresno Soap Co.
Frisco Cash Market.
San Joaquin Reduction Co., P. O. Box 823.
HANFORD—J. P. DeMont Hanford Tallow Works.
HAYWARD—Hellwig & La Grave.
JACKSON—Geo. L. Thomas.
LANKERSHIM—Lankershim Pkg. Co.
LOS ANGELES—Los Angeles Fertilizer Co., 2643 E. 25th St.
California Rendering Co.
MONTEREY—Mammalian Fish Products Co.
Monterey Fish Products Co., Ocean Ave.
LOS ANGELES—Pacific Reduction Co.
Union By-Products Co.
MODESTO—Modesto Tallow & Reduction Works.
MOSS LANDING—California Sea Products Co.
NEWMAN—J. P. De Mont.
OAKLAND—W. Coast Soap Co., 26th & Poplar Sts.
Oakland Reduction & Fertilizer Works, 2228 Rosedale Ave.
SACRAMENTO—Sacramento Reduction Works.
SAN DIEGO—Chas S. Hardy.
SAN FRANCISCO—California Fertilizer Works, 1332 Evans Ave.
California Tallow Works, 214 Front St.
Hilson & Hildebrand.
Rathjens & Kupfer.
Royal Tallow Works, 1490 Evans Ave.
So. San Francisco Tallow Works, 1420 Evans Ave.
Western Reduction Company.
Western Tallow Co., 1499 Evans Ave.
Union Products Corp., Evans Ave. and Keith St.
SAN JOSE—B. La Clergue & Co., R. F. D. Box 471.
SAN PEDRO—Nielsen & Kittle Canning Co.
SAN RAFEL—M. N. Schaefer Co.
STOCKTON—Stockton Tallow Works Co.
TURLOCK—Tom Bozinni.

COLORADO

DENVER—Capitol Rendering Company, Stock Yards Station.
Denver Soap & Manufacturing Co.
Ruddy Rendering Company, Stock Yards Station.
Union Rendering Company.
FT. COLLINS—Lormie Company.
GRAND JUNCTION—W. L. Peach Rendering Company.
GREELEY—Greeley Rendering Company.
MONTE VISTA—Busch-Nelson Rendering Company.
PUEBLO—Wm. Comerford Rendering Co.
Pueblo Rendering Company.
ROCKYFORD—Stauffer Pkg. Co.
STERLING—A. Schmidt Rendering Co., 213 So. Front St.

CONNECTICUT

ALLINGTOWN—Connecticut Fat Rendering & Fertilizer Corp.
BRIDGEPORT—Bridgeport Tallow Company.
Chas. Fischer & Company.
BRISTOL—R. W. Williams.
MIDDLETOWN—Wm. Allison.
Allison Bros. Soap Co.
Curtis C. Camp.
Rogers & Hubbard Company.
MYSTIC—Mystic Rendering Co.
Wilcox Fertilizer Company.
NEW HAVEN—Frank S. Platt Co.
L. T. Frisbie Company.
New Haven Rendering Co.
NEW LONDON—J. B. Brigham.
NORWICH—M. E. Morse.
ORANGE—New Haven Rendering Co.
S. W. Woodruff & Sons.
ROCKFALL—Rogers Mfg. Co.
STRATFORD—Jos. A. Allard.
TORRINGTON—Daniel Pullin, So. Main St.
WARRENVILLE—E. L. James.
WATERBURY—Waterbury Rendering Co.
WILLIMANTIC—H. A. Bugbee.

WASHINGTON, D. C.

Milton Hopfenmaier.
Norton & Company.
Washington Abattoir Co.

DELAWARE

FREDERICA—L. C. Rogers.
LEWES—Lewes Fisheries Co.
MILFORD—Draper Davis Co.
MILTON—Milton Fertilizer Co., Inc.
SMYRNA—Lewis M. Price.
WILMINGTON—James Finan.
Wilmington Sanitary Co.
Oil Seeds Company.

FLORIDA
TAMPA—Southern Tallow Co.

GEORGIA
ATLANTA—Schoen Bros., Inc.

IDAHO
KIMBERLEY—Jack France & Sons.

ILLINOIS
AURORA—A. Rogers.
BELLEVILLE — Belleville Rendering Company.
  Johnson Rendering Company.
BELVIDERE—Frank Rogers.
BLOOMINGTON—Geo. Agle & Sons, Inc.
CHICAGO—A. M. Adler & Co., 11 So. La Salle St.
  Adler & Oberndorf.
  John S. Camrell & Co., 32nd and So. Robey Sts.
  Darling & Co., U. S. Yards.
  Fitzpatrick Bros., 1319 W. 32nd Place.
  John Fitzpatrick & Co., 2800 So. Western Ave.
  Globe Rendering Co., U. S. Yards.
  General Rendering Company, 4100 So. Ashland.
  Hines & Co., U. S. Yards.
  Hine Bros. & Co.
  Michael Kirchhoff & Son.
  J. Scannell & Company.
  Wm. D. Smith Co., 4419 So Ashland Ave.
DANVILLE—Danville Rendering Company.
  Max Hodges.
DE KALB—Wm. Ballon.
DIXON—Peter McCoy.
EARLVILLE—Bert. Gould.
EAST ST. LOUIS—East St. Louis Rendering Company.
FREEPORT—Freeport Oil & Rendering Works.
  Jno. Hartmann.
  R. H. Hoover.
  Hoover & Isaac.
GALESBURG—Carl Haggenjos.
  Knox Rendering Company.
GENESEO—Fred R. Waters.
HENRY—Henry Rendering Co.
JACKSONVILLE—Ben Cohen.
JOLIET—Goss & Company.
  Joliet Rendering Co.
KANKAKEE—Geo. Drummond.
  Kankakee Rendering Wks.
KEWANEE—Cloustoun Rendering Co.
  Kewanee Rendering Company.
MONMOUTH—Monmouth Rendering Co.
MENDOTA—Rogers & Company.
OREGON—W. J. Salisbury.
OTTAWA—Ottawa Rendering Co.
  Frank Peltillion.
PARIS—Dr. Wm. Goff.
  W. H. Hoff Fertilizer Co.
PEORIA—Faber & Company.
PONTIAC—C. R. Tracy.
QUINCY—Quincy Soap Co.
ROCHELLE—John La Forge.
ROCKFORD—J. T. La Forge & Sons.
  Rockford Rendering Wks.
  Trenholm Rendering Works.
ROCK ISLAND—Twin City Rendering Company.
SANDWICH—E. Eva.
SAVANNAH—W. H. Griffith.
SPRINGFIELD—Thos. Walls Sons.
SPRING VALLEY—Jno. Wishnesky.
STREATOR — Klein Bros. Rendering Works.
UTICA—A. R. McDowell.
VILLA GROVE—Chas. Anderson.
WILMINGTON — Wilmington Hide & Tallow Works.

INDIANA
ALBANY—Albany Tanning Co.
ALEXANDRIA — Alexandria Fertilizer Works.
ANDERSON—Anderson Fertilizer Company.
ANGOLA—Angola Reduction Co.
AUBURN JUNCTION—DeKalb Tanking Company.
BAINBRIDGE—Collings Reduction Company.
BATESVILLE—E. F. Garinger.
BEDFORD—Bedford Reduction Co.
BICKNELL—Bicknell Abattoir Fertilizer Co.
BLOOMINGTON — Bloomington Reduction Company.
BRAZIL—Brazil Tankage Co.
  Stephenson & Armstrong.
BROOKSTON — Brookston Fertilizer Company.
BROOKVILLE—John Bunz.
CASTLETON—S. E. Test.
CAYUGA—Cayuga Tankage Co.
COLUMBUS—Columbus Sanitary Reduction Company.
CLINTON—John Gill.
CONNERSVILLE—Morris Cohen.
  Connersville Reduction Co.
CONVERSE — Louis Price Fertilizer Company.
CORYDON—Louis Quibbeman & Son.
CRAWFORDSVILLE—G o l d b e r g & Pearlman.
DANA—R. L. Cormack.
  Thos. Paul.
DECATUR—Decatur Fertilizer Co.
DELPHI—Barnhard Fertilizer Co.
EDINBURG—Blue River Reduction Company.
  Essex Bros.
ELKHART—Otis High.
EVANSVILLE — Interstate Rendering Company.
FARMERSBURG — Westfort Fertilizer Company.
FLORA—Flora Fertilizer Company.
FORTVILLE—Fortville Rendering Company.
FORT WAYNE—Farmers Chemical Fertilizer Company.
  Fort Wayne Rendering Co.
  Maier Hide & Fur Co.
  J. & H. Stadler Rendering Company.
  Weil Bros & Co.
FRANCISVILLE — Hubbell Fertilizer Company
FRANKFORT—Barnhard Fertilizer Company.
  Clinton Mfg. Company.
FRANKLIN—J. E. Walker Fertilizer Plant.

GENEVA—Zaggal Fertilizer Plant.
GOSHEN—Gerrit W. Clason.
GREENBURG—Greenburg Fertilizer Co.
  Geo. S. Littell.
  Robert A. Roberts.
GREENFIELD—Hancock Fertilizer Co.
HAGERSTOWN—Augustus Weidman.
HARRISON—Ed. Schwing.
HENRYVILLE—Cummings Bros.
HUNTINGTON—Huntington Fertilizer Company.
INDIANAPOLIS—R. R. Belt & Stock Yard Co.
  M. L. Goldberg & Son.
  Indianapolis Reduction Co.
  Pitman-Moore Co.
  E. Rauk & Sons Co.
  J. Wachtel Rendering Wks.
JEFFERSONVILLE—Mortz Willinger.
JONESBORO—C. L. McDonald.
JUDSON—Clare Connelly.
KLENDALLVILLE—Klendallville Fertilizer Company.
KENTLAND—Interstate Rendering Company.
  Newton Reduction Company.
KOKOMO—Barnhard Fertilizer Co.
  Kokomo Fertilizer Co.
KOUTS—Watson & Herring.
LAFAYETTE—Barnhard Fertilizer Company.
  M. & J. Schnaible Co.
LEXINGTON—Walter Parks & Son.
LIBERTY—Drook Rendering Co.
  Liberty Rendering Plant.
LOGANSPORT—The Heppe Co.
LOOGOOTEE—C. C. James & Co.
LOWELL—Kenney Bros. Reduction Co.
LYNN—S. O. Adams.
MARION—Goldreich Fertilizer Company.
MARKLE—Markle Feeding & Rendering Company.
MARTINSVILLE—C. F. Schnaiter.
MICHIGAN CITY—Michigan City Reduction Company.
MIDDLETON—F. B. Huff.
  Middleton Reduction Company.
MONTICELLO—Barnhard Fertilizer Company.
MONTPELIER—Montpelier Fertilizer Company.
MORGANTOWN—J. Haggard & Son.
MUNCIE—Caldwell Tankage Co.
  Muncie Tanking Company.
  Tanking & Fertilizer Company.
NEW ALBANY—Conrad Kammerer Glue Co.
NEW CASTLE—Hansard & Pickle.
  C. M. Hauser & Son.
NOBLESVILLE—C. C. James & Co.
  Wilson Bros. Fertilizer Co.
NORTH MANCHESTER—North Manchester Fertilizer Company.
ODON—Odon Reduction Co.
ORESTES—Orestes Fertilizer Co.
PERU—Peru Fertilizer Co.
PLAINFIELD—Plainfield Reduction Company.
  Verl Crews.
PLYMOUTH—Plymouth Fertilizer Company.
PORTLAND—R. F. Bone.
  Portland Fertilizer Company.
RENSSELAER—Jasper Reduction Co.
RICHMOND—Clendenin & Co.
ROCHESTER—Rochester Fertilizer Company.
RUSHVILLE—Bursback Fertilizer Co.
SCIPIO—Scipio Fertilizer Co.
SEYMOUR—F. F. Buhner Fertilizer Co.
SHELBYVILLE—Robert Bardback Sons Co.
SHIRLEY—Shirley Fertilizer Company.
SOUTH BEND—S. W. Lippman.
  Elmer Strayer.
SULLIVAN—Sullivan Reduction Company.
TERRE HAUTE—Terre Haute Abattoir Company.
  Terre Haute Hide & Fertilizer Company.
  Terre Haute Tallow & Grease Company.
THORNTON—Farmers Reduction Co.
TIPTON—Tipton Reduction Plant.
TRAFALGAR—W. C. Tucker.
UNDERWOOD—Winnie Stewart.
VEEDERSBURG—Fountain Fertilizer Co.
WABASH—Joe Feighner Fertilizer Co.
WAKRUSA—Elkhart County Fertilizer Company, R. F. D No. 3.
WARSAW—Warsaw Fertilizer Company.
WAYNETOWN—Chas. Dwiggins & Sons.
WESTPORT—Westport Fertilizer Company.
WINAMAC—O. W. Crawford.
WINCHESTER—Roby Bros.
ZIONSVILLE—Zionsville Tankage Company.

IOWA

ALGONA—Algona Rendering Works.
  F. C. Pohlman.
AMES—Ames Disposal Works.
  C. S. La Forge.
AVOCA—H. M. Deeds.
BATTLE CREEK—Battle Creek Rendering Company.
BOON—Boon Rendering Works.
CARROLL—O. C. Clausen.
CEDAR RAPIDS—Will Ater.
CHEROKEE—Cherokee Rendering Company.
CLINTON—Clinton Rendering Co.
CUMBERLAND—Cumberland Rendering Works.
DAVENPORT—Davenport Slaughtering & Rendering Company.
DES MOINES—D. M. Hide & Rendering Company.
  Iowa Reducing Company.
  T. T. La Forge.
DUBUQUE—Friths Rendering Works.
DYSART—Dysart Rendering Co.
EAGLE GROVE—Eagle Grove Rendering Co.
FAIRFIELD—MaGees Rendering Co.
FONDA—Fonda Rendering Co.
FORT DODGE—E. D. Clagg, 301 Central Ave.
GALVA—Galva Rendering Co.
GRUNDY CENTER—Grundy Center Iron & Metal Co.
IDA GROVE—Ida Grove Rendering Works.
IOWA CITY—H. Scholman.
JEFFERSON—Jefferson Produce Company.
KOLONA—J. W. Kelly.
  Kolona Rendering Works.

MACON CITY—Macon City Rendering Company, 19 N. Federal St.
MANCHESTER—T. B. Hawkes.
MANNING—Wenzel Rendering Co.
MARENGO—Hawkeye Rendering Co.
MAPLETON—F. W. Hough.
MARSHALLTOWN—Marshalltown Rendering Company.
C. W. Kirk.
MASON CITY—Burwell & Lund, 671 1st St., S. E.
MINDEN—D. D. Addison.
MONTEZUMA—C. F. Lightlinger.
Montezuma Rendering Co.
MOVILLE—W. A. Vigars.
MUSCATINE—C. E. Richard Sons Co.
NEW LONDON—New London Rendering Company.
NORTH ENGLISH—Otto Kucera.
OAKLAND—Geo. Addison.
OSKALOOSA—Otis Taylor.
OXFORD JUNCTION—Oxford Junction Rendering Co.
ROCKFORD—Urdangen & Weismann.
ROCK VALLEY—Rock Valley Rendering Works.
SAC CITY—Sac City Rendering Works.
A. F. Witte.
SIGOURNEY—Sigourney Hide & Tallow Company.
SIOUX CITY—Farmers Rendering Company.
Iowa Rendering Company.
TAMA—J. J. Hall.
TRAER—Alex Thompson.
Traer Rendering Company.
VAIL—Vail Rendering Co.
WALL LAKE—Wall Lake Rendering Company.
WALNUT—Chas. L. Addison.
WATERLOO—Cole Rendering Works.
WILLIAMSBURG—Manor & O'Donnell.
WEST LIBERTY—West Liberty Hide & Rendering Company.
WILTON JUNCTION—Wilton Hide & Rendering Co.

KANSAS

KANSAS CITY—Morris & Co.
Swift & Company.
Wilson & Company.
ROSEDALE—Far Famed Mt. Sausage Company.
TOPEKA—Topeka Rendering Co., 819 E. 6th St.

KENTUCKY

CENTRAL CITY—Moulden & Hughes.
HOPKINSVILLE—I. L. Freedman.
LOUISVILLE—American Hide & Tallow Company.
Fred Krauth.
Louisville Hide & Tallow Company.
Marx Hide Co., 816 W. Main St.
OWENSBORO—Field & Company, Box 413.
PADUCAH—Thomas Challenor.
S. A. Jones & Sons.
Kolb & Sons.

LOUISIANA

ALEXANDRIA—Rapides Pkg. Co.
OPELOUSAS—J. M. Boagni.
SHREVEPORT—F. Noeth & Co.
NEW ORLEANS—St. Bernard Rendering & Fertilizer Co., 1003 Title Guarantee Bldg.

MAINE

BANGOR—S. A. Maxfield Co.
BIDDEFORD—Eastman P. Seavey Tallow Works.
Willard Carville.
BERWICK—O. H. Butler Soap Factory.
EAST DOERING—Portland Rendering Company.
NEWPORT—Weymouth Wool Co.
PORTLAND—J. Carney & Co.
John Kern, 901 Washington Ave.
Portland Rendering Co.
ROCKLAND—Josiah Torrey.

MARYLAND

BALTIMORE—Agriculture Mfg. Co.
Baltimore Fertilizer Co.
Baltimore Oil Company.
Garbage Reduction Co.
Jacob W. Hook & Company.
Chas. Klemm & Co.
Levering Fertilizer Company.
Miller Fertilizer Company.
J. Moores & Co., 920 E. 4th St.
CUMBERLAND—Keystone Hide Co.
Hirsch Brothers, Inc.
FREDERICK—Ramsburg Fertilizer Co.
HAGERSTOWN—Wm. H. Bixler.
D. A. Thomas & Company.
LINWOOD—John A. Engler.
RECKFORD—Reckford Fertilizer Company.
SNOW HILL — Worchester Fertilizer Company.
TANEYTOWN—Reindollar Company.
WESTPORT—McNamara Bros.

MASSACHUSETTS

ATTELBORO—N. Roy & Son.
AUBURN—Worchester Rendering Co.
BILLERICA—Lowell Rendering Company.
BOSTON—Consolidated Rendering Company, 40 Market St.
J. C. Dow Company.
Eastern Oil & Rendering Co., 92 State St.
J. F. Morse & Co.
New England Fertilizer Co.
New England Rendering Co.
N. Ward Company.
CAMBRIDGE—John Reardon & Sons Company.
CHELMSFORD—Clarence Reemcs.
Whitman & Pratt Rendering Co.
CHELSEA—A. Lord & Company.
FALL RIVER — Butchers Rendering Company.
FITCHBURG—Fitchburg Rendering Company.
HOLYOKE—Abbott Soap & Fertilizer Company.
LANCASTER—Jos. H. Whelan.
LAWRENCE — Beach Soap Company, Lawrence and Maple Sts.
LONDON—Deer Creek Fertilizer Company.
LOWELL—Lowell Fertilizer Co.
Lowell Rendering Company.
Whitman & Pratt Rendering Company.
LYNN—Lynn Grease Extracting Company.
George E. Marsh Co.
Town Extract Company.

MANISTEE—L. M. Roussin.
MILLBURY—Home Soap Company.
MONSON — East Moulton Rendering Works.
NEW BEDFORD—New Bedford Extraction Company.
 Thomas Herson & Company.
NORTH ANDOVER—John Glennis.
PEABODY—American Degreasing Company.
 Newell & Knowlton, Inc.
PITTSFIELD—Owen-Coogan & Son, 39 Elm St.
PLAINFIELD — Springfield Rendering Company.
SOMERVILLE — Hinckley Rendering Company.
 George W. Norton.
SHERBORN—Sherborn Rendering Company.
SPRINGFIELD—Springfield Rendering Company.
W. BRIDGEWATER—S. Winter Company.
ROXBURY—Jas. F. Morse & Co., 66 Norfolk Ave.

MICHIGAN

BATTLE CREEK—Stewart Bros.
DETROIT—Detroit Reduction Co.
 C. E. North, 2412 20th St.
 Schulte Soap Company.
ECORSE—Millenbach Bros. Co.
GRAND RAPIDS—Grand Rapids Glue Company.
KALAMAZOO — Kalamazoo Rendering & Fertilizer Company.
ST JOE—E Burton Rendering Plant.
JAMESTOWN — Jamestown Tankage Works.
LANSING—The Prugleman Rendering Works.

MINNESOTA

ALPHA—Alpha Rendering Works.
DULUTH—T. E. Halford Co.
MINNEAPOLIS—Hy. Mengelkoch, 1009 Main St., N. E.
MOORHARD—G. Zewas.
NEW BRIGHTON—Hogland Bros.
 Mengelkoch Tallow Mfg. Co.
 Minneapolis Hide & Tallow Company.
 Minnesota Rendering Co.
 Northern Rendering Co.
 Olson Martin Rendering Plant.
 Van Hoven Company.
ROCHESTER—Rochester Rendering Works.
ST. PAUL—Luley Abattoir Co.
S. ST. PAUL—I. T. McMillan Company.
 Union Rendering Company.
 D. Bergman & Co.
S. STILLWATER—August Utecht.
 St. Croix Rendering Works.

MISSISSIPPI

GREENWOOD—Peltz & Son.

MISSOURI

KANSAS CITY — Standard Rendering Company.
ST. JOSEPH—St. Joseph Rendering Works.
S. ST. JOSEPH—Union Rendering Company.
ST. LOUIS—Bell Oil Company.
 Binz Hide & Tallow Co.
 J. B. Dick & Co.
 Leonard Haefele, 825 Tesson St.
 Holste Grease & Tallow Co.
 St. Louis Hide & Tallow Works.
 United Hide & Fur Co., Inc., 108 Commercial St.
 Chas. G. Ziegenbalg.
SEDALIA—John T. Hoffman.
 P. O. Sedalia Farmers Co.
SPRINGFIELD—Walsh Pkg. Co.

NEBRASKA

LA PLATTE—Union Rendering & Refining Company.
LINCOLN—C. W. Swingle.
OMAHA—Omaha Rendering & Feed Company.
 S. Frank & Co.
 Union Stock Yards Co.

NEW HAMPSHIRE

MANCHESTER—Manchester Rendering Company.
PORTSMOUTH—Eastern Oil & Rendering Company.

NEW JERSEY

ASBURY PARK—Flavell Company.
BRIDGETON—John A. Minch.
CARTERET — American Agricultural Chemical Co., Liebig Wks.
 Consumers Chemical Corp.
CHEEKTOWAGA—Fred L. Nehbrass Soap Manufacturing Co.
DUNDEE LAKE — Somers Rendering Company.
ELIZABETH—Edward Glaser, 664 Jefferson Ave.
 Robert C. Mauers Sons.
GLOUCESTER—Chas. Craig.
 P. Maeley Sons.
JERSEY CITY—Atlan Soap Wks., Inc., 142 Logan Ave.
 Lafayette Provision Co.
KEARNEY — Schwarz Brothers, 1100 Harrison Ave.
 Harry J. Theobald, 188 Schuler Ave.
MERCHANTSVILLE—Collins & Panscoat.
MIDDLETON—Milo Rendering Factory.
NEWARK—Adam Sadowasky.
 Albert Mertz, 578 S. 20th St.
 American Tallow Co., Plum Point Lane.
 Carstons Mfg. Co.
 Independent Tallow Co.
 Lister Agricultural Chemical Works, Lister St.
 Mapes Formula & Peruvian Guano Company, Ferry St.
 Noll & Fischer.
 Standard Tallow Company.
NEW BRUNSWICK—Samuel Lederer & Sons.
NORTH PATERSON—Stonemeal Fertilizer Company.
PAULSBORO—I. P. Thomas & Sons Company.
PERTH AMBOY—Martin Ortel.
PLAINFIELD — Middlesex Fertilizer Company.

PORT MONMOUTH—Monmouth Oil & Guano Co.
SECAUCUS—Hy. Beekman.
  Leonard Heflixh.
  Margaret Heitzemann.
  John Hinesee, Jr.
  Henry Henkel.
  Ida Hummel & C. Kuscharft.
  Fritz Koenemund.
  Robt. Koenemund.
  John Lasky.
  National Agric. Chemical Co. of New Jersey, County Ave.
  R. Prahm.
  Schaffner & Fox.
  Anthony Schmidt.
  Wm. Seidel.
  Louis Stern.
  R. & O. Tischmann.
  Gustav Wogisch.
  Adam Zengal.
  Wm. Zengal.
  Louis Zurig.
TEANECK—Hy. Claussen.
TRENTON—O. F. Neidt & Co.
  Trenton Bone Fertilizer Co.
JERSEY CITY—Butchers Fat Rendering Company, 665 Newark Ave.
  United Butchers' Fat Rendering Company.

NEVADA

RENO—Reno Tallow & Fertilizer Company.

NEW YORK

BALDWINSVILLE—Burt Giddings.
BROOKLYN—Barr Bros.
  Long Island Soap Company, Meeker Ave.
  Abraham Nachman, 177 Greenpoint Ave.
  Products Mfg. Company.
  Joseph Rosenberg's Sons.
  G. Weiss & Son, Cherry St. and Gardner Ave.
BUFFALO—American Agricultural Chemical Company.
  Bowker Fertilizer Co.
  Buffalo Fertilizer Co.
  A. H. Case & Company.
  Crocker Fertilizer & Chemical Company.
  International Agricultural Corporation.
  Leo Kraus.
  Milsom Rendering Works.
  Schaal Sheldon Fertilizer Company.
  Wolf Hide Company.
CHEEKTOWN—Milsom Rendering Works.
DUNKIRK—Joseph Gostomski, 424 Lord St.
  Joseph J. O'Brocta, 2 Willow Road.
ELMIRA—Frank Stadelmaier.
JAMESTOWN—Dayle & Maie.
GOWANDA—Eastern Tanners' Glue Co.
HUDSON—Smith-Patterson.
  Thomas Stackpool.
KINGSTON—Fischang Bros.
  Roach Bros., 38 Ann St.
LOCKPORT—S. T. Argue.
LONG ISLAND CITY—Haberman Co., Inc., 315-23 Borden Ave.
  Preston Works.
  Van Iderstine Company.
MIDDLETOWN—Theodore Leidy, 99 Fulton St.
MOUNT VERNON—Adolph Isaac, 619 S. 5th Ave.
NEWBURGH—Newburgh Rendering Company.
NEW YORK—Herman Brand, Inc., 404 E. 48th St.
  General Rendering Co., Inc., 816 First Ave.
  Fred Lesser, 754 First Ave.
  Jacob Levy & Co., 765 First Ave.
  Katzenstein Bros., 49 E. 135th St.
NIAGARA FALLS—Niagara Falls Reduction Company.
ONEIDA—Robt. Paul.
PENFIELD—Wm. Stappenbeck & Son.
PLATTSBURG—B. Tierney & Sons.
PORT CHESTER—Smith, Angevine & Co., Inc.
ROCHESTER—Genesee Reduction Company.
  Herbert Price.
  Rochester Hide & Tallow Co.
  Rochester Tallow Company.
  Stappenbeck Brothers.
ROME—Rome Rendering Works.
ROTTERDAM—H. M. Stangon.
SOUTHPORT—Frank Stadelmaier.
SYRACUSE—G. W. Finn.
  Syracuse Rendering Co.
TROY—Fat Melting & Calf Skin Association.
WATERLOO—Waterloo Soap Co.
WESTCHESTER—R. Monti & Son, 1087 Sackett Ave.
WHITESBORO—Stappenbeck & Son.
WILSON—James Haberman.

OHIO

AKRON—Akron Soap Company.
  C. A. Schell Prov. Co.
ALLIANCE—Alliance Fert. Co.
AMANDA—Amanda Tankage Wks.
ASHTABULA—John Dhen, Fairview Ave.
BELLEFONTAINE—Bellefontaine Mfg. Co.
BLUFFTON—Putnam Tanking Co.
CANTON—Canton Fertilizer & Chemical Company.
  F. W. Renner & Sons.
  Rennes & Stein.
CINCINNATI—American Agricultural Chemical Company.
  Cincinnati Hide Co.
  Cinn. Phosphates Company.
  Cinn. Reduction Co.
  Conway Tallow Company.
  Jacob Freiser.
  Chas. Hulsenmann.
  Kaufmann Fertilizer Company.
  John Mitchell.
  P. L. Neville.
  Union Reduction Company.
  Virginia Chemical Co.
CLEVELAND—Cleveland Garbage Disposal Company.
  Cuyahoga Rendering & Soap Company.
  Farmers' Chemical & Fertilizer Company.
  Independent Glue Company.
  C. Masek Glue Company, 4076 Jenning Road.
  J. L. & H. Stadler Rendering & Fertilizer Co., 908 Dennison Ave.
  W. H. Teare & Company.
COLUMBUS—John B. Bass Rendering Company.

E. G. Buchsieb.
Columbus Rendering Co.
Farmers' Fertilizer Co.
CAMDEN—Edw. Slover Fert. Co.
CONNEAUT—Conneaut Reduction Co.
ELYRIA—Mendelson Reduction Company.
FOSTORIA—A. Wernick.
GREENVILLE—Winchet Fertilizer Company.
HAMILTON—Miami Fertilizer Company.
Edward Motzer.
HARRISON—Schwing Fertilizer Company.
HARVESBURG—Harvesburg Fertilizer Company.
KENTON—Kenton Reduction Co.
LOCKLAND—Elmwood Rendering Co.
MANSFIELD—The Richland Fertilizer Company.
MARIETTA—Marietta Bone & Phosphate Company.
NEW PHILADELPHIA—New Philadelphia & Dover Fertilizer Company.
PAINESVILLE—Charles Massena.
PIQUA—Wm. Rhodehamel.
PORTSMOUTH—Portsmouth Chemical Company.
SANDUSKY—Musson Soap & Tallow Company.
TRIFFIN—Seneca Fertilizer Co.
WEST TOLEDO—N. Rassel Soap Company.
YOUNGSTOWN—Youngstown Fertilizer Company.
Youngstown Hide & Tallow Company.
GREENVILLE—The Greenville Fertilizer Company, P. O. Box 125.

OREGON

ASTORIA—DeForce Oil Works.
PORTLAND—Allen & Hendrickson Packing Co., 602 Lewis Bldg.
David M. Dunne.
Oregon Bone Meat Works.
Powell Valley Rend. Co., 100 S St.
Portland Rendering Co.
Superior Oil & Process Co.

PENNSYLVANIA

ALLENTOWN—R. A. Reichard.
ALTOONA—Monongahela Melting Co.
BETHLEHEM—Alvin Hill & Son.
BRADDOCK—Monongahela Melting Company, Ltd.
BRISTOL—John R. Williams, R. F. D.
CARLISLE—S. B. Romberger & Sons.
CHARLEROI—Thos. Lowstetter.
CHESTER—Henry V. Baxter.
CHESWICK—C. Mardorf & Sons.
COATESVILLE—J. C. Downward Co.
Frank Hartranft, 385 Chas. St.
DU BOIS—Swacks Fert. Wks., Ltd., Quarry Ave.
EASTON—Berger Brothers, R. F. D. No. 6.
ELDORADO—Monongahela Melting Company, Ltd.
ELIZABETHTOWN—Farmers' Fertilizer Works, Bldg. St.
ERIE—Erie Production Co.
ERWIN—Monongahela Melting Company, Ltd.
ESPY—Espy Humus Fertilizer Company.
FAIRVILLE VILLAGE—A. Bean.
FRANKVILLE—Wagner Bros.
FRANKLIN—Franklin Reduction Works, R. F. D. No. 6.
FORD CITY—Ford City Lime Fertilizer Company.
GETTYSBURG—Oyler & Spangler, R. F. D. No. 48.
HANOVER—A. F. Rees, P. O. Box 266.
HARRISBURG—Harrisburg Rendering & Hide Company, 11th and Walnut Sts.
Penna. Reduction Co., Cameron and Seneca Sts.
HARTFIELD—James Romig.
HAZELTON—Anson B. Schoemaker.
JOHNSTON—Caples & Moore, 340 Stony Creek St.
KENNETT SQUARE—Joseph R. Gawthrop, 517 Broad St.
LANCASTER—Conestoga Glue Wks.
Keystone Hide Company.
Geo. Lamparters Sons, Rockland St.
Lancaster Chemical Co., Box 184.
Lancaster Glue Works.
Leonard Stapf.
LEHIGHTON—John Lechler.
LEWISTON—Ben Wollner & Bro.
LEBANON—Lebanon Fertilizer Works, Hoffman and Green Sts.
LEHIGH GAP—Enox Roseberry.
LINCOLN UNIVERSITY—Henry Cope & Company.
LINFIELD—Jacob Trinley & Sons.
LORAINE—Ahrens Fertilizer Company.
MILESBURG—Smith's Rendering Works.
MILLERSBURG—J. E. Kahler.
MILROY—Mifflin County Rendering Works.
McKEESPORT—Monongahela Melting Company.
NEW CASTLE—Joseph Cohen, 413 E. Reynolds St.
Fazzoni Brothers, P. O. Box 517.
Knoblook Brothers, S. Mercer St.
John Rentz, R. F. D. No. 2, Harbor Rd.
NEW KENSINGTON—Freedom Oil Works Co.
NEWTON—T. S. Kinderdine & Sons.
N. E. PITTSBURGH—Staab Soap Company.
NORTH WALES—Union Chemical Co., Inc., Rorer Bldg.
OXFORD—Oxford Pkg. Co.
PECKVILLE—Maines Rendering Company.
PHILADELPHIA—American Agricultural Company, 897 Drexel Bldg.
Berg Company, Ontario and Richmond Sts.
Bough & Sons Co., 20 S. Delaware Ave.
Charles A. Green, 1214 Girard Ave.
General Mfg. Co., 30th and Market Sts.
Arthur Gore, 3rd and Curtin Sts.
Grays Ferry Abattoir Co., 36th St. and Grays Ferry.
Heinemann & Company.
Independent Mfg. Co., Wheat Sheaf Lane and Armingo Ave.
Keystone Bone Fertilizer Co.
Adam Marker.
Phila. Animal Product Co., 30th and Race Sts.
Phila. Fertilizer Co., 36th and Gray Sts.
F. J. Quigley Tallow Co.

W. L. Schoemaker & Co., Ltd.
F. W. Tunnell & Co., Inc., 15 N. 15th St.
PHOENIXVILLE—Chas. F. Bader.
Wm. M. Deger.
PITTSBURGH—Falk Co., 1217 Farmers' Bank.
Standard Animal Product Co., 326 Diamond St.
Walker, Stratman & Co., Herrs Island.
World Fertilizer Process Co., 20th and Chapment Sts.
PITTSTON—Exeter Rendering Works, 604 Wyoming Ave.
PLYMOUTH—Sickler Fertilizer Works.
POTTSTOWN—Harry Moyerman.
POTTSVILLE—O. L. Romberger.
PROVIDENCE SQ.—G. W. Schweiker.
READING—Globe Rendering Co., 535 Court St.
Reading Bone Fertilizer Co., P. O. Box 832.
RIDGWAY—Keystone Hide Co., Masonic Temple.
SCRANTON—Dickinson Rendering Company.
Hewett Fertilizer Co., 215 Pauli Bldg.
SHARON—Sharon Rendering & Fertilizer Works, R. F. D. No. 57.
SHENANDOAH—Shenandoah Abattoir Company.
SOUTHAMPTON—S. W. Dannehower.
SPRINGTOWN—Frank P. Miller.
STEWARTSTOWN—Abram Waltemeyer.
TRUCKVILLE—B. G. Laskowski.
VILLAGE KING OF PRUSSIA—Anderson & Walker.
WHITE OAK—Mrs. Hiram W. Diehm.
WILKES-BARRE—Bomberger Rendering Co., Franklin Junc.
Sickle's Fertilizer Works, 387 S. River St.
WILLIAMSPORT—Keystone Glue Company.
WOMELSDORF—Farmers' Fertilizer Seed & Hay Co.
WORCHESTER—Geo. Veits.
YORK—York Chemical Works.
York Sanitary Reduction Company.
INDIANA—I. C. Rendering Co., Inc.
PHILADELPHIA—Mutual Rendering Company, Ontario and Brabant Sts.
SHAMOKIN—Shamokin Fertilizer Works.

RHODE ISLAND

NEWPORT—John H. Donovan, 84 Broadway.
PAWTUCKET—L. B. Darling Fertilizer Company.
Whatcheer Chemical Co.
Wrensch Mfg. Company.
PROVIDENCE—E. Wolf & Co.
WOONSOCKET—P. J. O'Donnell & Son.

SOUTH CAROLINA

GREENVILLE—M. H. Goodlett.

SOUTH DAKOTA

MONTROSE—Montrose Rendering Works.
SIOUX FALLS—Sioux Falls Rendering Works.

TENNESSEE

MEMPHIS—Steinberg & Co.
NASHVILLE—Nashville Hide & Melting Ass'n.

TEXAS

DALLAS—Dallas Fert. & Reduction Company.
HOUSTON—Texas Chemical Co.

UTAH

SALT LAKE CITY—C. V. Hepworth.
John Skola.

VERMONT

BURLINGTON—Burlington Rendering Company.

VIRGINIA

ALEXANDRIA—J. Dreifus & Sons.
CHERRYHILL—Washington Fertilizer Company.
NORFOLK—Kenone Rendering Co.
Norfolk Tallow Co., 603 Union St.
Whitman & Pratt Rendering Company.
E. C. Dunn.
REEDVILLE—C. E. Davis Pkg. Co.
RICHMOND—Richmond Refining Company.
NORFOLK—Norfolk Tallow Co.
Wynne Lard & Prov. Company, 1141 May Ave.

WASHINGTON

ANACORTES—Marani Products Company.
BLAINE—Wannenweitch Reduction Company.
SEATTLE—Mutual Meat Company, 1225 Burns St.
N. Pacific Sea Products Co., 1820 L. C. Smith Bldg.
N. Western Products Co., F. R. D., Renton.
Universal By-Products Co.
TACOMA—Tacoma Tallow Works.

WEST VIRGINIA

CHARLESTON—Fisher & Fruth.
HUNTINGTON—Thomas & Allen.
WHEELING—Wheeling Butchers' Association.

WISCONSIN

KENOSHA—U. S. Koos & Son Co.
LA CROSSE—La Crosse Rendering Works.
MADISON—Capital City Soap Co.
MARTINTOWN—Fosdick & Higley.
MILWAUKEE—Milwaukee Tallow & Grease Co.
United States Glue Co.
OSHKOSH—Oshkosh Soap Co.
SEYMOUR—Seymour Rendering Co.
WATERTOWN—Robert Borkenhagen.

REFINERS OF EDIBLE VEGETABLE OILS

American Cotton Oil Co.—Pres., W. O. Thompson; Vice-Pres., Lyman H. Hine. Offices: 65 Broadway, New York, N. Y. Refineries: Chicago, Ill.; Cincinnati, O.; Gretna, La.; Guttenburg, N. J.; Houston, Tex. (Industrial Cotton Oil Properties); Memphis, Tenn.; Providence, R. I.; St. Louis, Mo.

American Cocoanut Butter Company.—Offices, 233 Broadway, New York, N. Y.

Arizona Egyptian Cotton Co.—Pres., Herbert B. Atha; Vice-Pres., F. R. Behrends. Offices and Refinery, Phoenix, Ariz.

Armour & Co.—Mgr., Refining Dept., G. G. Fox. Offices, Chicago, Ill. Refineries: Chicago, Ill.; Kansas City, Kan.; Fort Worth, Tex.; Chattanooga, Tenn.; Jersey City, N. J.

Aspegren & Co., Inc., New York.—(See Portsmouth Cotton Oil Refining Corp.)

Atlanta Refining & Mfg. Co.—Offices and Refinery, Atlanta, Ga.

Baker Cocoa Company—Offices, New York, N. Y. Refineries: Newark, N. J.; Philadelphia, Pa.

Berlin Mills Co.—Offices and Refinery, Berlin Mills, Vt.

Blanton-Sims Co.—Pres., D. A Blanton; Mgr., Harvey Sims. Offices and Refinery, St. Louis, Mo.

Boyer Oil Mfg. Co.—Pres., J. R. C. Boyer. Offices, 29 Broadway, New York, N. Y. Refineries, Indianapolis, Ind.; Singac, N. J.

Brenham Compress, Oil & Mfg. Co.—Pres., R. P. Thompson. Offices and Refinery, Brenham, Tex.

Buckeye Cotton Oil Co.—Pres., W. E. McCaw. Office and Refinery, Cincinnati, O.

Capitol Refining Co.—Pres., J. C. Dold. Offices, Washington, D. C. Refinery, South Washington, Va.

Capital City Dairy Co.—Pres., E. P. Kelly. Office and Refinery, Columbus, O.

Chickasha Cotton Oil Co.—Pres., R. K. Wootten; Vice-Pres., Joab Mulvane. Offices and Refinery, Chickasha, Okla.

Continental Cotton Oil Co.—Pres. and Mgr., John Guitar. Office and Refinery, Abilene, Tex.

Columbia Cotton Oil Co.—Pres., J. O. Hutcheson. Office and Refinery, Magnolia, Ark.

Cooknut Corporation.—Offices and Refinery, Baltimore, Md.

Corn Products Refining Co.—Offices, 26 Broadway, New York, N. Y. Refineries: Argo, Ill. (Corn Oil); Shadyside, N. J. (Corn Oil).

Cudahy Packing Co.—Vice-Pres., G. C. Shepard. Offices, Chicago, Ill. Refineries: East Chicago, Ill.; Memphis, Tenn.; Kansas City, Kas.; Omaha, Neb.

Cuero Cotton Oil & Mfg. Co.—Pres. and Mgr., Thornton Hamilton. Offices and Refinery, Cuero, Tex.

Dallas Oil & Refining Co.—Pres. and Mgr., J. S. LeClercq. Offices and Refinery, Dallas, Tex.

Dixie Cotton Oil Co.—Pres., W. P. De Jarnette; Vice-Pres., Michael Loeb. Offices and Refinery, Montgomery, Ala.

Dixie Refining Co.—Pres., Alton Boyd. Offices, Memphis, Tenn. Refinery, New Orleans, La.

Eagle Cotton Oil Co.—Pres., E. Cahn, Sr.; Mgr., E. Cahn, Jr. Offices and Refinery, Meridian, Miss.

Edison Oil Co.—Mgr., N. B. Solomon. Offices and Refinery (peanut oil), Edison, Ga.

El Paso Refining Co.—Pres., J. B. Dala. Offices and Refinery, El Paso, Tex.

Electrox Company—Pres., W. H. Ballance, Jr. Offices and Refinery, Peoria, Ill.

Fidelity Products Co.—Pres. and Mgr., W. H. Jasspon. Offices and Refinery, Houston, Tex.

Florida Cotton Oil Co.—(See Osage Cotton Oil Co.)

Forney Cotton Oil & Ginning Co.—Pres., W. A. Brooks. Offices and Refinery, Forney, Tex.

Gayoso Oil Works.—Pres., Milton Anderson. Offices and Refinery, Memphis, Tenn.

Globe Cotton Oil Mills.—Pres., Will E. Keller. Offices, Los Angeles, Calif. Refineries: El Centro, Calif.; Los Angeles (Vernon), Calif.

Gulf & Valley Cotton Oil Co.—Pres., John Aspegren; Mgr., H. P. Sanchez. Offices and Refinery, New Orleans, La.

Hancocks Products Co.—Offices and Refinery, Kingston, N. Y.

Hightower Oil Co.—Mgr., E. M. Hightower. Offices and Refinery (peanut oil), Brundidge, Ala.

Hodgson Oil Refining Co.—Pres., Harry Hodgson. Offices and Refinery, Athens, Ga.

Houston Packing Co.—Pres., R. E. Paine. Offices and Refinery, Houston, Tex.

Imperial Valley Oil & Cotton Co.—Pres., J. D. Dale. Offices, Calexico, Calif. Refinery, El Centro, Calif.

India Refining Company—Offices and Refinery, Philadelphia, Pa.

Industrial Cotton Oil Properties—(See American Cotton Oil Co.)

International Refining Co.—Pres., H. H. Coleman. Offices and Refinery, San Antonio, Tex.

International Vegetable Oil Co.—Pres., W. W. Banks. Offices, Atlanta, Ga. Refinery, Savannah, Ga.

Interstate Cotton Oil Refining Co.—Pres. and Mgr., C. A. Sanford. Offices and Refinery, Sherman, Tex.

Jones Packing & Provision Company, Ltd.—Offices and Refinery, Smith Falls, Ontario, Canada. A corporation. Capital, $40,000; issued, $35,000. Stockholders, 5. President, M. A. Jones; Secretary and Treasurer, A. M. Jones; Joint Manager, A. M. Jones and C. G. Jones.

Lange Soap Co.—Offices and Refinery, San Antonio, Tex.

Leder Oil Co.—Pres., Isidore Bley. Offices and Refinery, Demopolis, Ala.

Liberty Cotton Oil Co.—Pres., Joseph F. Rumsey. Offices and Refinery, Oklahoma City, Okla.

Lookout Oil & Refining Co.—Mgr., Mercer Reynolds. Offices and Refinery, Chattanooga, Tenn.

Los Angeles Soap Co.—Offices and Refinery, Los Angeles, Calif.

Louisville Food Products Co.—Offices, Chicago, Ill. (Van Camp Co.). Refinery, Louisville, Ky.

Magnolia Provision Co.—Pres., E. H. Astin; Vice-Pres. and Mgr., B. D. Cash. Offices and Refinery, Houston, Tex.

McGregor Cotton Oil Co.—Pres., J. F. Cavitt. Offices and Refinery, McGregor, Tex.

Merchants & Planters Oil Co.—Pres., W. M. Rice. Offices and Refinery, Houston, Tex.

Morris & Co.—Mgr., Refinery Dept., Ernest Kissling. Offices, Chicago, Ill. Refineries: Chicago, Ill.; East St. Louis, Ill.; Kansas City, Mo.; Oklahoma City, Okla.

Mutual Refining Co.—Pres., J. A. Underwood. Offices and Refinery, Sherman, Tex.

National Oil Treating Co.—Offices and Refinery, Chicago, Ill.

Nucoa Butter Company—Offices, New York, N. Y. Refinery, Bayonne, N. J.

Oil Seeds Co.—Pres., M. B. Snivily. Offices, 35 William St., New York, N. Y. Refinery, New York, N. Y.

Osage Cotton Oil Co.—Pres., W. B. Riddell. Offices, Chattanooga, Tenn. Refinery, Jacksonville, Fla.

Pacific Trading Corporation of America.—Pres., A. Gjessing. Offices, 90 West St., New York, N. Y. Refinery, Dobbs Ferry, N. Y.

F. S. Perry Gin & Mill—Pres. and Mgr., F. S. Perry. Offices and Refinery, Gorman, Tex.

Phoenix Cotton Oil Co.—Pres., J. H. Du Bose. Offices and Refinery, Memphis, Tenn.

Portsmouth Cotton Oil Refining Corp.—Pres. and Gen. Mgr., John Aspegren. Offices, Produce Exchange Bldg., New York, N. Y. Refinery, Portsmouth, Va.

Procter & Gamble Co.—Office, Cincinnati, Ohio. Refineries: Dallas, Tex.; Kansas City, Kan.; Los Angeles, Calif.; Port Ivory, N. Y.

Rambo & Sealey.—Mgr., C. J. Rambo. Offices and Refinery (peanut oil), Edison, Ga.

H. Schumacher Oil Works—Pres., E. H. Terrell. Offices and Refinery, Navasota, Tex.

Seaboard Refining Co., Ltd.—Pres., A. P. Sauer; Vice-Pres. and Gen. Mgr., E. T. George. Offices and Refinery, New Orleans, La.

Southern Cotton Oil Co.—Pres., C. G. Wilson; Vice-Pres., G. F. Tennille. Offices, 120 Broadway, New York, N. Y. Refineries: Atlanta, Ga.; Augusta, Ga.; Bayonne, N. J.; Charlotte, N. C.; Chicago, Ill.; Gretna, La.; Little Rock, Ark.; Memphis, Tenn.; Montgomery, Ala; Savannah, Ga.

Southland Cotton Oil Co.—Pres., A. J. Buston. Offices, Paris, Tex. Refineries: Corsicana, Tex.; Paris, Tex.; Shreveport, La.; Temple, Tex.

Southport Mill, Ltd.—Pres. and Mgr., A. D. Geoghegan. Offices and Refinery, New Orleans, La.

Spencer Kellogg Co.—Pres., Spencer Kellogg, Sr.; Vice-Pres., Spencer Kellogg, Jr. Offices, Buffalo, N. Y. Refineries: Buffalo, N. Y. (Kellogg Products Co.); Shadyside, N. J.

Stevenson & Co., Inc.—Offices, 44 Whitehall St., New York, N. Y. Refinery, Boonton, N. J.

Swift & Co.—Mgr., Refinery Dept., J. F. Smith. Offices, Chicago, Ill. Refineries: Chicago, Ill.; Atlanta, Ga.; Charlotte, N. C.; Kansas City, Kan.; Fort Worth, Tex.; Houston, Tex.; Toronto, Ont.; Winnipeg, Man.

Taft Oil & Gin Co.—Gen. Mgr., H. V. Fotick. Offices and Refinery, Taft, Tex.

Tar River Oil Co.—Pres., Dr. L. L. Staton. Offices and Refinery, Tarboro, N. C.

Terrell Cotton Oil Co.—Pres., W. P. Allen. Offices and Refinery, Terrell, Tex.

Terminal Oil Mill Co.—Pres. and Mgr., J. E. Quarles. Offices and Refinery, New Orleans, La.

Texas Refining Co.—Pres., F. J. Phillips; Vice-Pres., J. B. Clayton. Offices and Refinery, Greenville, Tex.

Trinity Cotton Oil Co.—Pres., E. R. Callier; Vice-Pres., F. C. Callier. Offices and Refinery, Dallas, Tex.

Tuscaloosa Cotton Oil Co.—Pres. and Mgr., E. B. Nuzum. Offices and Refinery, Tuscaloosa, Ala.

Many soap manufacturers and all large packers have their own refineries for both animal and vegetable oils.

MANUFACTURERS OF OLEOMARGARINE
(Licensed by the U. S. Bureau of Internal Revenue)

CALIFORNIA
WESTERN MEAT COMPANY, Walker Ave., Sixth and Townsend Sts., So. San Francisco.
NUCOA BUTTER CO., 1864 Bryant St., San Francisco.
PACIFIC FOOD PRODUCTS CO., San Francisco.
DANIEL M. HERRIN, Exeter.
MORRIS & CO., 734 Terminal St., Los Angeles.
WILSON & CO., 1000 Lyons St., Los Angeles.

COLORADO
SWIFT & COMPANY, Denver.

ILLINOIS
FRIEDMAN MANUFACTURING CO., Packers and Transit Ave., U. S. Yards, Chicago.
SWIFT & CO., Exchange Ave., Union Stock Yards, Chicago.
WILLIAM J. MOXLEY, INC., Randolph and Clinton Sts., Chicago.
JOHN F. JELKE CO., Western Ave., Chicago.
ARMOUR & CO., 4320-4340 Central Ave., Chicago.
G. H. HAMMOND CO., 45th and Center Ave., Chicago.
MORRIS & COMPANY, 42d and Laflin Sts., Chicago.
WILSON & COMPANY, 41st and Ashland Ave., Chicago.
DOWNEY-FARRELL CO., 241-255 E. Illinois St., Chicago.
B. S. PEARSALL BUTTER CO., Elgin.
ED. S. VAIL BUTTERINE CO., 4532-6 Gross Ave., Chicago.
TROCO NUT BUTTER CO., 220 E. Superior St., Chicago.
NUCOA BUTTER CO., 2802 S. Kilbourne Ave., Chicago.
GLIDDEN NUT BUTTER CO., 2670 Elston Ave., Chicago.
SWIFT & CO., East St. Louis.
MORRIS & CO., East St. Louis.

INDIANA
STANDARD NUT MARGARINE CO., Indianapolis.
KINGAN & CO., Indianapolis.

KANSAS
SWIFT & CO., Kansas City.
MORRIS & CO., Kansas City.
WILSON & CO., Kansas City.

MARYLAND
A. H. KUHLEMANN CO., Baltimore.
J. H. FILBERT, INC., 804 Calverton Road, Baltimore.
BALTIMORE BUTTERINE CO., Baltimore.

MASSACHUSETTS
SWIFT & CO., Cambridge.
SWEET NUT BUTTER CO., Boston.

MICHIGAN
SHEDD CREAMERY CO., 660 Vinewood Ave., Detroit.

MINNESOTA
SWIFT & CO., So. St. Paul.
NORTHERN COCOANUT BUTTER CO., 67-73 Nicollet Ave., Minneapolis.

MISSOURI
BLANTON-SIMS CO., St. Louis.
ST. LOUIS INDEPENDENT PACKING CO., trading as Missouri Butterine Co., St. Louis.
CROWN MARGARINE CO., St. Louis.
OTTO F. STIFELS UNION BREWING CO., 3145-51 Michigan Ave., St. Louis.
HARROW-TAYLOR BUTTER CO., 614 Broadway, Kansas City.

NEBRASKA
SWIFT & CO., South Omaha.

NEW JERSEY
SWIFT & CO., Jersey City.
HOLLAND BUTTERINE CO., Jersey City.
KEYSTONE CHURNING CO., Jersey City.
AMERICAN BUTTERINE CO., Jersey City.
E. A. STEVENSON & CO., Boonton.
NUCOA BUTTER CO., Bayonne.
HAUCK NUT BUTTER CO., Newark.

NEW YORK
PALMINE MANUFACTURING CO., INC., Dobbs Ferry.
PAWLING PRODUCTS CO., Pawling.
NUT GROVE BUTTER CO., 106 Noxon St., Syracuse.
KELLOGG-PRODUCTS, INC., Buffalo.

OHIO
OHIO BUTTERINE CO., Cincinnati.
MIAMI BUTTERINE CO., 107-9 E. Pearl St., Cincinnati.
PROCTER & GAMBLE CO., Cincinnati.
ROCK ISLAND BUTTER CO., Toledo.
OHIO DAIRY CO., Toledo.
CAPITAL CITY PRODUCTS CO., Columbus.

OREGON
SWIFT & COMPANY, Fourth and Hoyt Sts., Portland.

RHODE ISLAND
OAKDALE MANUFACTURING CO., Providence.
PROVIDENCE CHURNING CO., Providence.
RUMFORD CHURNING CO., E. Providence.
NUT GROVE BUTTER CO., Providence.
NATIONAL DAIRY CO., Providence.
VERMONT MFG. CO., Providence.

TEXAS
SWIFT & COMPANY, Fort Worth.

WISCONSIN
WISCONSIN BUTTERINE CO., 596-8 Clinton St., Milwaukee.
D. E. WOOD BUTTER CO., Evansville.
WONDER-NUT FOOD PRODUCTS CO., Jefferson.

BROKERS

(This list includes packinghouse products and animal and vegetable oils)

ALABAMA

Birmingham

ALLEN & CO.
HACKNEY & CO., J. T.—5½ N. 20th St.
HARRIS, O. H., & CO.—Lyric Bldg.
KELLY-MURPHY—601 American Trust Bldg.
STEELE BY-PRODUCTS CO.
SMITH, T. L.—402 Farley Bldg.
SCHAEFER & JAMES.

Dothan

SMITH, HERBERT H.
FRAZIER BROKERAGE CO.—Rooms 1 and 2, Malone Grocery Bldg. Private ownership, W. H. Frazier. Codes—Robinson, Armsby, Yopp's 7th, Modern Economy and private. All packinghouse products, lard and compound, vegetable oils, S. P. meats, tallows and greases.

Huntsville

LYLE & LYLE.

Mobile

BLACK, S. C., CO.
BROWN & BROWN.
CLEVELAND BROS.
McMILLAN & HARRISON.
ZIMMERMAN, J., & CO.

Montgomery

CANTELYOU BROS.
HEWITT, C. G., & SON—1123 Bell Bldg., P. O. Box 646. Partnership. Codes—Cross, Robinson and Yopps. Vegetable oils, fertilizers and fertilizer materials.

ARKANSAS

Little Rock

W. F. BRIDEWELL CO.—909-10 Southern Trust Bldg. A corporation. Codes—Yopps. President, W. F. Bridewell; Vice-President, I. L. Hathoway; Secretary and Treasurer, Sam. A. Sanders. Vegetable oils and cotton seed products.
CAPLE & STOCKTON.
CAMPBELL CO.—202 A. O. U. W. Bldg.
DAVIS, S. P.—207 Southern Trust Bldg.
FARMER-SILSON CO.—215 Bankers Trust Bldg.
HAYES GRAIN & COM. CO.
MUNN BROKERAGE CO.

CALIFORNIA

San Francisco

HENRY W. PEABODY & CO.—64 Pine St. Partnership. Codes—Yopps and Robinson. Vegetable oils and tallow and greases.
ROLPH, MILLS & CO.—149 California St. A corporation. President, Thomas Rolph; Vice-President, H. B. Mills. Codes—All standard codes; code word, "Patnoble." Vegetable oils.

ANDERSON, A. O., & CO.
AMERICAN FINANCE & COMMERCE CO.
BERESFORD, JOHN B., & CO.—Merchants Exchange Bldg.
BURKE, R. C., & CO.
CHINA AGENCY & TRADING CO.—519 California St.
CHRISTENSON, HANIFY & WEATHERWAX—210 California St.
DUREL & DODGE—255 California St.
HARRY GREEN & CO.—216 Pine St.
RALPH HIND & CO.—230 California St.
WALTER R. KIRK—15 California St.
MONROE, LEON & TEES, INC.—311 California St.
ORIENTAL VEGETABLE OIL CO.—114 Sansom St.
C. B. PETERS & CO., INC.—260 California St.
RUTGERS, BLECKKER & CO.—200 California St.
WILLITTS & PATTERSON—1 Drumm St.
HUIE & BOLTON—485 California St. Partnership. Codes—Yopps and Greigs. All packinghouse products, lard and compound, casings, vegetable oils, copra and oil seeds, and tallow and greases.
FRED L. KING—149 California St. Individual ownership. Codes—Western Union; code word, "Kingoil." Vegetable oils, tallow and greases and fertilizers.
HARVEY J. BOUTIN—681 Market St. Individual ownership. Codes—All standard; code word, "Boutin." Vegetable oils, tallow and greases, neatsfoot oil, lard oil and oleo stearine.
E. H. OTTO & CO.—214 Front St. Individual ownership. Codes—All standard; code word, "Emotto." Packinghouse products, vegetable oils, oil crushing seeds and oriental products.
W. T. PIDWELL—112 Market St. Individual ownership. Codes—Yopp and Bentley. Oil cake and meal—also Selling Representative of Southern Cotton Oil Trading Co.

Los Angeles

FRANK D. SAWYER—Higgins Bldg. Individual ownership. Codes—Cross and Yopps. All packinghouse products, lard and compound, vegetable oils, S. P. meats, tallow and greases.
BALFOUR-GUTHRIE & CO.—615 H. W. Hellena Bldg., 4th and Spring Sts. Partnership. Codes—Bentley, Scotts, 10th edition; code word, "Balfour." Casings, vegetable oils, tallow and greases. Mostly for import and export.
BRIDGES, H. E., & CO.
MADISON, A. J., & CO.
STONE COMPANY, JOHN A.—401 S. Grand Ave.
WILLITS & GREEN.

GEORGIA

Atlanta

ASHCRAFT-WILKINSON CO.—Candler Bldg. Codes—Yopp's and Robinson's.
W. M. HUTCHINSON—Atlanta Trust Co. Bldg. Codes—Yopp's, A. B. C., Robinson's.
JOHNSON-MORRISON CO.—610-611 Citizens & Southern Bank Bldg. Codes—Yopp's, Robinson, Bentley, A. B. C., 5th Edition, Cross.
M. C. KING CO. Codes—Western Union, Yopp's, and A. B. C., 4th and 5th Edition.

PALMER-MURPHY CO.—817-818 Atlanta Thust Bldg.
PINSON BROKERAGE CO.—1126 Healey Bldg. Codes—Yopp's and Robinson's.
R. M. SIMS CO.—1316-17 Atlanta Trust Co. Bldg. Code—Yopp's.
A. A. SMITH. Codes—Lieber's, A. B. C., 5th Edition, Robinson's and Yopp's.
TAYLOR COMMISSION CO. Code—Yopp's.
A. P. TREADWELL & CO.—918 Healey Bldg.

Augusta

R. E. BARINOWSKI—Lamar Bldg. Codes—Yopp's and Robinson's. Individual ownership.
HEATH, MALLARD & DANIEL. Codes—Yopp's and Robinson's.

Savannah

WALLACE & IVERSON—27 Whitehall St. All packinghouse products.

ILLINOIS

Chicago

A. L. WEBSTER & CO.—111 W. Washington St. Individual ownership. Codes—Cross; code word, "Alwebco." Tallow and greases and hides and skins.
TOMKINS, PHEE & CO.—401 Webster Bldg., 327 S. LaSalle St. Partnership. Codes—Cross. Tallows, greases, oils and tankage.
STEITZ BROKERAGE CO.—111 W. Washington St. Individual ownership. Codes—Bentley's. Lard and compound, vegetable oils and tallow and greases.
KELLY BROKERAGE CO.—434 Postal Telegraph Bldg. Partnership. Codes—Cross and Western Union; code word, "Kerbroco." All packinghouse products, lard and compound, S. P. meats and tallow and grease.
DAVIDSON COMMISSION CO.—140 W. Van Buren St. A corporation. President, A. A. Davidson; Secretary, F. H. Harrison; Treasurer, W. White. Codes—Cross, Robinson and Yopp. All packinghouse products, lard and compound, vegetable oils, S. P. meats and tallow and greases.
WALTER R. KIRK, 327 S. La Salle St. Individual ownership. Codes—Lieber's, A. B. C., 4th and 5th, Yopp's, Cross, Robinson's, Yopp's Revised, W. U. and Bentley's; code word, "Kirkcliffe." Vegetable oils, tallow and greases.
JOHN W. HALL—327 S. La Salle St., Webster Bldg. Individual ownership. Codes—Cross, Yopp, Robinson, Bentley and private. All packinghouse products, lard and compound, vegetable oils, S. P. meats, tallows and greases.
HERBERT BROKERAGE CO.—713 Postal Telegraph Bldg. Packinghouse products, tallow, greases, sausage material.
E. G. JAMES—309 S. La Salle St. All packinghouse products.
ROY L. NEELY—327 S. La Salle St. Casings.
J. F. NICOLAS—327 S. La Salle St. All packinghouse products, hides, sheepskins, etc.
McPHERSON-CARNEGIE CO.—168 N. Michigan Ave. Partnership. Code—Cross. All packinghouse products.
CHAS. SINCERE & CO.—141 W. Jackson Blvd. Provisions.

RUMSEY & CO.—91 Board of Trade Bldg., 141 Jackson Blvd. A corporation. President and Treasurer, H. A. Rumsey; Secretary, H. B. Godfrey. Codes—Utility, Jacobian and Private; code word, "Chappins & Rumsey." All packinghouse products.

H. H. MOORE—327 South La Salle St. Individual ownership. Codes—Cross, Robinson and Yopp's; code word, "Galmoore." Vegetable oils and tallow and greases.

GEORGE SUNDERLAND—332 South La Salle St., Postal Telegraph Bldg. Codes—Cross. All packinghouse products.

GEO. TSCHAPPAT & SON—327 S. La Salle St. Partnership. Codes—Robinson and Cross. Tallow and greases.

W. L. GREGSON & CO.—327 S. La Salle St. Individual ownership, W. L. Gregson. Codes—Cross; code word, "Andgregson." All packinghouse products.

ZIMMERMAN-ALDERSON-CARR CO.—111 West Monroe St. A corporation. President, W. L. Alderson; Vice-President, W. D. Carr; Secretary, W. B. Burr; Treasurer, W. L. Alderson. Codes—Yopp's, Cross and Robinson. Vegetable oils, tallow and greases.

WILLIAMS COMMISSION CO.—327 S. La Salle St. A corporation. President, Geo. A. Williams; Vice-President, S. A. Corker; Secretary and Treasurer, H. S. Haze. Codes—W. U. 5 letter, W. U., Universal, A. B. C., Bentley's and Lieber's; code word, "Wiloilcom." All packinghouse products, lard and compound, vegetable oils, tallow and greases and a full line of chemicals.

STERNE & SON CO.—150 W. Van Buren St. A corporation. President, C. B. Martin; Vice-Presidents, C. H. Sterne and D. P. Cosgrove; Secretary and Treasurer, J. G. Gilkison. All packinghouse products, vegetable oils, tallow and grease. Codes—Cross, Yopp, Lieber's, Bentley and Private; code word, "Gemsterne." Offices at New York, Memphis, Dallas, Houston and San Francisco.

J. C. WOOD & CO.—70 Board of Trade Bldg. Partnership. Codes—Cross. Provision brokers only.

JOHN S. CAMPBELL & CO.—140 S. Dearborn St.

PROCTOR & JOHNSON—327 S. La Salle St.

SUZUKI & CO.—821, 28 E. Washington St. (V. O.)

GALLAGHER, D. J.—4209 S. Racine Ave.

HATELY BROS. CO.—Board of Trade Bldg. Provisions.

RASCHKE BROKERAGE CO.—4148 S. Halsted St. Codes—Cross; code word, "Itradco." Packinghouse brokerage.

HENRY DUMMERT—118 N. La Salle St. Individual ownership. Tallow and greases.

CARRUTHERS BROKERAGE CO.—332 S. La Salle St. Individual ownership. Codes—Yopp's and Cross. Tallow, greases and all packinghouse products and by-products.

CROSS, ROY & SAUNDERS—140 W. Van Buren St. Provisions.

M. K. PARKER & CO.—Webster Bldg.

SCHWARTZ & CO.—305 S. La Salle St.

J. TREDWELL & CO.—327 S. La Salle St.

LOUISIANA

New Orleans

E. D. CAMBON—613 Poydras st.

W. D. COOPER—867 Magazine St.

MARYLAND
Baltimore

KAUFMAN PACKING CO.—Sixth St. A corporation. President, H. J. Kaufman; Vice-President, W. C. Kaufman; Secretary, Halver B. Kaufman; Treasurer, J. L. Kaufman.

SAM'L KRAUS & SON—316-318 S. Eutaw St. Partnership. Codes—Cross and Yopp. All packinghouse products, vegetable oils.

MASSACHUSETTS
Boston

GEORGE W. KING—502 Board of Trade Bldg. Individual ownership. Codes—Cross, Yopp's and Robinson. All packinghouse products, lard and compound and vegetable oils.

J. R. POOLE CO.—11 S. Market St. A corporation. President, J. R. Poole; Vice-President, W. G. Joyce; Secretary, H. S. Fraser and F. C. Clark. Codes—Cross and Robinson; codeword, "Jonarpool." All packinghouse products, lard and compound, casings, vegetable oils, S. P. meats and tallow and greases.

HERMON A. FLEMING CO.—88 Board St. Individual ownership. Codes—Cross, Robinson and Yopp's; code word, "Flemco-Boston." All packinghouse products.

H. P. HALE CO.—126 State St., Cunard Bldg. Individual ownership. Codes—Cross and Robinson. All packinghouse products, lard and compound, S. P. meats and tallow and grease.

P. G. GRAY CO.—Fidelity Bldg. Partnership. Codes—Cross, Griffin and Utility. All packinghouse products, lard and compound and S. P. meats. Specialize in dressed beef, lamb and veal, fresh pork loins, green and S. P. meats and pure lard.

WILLIAM G. JOYCE—Boston Fruit & Produce Exchange. All packinghouse products.

WILLIAM M. WARE & CO.—88 Broad St. Partnership. Codes—Bentley's and A. B. C., 5th edition; code word, "Wareeo." Oleo oil, stearine, cracklings, bones and tankage, rosin. Branch office at 351 Produce Exchange, New York City, and 90 St. James St., Montreal, Canada.

KENTUCKY
Louisville

HANKINS & ALLEN—31-32 Our Home Life Bldg. Partnership. Codes—Cross and Robinson. All packinghouse products.

MICHIGAN
Detroit

JOHN J. HAMEL—Room 610, 213 State St. Individual ownership. Codes—Cross and Robinson. Tallow and greases.

C. W. PERGANDE—4833 St. Clair Ave. Individual ownership. Tallow, greases, hides.

MINNESOTA
South St. Paul

THE HAAS COMMISSION CO.—201 Exchange Bldg. Individual ownership. President, C. L. Haas; Vice-President, G. M. Fitzgerald; Secretary, A. W. Thomas; Treasurer, W. F. Aull.

Minneapolis

WHEELER YOUNG CO.—655 Temple Court. A corporation. President, Harry K. Wheeler; Vice-President, E. M. Young; Secretary, E. M. Young; Treasurer, H. K. Wheeler. All pork products.

MISSOURI
St. Louis

JOHN RING—508 Merchants Exchange Bldg. Individual ownership. Codes—Cross and Robinson; code word, "Ring." Also order buyer and commission man. All packinghouse products, lard and compound, vegetable oils, S. P. meats and tallow and greases.

J. B. DICK & COMPANY.

PARKER-SAUNDERS—340 Pierce Bldg.

Kansas City

J. E. CHALLINOR—556 Live Stock Exchange Bldg. Individual ownership. Codes—Yopp's, Robinson, Cross and A. B. C., 5th edition; code word, "Challinor." All packinghouse products and cottonseed and linseed products.

NEW YORK
New York City

C. A. BECHSTEIN—443 W. 13th St. All packinghouse products.

ELBERT & CO.—27 William St. A corporation. President, A. Elbert; Vice-President, W. H. Beall; Secretary, M. Hallwood; Treasurer, B. Elbert. Codes—A. B. C., 4th, 5th and 6th editions, Bentley's, Lieber's, Robinson, Cross and Yopp's; code word, "Elbertonia." Lard and compound, vegetable oils and tallow and greases.

D. GECK, INC.—80 Maiden Lane. Tallow, grease, cracklings, tankage, hides, bones, etc.

JOHN THALLON & CO., INC.—New York Produce Exchange Bldg., 8 Broadway. A corporation. President, Monroe Washer; Secretary and Treasurer, D. W. Frazer. Codes—Bentley's, A. B. C. and Private; code word, "Thallon." Exporters of packinghouse and dairy products.

H. W. CALEF, INC.—331 Produce Exchange. Codes—Lieber's and Private. Vegetable oils and tallows and greases.

ASPEGREN & CO.—Produce Exchange. A corporation. Lard and compound and vegetable oils. Sole selling agents for Portsmouth Cotton Oil Refining Corp., Portsmouth, Va., Gulf & Valley Cotton Oil Co., Ltd., New Orleans, La., The International Vegetable Oil Co., Savannah and Atlanta, Ga.

WELCH, HOLME & CLARK CO.—383 West St. A corporation. President and Treasurer, A. M. Sherrill; Vice-President, How. Sherrill; Secretary, M. E. Clark. Codes—A. B. C., 5th edition (Imp.), Lieber's, Imp., W. U. and Cross; code word, "Ordeal." Tallow and greases and soap makers' supplies.

FREDERICK B. COOPER—228 Produce Exchange. Individual ownership. Codes—Cross and Griffin's. All packinghouse products, lard and compound and S. P. meats.

YOUNG COMMISSION CO., INC.—25 Broadway. A corporation. President, S. C. Young; Vice-President, F. S. Young, Jr.; Secretary and Treasurer, W. M. Sloan. Vegetable oils and tallow and greases.

DAVIS & BASSETT, INC.—299 Broadway. A corporation. President and Treasurer, J. L. Davis; Vice-President, H. D. Bassett;

Secretary, Frothingham. Codes—Bentley's, A. B. C., 5th edition, and Marconi; code word, "Davett." Purchase outright for export.

WILLIAM M. WARE & CO.—Produce Exchange Bldg. Partnership. Codes—Cross, Robinson and Private; code word, "Wilmware." Vegetable oils, tallow and greases, fertilizer, oleo oil and stearine.

EDWARD FLASH CO.—29 Broadway. A corporation. President and Treasurer, E. Flash, Jr.; Vice-President, W. A. Storts; Secretary, O. S. Flash. Codes—Yopp's, Lieber's and A. B. C., 5th; code word, "Ixofoldero-New York." Vegetable oils.

WORTHEN, TROTT & SULLIVAN—200 Produce Exchange. Partnership. Codes—Private, A. B. C., 5th, and Bentley's; cable address, "Longanka-New York." Lard and compound and vegetable oils, also oleo oil and stock, neutral and refined lard and choice white grease.

CHAS. HOLLINSHED CO., INC.—Produce Exchange. A corporation. President, Chas. Hollinshed; Vice-President, Secretary and Treasurer, W. J. Fischer. Codes—Private. All packinghouse products, lard and compound, vegetable oils and tallow and greases.

H. C. ZAUN—410 Produce Exchange. Individual ownership. Codes—Cross, Robinson and Utility. All packinghouse products, lard and compound, vegetable oils, S. P. meats and tallow and grease.

J. C. FRANCESCONI & CO.—25 Beaver St.; branches, 327 S. La Salle St., Chicago, and 486 California St., San Francisco. A corporation. President and Treasurer, J. C. Francesconi; Vice-President, W. A. Henson; Secretary, C. Francesconi. Codes—Cross, Robinson's and Yopp's. All packinghouse products and vegetable oils.

ARTHUR COMPANY—403 Produce Exchange. Individual ownership. Codes—Bentley's, Lieber's and A. B. C., 5th; code word, "Arthbroko." All packinghouse products and vegetable oils.

FONTANA BROS., INC.—424 Produce Exchange. A corporation. President, A. G. Fontana; Vice-President and Treasurer, F. G. Fontana; Secretary, G. A. Giglioli. Codes—Bentley's, Lieber's, A. B. C. and W. U.; code word, "Oderfla." All packinghouse products and vegetable oils.

W. B. CASSELL & CO.—1201-25 Beaver St. Partnership. Codes—Cross and Griffin. All packinghouse products and vegetable oils.

ARTHUR DYER—438 Produce Exchange. Individual ownership. Codes—Cross, Griffin and Armsby. All packinghouse products.

A. F. LOPEZ & CO., INC.—120 Produce Exchange. Exporters packinghouse products.

MacDOWELL-PETERMAN CO., INC.—15 William St. A corporation. President, C. E. MacDowell; Vice-President and General Manager, Andrew Peterman; Secretary, G. W. Smyth; Treasurer, A. Peterman. Codes—Cross, Robinson, Private and A. B. C., 4th, 5th and 6th editions; code word, "Macpete." All packinghouse products. Make a specialty of export trade.

SUZUKI & CO.—220 Broadway. Partnership. $25,000,000 paid capital. Codes—All standard codes; code word, "Kanetatsu New York." Suzuki & Co., New York, is the headquarters in the United States of Suzuki & Co., Kobe, Japan, being engaged in general export, import, manufacturing, and steamship operations, with branch offices in Seattle, Wash., San Francisco, Cal., Portland, Ore., and Fort Worth, Texas.

M. E. CLARENDON & SONS COMPANY—78 Gold St. A corporation. President, J. P. Clarendon; Secretary and Treasurer, L. F. Clarendon. Codes—A. B. C., Lieber's and Widebrook; code word, "Clarenhide." All packinghouse by-products, tallow and greases.

FINN & FINN—24 Stone St. Partnership. Codes—Cross and A. B. C., 5th edition. All packinghouse products and vegetable oils.

JOHN GOGGIN—Room 410 Produce Exchange. Lard and compound, vegetable oils, tallows and greases.

J. P. GRANT—25 Broadway. Codes—Cross, Yopp's, Lieber's, A. B. C., 5th edition. All packinghouse products, vegetable oils and S. P. meats.

JACKSON HATHAWAY—445 Produce Exchange Bldg. Codes—A. B. C., Lieber's and Private; code word, "Ballery." Vegetable oils, and tallow and grease.

TOBIAS T. PERGAMENT & CO.—New York Produce Exchange. Tallows and greases.

F. C. ROGERS—431 West 14th St. A. H. Olton, Manager. Provisions and packinghouse products.

W. D. VANDERHOVE & CO., INC.—309-311 New York Produce Exchange, New York City. A corporation. President, G. A. Dausey; Secretary, R. F. Biedermann. Codes—Cross. Tallows and greases.

C. W. ANDRUS & SON—350 Produce Exchange.
BAKER, CARVER & MORRELL CO.—39 Water St.
ROBERT G. BRANDT—C 20, Produce Exchange.
H. J. CANTRELL—59 Pearl St.
B. FRANKFELD & CO.—211 Produce Exchange.
CHAS. F. GARRIGUES CO.—54 Wall St.
C. C. HEIDELBERGER'S SONS—1085 Manhattan Ave.
S. HENLE—25 Beaver St.
H. HENTZ & CO.—22 William St.
H. P. KIDD—127 Produce Exchange.
E. S. KUH & VALK CO.—456 Produce Exchange.
KULLMAN & CO.—339 Produce Exchange.
D. C. LINK & CO.—118 Produce Exchange.
McGUIRE & JENKINS—16 Beaver St.
MORRIS & WILMARTH—217 Produce Exchange.
PRITCHARD & CO.—323 Produce Exchange.
RONEY & CO., INC.—81 Broad St.
ROUNDEY, F. B.—225 C, Produce Exchange.
W. H. STORY & CO.—213 Produce Exchange.
WILLIAMS COMMISSION CO.—23 Beaver St.

East Buffalo

WILLIAM LANSILL—963 William St. Individual ownership. Codes—W. U., Universal edition, Robinson, Yopp and Cross; code word, "Lansill-Buffalo." Vegetable oils and tallow and greases.

OHIO
Cincinnati

C. W. RILEY, JR.—2109 Union Central Bldg. Codes—Cross, Robinson and Miller's. All packinghouse products and vegetable oils.

Cleveland

ARTHUR CO.—879 The Arcade. All packinghouse products.

CHARLES A. STREETS—824 Engineers Bldg. Codes—Cross and Robinson. All packinghouse products, fertilizer materials and vegetable oils.

E. R. SMEAD CO.—327 S. La Salle St., Chicago; 1262 Hanna Bldg., Cleveland; 2 Rector St., New York City. A corporation. President and Treasurer, E. R. Smead; Vice-President, T. R. Proctor; Secretary, C. L. Small. Codes—Yopp, 7th edition. Vegetable oils.

PENNSYLVANIA

Philadelphia

W. T. RILEY—61-63 Main Floor Bourse Bldg. Individual ownership. Codes—Cross, Griffin and Private; code word, "Rileken." All packinghouse products and vegetable oils.

EDWARD B. CONLEY—5160 Pulaski Ave. Individual ownership. Codes—Cross, third edition. All packinghouse products.

FATS AND OILS SERVICE CO.—Commercial Trust Bldg.

F. C. ROGERS—267 North Front St. Provisions and packinghouse products.

Pittsburgh

H. R. SMITH CO.—Jenkins Arcade. A corporation. President, H. R. Smith; Vice-President, Wm. F. Gottschalk, Jr.; Secretary and Treasurer, W. H. Hervey. Codes—Cross and Robinson. All packinghouse products, grocer specialties and vegetable oils.

WM. F. HAMEL—331 Fourth Ave. Individual ownership. All packinghouse products, lard and compound, and tallow and greases. Codes—Robinson and A. B. C., 5th edition.

J. S. TAYLOR.

TENNESSEE

Memphis

A. G. PERKINS—Porter Bldg. Individual ownership. Codes—Yopp's; code word, "Siquis." Vegetable oils, cottonseed products and also press cloths.

GILES B. BONE CO.—1030 Falls Bldg. Individual ownership. Codes—Yopp and Robinson. Cottonseed products.

L. C. BARTON—Falls Bldg. Individual ownership. Codes—Cross, Sardy, Robinson and Yopp. Vegetable oils.

W. B. DASHIELL—1025 Falls Bldg. Individual ownership. Codes—Robinson and Yopp. Vegetable oils.

SCRUGGS-CARTER CO.—1512 Bank of Commerce Bldg. Partnership. Codes—Yopp's and Robinson's. Vegetable oils, cottonseed and peanut products.

W. C. NORTHERN—313 Central State Bank Bldg. Individual ownership. Codes—Yopp and Robinson. Wholesale cottonseed products.

STERNE-BARTON CO.—Falls Bldg. Codes—Yopp's, Cross, Sardy and Robinson. A corporation. Cottonseed products, vegetable oils and packinghouse products.

SCRUGGS-HILDEBRAND CO.—22 North Front St., 208 Falls Bldg. Partnership. Codes—Yopp's and Robinson. Vegetable oils, cottonseed products, mill feeds and fertilizer materials.

W. P. BATTLE & CO.—56 Porter Bldg. Individual ownership. Codes—Yopp's; code word, "Nub." Vegetable oils.

ROBERT RUFFIN—56 Porter Bldg. Individual ownership. Codes—Yopp's. Vegetable oils. (With W. P. Battle & Co.)

F. W. BRODE CORPORATION—119 Madison Ave. President, F. W. Brode; Vice-President, J. L. Brode; Secretary, B. D. Brode; Treasurer, J. H. Mangum. Codes—Yopp's, Robinson, Private, A. B. C., 5th edition, Hinrick's Baltimore, 2nd edition. Cable address, "Brode." Dealers in cottonseed and peanut products.

MANIRE BROKERAGE COMPANY.

TEXAS

Houston

L. A. HAMMER & CO.—415 Commercial Bank Bldg.

Dallas

FRED C. TONGUE & CO.—327-328 Slaughter Bldg. Individual ownership. Codes—Cross, Robinson, The Miller's. All packinghouse products, lard and compound, vegetable oils, S. P. meats. Specializing in Mexican business.

WASHINGTON

Seattle

FARRELL-BOSWELL & COMPANY—907 Hoge Bldg. Partnership. Codes—Cross, Robinson, Armsby and Yopp's. General brokerage edible and inedible oils, all kinds, both domestic and oriental; fish and wholesale fertilizers, packinghouse products; especially by-products.

GEO. W. BOWERS—Thompson Bldg.

W. J. LAKE & COMPANY, LTD.—600-A Central Bldg. Packinghouse products, edible oils, provisions. Codes—Western Union, A. B. C., 5th edition, Robinson, Cross, Yopp's. Cable address, "Lakeco."

LIVE STOCK ORDER BUYERS

ALABAMA

Montgomery

ABRAHAM BROS.—Cor. Perry and Jefferson Sts., P. O. Box 335, Partnership. Hogs for export only.

KENNETT, P. C., & SON—Union Stock Yards Company.

COLORADO

Denver

AMERICAN COMMISSION CO.
HAVENS LIVE STOCK CO.
HOUSTON COMMISSION CO.
LESLIE COY.
CLAYTON & MURNAN.
DEGEN BROS.
ESSER, HENRY.
FLETCHER & DODD.
HILL, W. S.
HALL, J. O., & SON.
HENDERSON, PAUL.
JAMISON, HARRY.
LEVY, ED., & SON.
LOWELL PURE BRED CATTLE CO.
MILLER, J. A., & SON.
MOOK BROTHERS.
MARLEY & DODD.
MILLER, JOE., & CO.
MILLER & MELNICK.
NUNN, JACK.
PEPPER, JOE.
RICHARDS & MITCHELL.
SINGER, HENRY.
SMITH, F. E.
SCHAEFFER, ARTHUR.
WYATT, CLIFF.
WHEELER, JAMES.
WILKERSON, B. R.

All Union Stock Yards, Denver.

GEORGIA

Atlanta

SHIPPEY, J. K., & BRO.—Miller Union Stock Yards.

ILLINOIS

Chicago

BOWLES LIVE STOCK COMMISSION CO.—U. S. Yards. A corporation. President, J. P. Bowles; Vice-President, Chas. F. Goepper; Secretary and Treasurer, L. F. Weeks.

BAKER HEYNE COMPANY—U. S. Yards. Partnership.

ADLER SONS & CO.—Exchange Bldg., U. S. Yards. Partnership.

REEVES, R. R., & CO.—U. S. Yards. Individual ownership.

CLARK, BOWLES & CO.—107 Exchange Bldg., U. S. Yards. Individual ownership, E. A. Clark.

WHITE & CO., C. E.—79 New Exchange Bldg., National Stock Yards. A corporation. President, C. E. White; Vice-President, F. M. Rook; Secretary and Treasurer, A. J. Hallows.

OWEN, C. EGAN & CO.—U. S. Yards. Codes—Cross.

MOOG & GREENWALD—28 Exchange Bldg., U. S. Yards. Individual ownership.

WOOL GROWERS COMMISSION COMPANY—188 Live Stock Exchange Bldg., U. S. Yards.

NIXON, C. E., & CO.—167 Exchange Bldg., U. S. Yards. Partnership.

GOOGINS & WILLIAMS—134 Live Stock Exchange Bldg., U. S. Yards. Individual ownership. Codes—Utility.

DOUD, KEEFER & ETTINGER, U. S. Yards. Partnership.

PRITCHARD COMMISSION CO.—Exchange Bldg. Partnership.

HARRY SHERWOOD & SON—27 Exchange Bldg., U. S. Yards. Partnership. W. H. Sherwood and R. H. Sherwood.

Following order buyers all care Union Stock Yards, Chicago:

KING, A. S.	NEWTON, THOS. D.
SINCLAIR, GILLETT & CO.	EGAN, E., & CO.
BRADBURN, W. J.	BROWN, KENNEDY & CO.
DARLINGTON & CO.	MARLOW, A.
FONLE BROS.	GANEY BROS.
LAWLER, J. J.	BRENNAN, F. J.
LETT, JOHN R., & CO.	L. W. BUDD.
NICHOLS, H.	FELIX GEHRMANN.
FOSTER, R. C.	HUFFMAN & HUTCHISON.
BELL, H. L.	KELLER, J. B.
HERBERT, G. R., & CO.	HARRY LOEWELL.
DANIEL, WM., & BRO.	MARTIN, C. H.
RONAN, T. J.	A. TREVELLYAN.

East St. Louis

MORRIS BROS. & DUNHAM.	ROUNTREE BROS.
NIFONG COMMISSION CO.	ARNOLD, HENRY.
COY & DELMORE.	GRIMES, I. C.
CAUDLE COMMISSION CO.	RICE, W. J., CO.
HOLLOWAY & CO.	HENSLEY, W. R., & CO.
HILTON, W. E., & CO.	SUNDHEIMER, BOB, & CO.
HUGHES, WM. S., CO.	MEYER, ABE.
WALKER-WATKINS.	SLOAN, JOE.
WATKINS & COMPANY.	YOUNG, W. H., & CO.
TIPPETT & CO., W. S.	SNIDER BROS.
KENNETT-SPARKS & CO.	PERSHALL, A. J.
CASH, E. T., & CO.	

All National Stock Yards, Ill.

Peoria

CHAS. F. HILL—Peoria U. S. Yards.
EVERETT BUSTER—U. S. Yards.
WHITE, F. E., & CO.—U. S. Yards.
PIERSON, C. A., & CO.—Foot of South St.

INDIANA

Indianapolis

ACKLEN, S. J.	REYNOLDS, FRANK R., CO.
LICHTENBERG CATTLE CO.	KENNETT, MURRAY & DARNELL.
GARDNER & HUSSEY.	
SCHULL, JOHN K.	HILL, J. W.
THOMPSON, J. W., & CO.	McMURRAY & JOHNSTON.
HERBERT SAWYER & CO.	KRAMER, C. F., CO.
POWELL, HARTING & CO.	KAHN, D. A.

All Union Stock Yards, Indianapolis.

Lafayette

KENNETT, MURRAY & CO.—Union Stock Yards.

IOWA

Sioux City

HEFNER BROS. & KLOEK.	BALDWIN, KITSELMAN & TIMMEL.
DONOHUE, J. J.	
RICE BROTHERS.	CLAY, ROBINSON & CO.

MIDWEST LIVE STOCK COM. CO.
IOWA COMMISSION CO.
HERMAN & McGLAUELIN.
STEELE-SIMAN & CO.
FITZSIMMONS-PIERCE-FRICK COMMISSION CO.
HUDSON-GIBBS COMMISSION CO.
LONG & HANSEN.
SWANSON, GILMORE & WALSH.
WAGNER, GARRISON & ABBOTT.
WAITT & LAKE COM. CO.
BIRMINGHAM, E. H.
FRANK E. SCOTT COMMISSION CO.
FLYNN COMMISSION CO.
GEORGE M. VICKERS COMMISSION CO.
GILMAN COMMISSION CO.
HIGGINS SHEEP COMMISSION CO.
INGWERSEN BROS.
LEE LIVE STOCK COMMISSION CO.
LYNCH & GAMET.
ROSENBAUM BROS. & CO.
SIOUX CITY LIVE STOCK COMMISSION CO.
WOOD BROS. & CO.

All Stock Yards, Sioux City, Iowa.

KANSAS
Wichita

FLINT HILLS LIVE STOCK COMMISSION CO.—47 Live Stock Exchange Bldg. Partnership.

STUART, C. A., LIVE STOCK COMMISSION COMPANY—Union Stock Yards.

SWANFELKT, E. S.—Wichita Union Stock Yards.

KENTUCKY
Louisville

COLLINS, RUSSELL, BAILEY & CO.
HENRY KNIGHT & SON, INC.
BOWLES, J. T.
WATKINS, CARRUTHERS & CO.
KENNETT, P. C., & SON.
GARRIOTT, T. L., & CO.
ROGERS & CO.

All Exchange Building, Bourbon Stock Yards.

MARYLAND
Baltimore

SUNDHEIMER, S.—2423 Madison Ave.

MICHIGAN
Detroit

SANDEL, STACY, BEADLE & GREEN.
BISHOP, HOLMES, HAMMOND & JACKSON.
JOHNSON, PRINCE, HAMMOND & HALL.
BRESNAHAN & SONS, T.
KENNETT, MURRAY & COLINA.

All Detroit Stock Yards.

MISSOURI
Kansas City

SCHWARTZ, BOLEN & CO.
CRIDER BROS. COM. CO.
CALLAHAN, J.
GRAYBILL & STEPHENSON.
M. K. & T. COMMISSION CO.
CHERRY-TILDEN LIVE STOCK COMMISSION CO.
RICE & KIRK.
FREED ORDER BUYING CO.
B. BALLING.
CLATTERBUCK, WILSON & LANGFORD.
GOODSON GREEN & CO.
L. LEVY.

LIVESTOCK ORDER BUYERS

CURTIS & WRIGHT.
ROBINSON-HOOVER COM. CO.
BOWLES LIVE STOCK COM. CO.
RECORD ORDER BUYING CO.
GEORGE W. SEARLS.
H. M. SPARROW.
H. STEINFELS & CO.
H. C. WILLIAMS.

All Live Stock Exchange Bldg., Kansas City Stock Yards.

St. Joseph

WOOD LIVE STOCK COM. CO.
HOLTMAN, J. W.
AUSTIN-HAMILL-DIXON LIVE STOCK COMMISSION CO.
CLAY ROBINSON & COMPANY.
GREAT WESTERN LIVE STOCK COMMISSION CO.
WHEELER & SONS.
VALLERY-BAKER-JACKSON-JORDON.
MORLOCK, W. H.
JONES, C. H.
ROUNDTREE, W. R.
TRIPLETT & SON.
JAMES STROCK.
TRAMP, CHAS.
HRENCHIR BROS.
AIKENS, J. V.
VENCILL & SON.

NEBRASKA

Omaha

J. W. MURPHY.
FRANK ANDERSON & SON.
ROBERTS BROS. & ROSE.
LAIRD LIVE STOCK COMMISSION CO.
GEO. M. WOOD SHEEP COMMISSION CO.
DONAHUE-RANDALL & CO.
JOHN HARVEY & CO.
ALLEN DUDLEY & CO.
J. H. LAWRENCE.
BYER BROS. & CO.
LLOYD McADAMS.
SWARTZ & COMPANY.

All Live Stock Exchange Bldg., Stock Yards Station.

NEW JERSEY

Jersey City

MYERS & HOUSEMAN—Jersey City Stock Yards.

Newark

CANFIELD COMMISSION COMPANY—32 Plane St., Newark Stock Yards.

NEW YORK

New York

McCABE, DREELAN & McCABE—312 West 60th St.

Buffalo

GEORGE C. EIRICK.
STACY, BEMENT & BEADLE, INC.
J. A. GRUNDTISCH.
LAMBERT CANNON.
ZIMMER BROS.
JOHNSTON, E. E.
FORD & HOLLOWAY.
CLAY, ROBINSON & CO.
DUNNING & STEVENS, INC.
SADLER, RORICK & CO.
MEEKS, BOREN & MILLER.
SWOPE, HUGHES, WALTZ & BENSTEAD.
RICE & WHALEY CO.
DALTON MEEKS CO.
IMHOFF COMMISSION CO.
WINDSOR BROS.

All Live Stock Exchange Bldg., East Buffalo.

OHIO
Cincinnati
KENNETT, COLINA & CO.—Union Stock Yards.

Cleveland
BENSTEAD, BRYANS & CO.
BOWER & BOWER.
NATIONAL LIVE STOCK COMMISSION CO.
SHIPPERS COMMISSION CO.
GREENE, EMBRY CO.
MEEKS, BOREN & THOMPSON.

All Cleveland Union Stock Yards.

Dayton
GREENE, EMBRY & PETERSON COMMISSION CO.—Union Stock Yards.
McLEAN & COMPANY, Union Stock Yards.

OKLAHOMA
Oklahoma City
LIBERTY LIVE STOCK COMMISSION CO.
BARFOOT-VINSON COMMISSION CO.
CASSIDY SOUTHWESTERN COMMISSION CO.
ENGLISH-LOBB COMMISSION CO.
FUSON COMMISSION CO.
HEALY & COMPANY.
MALOY COMMISSION CO.
NATIONAL LIVE STOCK COMMISSION CO.
OKLAHOMA LIVE STOCK COMMISSION CO.
SCANNELL-SLITT COMMISSION CO.
STRIBLING COMMISSION CO.
TATE-INGRAM LIVE STOCK COMMISSION CO.
WITHERSPOON-McMULLEN LIVE STOCK COMMISSION CO.
T. B. SAUNDERS & CO.
W. H. MAY.
HUGO KAPF.
WRIGHT, R. O.
CASH, JAMES.
LACKEY & PRYOR.

All Exchange Bldg., National Stock Yards.

PENNSYLVANIA
Pittsburgh
WILLIAM A. MERRITT—Exchange Bldg., U. S. Yards.
UNION LIVE STOCK COMPANY, U. S. Yards.
REEVES, R. R., & CO.—1210 Esplanade St.

Lancaster
MYERS, HOUSEMAN, RICE & WHALEY.
L. D. HIMMELBERGER.
J. H. & ELIAS STAUFFER.
H. H. SNAVELY.
KNOX, R. N.
MUSSER, H. C.
MINNICH & MYERS.
BUSH & BRUBAKER.
HOMSHER BROS.
LIED, WM.

All Union Stock Yards.

Philadelphia
HOLMES & CLARK.
PHILADELPHIA SHEEP CO.
LENAHAN, SMYTH & CO.
THEO. B. LANDIS.
B. F. BEAR.
CHRISTY & COMPANY.
HEILBRON & LOEB.
JOHN GARTLAND.
LAMAR HUTTON.
HENRICKSON & CO., J. E.
COULBOURN & NOBLE.
RICE, C. J.
GROSS, C., & BROTHER.
MONK, ARTHUR.

All West Philadelphia Stock Yards.

SOUTH DAKOTA
Sioux Falls

SAVAGE, E. W., & CO.
BIG SIOUX COMMISSION CO.
COOLEY, B. H.
TRI-STATE FARMERS COMMISSION CO.

All Live Stock Exchange Bldg., Stock Yards.

TENNESSEE
Nashville

MURRAY, F. L., & COMPANY—U. S. Yards.
KENNETT, P. C., & SON—U. S. Yards, 900 Second Ave., North.
BURNETT-WILSON-LA CROIS CO.—South Memphis Stock Yards.

WASHINGTON
Spokane

OVERMAN, A. V., & COMPANY.
MURPHY, P. W., COMMISSION CO.
HISLOP SHEEP COMPANY.
CONDON COMMISSION COMPANY.
PRIEST, ED.

All Spokane Union Stock Yards.

WEST VIRGINIA
Wheeling

TAVINER, RAY—Union Stock Yards.
PATTERSON, B. B.—Union Stock Yards.
SAX BROS.—Union Stock Yards.

WISCONSIN
Milwaukee

BOOTH & McDERMOTT.
BRUEMMER, J. P., & SON.
CLOUGH-COOK & CO.
HOLMES & ROBINSON, INC.
SPENCER, L., & COMPANY.
KANE, STRAUSS & WINKELEMANN.
MORRIS, BEGEL.
ISAAC, SIMON.
TERWILLEGER, L.
VAN NORMAN, G. B., & CO.

All Milwaukee Stock Yards.

THE NATIONAL PROVISIONER

ESTABLISHED IN 1889

Accepted at home and abroad as the authoritative source of information on all matters pertaining to the Meat Packing and Allied Industries. Official publication of the Institute of American Meat Packers and other trade associations.

Executives, officials, department heads, superintendents, foremen, branch house managers and salesmen are able to keep closely in touch with the various activities of the industry by reading THE NATIONAL PROVISIONER every week.

If you are not a subscriber, do not miss another issue. If there is anything you want to know about that you do not find in "The Packers' Encyclopedia," write to

THE NATIONAL PROVISIONER

Old Colony Bldg.
CHICAGO

15 Park Row
NEW YORK

Subscription in the U. S., $3.00.
Canada, $4.00. Foreign, $5.00
per year

Mr. American Meat Packer!

Are you affiliated with the one trade organization representing the Meat Packing industry? If not, you should know of our work and the benefits that come from such a membership.

OBJECTS OF THE ORGANIZATION

THE INSTITUTE OF AMERICAN MEAT PACKERS is an incorporated national organization composed of hundreds of packers located throughout the United States, Canada and foreign countries. The Institute is organized—

- a. to secure co-operation among the meat packers of the United States in *lawfully* furthering and protecting the interests and general welfare of the industry;
- b. to afford a means of co-operation with the federal and state governments in all matters of general concern to the industry;
- c. to promote and foster domestic and foreign trade in American meat products;
- d. to promote the mutual improvement of its members and the study of the arts and sciences connected with the meat-packing industry;
- e. to inform and interest the American public as to the economic worth of the meat-packing industry;
- f. to encourage co-operation with live stock producers and distributors of meat-food products.

HOW THE ORGANIZATION FUNCTIONS

The work of the Institute is largely handled under the guidance of committees appointed to consider, investigate and report on matters referred to them. The following is a partial list of permanent committees, and new ones are appointed from time to time:

- Committee on Eradication of Live Stock Diseases.
- Committee on Foreign Relations and Trade.
- Committee on Improved Live Stock Breeding.
- Committee on Industrial Relations.
- Committee on Live Stock Shipping Losses.
- Committee on Local Deliveries.
- Committee on Nutrition.
- Committee on Packing House Practice.
- Committee on Public Relations.
- Committee on Soft and Oily Hogs.
- Committee on Standardized Accounting.
- Committee on Standardized Containers.
- Committee to Confer with Government Officials.
- Committee to Confer with Live Stock Producers.
- Committee to Confer with Retail Dealers and Trade Associations.
- Finance Committee.
- Legal Committee.
- Traffic Committee.

Membership in these committees is a recognition of efficiency in a particular line of work. Only the most capable men are selected.

WASHINGTON SERVICE

With an efficiently managed Washington office all matters are promptly handled at a minimum of expense. With the industry practically controlled under federal laws, this represents a tremendous advantage to members located throughout the country.

WE INVITE INVESTIGATION

We take pride in the work we have accomplished and look forward to even greater success. We invite careful investigation, and welcome inquiries from members as well as non-members.

Our dues are low and are the smallest thing about the organization. Let us send you an application blank.

Institute of American Meat Packers
509 South Wabash Ave., Chicago, Illinois.

IN every community, one industry and one company is the acknowledged leader because of its integrity, its fair dealing, its progressiveness and because of invariably satisfactory service and high quality merchandise.

It is the aim of **Allied Packers Incorporated** to occupy this position in the meat food industry, and we are striving daily—yes, hourly—to maintain a definite, consistent policy which will achieve this ambition.

Our plants are modern in every respect and conveniently located to serve you at any point.

Chicago—Western Packing & Prov. Co.
Detroit—Parker Webb Co.
Buffalo—Klinck Packing Co.
Wheeling—F. Schenk & Sons Co.
Topeka—Chas. Wolff Packing Co.
Richmond—W. S. Forbes & Co., Inc.
Toronto—Canadian Packing Co.
Montreal—Canadian Packing Co.
Peterboro—Canadian Packing Co.
Hull—Canadian Packing Co.
Brantford—Canadian Packing Co.

Send in your inquiries. Our Products will please you.

ALLIED PACKERS
INCORPORATED

General Offices
CHICAGO—621 Postal Telegraph Bldg.

New York Office
40 Tenth Avenue

Armour's
Lighthouse Cleanser

An Invaluable Aid in The Packing House or Factory

Whatever you ask of a cleanser can be done with Lighthouse. It is a high grade scouring powder — a combination of natural cleansing agents, refined and powdered.

For removing dirt from tiling, concrete, metal or wood, Lighthouse Cleanser is unexcelled.

If you buy it in bulk—by the barrel, as many of our customers do—you'll find it remarkably convenient and economical. The small hand package is very convenient when distributing it to various departments.

Prompt delivery from any of our branch houses. Write for prices and information.

ARMOUR AND COMPANY
DEPT. OF SOAP SALES
CHICAGO

6357

THE CUDAHY PACKING CO.
GENERAL OFFICE:
111 WEST MONROE ST.
CHICAGO, ILLINOIS, U. S. A.

Old Dutch Cleanser

Manufacturers and Proprietors of

"Old Dutch Cleanser"
"Solvene"—A superior soap in shredded form
"Puritan" Hams—Bacon—Lard—Sausage
"White Ribbon"—A choice lard substitute
"Rex" Hams—Bacon—Lard—Canned Meats
"Sunlight" Butter—Eggs—Margarine
"Meadow Grove" Cheese

We Solicit Your Inquiries

concerning our facilities for supplying products for both export and domestic usages. We can furnish anything in beef, pork or mutton, or their by-products, fancy meat and sausage specialties, produce, soaps, glues, glue material, fertilizers, hides, skins, pelts, hog hair, wool, bones, industrial oils, hog and beef casings, oleo oil, oleo stock, neutral lard, barreled pork and beef.

TRADE MARK
Puritan
Hams and Bacon

Plants and Factories

Kansas City, Mo.	Wichita, Kans.	Los Angeles, Cal.
Sioux City, Iowa	Omaha, Nebr.	Salt Lake City, Utah
Memphis, Tenn.	Toronto, Canada	Calumet, Indiana

MEMBERS INSTITUTE OF AMERICAN MEAT PACKERS

Reg. U. S. Pat. Off.

Producers of

Niagara Hams and Bacon

White Rose Pure Refined Lard

Complete line of

Packing House and Food Products

■ ■ ■

Exporters of

Lard, Canned Meats, Casings
and
All Foreign Cuts of Meats

■ ■ ■

JACOB DOLD PACKING COMPANY
BUFFALO, N. Y.

Wichita, Kans.　　　Omaha, Neb.　　　Washington, D. C.
　　　　Liverpool, Eng.　　London, Eng.

MORRIS' Supreme FOODS

ON an infinite variety of foods, you will find the Yellow and Black label. It always means all that the name implies—*Supreme!*

MORRIS & COMPANY

PACKERS ... PROVISIONERS

Chicago   E. St. Louis   St. Joseph   Kansas City   Omaha   Oklahoma City

ROHE & BROTHER

527-543 West 36th Street
New York, N. Y.

Export Offices: 344 Produce Exchange

=== ESTABLISHED 1857 ===

PORK and BEEF PACKERS

LARD REFINERS

Provisions for Export and Home Trade in Any Desired Package

All the world eats meat

FUMBAR, GERMAN WEST AFRICA. Slaughtering here is rather ceremonious but unsanitary. The gentleman in the robes of office is a native slaughterer. Others in the picture comprise helpers, customers, and those who hang around the local market in Africa as in less remote places.

JERUSALEM. The equivalent of a "car load order" in the Holy Land.

CONSTANTINOPLE, TURKEY. A youthful meat trader on the way to his father's stall in the Constantinople market. Sanitation as a factor in distribution is an unconsidered subject.

REVAL, ESTHONIA. The temperature of this Baltic state is favorable to unrefrigerated handling in the out-of-doors. The Baltic temperament is accustomed to markets such as this, never having had the improvements that we take so much for granted.

(Photographs copyrighted by International Film Service Co., Inc.)

But it is prepared under widely different conditions.

Contrast the modern, clean, orderly Swift & Company packing plant with the unkempt ceremonial butcher of West Africa, killing clumsily in the open.

Compare the Swift refrigerator cars and branch houses carrying meat, refrigerated and carefully protected, from packing plant to dealer, with the boy in Jerusalem delivering on his shoulder meat that is unprotected from insects, dust and heat.

Swift & Company delivers meat products to towns and cities not served by branch houses through a system of car routes. Compare this with the Constantinople pack animal, sweaty and fly-bitten, conveying the unprotected meats on its back.

Look at your own retail dealer's modern service and equipment and compare that with the crude, outdoor market of Esthonia.

Consider the meager meat allowances some of these foreign people enjoy, and the 2¾ pounds per week of the average American.

Think also of the sort of meat the pictures suggest; compare the quality with that found in such delicate, delicious products as Swift's Premium Ham and Bacon and Swift's Fresh U. S. Inspected Meats.

American meat packing, as exemplified in Swift & Company, is both a result of, and a contribution to, civilization.

Swift & Company, U. S. A.
Founded 1868

WILSON'S

Certified FOOD PRODUCTS

Represent the Highest Quality that Choice Materials and Expert Preparation can Produce.

HAM	MARGARINE	PEANUT BUTTER
BACON	PORK SAUSAGE	SALAD OIL
LARD	SUMMER SAUSAGE	SLICED BACON
	SHORTENING	BOILED HAMS
	CANNED MEATS	

WILSON & CO.

GENERAL OFFICES: CHICAGO, ILL.

The Wilson label protects your table

Fred J. Anders   Chas. H. Reimers

ANDERS & REIMERS

Architects
Engineers

Specialists in Packing Plants
and Allied Industries

Erie Building   Cleveland, O.

Chemical & Engineering Company

431 South Dearborn Street
Chicago, Illinois

Chemists

Architects

Engineers

Practical Experts

We specialize in solving all Technical and Practical Problems in the Packing Industry

Henschien & McLaren

Packing House Architects and Engineers

1637 Prairie Ave.   Chicago, Illinois

We have in recent years done work for the following well-known packers:

Jacob Dold Packing Co.	Buffalo, N. Y.
Nuckolls Packing Co.	Pueblo, Colo.
William Davies Co., Inc.	Chicago, Ill.
William Davies Co., Ltd.	Toronto, Can.
Gunn's Limited	Toronto, Can.
John Morrell & Co.	Ottumwa, Ia.
John Morrell & Co.	Sioux Falls, S. D.
Rath Packing Co.	Waterloo, Ia.
Geo. A. Hormel & Co.	Austin, Minn.
White Provision Co.	Atlanta, Ga.
Neuhoff Packing Co.	Nashville, Tenn.
Louisville Provision Co.	Louisville, Ky.
East Tennessee Packing Co.	Knoxville, Tenn.
J. H. Allison & Co.	Chattanooga, Tenn.
Evansville Packing Co.	Evansville, Ind.
Birmingham Ice & Cold Storage Co.	Birmingham, Ala.
P. P. Williams & Co.	Vicksburg, Miss.
Independent Packing Co.	Chicago, Ill.
Louis Pfaelzer & Sons	Chicago, Ill.
City Abattoir Co.	Chicago, Ill.
Allied Packers, Inc.	Chicago, Ill.
Brennan Packing Co.	Chicago, Ill.
Agar Provision Co.	Chicago, Ill.
Oscar Mayer & Co.	Chicago, Ill.
Illinois Packing Co.	Chicago, Ill.
David Levi & Co.	Chicago, Ill.
Siegel-Hechinger Packing Co.	Chicago, Ill.
Kerber Packing Co.	Elgin, Ill.
Ogden Packing & Provision Co.	Ogden, Utah
Ogden Stock Yards Co.	Ogden, Utah
Yorkshire Creamery Co.	Ottumwa, Ia.
Universal Packing Co.	Fresno, Cal.
A. D. Davis Packing Co.	Mobile, Ala.
Belle Mead Farms	Belle Mead, Va.
Wheatfield Farms	Niagara, N. Y.
Irish Co-Operative Meat, Ltd.	Waterford, Ireland
Capitol Refining Co.	Washington, D. C.
Chas. Wolff Packing Co.	Topeka, Kan.
Figge & Hutwelker Co.	New York, N. Y.
C. A. Durr Packing Co.	Utica, N. Y.
Manitoba Cold Storage Co.	Winnipeg, Can.
S. S. Price	Lexington, Ky.
Field & Company	Owensboro, Ky.
Hammond Packing Co.	Cheyenne, Wyo.
Campbell Brothers Co.	Danville, Ill.
Brooklyn Retail Butchers Corp.	Brooklyn, N. Y.
Louis Meyer Company	Brooklyn, N. Y.
Rochester Packing Co.	Rochester, N. Y.
A. C. Hofmann & Sons	Syracuse, N. Y.
Columbus Packing Co.	Columbus, Ohio
Danahy Packing Co.	Buffalo, N. Y.
Emmart Packing Co.	Louisville, Ky.
Vissman Packing Co.	Louisville, Ky.

Rollins Burdick Hunter Co.
INSURANCE
In All It's Branches

We guarantee lowest rates.

We safeguard your interest.

We prevent losses, by eliminating their causes.

We audit your insurance accounts.

Charles E. Rollins, Jr.   John C. Pitcher
Arch O. Burdick   Raymond Kirk
Robert H. Hunter   Harry F. Thomas
Thomas J. Prindiville   Arthur Croxson

CHICAGO

New York   San Francisco
London   Detroit   Seattle

THE NATIONAL PROVISIONER

is the medium through which manufacturers of machinery, equipment and supplies for the meat packing and allied industries can establish their name, familiarize the trade with their product and secure the good will necessary to repeat orders.

As the recognized authority in the meat packing and allied industries it assures reputable concerns an opportunity to bring their sales appeal before the buying power in this field.

Published every Saturday for 33 years and read all over the world.

THE NATIONAL PROVISIONER

Old Colony Bldg.
CHICAGO

15 Park Row
NEW YORK

SIXTY-NINE years ago THE BRECHT COMPANY was founded by Mr. Gus V. Brecht. The business has since grown to international proportions, but the ideals bred into the organization by the Founder have come down through the years. The same spirit which helped to fabricate our early success continues with us. The same quality of service and merchandise, the same standards of men, the same ideals and purposes still dominate this organization.

The Brecht Company
ESTABLISHED 1853 ST. LOUIS

| NEW YORK | LIVERPOOL | CHICAGO | SAN FRANCISCO | MADRID |
| BUENOS AIRES | SYDNEY | SHANGHAI | CAPETOWN | PARIS |

MAIN OFFICES AND FACTORIES, ST. LOUIS, U. S. A.

Complete

Packing House machinery of every description for both large and small plants.

Offal machinery and equipment for the recovery of packing House by-products.

Canning machinery for every purpose allied to the meat packing industry.

Lard pails and cans for both animal and vegetable oil products.

Lard machinery, large and small. Complete lard refineries designed and manufactured.

Sausage casings carefully graded and packed. We are large exporters and importers of sausage making materials.

The BRECHT

12th and Cass Ave.

"All machinery, tools and the meat packing and

Branches:
NEW YORK
CHICAGO
SAN FRANCISCO
LIVERPOOL

The Brecht Company
ESTABLISHED 1853 ST·LOUIS

Equipment

Sausage making equipment. Complete sausage kitchens designed and manufactured.

Oil refining plants for animal and vegetable oils.

Oleo machinery for every purpose.

Refrigerators and specially designed cooling rooms for packing plants.

Market fixtures including counters, display cases, blocks and meat racks as well as small tools and equipment.

COMPANY

St. Louis, Mo.

"equipment used in allied industries"

Branches:
BUENOS AIRES
MADRID
SYDNEY
SHANGHAI

EVERYTHING for the sausage maker, meat packer, oil refiner and retail meat market man. THE BRECHT COMPANY has ong been known as the foremost manufacturer of equipment for use in the meat packing and allied industries. Through years of close association with the problems of the trade our Packing House engineers have gained a fund of knowledge which is valuable to YOU. They will be glad to help with your problem.

THE Brecht COMPANY
ESTABLISHED 1853 ST·LOUIS

MAIN OFFICES AND FACTORIES, ST. LOUIS, U. S. A.

Airoblast

PATENTED

The
A. B. C.
of
Meat Smoking

Saves Time
Saves Labor
Reduces Cost
Increases Profits

Recognized today as the most perfect method of smoking meat

Address

Airoblast Corporation

1807-1809 So. Clark St.   CHICAGO, ILL.

The Allbright-Nell Co.

5323 South Western Boulevard

Chicago, Illinois

LEADERS in the manufacture of packing house equipment.

ORIGINATORS of the present method of dehairing hogs by massaging with beaters.

INVENTORS of the lard cooling cylinder.

WE have maintained the leadership in the manufacture of all equipment necessary for the production of lard, lard compound and the refining of edible oils.

ORIGINATORS of Hog Dehairing Machines, Viscera Inspection Tables, Reversible Trolley Conveyors, Lard Cooling Cylinders, Deodorizing for Fats and Oils, Lard Crackling Flour Process.

Some of Our Specialties

Filter Presses, Tank Water Evaporator, Fertilizer Dryer.
Eldredge Patent Blow-up System for Tankage.

Overhead Tracking System. Baragwanath Barometric Condenser. Soap Machinery.

A Continuous Crackling Press

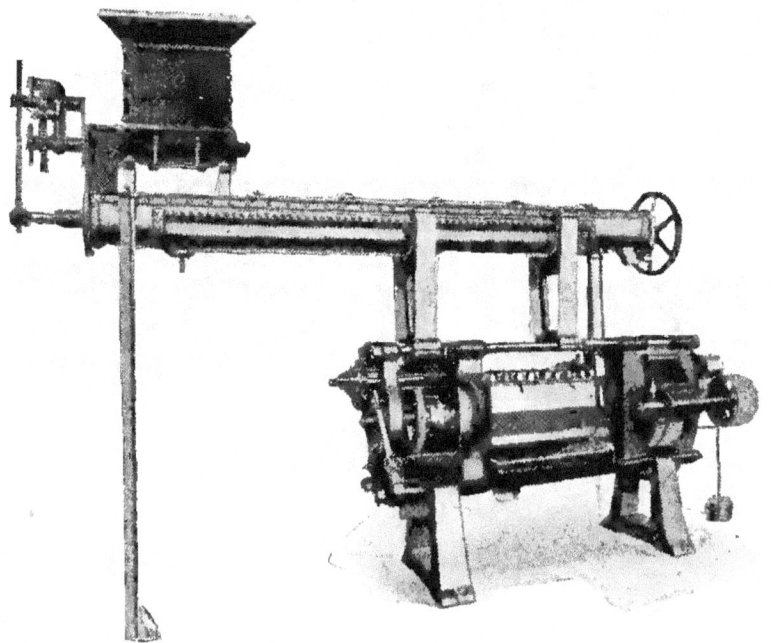

The **Anderson Expeller** will extract 25 per cent grease from the average hydraulic pressed cracklings.

Write for complete information.

Manufactured by

The V. D. Anderson Co.
1935 W. 96th Street

Cleveland   Ohio

ARCTIC

Ice Making
and
Refrigerating Equipment

In Sizes from One to Five Hundred Tons
Daily Capacity

Write Us for Estimates and Literature

The
Arctic Ice Machine Co. CANTON, OHIO, U.S.A.

BRINE SPRAY REFRIGERATION

THE choice of progressive packing house men—one experience with a Webster System and nothing else will do. We have convinced the most conservative. Proper nozzles *properly applied* is the secret of success. Our methods are original and practical—the product of patient scientific development on a large scale—proven in the manifold uses to which they have been put. In 1921 over 100,000 Webster Brine Spray Nozzles were at work in the World's largest and most modern packing houses.

Vigorous air circulation—without fans—dry ceilings, and quick, thorough chilling mark our systems. Products acquire and hold a finish which cannot be excelled. These results are obtained using weak, high-temperature brine—a combination which reduces shrinkage. Pressures required on Webster Sprays are exceptionally low, ranging from 3 to 9 lbs.—reducing pumping cost.

Webster Spray Nozzles—*as we apply them*—will meet your requirements' no matter how severe or unusual. We install complete brine spray systems for any service. Our broad experience at your disposal. Why not use it?

Summer Sausage Drying—with Webster Air Conditioning Apparatus—can be done at all seasons, and a product unparalleled in color and uniformity assured. We can show you working installations of all sizes. We design, manufacture and install complete air conditioning systems for this purpose, and a variety of others.

"THE SUCCESSFUL SYSTEMS ARE WEBSTER SYSTEMS"

Atmospheric Conditioning Corporation

MONADNOCK BLDG., CHICAGO — LAFAYETTE BLDG., PHILA.

For Drying or Rendering

Triumph Steam Dryers are used both for drying and rendering. The capacity is large and the power and steam consumption low. Nearly 1000 now in operation.

BARTLETT & SNOW DRYERS

comprise thirteen distinctly different types, each of which has a particular field of application. Nearly every drying requirement can be properly met by one of these thirteen types.

The C. O. Bartlett & Snow Co.
Cleveland, Ohio, U. S. A.

Style S Dryer—one of the thirteen different types. Used for drying tankage and similar materials on large scale.

"Niagara Brand"

Genuine

Double Refined

Saltpetre

(Nitrate of Potash)

and

Double Refined

Nitrate of Soda

*Both Complying with all the Requirements
of the B. A. I.*

Manufactured by

Battelle & Renwick
Established 1840

80 Maiden Lane,   New York

THE CASING HOUSE

All kinds of
SAUSAGE CASINGS
carefully cleaned and graded.

BERTH. LEVI & CO., INC.

ESTABLISHED 1882

NEW YORK   CHICAGO

LONDON

FILTER-CEL

TRADE MARK REGISTERED U.S. PATENT OFFICE

A CELITE PRODUCT

In the *Refining of Lard and Oils*

Filter-Cel is a very light-weight, highly porous siliceous powder, made from the mineral Celite, and especially prepared for facilitating filtration processes. It is used to advantage in filtration of all kinds of liquids, in conjunction with all types of filters.

Process:

Filter-Cel is added to the heated oil, in the clay kettle. The mixture is then thoroughly agitated by the paddles or air, and pumped through the filter press in the usual way, returning the oil from the press to the clay kettle until the oil is brilliant and clear.

The cloth in the filter press acts as a retaining medium for the Filter-Cel. A porous film of the filter aid is formed on the cloth, which retains and removes completely all of the most finely divided gummy and gluey substances, but allows the clear oil to pass.

Application:

The Filter-Cel treatment is applied to any raw fat or oil that comes from rendering tanks, expellers or hydraulic presses, such as: Lard, tallow, cottonseed oil, linseed oil, kernel oils, peanut oil, corn oil, olive oil and soya bean oil.

Keeping Qualities of Final Product:

The gummy and gluey impurities, along with the moisture, are completely removed by the Filter-Cel treatment. The medium, therefore, in which the ferments propagate best, has been eliminated. The resulting product has excellent keeping qualities.

Further Information:

The services of our technical staff are at the disposal of any one desiring help or information on any filtration problem. We will gladly submit filtered samples, together with complete data on the process. Samples of Filter-Cel for testing and experimental use will be forwarded on request.

Celite Products Company

Chicago—Monadnock Bldg.

New York—11 Broadway
Boston—79 Milk St.
Denver—Symes Bldg.
Cincinnati—Union Central Bldg.
Los Angeles—Van Nuys Bldg.
New Orleans—Whitney Central Bank Bldg.
San Francisco—Monadnock Bldg.
Buffalo—Mutual Life Bldg.
Detroit—Book Bldg.
Cleveland—Bulkley Bldg.
Philadelphia—Bulletin Bldg.
St. Louis—Railway Exchange Bldg.

The First Trade-Marked Tight Oak Barrel!

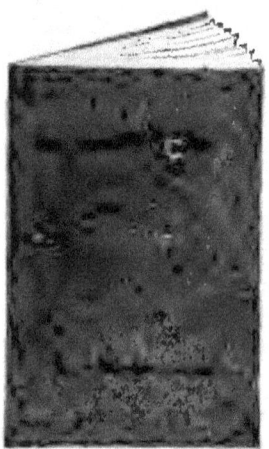

The First Tight Oak Barrel Catalog. Executives of leading packing concerns have written us personal letters of thanks for this book.

WHEN a Cleveland Cooperage Barrel successfully passes the last of nine rigid inspections, then, and not until then, is it branded with the "Triangle C" trade-mark,—a signature to our pledge of unqualified merit.

A Text Book for the Cooperage End of the Packing Industry!

32 pages of Barrel Facts. The first book to present in a single volume complete barrel specifications, barrel-buying data and barrel-handling pointers. Careful reading by an executive is virtually certain to result in greater packing economy for you. Gladly given to executives on request. *Write*.

The Cleveland Cooperage Company
2229 W. 61st ST., CLEVELAND, OHIO

"Boss" Packing House Equipments

Grate and U
Hog Dehairers
Jerkless Hog Hoists

Patented in United States and Canada

The Great Winners on Merits

"Boss" Meat Cutter

Use "Boss"
Hog and Beef Killing
Sausage and Lard
Tanking and Drying
Outfits

Consult us about equipment before building or altering your plant

The Cincinnati Butchers' Supply Co.
Manufacturers

Branch Office
975 Old Colony Bldg.
Chicago, Ill.

Main Office and Factory
1972-2008 Central Ave.
Cincinnati, Ohio

The "Dayton" System
of
Rendering and Drying Tanks
(Combination Type)

Will give you

Quick Results and Uniform Tankage

❖ ❖

Descriptive Bulletin on Request

❖ ❖

Dayton Beater & Hoist Co.
Dayton, Ohio

PRESSES

for
Tankage, Glue
Oleo, Crackling,
Stearic Acid,
Tanneries, etc.

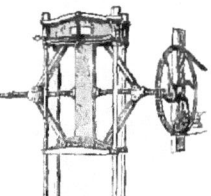
Oleo Press

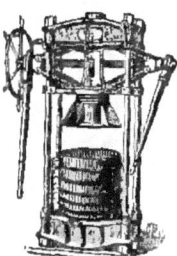
Scrap Press

Our Knuckle Joint Oleo Presses have been in use for many years by all the leading manufacturers of Oleo oil in this and other countries. The slow, steady pressure exerted on the material gives results unexcelled both as to quality of product and quantity produced.

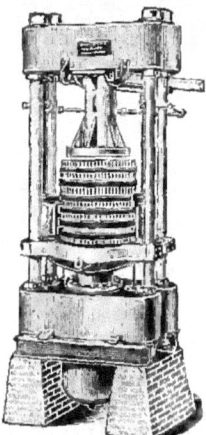
Scrap Press

We make Scrap Presses of various sizes and with pressures of from 60 tons to 900 tons to meet the requirements of the user. The heavier pressures give surprising results both a to economy of labor and small percentage of grease left in the crackling. It will pay you to investigate.

Our Tankage and Glue Stock Presses are "Standard," and their reputation for excellence established by years of use in both large and small Packing Houses. If you are interested in presses for this or any other purpose, let us tell you more about them.

We also make Pressure Pumps, Accumulators, Operating Valves, Fittings, etc.

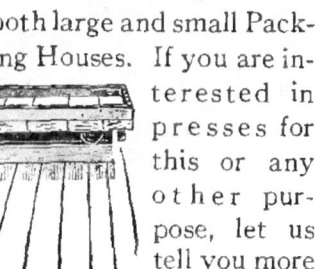
Tankage Press

Tankage Press

Dunning & Boschert Press Co., Inc.
362 West Water St.　　*Established 1872*　　Syracuse, N. Y.

"ENTERPRISE"
New Power Choppers—Tinned

No. 166

Height.....................37½ inches	Speed of Pulleys, 200 revolutions per minute.
Length.....................49 inches	Power 7 to 10 horse.
Diameter of Hopper.............20 inches	Capacity per hour, 1500 lbs. of Beef cut three times; 3000 lbs. Pork cut twice.
Diameter of Throat.............7 inches	
Diameter of Plate..............8⅝ inches	Weight of Machine complete, 737 lbs.
Extra Heavy Pulleys.......28x4¼ inches	

Double Thickness Belt must be used

Greater Capacity can be secured by increasing speed of pulleys. This can be done up to 30 per cent without injury to machine.

Packed 1 in a crate, weight 1000 lbs.

No. 186

Fitted with 15 H. P. Motor
For Direct or Alternating Current

Height, 41 in. Length, 82 in. Width, 36 in. Weight, 3,000 lbs. Cylinder with or without Steam Jacket as desired. The Gears are cut, the smaller one being made of Raw Hide.

Capacity

Beef (first cutting) ....5,000 lbs. per hour
Beef (second cutting).....................2,500 lbs. per hour

Four Plates are furnished with each machine, one fine (⅛-inch holes), one medium (¼-inch holes), one coarse (⅜-inch holes) and one Fat (1½-inch holes), also three knives and one Fat Knife.

Packed 1 in a crate, weight 3,250 lbs.

THE ENTERPRISE Mfg. Co. of Pa.
Phila., U. S. A.

FRICK
Ammonia Compressors

Vertical Single-acting and Horizontal Double-acting Medium Speed Types.

Made in sizes to meet the requirements of the Meat Packing and allied industries.

Ask for Catalogs A-9 and A-10.

Branch Offices and Sales Agents in all principal cities throughout the world

Gruendler Heavy Duty Crushers and Pulverizers

This machine is especially adapted for crushing and grinding green bones, steamed bones, tankage, manure, acid cake, limestone and various other materials.

Write Us for Full Particulars

Gruendler equipment is built right, our engineering department has had years of experience in the meat and allied industry field and we are ready to serve you.

We also manufacture disintegraters, hammer crushers, shredders and pulverizers for all purposes.

Gruendler Patent Crusher & Pulverizer Co.
942 N. Main St.   St. Louis, Mo.

TYPE B BOILER Made in 6 Sizes

The Latest Ham Containers

The latest ham boilers with the yielding spring pressure attachment and, therefore, the only ones that will reduce the shrinkage in boiling. They are cheaper in the end than any boiler on the market.

No power-press needed.
No string needed for tying ham.
No cloth wrapper while boiling the ham.
Best quality ham. It cooks in its own juice, thus retaining its flavor and nourishing qualities.
Holds together firmly under any conditions.
Boilers—Made of cast aluminum. No rust spots. Always sanitary.
Based on simple common sense principles, they can be worked by anybody.

Beware of infringements. Infringements will be prosecuted.

Write for details to

The Ham Boiler Corporation

1762 Westchester Ave.   New York City

Meat Curing Hogsheads for Any Capacity

Our facilities are unexcelled for manufacturing hogsheads from strictly genuine quartered white oak, 1 inch thick before dressed for the staves, and white pine or genuine long leaf yellow pine 2 inches thick before dressed for the bottoms.

We use standard hooping of five galvanized hoops, 2 inches wide, No. 14 gauge. Standard size is 1500 lbs. capacity, but hogsheads of any capacity less than 1500 lbs., with or without covers, made to special order.

We quote prices without obligating our client.

The Hauser-Stander Tank Co.
"We Win With Quality"
Cincinnati, O., U. S. A.

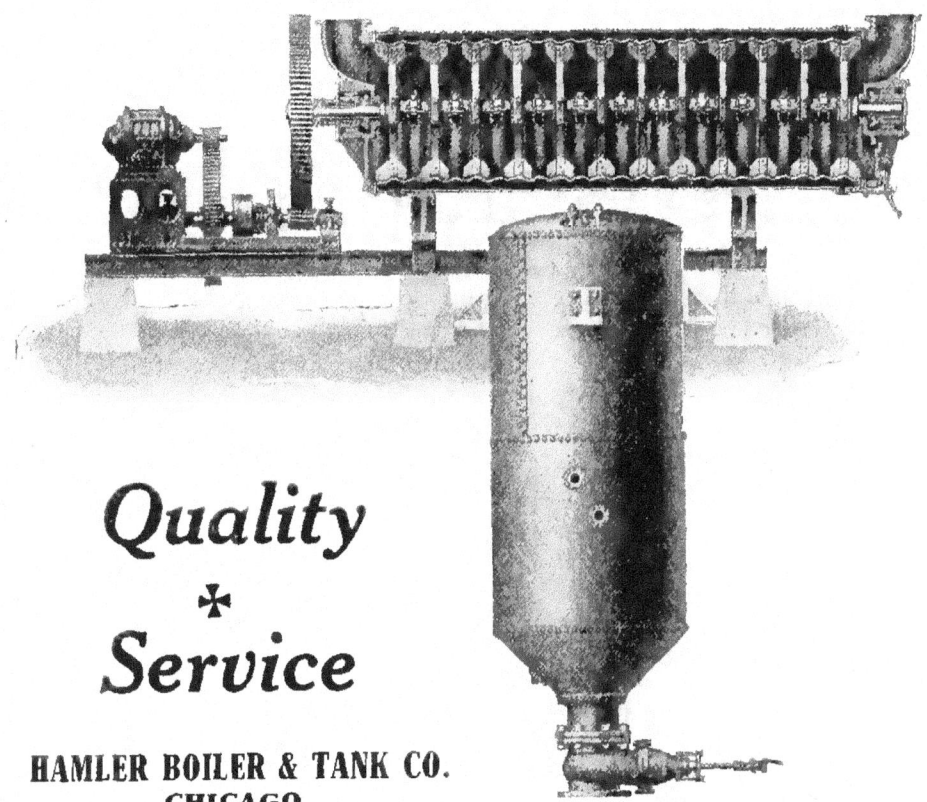

Tierces
Oil Barrels
Pork Barrels
Half Barrels
and Kegs

For Domestic and Foreign Trade

❖

Tierces a Specialty
Capacity 2,000 Packages per Day.

❖

C. G. Hopkins Cooperage Co.
Established 1902
Joplin, Mo.

Lard Presses
Sheep Skin Presses
Stearic Acid Presses
Oleo Presses
Tankage Press Cloth
Tankage Racks
Tankage Trucks
Garbage Presses
Fertilizer Presses
Hydraulic Pumps
Hydraulic Valves
Hydraulic Fittings

"For Your Pressing Needs"

H - P - M
HYDRAULIC

Tankage Presses

There is no shoddy material used in the construction of H-P-M Presses. All material and workmanship is of the highest quality. All parts are rigidly inspected and tested. Every press sold under the H-P-M Guarantee. Pressure capacities 15 to 500 tons. Single or double transfer car systems.

Ask your Jobber about H-P-M products or write direct to factory.

The HYDRAULIC PRESS MFG. CO.
45 Lincoln Avenue, Mount Gilead, Ohio

BRANCH OFFICES:
NEW YORK   BUFFALO   CLEVELAND   SAN FRANCISCO

Packers

wanting to extend their business to

Scandinavian and Baltic Markets

are offered the services of

Johs. P. Larsen

The Danish Oleo Works

10 Dortheavej

Copenhagen · L

DENMARK

Packing House Products · Oleo Oil
Oleo Stock   Neutral Lard
Tallow for Edible and Technical Purposes
Vegetable Oils for Margarine Making

∎∎∎

Cable Address: "NIRAGRAM," Copenhagen
Code: A.B.C. 5th Improved
Bank: Kjobenhavns Handelsbank, Copenhagen

TANKHOUSE, ABATTOIR
RENDERING PLANT
PACKING HOUSE

Odors eliminated
12% ammonia tankage

Can be made from waste hair

The **MACLACHLAN PROCESS** not only cuts odors but saves time and money in the cooking and drying of tankage. It decreases the amount of retained grease and gives a higher ammonia content.

The process can be added to any standard tankhouse layout and results in economies that pay for the installation many times over.

For a full description of the MACLACHLAN PROCESS see chapter on tankage.

Details of its application to your plant will gladly be sent. Our engineering advice is free.

MACLACHLAN REDUCTION PROCESS CO., Inc.
40 RECTOR ST., NEW YORK CITY

SPICES
HERBS—SEEDS

We have studied the spice requirements of the Meat Packing Trade for many years. Our long experience, and the fact that we are direct importers and grinders, places us in position where we can serve you best on the entire line of spices, condimental seeds and herbs.

We specialize in—

Allspice	Pepper—Black	Caraway Seed
Cinnamon	Pepper—Red	Cardamon Seed
Cloves	Pepper—White	Celery Seed
Ginger	Sausage Seasonings	Coriander Seed
Mace	Marjoram	Powdered Onion
Mustard	Sage	" Garlic
Nutmegs	Savory	Saltpetre
Paprika	Thyme	

Our spices are carefully sifted and cleaned, and are ground by special improved processes retaining their original pungency and flavor.

SAUSAGE SEASONINGS

We can supply our own combinations, or will compound goods in conformity with your private formulae.

We solicit the opportunity of figuring on your needs, and will be glad to submit samples, prices, etc.

McCORMICK & CO., Inc.
Spice Importers and Grinders
BALTIMORE, U. S. A.

MECHANICAL

PACKING HOUSE MACHINERY and EQUIPMENT

In almost every civilized country in the world "**Mechanical**" machinery and equipment are used, reflecting in their consistent daily service credit that is far greater than words or pictures can describe.

We also manufacture special equipment for the Fertilizer, Sausage, Oleomargarine, Creamery, Canning, Lard and Vegetable Oil industries, as well as railway bumping posts, boiler and tank work, etc.

Our extensive manufacturing and engineering facilities are at your service.

Let us assist you in solving your equipment problems.

We invite your inquiries.

THE MECHANICAL MFG. CO.

Established 1889

Pershing Road and Loomis Street

Chicago

Products of Quality and Distinction

Morrison's Machinery

for

Butchers, Packers, Renderers, Garbage Reduction and Fertilizer Manufacturers

❖ ❖

Our Specialty: Combined Rendering and Drying Tanks, for the building up of low grade Tankage for Hog and Poultry Feed, for adding blood and Concentrated Tank Water by vacuum producing a uniform analysis. They are best for converting garbage into a high grade vegetable feed. Absolutely *Sanitary* and *Odorless* in operation. Personal supervision of all installations is a guarantee of highest results.

Horizontal Tankage Dryers, Vertical Tankage Dryers, Crackling Breakers, Cookers, Rendering Tanks, Tank Water Evaporators, Vacuum Pumps, Engines, Boilers, Steam and Vacuum Gauges, Pressure Regulators and Reducing Valves.

William G. Morrison
1-3 North Main Street,   Dayton, O., U. S. A.

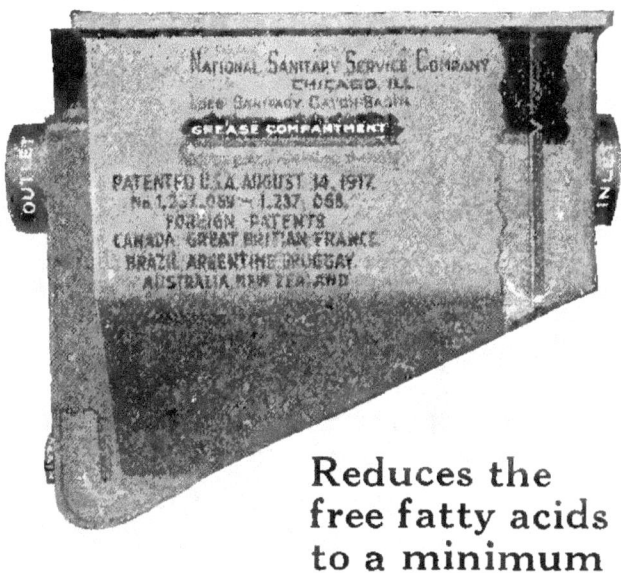

A CASING INSTITUTION
guaranteed by
A WORLD WIDE REPUTATION

for

INTEGRITY
QUALITY
SERVICE

E SOLICIT the patronage of every sausage manufacturer who desires casings a little better than his competitor ❦ ❦ ❦

Oppenheimer Casing Co.
1016-26 W. 36th Street   CHICAGO, ILL.

Also at

New York : London : Toronto : Buenos Aires : Wellington

Factories and Agencies throughout the World

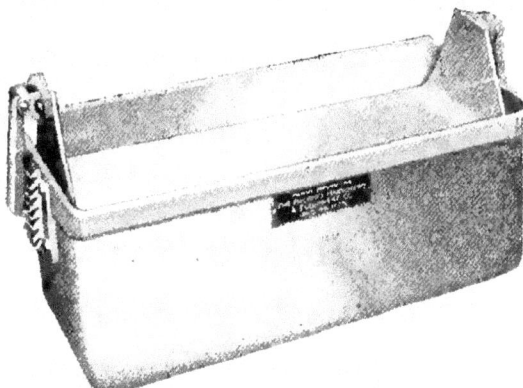

Conveyors for every purpose	Machinery
Overhead Tracking	Lard Tanks and Kettles
Iron Doors and Frames	Smoke House Equipment
Sanitary Tables	Filter Presses
Sanitary Trucks	Hog Dehairing Machines
Tank Water Evaporators	Packing House and Abattoir Specialties
Fertilizer Dryers	

SQUARE FROM END TO END

ARE THE HAMS PRODUCED BY THE

JORDAN

Square Ham Retainer and Ham Boiler

"Jordan" SQUARE Ham Retainer and Boiler is the only Retainer that will produce an absolutely square ham from end to end.

Above illustration shows design of "Jordan" Retainer, which produces absolute square end ham after cooking.

Cover designed in such a manner to properly curve ham and permit the use of a power or hand press to insure solid formed ham.

Lever ratchet absolutely cannot become loose, as same catches automatically and holds without the use of spring.

Eliminates cloth wrappers while boiling ham. Has no springs to crystallize or weaken. Cooks ham in its own juice, thus retaining all its natural flavor and nourishing qualities. One slice makes two perfect sandwiches.

Manufactured in three sizes—

8 to 12 lb. boned and fatted hams......Stock No. A-1
12 to 16 lb. boned and fatted hams......Stock No. A-2
16 lbs. and over..........................Stock No. A-3

THE PACKERS MACHINERY & EQUIPMENT CO.

Manufacturers of
PACKING HOUSE MACHINERY

1400-10 W. 47th St., CHICAGO, Ill., U.S.A. Phones Boulevard 1605-6
Baltimore Office, 620 Dennison St.

*Manufacturers of famous "ECONOMY" Pan Type Viscera Table.
"Economy" Rapid Meat Cutter and "Hildebrandt Revolving Smoke House."*

Patent Casing Co.
Chicago

Sole licensed manufacturers
of all kinds of

Sewed Casings

Sewed under the
May Patents
Granted by the
United States - Canada
and
Principal Countries
of Europe

Patent Casing Co.
617-621 W. 24th Place
Chicago, Ill.

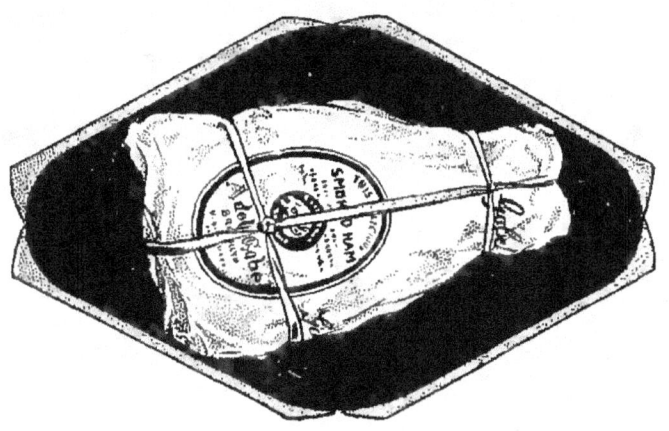

Wrapped for Protection

in

Paterson Vegetable Parchment
"Nearest to Perfection"

Established 1885   Incorporated 1891

The Standard for Nearly Forty Years

Manufactured Exclusively by

The Paterson Parchment Paper Co.

Office and Mills

Passaic, - New Jersey

Standard Results Automatically Assured in

→ Scalding and
→ Dehairing Tanks
→ Cooking Vats and
→ Retorts
→ Smoke Houses
→ Evaporators

or any other places where even temperature is essential, if that temperature is controlled by

Powers Automatic Thermostatic Regulators

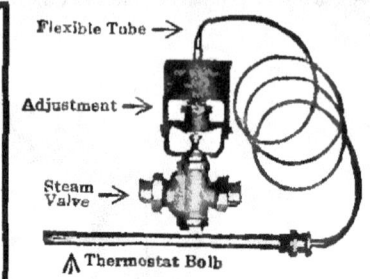

The Powers Regulator No. 11

Entirely automatic. Reliable. Accurate. Can be set for the desired temperature. Easily applied. Put thermostat bulb in liquid to be controlled and valve in steam supply.

Used on Hog Scalding and Dehairing Tanks, prevents mutilating or cutting of the skins.

Powers Regulator No. 16

Especially adapted for use in ham cookers, vats and open tanks. It is quickly and easily installed and operated.

It relieves your employee of the duty of constantly testing the temperature of the liquid. It saves time and labor and assures absolute uniformity in the product.

Many of the World's Largest Packers are using Powers Regulation—much to their profit.

Let us show *you* where you can save money, meat, men, and time with Powers Automatic Heat Control.

Heat Regulation has been our sole business for over thirty years—our wide experience is freely at your command.

THE POWERS REGULATOR CO
Specialists in Automatic Heat Control

R-903—126 East 44th St., New York
2784 Greenview Ave., Chicago
535 Boston Wharf Bldg., Boston
The Canadian Powers Regulator Co., Ltd., Toronto, Ont.
(1696)

"Pack it in Wood"

Since the beginning of time, down through all the ages, containers made of wood have been the most satisfactory for the packing of food products.

EDWIN C. PRICE COMPANY

408 Commerce Building
KANSAS CITY, MO.

1822 So. Clark Street
CHICAGO, ILL.

"Pack it in Wood" Whitewood

Double Refined
Nitrate of Soda

The same formulas are used with Double Refined Nitrate of Soda as with Saltpeter, except that 16% less Double Refined Nitrate of Soda should be used; the reason for this being that 84 parts of nitrate of soda are equivalent to 100 parts of saltpeter.

REX BRAND
The King of Nitrates

Complies with
B. A. I. Requirements

Write for Prices
Immediate Deliveries

∎ ∎ ∎

Stauffer Chemical Co.
Chauncey, New York

San Francisco Salt Refinery
San Francisco, Calif.

∎ ∎ ∎

Sales Agent—Edgar R. Adler,
4148 South Halsted St., Chicago, Ill.

Before You Purchase

Your

Sausage Machinery

be sure to investigate the merits
of the New Model

"BUFFALO" Silent Meat Cutters
"BUFFALO" Meat Mixers
"BUFFALO" Sausage Stuffers
"BUFFALO" Lard Mixers
"BUFFALO" Grindstones
"BUFFALO" Spice Mills
Kraut Cutters, etc.

❖ ❖

We specialize on the above machines. They are being *used successfully by thousands of sausage makers* all over the world.

❖ ❖

Write for catalog and full particulars

John E. Smith's Sons Co.

51 Broadway   Buffalo, N. Y.

Established 1868

S. OPPENHEIMER & CO.

Sausage Casings

IMPORT   EXPORT

96-100 Pearl St.
NEW YORK

2700-2706 Wabash Ave.	47-53 St. John St.
CHICAGO	**LONDON**
Luisenhof	73-77 Boulcott St.
HAMBURG	**WELLINGTON, N. Z.**

Stedman's Grinding and Screening Machinery

The Disintegrator illustrated is the only machine that will successfully grind Fertilizer and Hog Feed Tankage. *Ask for catalog No. 12.*

Our Hexagon Revolving Screen offers an efficient and economical solution to your screening problems, and particularly for materials of a wet or gummy nature which offer difficulties with other types of Screens.

A simple mechanical Tapping Device keeps the screen clear without the employment of labor to clean out clogged materials. The ideal Tankage Screen.

Write for bulletin No. 105.

Details of tapping device which keeps wire cloth clean.

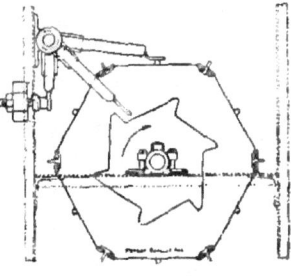

We make a specialty of designing Grinding and Screening Plants. Let us have your inquiries.

Stedman's Foundry & Machine Works
Founded 1834   Aurora, Indiana

Atlanta, Ga.: Hurt Bldg., Murphy-Rountree Co., District Sales Representative.

A Return of Over 100 Per Cent Annually

seems impossible, but we have over 450 Swenson's in packing houses and rendering plants that are getting that every year from the tankwater saved in our evaporators which are also used for making over 80% of all the glue and beef extract produced in this country.

The endorsement of our ideas on construction and design by such companies as Swift, Armour, Cudahy, Wilson, Morris, American Glue Company, U. S. Glue Company, Eastern Tanners Glue Company, etc., and the number of repeat orders from these people is proof enough why our equipment has been adopted as

"The Standard"

in animal product recovery processes.

Over 6,000,000 gallons of tank water are handled in Swenson's every day.

Mr. Manager! Can you afford to run this valuable waste material to the sewer any longer?

Write today for literature and complete data on any problem which interests you.

Swenson Evaporator Co.

Sales and Engineering Offices:
945 Monadnock Block, Chicago

Shops and General Offices:
Harvey, Illinois

Cable Address: "Evaporator Chicago," Western Union Code

Hydraulic Presses

for

TANKAGE :: LARD :: SCRAPS :: OLEO
STEARIC ACID :: LEATHER :: SHEEP-
SKIN AND HIDE BALING

Belt and Steam-Driven Pumps
Valves and Fittings

Cotton and Jute Press Cloth
Steel and Wood Press Racks

□ □ □

Write for catalogue

Thomas–Albright Company
Goshen, Indiana

Spices
Seeds
Herbs

Thomson & Taylor Company
Chicago

All Varieties of
Sausage Seasonings

Union Insulating & Construction Company

Great Northern Building, Chicago

S. E. McPartlin   J. H. Bracken   E. S. Main

Contractors for Corkboard Insulation

for every type of Refrigerated Building

Showing insulation and meat rails being applied
at the United States Cold Storage Co., Chicago

Contractors of Built-in Refrigerators

of every type in Hotels, Restaurants,
Provision Houses, Hospitals. etc.

Cork Board, Insulating Asphalt, Granulated Cork, Mineral Wool, Insulating Paper, Meat Rails, Scales, Tracking.

Celotex Insulating Lumber, especially designed for roof insulation and for the insulation of dwellings, apartments and garages. Celotex may be used in place of sheathing lumber and as a stucco and plaster base.

Members of this firm have specialized in Heat and Cold
Insulation for Twenty Years

☐ ☐ ☐

Estimates and Sketches cheerfully furnished

Quality
Capacity
Service

Warehouses 700 Ft. Long

Immediate Shipments
Carlots Only

The
Union Salt Company
Cleveland, Ohio

"IDEAL" STORAGE VATS
For Pickling   For Curing

Don't Give Up in Despair!

When you are looking for pickling vats, remember that we can supply them in lots from one to a carload—and we will save you at least 50% on the transaction.

Built of hardwood, bound with eight iron hoops; of 170 gallons capacity; and warranted to pass the U. S. Government inspection.

Write us for prices and delivery.

❖ ❖

United Cooperage Company
1115 Fullerton Avenue
Chicago, Illinois

United States Cold Storage Co.
2101 W. Pershing Road
Chicago
Chicago Junction Railway

Reasonable Storage Rates and Lowest Insurance Rates on all Packing House Products. Accurate Control of Temperature—as required from 15 degrees below zero to 50 degrees above. Prompt and efficient service. Our location is adjacent to the Union Stock Yards.

*Approved by the Board of Trade
and the Chicago Mercantile Exchange*

Write, Wire or Phone Lafayette 3060

United States Cold Storage Co.
Chicago

Horizontal Refrigerating Machine

Ice Making and Refrigerating Machinery

CORLISS AND POPPET
Valve Engines

Horizontal Ammonia Compressors from 6 Tons to 750 Tons Daily Capacity

Our Representatives are at your service to assist you in the selection of equipment which will be best suited for your particular requirements.

Twin Cylinder vertical ammonia compressors from ½ ton and up.

Vertical Twin Cylinder Refrigerating Machine

The Vilter Manufacturing Company

893 Clinton Street, Milwaukee, Wis.

Branch Offices in all Principal Cities

G. Van Gelder & Co.

Reguliersgracht 29

AMSTERDAM, Holland

Importer and Exporter of First Class
Selected and Unselected

Sheep Casings

∎ ∎ ∎

BRANCHES:

Chicago **Hamburg**
128 North Wells Street Grosse Johannisstrasse 15

∎ ∎ ∎

Cable address: All offices, "CATAI"
Codes: A.B.C. 5th and Impr., A.B.C. 6th Ed.
Bentley, Marconi, Western Union, Lieber.

The South's Largest Packing Corporation

The building of the new $2,000,000 plant of the Memphis Packing Corporation, Memphis, Tennessee, marks another great stride in the South's industrial development.

The personnel of this modernly equipped plant is composed of men who have had broad experience in other large packing establishments of our Country. Their exceptional work with these concerns puts them in the "Expert Class."

We are proud to say the officials in charge of the construction work in the Refrigerating Plant of the Memphis Packing Corporation specified

Vogt Ice-Making & Refrigerating Equipment

Absorption and Compression Systems

Write for Bulletins

HENRY VOGT MACHINE COMPANY
Incorporated
Louisville, Ky.

BRANCH OFFICES:
NEW YORK, CHICAGO, TULSA, DALLAS

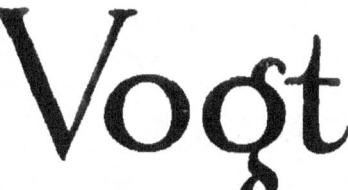

Ingersoll-Rand
AMMONIA COMPRESSOR IS USED IN THE VOGT COMPRESSION SYSTEM

MANUFACTURERS OF ICE-MAKING AND REFRIGERATING MACHINERY. DROP FORGED STEEL VALVES AND FITTINGS, WATER TUBE AND HORIZONTAL RETURN TUBULAR BOILERS, OIL REFINERY EQUIPMENT

Westinghouse offers a complete line of Specially Designed Motors and Control for the Packinghouse

Other
Westinghouse Products for the Packinghouse:

Synchronous Motors
Motor Generators
Turbines—Condensers
Transformers
Switchboards and Switching Equipments
Arc Welding Equipments
Electric Baking Ovens
Industrial Lighting
Space Heaters

*Westinghouse Electric and Manufacturing Co.
East Pittsburgh, Pa.*

Westinghouse

Crushers and Grinders
for Tankage, Bones, and All By-Products

Armour, Swift, Morris and hundreds of other meat packers both large and small are using Williams Crushers and Grinders because of their large capacity—small horse power requirements—low upkeep expense.

They are built in sizes and models to suit every purpose. Big green beef heads, knuckles, in fact, the largest bones in the carcass can be reduced to any fineness with the Williams Mogul Crusher, while smaller bones, tankage and similar materials can be ground with the "Regular," "No. 730," or "Little Giant" type, depending on local conditions. Furthermore, each of these types is built in from 4 to 6 sizes, permitting selection of exactly the proper machine for your use.

Williams Crushers and Grinders reduce by the Hinged Hammer Principle invented and developed by our company. This means that tankage need not be screened after leaving the machine, the screen being right in the grinder in the shape of the cage, which holds the material within the grinder until of the proper fineness.

The above photo shows a No. 730 tankage grinder with air equipment for handling ground material. For conservation of ground goods, economy and cleanliness in handling, this method cannot be equaled. If you have a crushing or grinding problem put it up to our engineering department. Our one-quarter of a century's experience is your guarantee of satisfaction. Free engineering service and plans.

WRITE TO DEPT. 37.

Williams Patent Crusher and Pulverizer Co.   Established 1871

Plant and General Offices 2701-2723 North Broadway

Chicago Office   St. Louis, Missouri, U. S. A.   San Francisco Office:
37 W. Van Buren Street   67 Second Street

Wolf, Sayer & Heller
Incorporated

The Leading Casing House of the World

Cleaners, Importers and Exporters of all kinds of

Sausage Casings

Also

Staple Sausage Room Supplies

Special Victor Flour

Best Seasoning

English Breakfast Fancy Pork Sausage Seasoning

Concentrated Garlic

Colors, Preservatives, Brines, etc.

∎ ∎ ∎

Wolf, Sayer & Heller, Inc.

Fulton and Peoria Sts., Chicago
437 W. 16th St., New York
76 St. Paul St. E., Montreal
7 and 8 Snow Hill, London, E. C.
153 Hereford St., Christchurch, N. Z.
76 Spaldingstr., Hamburg

An Ideal Unit for the Packing House

The YORK Semi-Enclosed Vertical Single-Acting Machine with Direct Motor Mounting

WHERE electric current is available at a reasonable cost, our Semi-Enclosed Machine, with direct motor mounting, makes a neat, clean and highly economical plant—no belts, no engine or steam lines.

The machine occupies a comparatively small floor space. All the power developed by the motor is delivered to the crank-shaft of the machine. These machines are built in sizes from 30 tons refrigerating capacity upwards.

Write for detailed information and prices.

YORK MANUFACTURING CO.
YORK   PENNA.

"You get what you give"

John W. Hall

Webster Building
327 So. La Salle St., Chicago

Broker

Packing House Products

SPECIALTIES:

Tallows—Oils—Greases—Tankage—Blood—Liquid Stick—Bone Meals—Bones—Glue and Gelatin Stocks—Pig Skins—Hog Hair—Cracklings—Horns—Hoofs—Cattle Switches Pure Animal Bone Black—Syrups and Sugars for Curing Purposes

---

CHAS. D. KOUTERICK, MGR.   D. GECK, PRESIDENT

"WE SERVE TO SATISFY"

BUYERS AND SELLERS
OF
GLUES
TALLOW—GREASE
CRACKLING—BLOOD—TANKAGE
HIDES—SKINS
BONES

We're here to produce results and work like "HELEN BLAZES" to get 'em! Wire or write your offerings.

D. GECK, Inc.

80 MAIDEN LANE   NEW YORK

PHONE JOHN 1519

Member New York Produce Exchange
Philadelphia Commercial Exchange

F. C. Rogers

Broker

Packing House Products

Beef and Small Stock

Philadelphia	New York
267 North Front Street	431 West 14th Street

Charles A. Streets

Engineers Building
Cleveland, Ohio

Broker

Packing House Products
Green and Cured Meats
Lard, Tallow, Greases, Oils
Fertilizer Materials

Codes: Cross, Yopp, Robinson

Packers–Renderers
Animal
Inedible By-Products
Buyers

Tallow	Bones
Grease	Hoofs
Cracklings	Hair
Tankage	Hides
Blood	Skins
Stick	Gluestock

Manufacturers
Darling & Company
Union Stock Yards, Chicago

TOPICAL INDEX

(Need for extended or cross references to subjects in this Packers' Encyclopedia is not so great as in many works of this character, because the arrangement of the text is itself in topical form. The subjects are treated in the order in which they come in packinghouse operations, and most details are easily located for that reason.)

A

Accounting, packinghouse costs and	182-195
Animal glands	167
Ammonia in tankage, calculating	138

B

Bacon, curing	91-97
Backs, English rib	87
Backs, English short clear	87
Bacon hogs	65
Back-packing pork cuts	94
Backs, rib	85
Backs, rough	85
Backs, short clear	85
Backs, short fat	85
Beef animals, grading of	3
Beef, baby	3
Beef casings, handling and grading:	
rounds	43-45
middles	45-46
bungs	46-47
weasands	47-48
bladders	48
calf bladders	48
Beef boning	29-30
Beef, boneless, compared to carcass cuts	32
Beef breeds described	1
Beef, carcass cuts, Chicago style	30
Beef coolers, temperatures in	24
Beef, curing barrelled	37
Beef, curing of corned	166
Beef, edible offal or fancy meats	41
Beef extract, manufacture of	49-52
Beef, grading carcass	24
Beef, handling for export	28
Beef hearts	41
Beef, loading in cars	25, 26, 27
Beef livers	41
Beef, manufacture of dried	37
Beef, mess	37
Beef offal, handling edible:	
head	38
viscera	38-39
liver	39
pluck	39
paunch	39
inedible offal	40
feet	40
Beef ox tails	41
Beef, plate	36
Beef, ribbing	25
Beef tongues	41
Beef tripe	41
Beef shrinkage	24
Beef slaughtering operations:	
driving to knocking pen	10
knocking	11
shackling	11
sticking, regular	13
heading	14
foot skinning	14
leg breaking	14
ripping open	14
raising gullet	14
floorsman	15
breastbone sawer	15
aitch bone opening	15
fell cutting	15
rumping	17
bung dropping	17
tail ripping	17
aitch bone sawing	17
fell pulling and beating	17
backing	17
clearing out	17
hide dropping	17
gutting	19
tail sawing	19

splitting 19
  neck splitting 19
  scribe sawing 19
Beef, wholesale cuts............ 35
Beef, wrapping for shipment..... 29
Beef yields in boning a carcass.. 31
Beef yields in cutting............ 33
Beef yields per cent from various
   grades 8
Bellies, clear 83
Bellies, clear, square cut and
   seedless 83
Bellies, English 85
Bellies, rib 83
Bleaching vegetable oils......... 211
Blood per head, calculating
   weight of 136
Blood, cooking and pressing..135, 136
Blood and tankage yields........ 135
Bones, horns and hoofs.......... 143
Bones, steamed 136, 137
Boning beef 29, 30
Borax, use in curing meats...... 92
Boxes for dry salt meats......... 89
Brine calculating table........96, 97
Brine systems, various.......... 202
Butcher hogs 65
Butts, jowl 85
By-products, chemical analysis
   of inedible 137
By-products, cost and return on. 146
By-product yield of 1,000-lb. steer 8

C

Calf skins, curing............... 62
Calves, blood and tankage yields
   from 125
Calves, dressing of veal.......... 125
Casings, calves and yearlings... 125
Casings, handling and grading
   beef 43-49
Casings, hog 115
Casings, sheep 125
Canned soups, formulas for...... 164
Canning meats, methods of...... 163
Catch basins 114
Caul-dressing sheep and lambs. 123
Cattle, breeds of................. 1
Cattle, dressing percentages of.. 8
Cattle, dual purpose............. 2
Cattle, figuring costs on......... 186
Cattle-dressing gang, labor schedule of 20-21
Cattle, market classes and grades
   of 3, 5, 6
Cattle switches, curing.......... 61

Chemistry, packinghouse 169-181
Compound lard, manufacture
   of 214-216
Construction of packing plants..
  196-199
Cooking rendering tanks......... 127
Coolers, air circulation in........ 203
Costs and accounting, packinghouse 182-195
Costs on cattle, figuring......... 186
Costs on hogs, figuring.......... 189
Corned beef, short cure for...... 166
Costs, "opportunity" 192
Cured meats, calculating weights
   of 95
Cured meats, drainage allowance
   on 95
Cured meats, overhauling dry
   salt 93
Cured meats, overhauling pickle. 93
Curing ages for meats........... 94
Curing, dry salt................. 93
Curing, formulas for dry sugar.. 93
Curing materials, chemical tests
   of 178-181
Curing meats for export......... 93
Curing periods for pork cuts.... 95
Curing pork cuts..............91-97
Curing temperatures for pork
   cuts 93

D

Deodorizing vegetable oils....... 213
Departmentization of packing
   business 190
Digester tankage or hog feed.138-139
Dry sausage 156
Dry salt meats, regulations for... 83

E

English meats in borax, packing 94

F

Feeds, specifications for animal.. 136
Feet, handling cattle............. 40
Fertilizer materials, chemical
   tests of 170
Fuller's earth for refining lard.. 107

G

Gallstones 168
Gas-fired apparatus for smoking
   meats 102
Glands, saving and uses of animal 167

TOPICAL INDEX

Glue, manufacture of.........141-142
Greases, chemical standards for. 129
Grease tests, inedible............ 130
Green or sweet pickle meats, regulations for 81

H

Hair from cattle ears, uses of... 14
Hair, handling hog............... 145
Ham boning and cooking....102, 103
Hams in freezer, carrying....... 96
Hams, containers for cooking.... 103
Hams, curing 91-97
Hams, long cut.................. 87
Hams, Manchester 87
Hams, shrinkages in cooking.... 103
Hams, skinned 81
Hams, South Staffordshire....... 87
Hams, standard 81
Heads, handling cattle........... 38
Hide allowance, grubby.......... 61
Hides, building the pack........ 59
Hides, grading 58
Hide grubbing seasons........... 62
Hide left on feet, head, etc..... 61
Hides, salting 59
Hide spread, measuring......... 62
Hides, taking out of pack....... 61
Hides, temperatures in pack.... 59
Hides, trimming 59
Hogs, bacon and lard types..... 63
Hog bladders 117
Hogs, breeds of................. 63
Hog bungs 116, 117
Hog carcass shrinkages in cooler 76
Hog casings, cleaning, grading, salting and packing......115-118
Hogs in cooler, hanging......... 76
Hog cooling temperatures....... 76
Hog coolers, methods of chilling. 76
Hogs, dressing yields of........69-71
Hog feed or digester tankage.138, 139
Hogs, figuring costs on......... 189
Hog hair, handling............. 145
Hog hoists, types of............ 75
Hog killing costs............... 75
Hog killing floor, height of..... 76
Hog killing operations:
    shackling 72
    sticking 72
    bleeding 72
    scalding and dehairing....... 72
    heading 73
    gutting 73
    splitting 73
    dressing 75
    hanging 75

Hogs, market classes and grades of 65
Hog middles 118
Hog offal or miscellaneous meats, edible 118, 119
Hog products, green, temperatures for carrying............ 76
Hogs, tests on..................69-71
Hogs, trade requirements in.... 63
Holding green hog meats........ 94
Hydrogenation of oils and fats... 218

I

Icing refrigerator cars, cost of.. 261
Inedible by-products 127
Inedible hog products, grading.. 128

K

Kosher killing 12

L

Lambs, market classes of........ 121
Lambs, tests on killing.......... 124
Lard agitator, use of............ 107
Lard, butchers' 112
Lard cooling cylinders, types of 108-109
Lard cracklings, method of pressing 111
Lard crackling flour, uses of.... 112
Lard definitions in trading...... 114
Lard filter press................ 108
Lard, kettle rendered, ingredients 110
Lard, kettle rendered, chilling... 110
Lard, kettle-rendered, hashing... 110
Lard, kettle-rendered, rendering. 110
Lard, kettle-rendered, settling and filling 111
Lard, kettle-rendered, yields.... 111
Lard manufacture, methods of.. 103-114
Lard, making compound......214-216
Lard, neutral, grades of......... 112
Lard, neutral, hashing and rendering 112
Lard, neutral, to get quality.... 113
Lard, neutral, yields............ 113
Lard oil and stearine.........113-114
Lard oils and greases........... 140
Lard operating precautions..... 114
Lard, prime steam, ingredients for 103
Lard, prime steam, rendering.105-106
Lard, prime steam, cooking...105-106
Lard, prime steam, drawing off.. 106

Lard, prime steam, refining or bleaching 107
Lard, prime steam, filtering...... 108
Lard, prime steam, cooling...... 108
Lard refining kettles............. 107
Lard rolls, use of................. 109
Lard, stiffening of 109
Lard, wet neutral................. 113
Lard yields from various fats.109-110

M

Margarin, manufacture of....... 219
Markets, influence on accounting 193
Meat canning 163
Middles, Dublin 87
Mutton fats in oleo products.... 55

N

Neatsfoot oil 140-141
Nitrate of soda................. 92

O

Offal, edible hog................ 118
Offal from sheep killing......123-124
Offal, handling edible beef......38-43
Offal test, butcher cattle......... 9
Offal tests, shipping cattle....... 9
Offal, tests on condemned........ 131
Oil refining, vegetable........205-214
Oleo fats, yields from........... 55
Oleo oil, grades of.............. 55
Oleo oil, handling............... 55
Oleomargarin, making of........ 219
Oleo products, fats included in.. 52
Oleo products, manufacture of... 52
Oleo products, yields of fats for. 56
Oleo stock, seeding process for... 54
Oleo stock, pressing............. 54
Oleo stocks, tests on............ 56
Oleo stearine, handling.......... 55

P

Packing plants, construction of 196-199
Packing plants, location of...... 195
Pickle, formulas for sweet....... 92
Pickle, method of making........ 92
Pickle, pumping 92
Pickle, recovery of after curing meats 95
Pickled meats, uniformity of.... 83
Picnics, standard 83

Pig's feet, preparation of........ 119
Pig's feet, yields from........... 119
Plates, regular 85
Plates, clear 85
Pork cuts, Board of Trade regulations for 79-91
Pork, barrelled 79
   mess 79
   back 79
   extra clear 79
   clear 79
   clear back 79
   fat back 79
   ham butt 79
   bean 81
   jowl 81
   clear plate 81
   plate 81
   shoulder butt 81
   clear shoulder 81
   loin 81
Pork barrels, specifications for.. 89
Pork cuts, branding............. 83
Pork cuts, range of weights in trade rules 89
Pork meats, grading of.......... 91
Provision plant, layout of....... 161

R

Refrigerator cars, icing 26, 261
Refrigeration systems, packinghouse 200-204
Rendering inedible products..... 129

S

Salt, use in curing meats........ 91
Saltpeter, use in curing meats... 92
Sausage, drying summer........ 157
Sausage, emulsion method....... 148
Sausage factory, arrangement of 161
Sausage formulas, cooked....... 154
Sausage formulas, dry or summer 157
Sausage formulas, fresh......... 149
Sausage formulas, smoked....... 150
Sausage, general directions for making 147
Sausage, moisture table for..... 148
Sausage, smoking summer...... 156
Sausage yields 147
Sawdust used for smoking meats. 102
Scalding tubs, dimensions of.... 75
Screening inedible products...... 137
Sheep and lambs, market classes of 121

Sheep killing methods..........121-123
Sheep and lambs, styles of dressing 123
Sheep carcass yields............ 124
Sheep casings 125
Sheep, blood and tankage yields from 125
Shipper pigs, method of dressing. 76
Shipper pigs, tests on.......... 77
Shoulders, New Orleans......... 85
Shoulders, New York............ 83
Shoulders, regular 85
Shoulders, square 87
Sides, Birmingham 87
Sides, Cumberland 87
Sides, extra short clear........ 83
Sides, extra short rib 83
Sides, Irish cut 87
Sides, long clear 85
Sides, long rib 87
Sides, short clear 83
Sides, short rib 83
Sides, South Staffordshire...... 87
Sides, Wiltshire 85
Sides, Yorkshire 87
Smoking, preparing meat for.... 98
Smoking, length of time required for 99
Smoking, hanging meat after.... 99
Smokehouses, construction of... 100
Smokehouses, various types of.. 100-101
Smoking meats, materials used in 102
Smoking sweet pickle and dry salt meats 98-102
Sugar, use in curing meats...... 91
Summer sausage 156

T

Tallows and greases, chemical tests of 174
Tallow and grease, refining inedible 139-140
Tallow grading standards 57
Tallow, products for prime 57
Tallow, raw products for edible. 57
Tallow yield per head 57
Tallow yields, inedible 131
Tankage, blow system for 128
Tankage, deodorizing 127
Tankage, digester or hog feed 138-139
Tankage, drying concentrated.133-134
Tanking hog hair................ 146
Tank house, inedible 127
Tank house odors, removing 127
Tankage method for small plants 134
Tankage, pressing inedible 130
Tankage values, calculating..... 138
Tank water value table 133
Tank water evaporation......131-132
Tankage yields, inedible 130-131
Tierces for pickled meats 89
Tongue, potted 166
Tongues, removing beef 14

V

Vegetable margarin, making ... 219
Vegetable oil refining.........205-214
Viscera inspection tables........ 73

W

Wiltshire sides, directions for curing 94
Winter oil, manufacture of..216-218

INDEX TO ILLUSTRATIONS

	Page
Beef carcass, standard method of dressing	18
Beef cuts, standard wholesale	34
Beef forequarter, how to hang in refrigerator	27
Beef killing equipment in large plant	11
Beef killing equipment in small plant	13
Beef killing floor, lay-out of modern	16
Beef, market grades of:	
Choice, good, medium and common steer sides	22
Prime baby beef, good heifer, good cow and good bull sides	23
Canning floor, lay-out of modern meat	165
Cattle business, plan for departmentizing	190
Cattle, market classes of:	
Baby beeve	4
Common cow	4
Common killing steer	2
Good fat cow	4
Medium beef steer	2
Prime killing steer	2
Coolers, types of overhead bunkers in	200-203
Hide, proper lay-out of packer	60
Hog business, plan for departmentizing	191
Hog carcass, domestic cuts	70, 78
Hog carcass, export cuts	80
Hog killing floor, lay-out of	74
Hogs, market classes and grades of:	
Choice, good fat, medium butcher	66
Common, skip or rough, rough packing sow	67
Hogs, typical bacon and lard	64
Lambs, market grades of	120
Lard plant, elevation of kettle-rendered	111
Lard refinery, elevation of	104
Offal floor, lay-out of an	42
Oleo plant, elevation of an	53
Oil refinery, elevations of typical vegetable	206, 208, 210, 212
Packing plants, typical lay-outs for	197-199
Pork cuts, Board of Trade regulations, domestic and export	82, 84, 86, 88, 90
Sausage factory and provision plant, lay-out of	162
Sheep killing floor, lay-lout of modern	122
Smoke house, circular	100
Smoke house, continuous operation	101
Smoke houses, layouts of	98-99
Slaughterings, chart showing classes of	7
Tank house, elevation of typical	126

INDEX TO ADVERTISEMENTS

Air Conditioning Apparatus
Atmospheric Conditioning Corp.. 471

Architects and Engineers
Anders & Reimers............... 458
Chemical & Engineering Co..... 459
Henschien & McLaren 460

Boilers
Hamler Boiler & Tank Co....... 485
Mechanical Manufacturing Co... 491
Morrison, Wm. G............... 492
Vogt Machine Co., Henry........ 513

Brokers
Geck, Inc., D. 518
Hall, John W................... 518
Johs. P. Larsen, Copenhagen.... 488
Rogers, F. C................... 519
Streets, Chas. A............... 519

Casings
Berth, Levi & Co., Inc.......... 474
Brecht Co., The...............463-466
Oppenheimer Casing Co.......... 494
Oppenheimer & Co., S........... 502
Patent Casing Co. 496
Van Gelder & Co., G............ 512
Wolf, Sayer & Heller, Inc....... 516

Catch Basins
National Sanitary Service Co.... 493

Chemicals
Battelle & Renwick............. 473
San Francisco Salt Refinery..... 500

Chemists
Chemical & Engineering Co..... 459

Cleaning Materials
Armour & Company............. 451
Cudahy Packing Co............. 452

Cold Storage
United States Cold Storage Co.. 510

Cooperage—Lard Tubs, Barrels, Kegs, Etc.
Cleveland Cooperage Co......... 476
Hauser-Stander Tank Co., The.. 484
Hopkins Cooperage Co., C. G.... 486
Price Co., Edwin C............. 499
United Cooperage Co........... 509

Crackling Presses
Anderson Co., The V. D........ 469
Dunning & Boschert Press Co... 479
Hydraulic Press Mfg. Co........ 487

Crushing and Grinding Machinery
Gruendler Patent Crusher & Pulv. Co. 482
Stedman's Foundry & Machine Works 503
Williams Patent Crusher & Pulv. Co. 515

Curing Materials
Battelle & Renwick............. 473
San Francisco Salt Refinery..... 500
Union Salt Co. 508

Dryers and Tanks
Allbright-Nell Co., The.......... 468
Bartlett & Snow Co., The C. O.. 472
Brecht Co., The...............463-466
Cincinnati Butchers' Supply Co.. 477
Dayton Beater & Hoist Co...... 478
Hamler Boiler & Tank Co....... 485
Mechanical Manufacturing Co.... 491
Morrison, Wm. G............... 492

Electric Motors & Packinghouse Equipment
Westinghouse Electric & Mfg. Co. 514

Evaporators
Allbright-Nell Co., The.......... 468
Morrison, Wm. G............... 492
Swenson Evaporator Co......... 504

Ham Containers
Ham Boiler Corporation, The... 483
Packers Machinery & Equip. Co., The 495

Hot Water Regulators
Powers Regulator Co., The...... 498

Insulation Material
Union Insulating & Construction Co. 507

Insurance
Rollins Burdick Hunter Co...... 461

Lard and Oil Filtration
Celite Products Co............... 475

Meat Curing Hogsheads and Vats
Hauser-Stander Tank Co., The.. 484
United Cooperage Co............ 509

Meat Packers
Allied Packers, Inc. 450
Armour & Company............. 451
Cudahy Packing Co. 452
Dold Packing Co., Jacob......... 453
Morris & Co. 454
Rohe & Brother 455
Swift & Company 456
Wilson & Co. 457

Motors and Control Apparatus
Westinghouse Electric & Mfg. Co. 514

Oils and Fats
Johs. P. Larsen, Copenhagen.... 488

Oil Refining and Margarin Machinery
Allbright-Nell Co., The.......... 468
Brecht Co., The...............463-466
Mechanical Mfg. Co., The....... 491

Oil Refining Materials
Celite Products Co............... 475

Packers' Association
Institute of American Meat Packers 448, 449

Packinghouse Machinery and Equipment
Allbright-Nell Co., The.......... 468
Brecht Co., The...............463-466
Cincinnati Butchers' Supply Co., The 477
Hydraulic Press Mfg. Co......... 487
Maclachlan Reduction Process Co., Inc. 489
Mechanical Mfg. Co., The....... 491
Morrison, Wm. G................ 492
Packers' Machinery & Equip. Co. 495
Smith's Sons Co., John E........ 501
Thomas-Albright Co............. 505

Parchment Paper
Paterson Parchment Paper Co., The 497

Presses for Lard, Oils, Acids, Fertilizers, Tankage, Etc.
Anderson Co., V. D............. 469
Dunning & Boschert Press Co... 479
Hydraulic Press Mfg. Co., The... 487
Thomas-Albright Co............. 505

Refrigerating and Ice-Making Machinery
Arctic Ice Machine Co........... 470
Brecht Co., The...............463-466
Frick Co. 481
Vilter Mfg. Co., The............. 511
Vogt Machine Co., Henry........ 513
York Manufacturing Co.......... 517

Refrigeration, Brine Spray
Atmospheric Conditioning Corp.. 471

Refrigeration Regulators
Powers Regulator Co., The...... 498

Renderers—Tallow, Oleo Oil, Etc.
Johs. P. Larsen, Copenhagen.... 488

Renderers—Inedible By-Products
Darling & Co. 520

Rendering Equipment
Allbright-Nell Co., The.......... 468
Bartlett & Snow Co., The C. O.. 472
Brecht Co., The...............463-466
Cincinnati Butchers' Supply Co., The 477
Dayton Beater & Hoist Co....... 478
Hamler Boiler & Tank Co...... 485
Maclachlan Reduction Process Co. 489
Mechanical Mfg. Co., The........ 491
Morrison, Wm. G................ 492

Salt
Union Salt Co. 508

Sausage Machinery
Brecht Co., The...............463-466
Cincinnati Butchers' Supply Co., The 477
Enterprise Manufacturing Co.... 480
Mechanical Manufacturing Co., The 491
Smith's Sons Co., John E........ 501

Sausage Seasonings
McCormick & Co., Inc........... 490
Thomson & Taylor Co........... 506
Wolf, Sayer & Heller............. 516

INDEX TO ADVERTISEMENTS

Screening Machinery
Stedman's Foundry & Mach. Works 503

Smokehouse Equipment
Airoblast Corporation 467

Spices, Herbs, Etc.
McCormick & Co., Inc........... 490
Thomson & Taylor Co........... 506

Tankage Process
Maclachlan Reduction Process Co. 489

Temperature Regulators
Powers Regulator Co., The....... 498

Trade and Market Information
The National Provisioner....447, 462

Valves, Gauges, Pumps, Etc.
Hydraulic Press Mfg. Co., The.. 487
Morrison, Wm. G................ 492
Powers Regulator Co., The...... 498
Thomas-Albright Co............ 505